Greek Alphabet

A	α	alpha	H	η	eta	N	ν	nu	T	τ	tau
B	β	beta	Θ	θ	theta	Ξ	ξ	xi	Y	υ	upsilon
Γ	γ	gamma	I	ι	iota	O	o	omicron	Φ	φ	phi
Δ	δ	delta	K	κ	kappa	Π	π	pi	X	χ	chi
E	ε	epsilon	Λ	λ	lambda	P	ρ	rho	Ψ	ψ	psi
Z	ζ	zeta	M	μ	mu	Σ	σ	sigma	Ω	ω	omega

Perspectives
of
Modern Physics

Perspectives of Modern Physics

Arthur Beiser
Formerly Associate Professor of Physics
New York University

INTERNATIONAL STUDENT EDITION

McGRAW-HILL KOGAKUSHA, LTD.
Tokyo Auckland Beirut Bogota Düsseldorf Johannesburg
Lisbon London Lucerne Madrid Mexico New Delhi Panama
Paris San Juan São Paulo Singapore Sydney

Perspectives
of
Modern Physics

INTERNATIONAL STUDENT EDITION

TOSHO PRINTING CO., LTD., TOKYO, JAPAN

Preface

My object in preparing "Perspectives of Modern Physics" was to provide an account of the origin of some of the basic properties of atoms, molecules, solids, and nuclei that would be useful both as an introduction to the ideas involved for those who intend to proceed further into any of the subject areas (for instance, students majoring in physics or chemistry) and as a terminal survey for those whose principal interests lie elsewhere (for instance, students majoring in biology or engineering). No attempt was made to provide either overwhelming rigor or encyclopedic coverage: I believe the former is best approached in successive stages so as not to obscure the physics of what is being considered, and that the student in search of the latter is best served by a secure grounding in fundamentals so that he can read the literature on his own.

While a number of topics are approached from the point of view of someone in the laboratory trying to make sense of his results, more often the point of view is that of someone at a desk trying to understand the structure and behavior of matter in terms of theories and models that experiment has shown to be valid. Although all of modern physics—indeed, all of science—rests upon an empirical foundation, it is frequently more economical of the student's time to begin with the concepts that unify the data rather than with the data themselves or the experimental methods by which they were obtained.

A special point is made of using basic ideas and methods of approach in several contexts whenever possible, both to emphasize their general applicability and to give the student insight into how they are actually incorporated into the theoretical fabric of physics. For instance, quite apart from its intrinsic interest, the theory of molecular spectra touches upon angular momentum quantization, the quantum-mechanical harmonic oscillator, and statistical mechanics, and thus, despite its specialized nature, this theory is entitled to the space devoted to it.

While my concern is more with the physics of each situation than with its mathematics, I have nevertheless felt it essential to show how numerical results are obtained and not merely to tell how this is done. Since this is a textbook and not a parlor game, I have not omitted alternate steps in the various arguments but have instead tried to make them reasonably complete. In this way I hope that readers whose mathematical facility is not as yet mature do not suffer any disadvantage; in fact, perhaps their attainment of such maturity will be aided by seeing the weaving process itself, as it were, as well as the final tapestry. However, the book is not meant as a text on quantum mechanics and statistical mechanics, but as one on modern physics—which means, in practice, that I have developed the notions of the former disciplines only as far as they are required for understanding the chief concepts of atomic, molecular, solid-state, and nuclear physics.

Arthur Beiser

Contents

CONTENTS

Perspectives
of
Modern Physics

Chapter 1
Special Relativity

Our study of modern physics begins with a consideration of the special theory of relativity. This is a logical starting point, since all of physics is ultimately concerned with measurement and relativity involves an analysis of how measurements depend upon the observer as well as upon what is observed. From relativity emerges a new mechanics in which there are intimate relationships between space and time, mass and energy. Without these relationships it would be impossible to understand the microscopic world within the atom whose elucidation is the central problem of modern physics.

1.1   The Michelson-Morley Experiment

The wave theory of light was devised and perfected several decades before the electromagnetic nature of the waves became known. It was reasonable for the pioneers in optics to regard light waves as undulations in an all-pervading elastic medium called the *ether*, and their successful explanation of diffraction and interference phenomena in terms of ether waves made the notion of the ether so familiar that its existence was accepted without question. Maxwell's development of the electromagnetic theory of light in 1864 and Hertz's experimental confirmation of it in 1887 deprived the ether of most of its properties, but nobody at the time seemed willing to discard the fundamental idea represented by the ether: that light propagates relative to some sort of universal frame of reference. Let us consider an example of what this idea implies with the help of a simple analogy.

Figure 1.1 is a sketch of a river of width D which flows with the speed v. Two boats start out from one bank of the river with the same speed V. Boat A crosses the river to a point on the other bank directly opposite the starting point and then returns, while boat B heads downstream for the distance D and then returns to the starting point. Let us calculate the time required for each round trip.

We begin by considering boat A. If A heads perpendicularly across the river, the current will carry it downstream from its goal on the opposite bank (Fig. 1.2). It must therefore head somewhat upstream in order to compensate for the current. In order to accomplish this, its upstream component of velocity should be exactly $-v$ in order to cancel out the river current v, leaving the component V' as its net speed across the river. From Fig. 1.2 we see that these speeds are related by the formula

$$V^2 = V'^2 + v^2$$

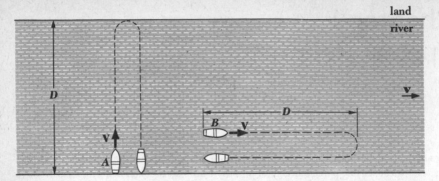

FIGURE 1.1 Boat A goes directly across the river and returns to its starting point, while boat B heads downstream for an identical distance and then returns.

so that the actual speed with which boat A crosses the river is

$$V' = \sqrt{V^2 - v^2}$$
$$= V\sqrt{1 - v^2/V^2}$$

Hence the time for the initial crossing is the distance D divided by the speed V'. Since the

FIGURE 1.2 Boat A must head upstream in order to compensate for the river current if it is to go directly across the river.

reverse crossing involves exactly the same amount of time, the total round-trip time t_A is twice D/V', or

1.1
$$t_A = \frac{2D/V}{\sqrt{1 - v^2/V^2}}$$

The case of boat B is somewhat different. As it heads downstream, its speed relative to the shore is its own speed V *plus* the speed v of the river (Fig. 1.3), and it travels the distance D downstream in the time

$$\frac{D}{V + v}$$

On its return trip, however, B's speed relative to the shore is its own speed V *minus* the speed v of the river. It therefore requires the longer time

$$\frac{D}{V - v}$$

to travel upstream the distance D to its starting point. The total round-trip time t_B is the sum of these times, namely,

$$t_B = \frac{D}{V + v} + \frac{D}{V - v}$$

Using the common denominator $(V + v)(V - v)$ for both terms,

$$t_B = \frac{D(V - v) + D(V + v)}{(V + v)(V - v)}$$

$$= \frac{2DV}{V^2 - v^2}$$

1.2
$$= \frac{2D/V}{1 - v^2/V^2}$$

FIGURE 1.3 The speed of boat B downstream relative to the shore is increased by the speed of the river current while its speed upstream is reduced by the same amount.

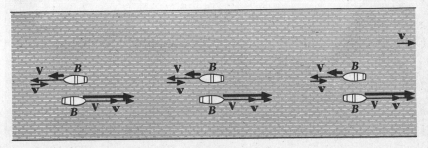

which is greater than t_A, the corresponding total round-trip time for the other boat. The ratio between the times t_A and t_B is

$$\frac{t_A}{t_B} = \sqrt{1 - v^2/V^2}$$

1.3

If we know the common speed V of the two boats and measure the ratio t_A/t_B, we can determine the speed v of the river.

The reasoning used in this problem may be transferred to the analogous problem of the passage of light waves through the ether. If there is an ether pervading space, we move through it with at least the 3×10^4 m/sec (18.5 mi/sec) speed of the earth's orbital motion about the sun; if the sun is also in motion, our speed through the ether is even greater (Fig. 1.4). From the point of view of an observer on the earth, the ether is moving past the earth. To detect this motion, we can use the pair of light beams formed by a half-silvered mirror instead of a pair of boats (Fig. 1.5). One of these light beams is directed to a mirror along a path perpendicular to the ether current, while the other goes to a mirror along a path parallel to the ether current. The optical arrangement is such that both beams return to the same viewing screen. The purpose of the clear glass plate is to ensure that both beams pass through the same thicknesses of air and glass.

If the path lengths of the two beams are *exactly* the same, they will arrive at the screen in phase and will interfere constructively to yield a bright field of view. The presence of an ether current in the direction shown, however, would cause the beams to have different transit

FIGURE 1.4 Motions of the earth through a hypothetical ether.

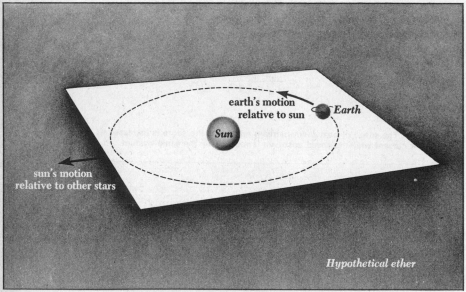

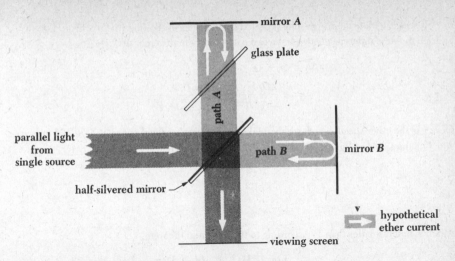

mirror A

glass plate

path A

parallel light
from
single source

path B

mirror B

half-silvered mirror

v
hypothetical
ether current

viewing screen

FIGURE 1.5 The Michelson-Morley experiment.

times in going from the half-silvered mirror to the screen, so that they would no longer arrive at the screen in phase but would interfere destructively. In essence this is the famous experiment performed in 1887 by the American physicists Michelson and Morley.

In the actual experiment the two mirrors are not perfectly perpendicular, with the result that the viewing screen appears crossed with a series of bright and dark interference fringes due to differences in path length between adjacent light waves (Fig. 1.6). If either of the optical paths in the apparatus is varied in length, the fringes appear to move across the screen as reinforcement and cancellation of the waves succeed one another at each point. The stationary apparatus, then, can tell us nothing about any time difference between the two paths. When the apparatus is rotated by 90°, however, the two paths change their orientations relative to the hypothetical ether stream, so that the beam formerly requiring the time t_A for the round trip now requires t_B and vice versa. If these times are different, the fringes will move across the screen during the rotation.

FIGURE 1.6 Fringe pattern observed in Michelson-Morley experiment.

The Michelson-Morley Experiment

Let us calculate the fringe shift expected on the basis of the ether theory. From Eqs. 1.1 and 1.2 the time difference between the two paths owing to the ether drift is

$$\Delta t = t_B - t_A$$

$$= \frac{2D/V}{1 - v^2/V^2} - \frac{2D/V}{\sqrt{1 - v^2/V^2}}$$

Here v is the ether speed, which we shall take as the earth's orbital speed of 3×10^4 m/sec, and V is the speed of light c, where $c = 3 \times 10^8$ m/sec. Hence

$$\frac{v^2}{V^2} = \frac{v^2}{c^2}$$

$$= 10^{-8}$$

which is much smaller than 1. According to the binomial theorem,

$$(1 \pm x)^n = 1 \pm nx + \frac{n(n-1)x^2}{2!} \pm \frac{n(n-1)(n-2)x^3}{3!} + \cdots$$

which is valid for $x^2 < 1$. When x is extremely small compared with 1,

$$(1 \pm x)^n \approx 1 \pm nx$$

We may therefore express Δt to a good approximation as

$$\Delta t = \frac{2D}{c} \left[\left(1 + \frac{v^2}{c^2} \right) - \left(1 + \frac{1}{2} \frac{v^2}{c^2} \right) \right]$$

$$= \left(\frac{D}{c} \right) \left(\frac{v^2}{c^2} \right)$$

In this expression D is the distance between the half-silvered mirror and each of the other mirrors. The path difference d corresponding to a time difference Δt is

$$d = c \, \Delta t$$

If d corresponds to the shifting of n fringes,

$$d = n\lambda$$

where λ is the wavelength of the light used. Equating these two formulas for d, we find that

$$n = \frac{c \, \Delta t}{\lambda}$$

$$= \frac{Dv^2}{\lambda c^2}$$

In the actual experiment Michelson and Morley were able to make D about 10 m in effective length through the use of multiple reflections, and the wavelength of the light they used was about 5,000 A (1 A = 10^{-10} m). The expected fringe shift in each path when the apparatus is rotated by 90° is therefore

$$n = \frac{Dv^2}{\lambda c^2}$$

$$= \frac{10 \text{ m} \times (3 \times 10^4 \text{ m/sec})^2}{5 \times 10^{-7} \text{ m} \times (3 \times 10^8 \text{ m/sec})^2}$$

$$= 0.2 \text{ fringe}$$

Since both paths experience this fringe shift, the total shift should amount to $2n$ or 0.4 fringe. A shift of this magnitude is readily observable, and therefore Michelson and Morley looked forward to establishing directly the existence of the ether.

To everybody's surprise, *no fringe shift whatever* was found. When the experiment was performed at different seasons of the year and in different locations, and when experiments of other kinds were tried for the same purpose, the conclusions were always identical: no motion through the ether was detected.

The negative result of the Michelson-Morley experiment had two consequences. First, it rendered untenable the hypothesis of the ether by demonstrating that the ether has no measurable properties—an ignominious end for what had once been a respected idea. Second, it suggested a new physical principle: the speed of light in free space is the same everywhere, regardless of any motion of source or observer.

1.2 The Special Theory of Relativity

We mentioned earlier the role of the ether as a universal frame of reference with respect to which light waves were supposed to propagate. Whenever we speak of "motion," of course, we really mean "motion relative to a frame of reference." The frame of reference may be a road, the earth's surface, the sun, the center of our galaxy; but in every case we must specify it. Stones dropped in Bermuda and in Perth, Australia, both fall "down," and yet the two move in exactly opposite directions relative to the earth's center. Which is the correct location of the frame of reference in this situation, the earth's surface or its center? The answer is that *all* frames of reference are equally correct, although one may be more convenient to use in a specific case. *If* there were an ether pervading all space, we could refer all motion to it, and the inhabitants of Bermuda and Perth would escape from their quandary. The absence of an ether, then, implies that there is no universal frame of reference, so that all motion exists solely relative to the person or instrument observing it. If we are in a free balloon above a uniform cloud bank and see another free balloon change its position relative to us, we have no way of knowing which balloon is "really" moving (Fig. 1.7). Should we be isolated in the universe, there would be no way in which we could determine whether we

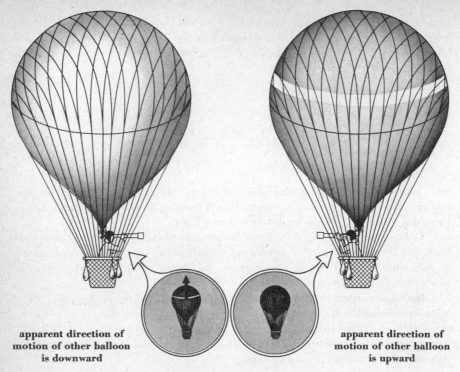

apparent direction of
motion of other balloon
is downward

apparent direction of
motion of other balloon
is upward

FIGURE 1.7 All motion is relative to the observer.

are in motion or not, because without a frame of reference the concept of motion has no meaning.

The theory of relativity resulted from an analysis of the physical consequences implied by the absence of a universal frame of reference. The special theory of relativity, developed by Albert Einstein in 1905, treats problems involving *inertial frames of reference*, which are frames of reference moving at constant velocity with respect to one another. The general theory of relativity, proposed by Einstein a decade later, treats problems involving frames of reference accelerated with respect to one another. An observer in an isolated laboratory *can* detect accelerations. Anybody who has been in an elevator or on a merry-go-round can verify this statement from his own experience. The special theory has had a profound influence on all of physics, and we shall consider it in this chapter and the first part of Chap. 2 before a brief glance at the general theory.

The special theory of relativity is based upon two postulates. The first states that **the laws of physics may be expressed in equations having the same form in all frames of reference moving at constant velocity with respect to one another.** This postulate expresses the absence of a universal frame of reference. If the laws of physics had different forms for different observers in relative motion, it could be determined from these differences which objects are "stationary" in space and which are "moving." But because there is no universal frame of reference, this distinction does not exist in nature; hence the above postulate.

The second postulate of special relativity states that **the speed of light in free space has the same value for all observers, regardless of their state of motion.** This postulate follows directly from the result of the Michelson-Morley experiment (and others).

At first sight these postulates hardly seem radical. Actually they subvert almost all of the intuitive concepts of time and space we form on the basis of our daily experience. A simple example will illustrate this statement. In Fig. 1.8 we have the two boats A and B once more, with boat A stationary in the water while boat B drifts at the constant velocity **v**. There is a dense fog present, and so on neither boat does the observer have any idea which is the moving one. At the instant that B is abreast of A, a flare is fired. The light from the flare travels uniformly in all directions, according to the second postulate of special relativity. An observer on either boat must see a sphere of light expanding with *himself* at its center, according to the first postulate of special relativity, even though one of them is changing his position with respect to the point where the flare went off. The observers cannot detect which of them is undergoing such a change in position since the fog eliminates any frame of reference other than each boat itself, and so, since the speed of light is the same for both of them, they must both see the identical phenomenon.

Why is the situation of Fig. 1.8 unusual? Let us consider a more familiar analog. The boats are at sea on a clear day and somebody on one of them drops a stone into the water

FIGURE 1.8 Relativistic phenomena differ from everyday experience.

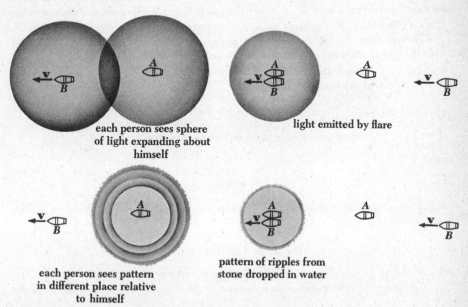

each person sees sphere
of light expanding about
himself

light emitted by flare

each person sees pattern
in different place relative
to himself

pattern of ripples from
stone dropped in water

The Special Theory of Relativity

when they are abreast of each other. A circular pattern of ripples spreads out, as at the bottom of Fig. 1.8, *which appears different* to observers on each boat. Merely by observing whether or not he is at the center of the pattern of ripples, each observer can tell whether he is moving relative to the water or not. Water is in itself a frame of reference, and an observer on a boat moving through it measures ripple speeds with respect to himself that are different in different directions, in contrast to the uniform ripple speed measured by an observer on a stationary boat. It is important to recognize that motion and waves *in water* are entirely different from motion and waves *in space;* water is in itself a frame of reference while space is not, and wave speeds in water vary with the observer's motion while wave speeds of light in space do not.

The only way of interpreting the fact that observers in the two boats in our example perceive identical expanding spheres of light is to regard the coordinate system of each observer, from the point of view of the other, as being affected by their relative motion. When this idea is developed, using only accepted laws of physics and Einstein's postulates, we shall see that many peculiar effects are predicted. One of the triumphs of modern physics is the experimental confirmation of these effects.

1.3 The Galilean Transformation

Let us suppose that we are in a frame of reference S and find that the coordinates of some event that occurs at the time t are x, y, z. An observer located in a different frame of reference S' which is moving with respect to S at the constant velocity v will find that the same event occurs at the time t' and has the coordinates x', y', z'. (In order to simplify our work, we shall assume that v is in the $+ x$ direction, as in Fig. 1.9.) How are the measurements x, y, z, t related to x', y', z', t'?

At first glance the answer seems obvious enough. If time in both systems is measured from the instant when the origins of S and S' coincided, measurements in the x direction made in S will exceed those made in S' by the amount vt, which represents the distance that S' has moved in the x direction. That is

1.4
$$x' = x - vt$$

There is no relative motion in the y and z directions, and so

1.5
$$y' = y$$

1.6
$$z' = z$$

In the absence of any indication to the contrary in our everyday experience, we further assume that

1.7
$$t' = t$$

The set of Eqs. 1.4 to 1.7 is known as the *Galilean transformation.*

To convert velocity components measured in the S frame to their equivalents in the

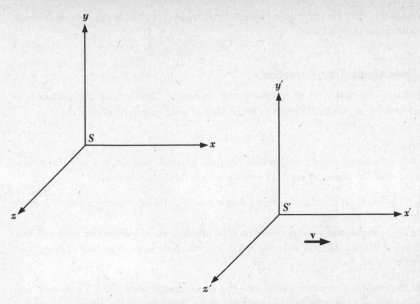

FIGURE 1.9 Frame S′ moves in the + x direction with the speed v relative to frame S.

S′ frame according to the Galilean transformation, we simply differentiate Eqs. 1.4 to 1.6 with respect to time:

1.8
$$v'_x = \frac{dx'}{dt'} = v_x - v$$

1.9
$$v'_y = \frac{dy'}{dt'} = v_y$$

1.10
$$v'_z = \frac{dz'}{dt'} = v_z$$

While the Galilean transformation and the velocity transformation it leads to are both in accord with our intuitive expectations, they violate both of the postulates of special relativity. The first postulate calls for identical equations of physics in both the S and S′ frames of reference, but the fundamental equations of electricity and magnetism assume very different forms when Eqs. 1.4 to 1.7 are used to convert quantities measured in one frame into their equivalents in the other. The second postulate calls for the same value of the speed of light c whether determined in S or S′. If we measure the speed of light in the x direction in the S system to be c, however, in the S′ system it will be

$$c' = c - v$$

according to Eq. 1.8. Clearly a different transformation is required if the postulates of special relativity are to be satisfied.

1.4 The Lorentz Transformation

We shall now develop a set of transformation equations directly from the postulates of special relativity. A reasonable guess as to the kind of relationship between x and x' is

1.11
$$x' = k(x - vt)$$

where k is a factor of proportionality that does not depend upon either x or t but may be a function of v. The choice of Eq. 1.11 follows from several considerations:

1. It is linear in x and x', so that a single event in frame S corresponds to a single event in frame S', as it must.
2. It is simple, and a simple solution to a problem should always be explored first.
3. It has the possibility of reducing to Eq. 1.4, which we know to be correct in ordinary mechanics.

Because the equations of physics must have the same form in both S and S', we need only change the sign of v (in order to take into account the difference in the direction of relative motion) to write the corresponding equation for x in terms of x' and t':

1.12
$$x = k(x' + vt')$$

The factor k must be the same in both frames of reference since there is no difference between S and S' other than in the sign of v.

As in the case of the Galilean transformation, there is nothing to indicate that there might be differences between the corresponding coordinates y, y', and z, z' which are normal to the direction of v. (We shall verify this assumption in Sec. 1.6.) Hence we again take

1.13
$$y' = y$$
1.14
$$z' = z$$

The time coordinates t and t', however, are *not* equal. We can see this by substituting the value of x' given by Eq. 1.11 into Eq. 1.12. We obtain

$$x = k^2(x - vt) + kvt'$$

from which we find that

1.15
$$t' = kt + \left(\frac{1 - k^2}{kv}\right)x$$

Equations 1.11, 1.13, 1.14, and 1.15 constitute a coordinate transformation that satisfies the first postulate of special relativity.

The second postulate of relativity enables us to evaluate k. At the instant $t = 0$, the origins of the two frames of reference S and S' are in the same place, according to our initial conditions, and $t' = 0$ then also. Suppose that a flare is set off at the common origin of S and S' at $t = t' = 0$, and the observers in each system proceed to measure the speed with which the light from it spreads out. Both observers must find the same speed c, which means that in the S frame

1.16
$$x = ct$$

while in the S' frame

1.17
$$x' = ct'$$

Substituting for x' and t' in Eq. 1.17 with the help of Eqs. 1.11 and 1.15,

$$k(x - vt) = ckt + \left(\frac{1 - k^2}{kv}\right)cx$$

and solving for x,

$$x = \frac{ckt + vkt}{k - \left(\dfrac{1 - k^2}{kv}\right)c}$$

$$= ct\left[\frac{k + \dfrac{v}{c}k}{k - \left(\dfrac{1 - k^2}{kv}\right)c}\right]$$

$$= ct\left[\frac{1 + \dfrac{v}{c}}{1 - \left(\dfrac{1}{k^2} - 1\right)\dfrac{c}{v}}\right]$$

This expression for x will be the same as that given by Eq. 1.16, namely, $x = ct$, provided that the quantity in the brackets equals 1. Therefore

$$\frac{1 + \dfrac{v}{c}}{1 - \left(\dfrac{1}{k^2} - 1\right)\dfrac{c}{v}} = 1$$

and

1.18
$$k = \frac{1}{\sqrt{1 - v^2/c^2}}$$

The Lorentz Transformation

Inserting the above value of k in Eqs. 1.11 and 1.15, we have for the complete transformation of measurements of an event made in S to the corresponding measurements made in S' the equations

1.19
$$x' = \frac{x - vt}{\sqrt{1 - v^2/c^2}}$$
Lorentz transformation

1.20
$$y' = y$$
Lorentz transformation

1.21
$$z' = z$$
Lorentz transformation

1.22
$$t' = \frac{t - \dfrac{vx}{c^2}}{\sqrt{1 - v^2/c^2}}$$
Lorentz transformation

Equations 1.19 to 1.22 comprise the *Lorentz transformation*. They were first obtained by the Dutch physicist H. A. Lorentz, who showed that the basic formulas of electromagnetism are the same in all frames of reference in uniform relative motion only when these transformation equations are used. It was not until a number of years later that Einstein discovered their full significance.

In order to transform measurements from S' to S, the only changes in the Lorentz transformation equations that need be made are to exchange primed for unprimed quantities and vice versa, and to replace v by $-v$. Thus the inverse Lorentz transformation is

1.23
$$x = \frac{x' + vt'}{\sqrt{1 - v^2/c^2}}$$
Inverse Lorentz transformation

1.24
$$y = y'$$
Inverse Lorentz transformation

1.25
$$z = z'$$
Inverse Lorentz transformation

1.26
$$t = \frac{t' + \dfrac{vx'}{c^2}}{\sqrt{1 - v^2/c^2}}$$
Inverse Lorentz transformation

Two obvious aspects of the Lorentz transformation are worth noting. The first is that measurements of time as well as of position depend upon the frame of reference of the observer, so that two events which occur simultaneously in one frame at different places need not be simultaneous in another. The second is that the Lorentz equations reduce to the ordinary Galilean equations 1.4 to 1.7 when the relative velocity v of S and S' is small compared with the velocity of light c. Therefore we may anticipate that the peculiar consequences of special relativity to be explored in the remainder of this chapter will not be apparent except under circumstances where enormous velocities are encountered.

1.5   The Lorentz-FitzGerald Contraction

A rod is lying along the x' axis of a moving frame of reference S'. An observer in this frame determines the coordinates of its ends to be x_1' and x_2', and he concludes that the length L_0 of the rod is

1.27
$$L_0 = x_2' - x_1'$$

L_0 is the rod's length as measured in a frame of reference in which it is at rest. Suppose the same quantity is determined from a frame of reference S relative to which the rod is moving with the velocity v: Will the length L measured in S be the same as the length L_0 that was measured in S'? In more familiar terms, if a yardstick is in a speeding car (the S' frame), does it seem to have the same length to someone standing beside the road (the S frame) as it does to someone inside the car?

In order to find L, we use the Lorentz transformation to go from the coordinates x_1' and x_2' in the moving frame S' to the corresponding coordinates x_1 and x_2 in the stationary frame S. From Eq. 1.19 we have, since the measurements of x_1 and x_2 are made at the same time t,

$$x_1' = \frac{x_1 - vt}{\sqrt{1 - v^2/c^2}}$$

$$x_2' = \frac{x_2 - vt}{\sqrt{1 - v^2/c^2}}$$

and so

$$L_0 = x_2' - x_1'$$

$$= \frac{x_2 - x_1}{\sqrt{1 - v^2/c^2}}$$

By definition the length L measured in the stationary frame S is given by

1.28
$$L = x_2 - x_1$$

which means that

$$L_0 = \frac{L}{\sqrt{1 - v^2/c^2}}$$

or

1.29
$$L = L_0 \sqrt{1 - v^2/c^2} \qquad \text{Lorentz-FitzGerald contraction}$$

The length of an object in motion with respect to an observer appears to the observer to be shorter than when it is at rest with respect to him, a phenomenon known as the *Lorentz-FitzGerald contraction*.

Because the relative velocity of the two frames S and S' appears only as v^2 in Eq. 1.29, it does not matter which frame we call S and which S'. If we find that the length of a spaceship is L_0 when it is on its launching pad, we will find from the ground that its length L when moving with the speed v is

$$L = L_0 \sqrt{1 - v^2/c^2}$$

while to a man in the spaceship, objects on the earth behind him appear shorter than they did when he was on the ground by the same factor $\sqrt{1 - v^2/c^2}$. The length of an object is a maximum when measured in a reference frame in which it is stationary, and its length is less when measured in a reference frame in which it is moving.

The ratio between L and L_0 in Eq. 1.27 is the same as that in Eq. 1.3 when it is applied to the times of travel of the two light beams, so that we might be tempted to consider the Michelson-Morley result solely as evidence for the contraction of the length of their apparatus in the direction of the earth's motion. This interpretation was tested by Kennedy and Thorndike in a similar experiment using an interferometer with arms of unequal length. They also found no fringe shift, which means that these experiments must be considered evidence for the absence of an ether with all this implies and not only for contractions of the apparatus.

The relativistic length contraction is negligible for ordinary speeds, but it is an important effect at speeds close to the speed of light. A speed of 1,000 mi/sec seems enormous to us, and yet it results in a shortening in the direction of motion to only

$$\frac{L}{L_0} = \sqrt{1 - \frac{v^2}{c^2}}$$

$$= \sqrt{1 - \frac{(1,000 \text{ mi/sec})^2}{(186,000 \text{ mi/sec})^2}}$$

$$= 0.999985$$

$$= 99.9985 \text{ percent}$$

of the length at rest. On the other hand, a body traveling at 0.9 the speed of light is shortened to

$$\frac{L}{L_0} = \sqrt{1 - \frac{(0.9c)^2}{c^2}}$$

$$= 0.436$$

$$= 43.6 \text{ percent}$$

of the length at rest, a significant change.

The Lorentz-FitzGerald contraction occurs only in the direction of the relative motion: if v is parallel to x, the y and z dimensions of a moving object are the same in both S and S'.

An actual photograph of an object in very rapid relative motion might reveal a somewhat different distortion, depending upon the direction from which the object is viewed and the ratio v/c. The reason for this effect is that light reaching the camera (or eye for that matter) from the more distant parts of the object was emitted earlier than that coming from the nearer parts; the camera "sees" a picture that is actually a composite, since the object was at different locations when the various elements of the single image that reaches the film left it. This effect alters the Lorentz-FitzGerald contraction by extending the apparent length of a moving object in the direction of motion. A three-dimensional body, such as a cube, may be seen as rotated in orientation as well as changed in shape, again depending upon the position of the observer and the value of v/c. This visual effect must be distinguished from the Lorentz-FitzGerald contraction itself, which is a physical phenomenon. If there were no Lorentz-FitzGerald contraction, the appearance of a moving body would be different from what it is at rest, but in another way.

It is interesting to note that the above approach to the visual appearance of rapidly moving objects was not made until 1959, 54 years after the publication of the special theory of relativity.

1.6   Time Dilation

Time intervals, too, are affected by relative motion. Clocks moving with respect to an observer appear to tick less rapidly than they do when at rest with respect to him. If we, in the S frame, observe the length of time t some event requires in a frame of reference S' in motion relative to us, our clock will indicate a longer time interval than the t_0 determined by a clock in the moving frame. This effect is called *time dilation*.

To see how time dilation comes about, let us imagine a clock at the point x' in the moving frame S'. When an observer in S' finds that the time is t'_1, an observer in S will find it to be t_1, where, from Eq. 1.26,

$$t_1 = \frac{t'_1 + \dfrac{vx'}{c^2}}{\sqrt{1 - v^2/c^2}}$$

After a time interval of t_0 (to him), the observer in the moving system finds that the time is now t'_2 according to his clock. That is,

1.30
$$t_0 = t'_2 - t'_1$$

The observer in S, however, measures the end of the same time interval to be

$$t_2 = \frac{t'_2 + \dfrac{vx'}{c^2}}{\sqrt{1 - v^2/c^2}}$$

so to him the duration of the interval t is

1.31
$$t = t_2 - t_1$$

$$= \frac{t_2' - t_1'}{\sqrt{1 - v^2/c^2}}$$

or

1.32
$$t = \frac{t_0}{\sqrt{1 - v^2/c^2}}$$ **Time dilation**

A stationary clock measures a longer time interval between events occurring in a moving frame of reference than does a clock in the moving frame.

The slowing down of clocks in motion is so remarkable a phenomenon that it deserves a further look. Instead of starting from the Lorentz transformation, let us directly examine the mechanism of a clock and see how relative motion affects what we measure.

A particularly simple clock consists of a stick L long with a mirror at each end (Fig. 1.10). A pulse of light is reflected up and down between the mirrors, and an appropriate device is attached to one of the mirrors to give a "tick" of some kind each time the pulse of

FIGURE 1.10 A simple clock. Each "tick" corresponds to a round trip of the light pulse from the lower mirror to the upper one and back.

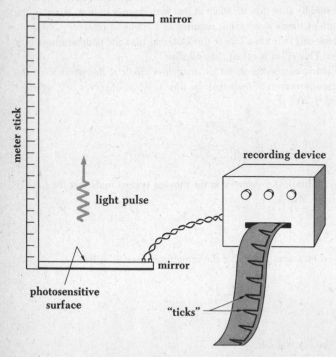

light strikes it. (Such a device might be a photosensitive surface on the mirror which can be arranged to give an electric signal when the light pulse arrives.) The time interval t_0 between ticks is

1.33
$$t_0 = \frac{2L}{c}$$

If the stick is 1 m long,

$$t_0 = \frac{2 \text{ m}}{3 \times 10^8 \text{ m/sec}} = 0.67 \times 10^{-8} \text{ sec}$$

and there are 1.5×10^8 ticks/sec. Two identical clocks of this kind are built, and one is attached to a spaceship mounted perpendicular to the direction of motion while the other remains at rest on the earth's surface.

The first question to ask is whether the length of the meter stick on the spaceship, as measured from the ground, is different when it is in motion. We assumed earlier in Eq. 1.20 that $y' = y$ and hence that there is no difference; now let us verify this statement by a hypothetical experiment. On the ground pencils are attached near the ends of the spaceship's meter stick, and by bringing this stick to the stick fixed on the ground, marks are made on the latter. The spaceship is then set in motion, and the same process is repeated with the pencils brushing against the meter stick on the ground as the moving stick speeds past. If the spaceship's meter stick is different in length from what it was at rest, the new pencil marks will not coincide with the old ones; if the moving stick is shorter, for instance, the new marks will be closer together than the old ones. The various marks are definite permanent physical things, and when the spaceship lands, its pilot can compare them and check whether his meter stick changed in length while in motion relative to the ground.

For the sake of argument, let us suppose a moving meter stick in the y direction does seem smaller to a stationary observer when it is moving in the x direction. An observer on the ground would therefore expect the new marks on the stationary stick to be closer together than the old ones. But to the pilot of the spaceship, the stick on the ground is in motion relative to him and hence is the shorter one, which means that the marks his pencils make on the latter ought to be *farther apart* than the marks made when both were at rest. There is no way to resolve this contradiction while retaining the idea that all motion is relative except by concluding that the new pencil marks must coincide with the old ones, in which case $y' = y$ as in Eq. 1.20.

(The preceding analysis holds equally well for a meter stick in the z direction, so $z' = z$ as in Eq. 1.21. However, the analysis cannot be applied to a meter stick in the x direction, which is the direction of motion, because the pencil marks would all fall on the same line and be indistinguishable. A different kind of experiment is needed here, and it will yield the Lorentz-FitzGerald contraction of Eq. 1.29.)

Having established that the distance L between the mirrors of the clock in the spaceship is unaffected by its motion, the next question to ask is how much time t elapses between ticks in the moving clock as measured by an observer on the ground with an identical clock

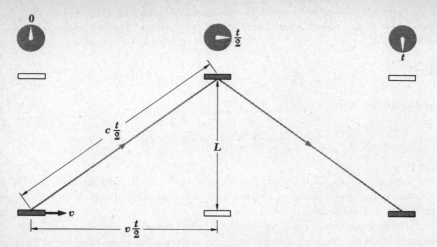

FIGURE 1.11 A light clock in the moving S′ system as seen by an observer in the stationary S system.

that is stationary with respect to him. Each tick involves the passage of a pulse of light at speed c from the lower mirror to the upper one and back. During this round-trip passage, the entire clock in the spaceship is in motion, which means that the pulse of light, as seen from the ground, actually follows a zigzag path (Fig. 1.11). On its way from the lower mirror to the upper one in the time $t/2$, the pulse of light travels a horizontal distance of $vt/2$ and a total distance of $ct/2$. Since L is the vertical distance between the mirrors,

$$\left(\frac{ct}{2}\right)^2 = L^2 + \left(\frac{vt}{2}\right)^2$$

$$\frac{t^2}{4}(c^2 - v^2) = L^2$$

$$t^2 = \frac{4L^2}{c^2 - v^2} = \frac{(2L)^2}{c^2(1 - v^2/c^2)}$$

and

$$t = \frac{2L/c}{\sqrt{1 - v^2/c^2}}$$

But $2L/c$ is the time interval t_0 between ticks on the clock on the ground, as in Eq. 1.33, and so

$$t = \frac{t_0}{\sqrt{1 - v^2/c^2}}$$

as before. The moving clock in the spaceship appears to tick at a slower rate than the stationary one on the ground, as seen by an observer on the ground.

Exactly the same analysis holds for measurements of the clock on the ground by the pilot of the spaceship. To him, the light pulse of the ground clock follows a zigzag path which requires a total time t per round trip, while his own clock, at rest in the spaceship, ticks at intervals of t_0. He too finds that

$$t = \frac{t_0}{\sqrt{1 - v^2/c^2}}$$

so the effect is reciprocal: *every* observer finds that clocks in motion relative to him tick more slowly than when they are at rest.

Our discussion has been based on a somewhat unusual clock that employs a light pulse bouncing back and forth between two mirrors. Do the same conclusions apply to more conventional clocks that use machinery—spring-controlled escapements, tuning forks, or whatever—to produce ticks at constant time intervals? The answer must be yes, since if a mirror clock and a conventional clock in the spaceship agree with each other on the ground but not when in flight, the disagreement between them could be used to determine the speed of the spaceship without reference to any other object—which contradicts the principle that all motion is relative. Detailed calculations of what happens to conventional clocks in motion—as seen from the ground—confirm this answer. For example, as we shall learn in Chap. 2, the *mass* of an object is greater when it is in motion, so that the period of an oscillating object must be greater in the moving spaceship. Therefore *all* clocks at rest relative to one another behave the same to all observers, regardless of any motion at constant velocity of either the group of clocks or the observers.

1.7 Meson Decay

A striking illustration of both the time dilation of Eq. 1.32 and the length contraction of Eq. 1.29 occurs in the decay of unstable particles called μ *mesons,* whose properties we shall discuss in greater detail later. For the moment our interest lies in the fact that a μ meson decays into an electron an average of 2×10^{-6} sec after it comes into being. Now μ mesons are created high in the atmosphere by fast cosmic-ray particles arriving at the earth from space, and reach sea level in profusion. Such mesons have a typical speed of 2.994×10^8 m/sec, which is 0.998 of the velocity of light c. But in $t_0 = 2 \times 10^{-6}$ sec, the meson's mean lifetime, they can travel a distance of only

$$y = vt_0$$
$$= 2.994 \times 10^8 \text{ m/sec} \times 2 \times 10^{-6} \text{ sec}$$
$$= 600 \text{ m}$$

while they are actually created at altitudes more than 10 times greater than this.

We can resolve the meson paradox by using the results of the special theory of relativity. Let us examine the problem from the frame of reference of the meson, in which its lifetime is 2×10^{-6} sec. While the meson lifetime is unaffected by the motion, its distance to the ground appears shortened by the factor

$$\frac{y}{y_0} = \sqrt{1 - v^2/c^2}$$

That is, while we, on the ground, measure the altitude at which the meson is produced as y_0, the meson "sees" it as y. If we let y be 600 m, the maximum distance the meson can go *in its own frame of reference* at the speed $0.998c$ before decaying, we find that the corresponding distance y_0 in *our reference frame* is

$$y_0 = \frac{y}{\sqrt{1 - v^2/c^2}}$$

$$= \frac{600}{\sqrt{1 - \dfrac{(0.998c)^2}{c^2}}} \; \text{m}$$

$$= \frac{600}{\sqrt{1 - 0.996}} \; \text{m}$$

$$= \frac{600}{0.063} \; \text{m}$$

$$= 9{,}500 \; \text{m}$$

Hence, despite their brief lifespans, it is possible for the mesons to reach the ground from the considerable altitudes at which they are actually formed.

Now let us examine the problem from the frame of reference of an observer on the ground. From the ground the altitude at which the meson is produced is y_0, but its lifetime in *our reference frame* has been extended, owing to the relative motion, to the value

$$t = \frac{t_0}{\sqrt{1 - v^2/c^2}}$$

$$= \frac{2 \times 10^{-6}}{\sqrt{1 - \dfrac{(0.998c)^2}{c^2}}} \; \text{sec}$$

$$= \frac{2 \times 10^{-6}}{0.063} \; \text{sec}$$

$$= 31.7 \times 10^{-6} \; \text{sec}$$

almost 16 times greater than when it is at rest with respect to us. In 31.7×10^{-6} sec a meson whose speed is $0.998c$ can travel a distance

$$y_0 = vt$$
$$= 2.994 \times 10^8 \, \text{m/sec} \times 31.7 \times 10^{-6} \, \text{sec}$$
$$= 9{,}500 \, \text{m}$$

the same distance obtained before. The two points of view give identical results.

1.8 Simultaneity

The relative character of time as well as space has many implications. For example, events that seem to take place simultaneously to one observer may not be simultaneous to another observer in relative motion, and vice versa. Let us consider two events—the explosion of a pair of time bombs, say—that occur at the same time t_0 to somebody on the ground but at the different locations x_1 and x_2. What does the pilot of a spaceship see? To him, the explosion at x_1 and t_0 occurs at the time

$$t_1' = \frac{t_0 - vx_1/c^2}{\sqrt{1 - v^2/c^2}}$$

according to Eq. 1.22, while the explosion at x_2 and t_0 occurs at the time

$$t_2' = \frac{t_0 - vx_2/c^2}{\sqrt{1 - v^2/c^2}}$$

Hence two events that occur simultaneously to one observer are separated by a time interval of

$$t_2' - t_1' = \frac{v(x_1 - x_2)/c^2}{\sqrt{1 - v^2/c^2}}$$

to another observer. Who is right? The question is, of course, meaningless: *both* observers are "right," since each simply measures what he sees.

Because simultaneity is a relative notion and not an absolute one, physical theories which require simultaneity in events at different locations must be discarded. The principle of conservation of energy in its elementary form states that the total energy content of the universe is constant, but it does not rule out a process in which a certain amount of energy ΔE vanishes at one point while an equal amount of energy ΔE spontaneously comes into being somewhere else with no actual transport of energy from one place to the other. Because simultaneity is relative, some observers of the process will find energy not being conserved. To rescue conservation of energy in the light of special relativity, then, it is necessary to say that, when energy disappears somewhere and appears elsewhere, it has actually *flowed* from the first location to the second. (There are many ways in which a flow of energy can occur, of course.) Thus energy is conserved *locally* in any arbitrary region of space at any time, not merely when the universe as a whole is considered—a much stronger statement of this principle.

Although time is a relative quantity, not all the notions of time formed by everyday experience are incorrect. Time does not run backward to *any* observer, for instance: a sequence of events that occur somewhere at t_1, t_2, t_3, . . . will appear in the same order to all observers everywhere, though not necessarily with the same time intervals $t_2 - t_1$, $t_3 - t_2$, . . . between each pair of events. Similarly, no distant observer, regardless of his state of motion, can see an event before it happens—more precisely, before a nearby observer sees it—since the speed of light is finite and signals require the minimum period of time L/c to travel a distance L. There is no way to peer into the future, although temporal (and spatial) perspectives of past events may appear different to different observers.

1.9   Space-Time

As we have seen, the concepts of space and time are inextricably mixed in nature. A quantity that one observer is able to measure with only a meter stick may have to be measured with *both* a meter stick and a clock by a different observer. A convenient and elegant way to express the results of special relativity is to regard events as occurring in a four-dimensional continuum called *space-time* in which the three coordinates x, y, z refer to space and a fourth coordinate ict refers to time ($i = \sqrt{-1}$). Such a coordinate system cannot be visualized, of course, but it is no more difficult to cope with mathematically than a three-dimensional system. The reason for choosing ict as the time coordinate instead of simply t is that the quantity

1.34
$$s^2 = x^2 + y^2 + z^2 - (ct)^2$$

is invariant under a Lorentz transformation. That is, if an event occurs at x, y, z, t in a frame of reference S and at x', y', z', t' in another frame S',

$$s^2 = x^2 + y^2 + z^2 - (ct)^2 = x'^2 + y'^2 + z'^2 - (ct')^2$$

Because s^2 is invariant, we can regard a Lorentz transformation merely as a rotation of the coordinate axes x, y, z, ict in space-time (Fig. 1.12).

The four cartesian coordinates x, y, z, ict define a vector in space-time, and this *four-vector* remains fixed in space-time regardless of any rotation of the coordinate system—that is, regardless of any shift in point of view from one inertial frame of reference S to another S'.

Another four-vector whose magnitude remains constant under Lorentz transformations is the four-vector momentum. This has the components p_x, p_y, p_z, iE/c, where p_x, p_y, p_z are the usual components of the linear momentum of a body and E is its total energy. Hence the value of

$$p_x{}^2 + p_y{}^2 + p_z{}^2 - \left(\frac{E}{c}\right)^2$$

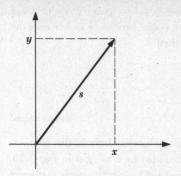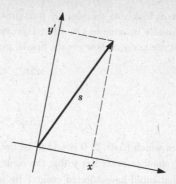

FIGURE 1.12 Rotating a two-dimensional coordinate system leaves unaltered the quantity $s^2 = x^2 + y^2 = x'^2 + y'^2$. This result can be generalized to the four-dimensional coordinate system x, y, z, ict.

is the same in all frames of reference even though p_x, p_y, p_z, and E separately may be different as measured in different frames. Electrodynamics, too, can be expressed in terms of four-vectors, though more elaborate mathematical quantities called *tensors* are also required here. For example, the observation in elementary physics that what appears to be a purely electric field to an observer at rest relative to a charge q appears as a combination of electric and magnetic fields to another observer in motion relative to q turns out to be perfectly natural in relativistic electrodynamics, where **E** and **B** are associated in a single field tensor.

While we shall go no deeper into this type of analysis, it is already evident that when the laws of physics are expressed in appropriate four-dimensional form, the result is a considerable simplification that clearly exhibits relationships which otherwise might be obscure or even overlooked. Historically, the incorporation of special relativity into physics via four-vectors and four-tensors has led both to a deeper understanding of natural laws already known and to the discovery of new phenomena and relationships.

Let us use some of the notions of space-time to confirm the statements made in the last paragraph of Sec. 1.8. Figure 1.13 shows two events plotted on the rectangular axes x and ct. Event 1 occurs at $x = 0$, $t = 0$ and event 2 occurs at $x = \Delta x$, $t = \Delta t$. The *interval* Δs between them is defined by

1.35
$$(\Delta s)^2 = (c\Delta t)^2 - (\Delta x)^2$$
Interval between events

The virtue of this definition is that $(\Delta s)^2$, like the s^2 of Eq. 1.34, is invariant under Lorentz transformations. If Δx and Δt are the differences in space and time between two events measured in the S frame and $\Delta x'$ and $\Delta t'$ are the same quantities measured in the S' frame,

$$(\Delta s)^2 = (c\Delta t)^2 - (\Delta x)^2 = (c\Delta t')^2 - (\Delta x')^2$$

Therefore whatever conclusions we arrive at in the S frame in which event 1 is at the origin hold equally well in any other frame in relative motion at constant velocity.

Now let us look into the possible relationships between events 1 and 2. Event 2 can be related causally in some way to event 1 provided that a signal traveling slower than the speed of light can connect these events, that is, provided that

$$c\Delta t > |\Delta x|$$

or

$$(\Delta s)^2 > 0 \qquad \text{Timelike interval}$$

An interval in which $(\Delta s)^2 > 0$ is said to be *timelike*. Every timelike interval that connects event 1 with another event lies within the *light cones* bounded by $x = \pm ct$ in Fig. 1.13. All events that could have affected event 1 lie in the past light cone; all events that event 1 is able to affect lie in the future light cone. (Events connected by timelike intervals need not *necessarily* be related, of course, but it is *possible* for them to be related.)

Conversely, the criterion for there being no causal relationship between events 1 and 2 is that

$$c\Delta t < |\Delta x|$$

FIGURE 1.13 The past and future light cones of event 1.

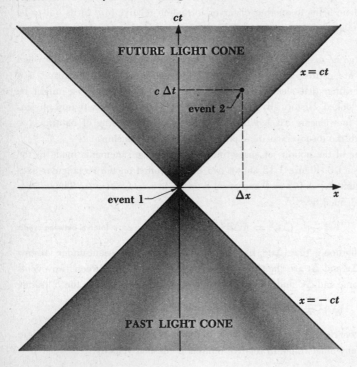

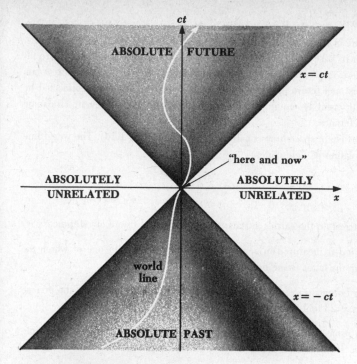

FIGURE 1.14　The world line of a particle.

or

$$(\Delta s)^2 < 0 \qquad \text{Spacelike interval}$$

An interval in which $(\Delta s)^2 < 0$ is said to be *spacelike*. Every event that is connected with event 1 by a spacelike interval lies outside the light cones of event 1 and neither has interacted with event 1 in the past nor is capable of interacting with it in the future; the two events must be entirely unrelated.

When events 1 and 2 can be connected with a light signal only,

$$c\Delta t = |\Delta x|$$

or

$$\Delta s = 0 \qquad \text{Lightlike interval}$$

An interval in which $\Delta s = 0$ is said to be *lightlike*. Events that can be connected with event 1 by lightlike intervals lie on the boundaries of the light cones.

These conclusions hold in terms of the light cones of event 2 because $(\Delta s)^2$ is invariant; for example, if event 2 is inside the past light cone of event 1, event 1 is inside the future light cone of event 2. In general, events that lie in the future of an event as seen in one frame of reference S lie in its future in every other frame S', and events that lie in the past of an event in S lie in its past in every other frame S'. Thus "future" and "past" have invariant meanings. However, "simultaneity" is an ambiguous concept, because all events that lie outside the past and future light cones of event 1 (that is, all events connected by spacelike intervals with event 1) can appear to occur simultaneously with event 1 in some particular frame of reference.

The path of a particle in space-time is called its *world line* (Fig. 1.14). The world line of a particle must lie within its light cones.

Problems

1. Is a laboratory at rest on the earth's surface really an inertial frame of reference?

2. The length of a rod is measured in several frames of reference, in one of which the rod is at rest. How can the latter frame be identified?

3. Starting from the Lorentz transformation equations 1.19 to 1.22, prove that the inverse transformation of Eqs. 1.23 to 1.26 is correct.

4. A meter stick is projected into space at so great a speed that its length appears contracted to only 50 cm. How fast is it going in miles per second?

5. A rocket ship is 100 m long on the ground. When it is in flight, its length is 99 m to an observer on the ground. What is its speed?

6. A rocket ship leaves the earth at a speed of $0.98c$. How much time does it take for the minute hand of a clock in the ship to make a complete revolution as measured by an observer on the earth?

7. A rocket ship travels away from the earth at 300 m/sec. How many years must elapse before a clock in the ship and one on the ground differ by 1 sec?

8. (a) The density of a substance is ρ in the S frame in which it is at rest. Find the density ρ' that an observer in the S' frame moving at a speed relative to S of v would determine. (b) Gold has a density of 19.3 g/cm³ when the sample is at rest relative to the observer. What is its density when the relative velocity is $0.9c$?

9. A certain particle has a lifetime of 10^{-7} sec when measured at rest. How far does it go before decaying if its speed is $0.99c$ when it is created?

10. Light of frequency ν is emitted by a source. An observer moving away from the source at the speed v measures a frequency of ν'. By considering the source as a clock that ticks ν times per second and gives off a pulse of light with each tick, show that

$$\nu' = \nu \sqrt{\frac{1 - v/c}{1 + v/c}}$$

This constitutes the *longitudinal doppler effect* in light. Why does the above formula differ from the corresponding one for sound waves in air?

11. The *transverse doppler effect*, which has no nonrelativistic counterpart, applies to measurements of light waves made by an observer in relative motion perpendicular to the direction of propagation of the waves. (In the preceding problem the observer moves parallel to the direction of propagation.) Show that in the transverse doppler effect

$$\nu' = \nu \sqrt{1 - v^2/c^2}$$

12. In 1903 Trouton and Noble performed an interesting experiment to detect the existence of an absolute frame of reference. What they did, in essence, was to place a charge $+q$ at one end of a rod L long and another charge $-q$ at the opposite end. The rod was suspended from its center with a tiny fiber so it could rotate in a horizontal plane. If it is supposed that all electromagnetic phenomena take place in a stationary frame of reference (the "ether"), then the charges are moving through the ether with the speed v, and each charge gives rise to a magnetic field. If the angle between the rod and the direction of v is θ, show that the torque on the rod produced by the action of these magnetic fields on the moving charges is $(q^2v^2 \sin 2\theta)/8\pi\varepsilon_0 Lc^2$.

Chapter 2
Relativistic Mechanics

We have seen that such fundamental physical quantities as length and time have meaning only when the particular reference frame in which they are measured is specified. Given this frame, we may compute what the values of these quantities are when measured in other reference frames in motion relative to the specified frame by applying the Lorentz transformation equations. An event in time and space—a collision between two bodies, for instance—will have a different appearance in different frames of reference. According to the postulates of special relativity, however, the laws of motion arrived at by observing events of this kind must have the same form in all frames of reference. In this chapter we examine how the principles of energy and momentum conservation and the theorem of velocity addition must be modified in the light of this requirement. Our discussion of relativity concludes with a glance at some of the ideas of general relativity.

2.1 Velocity Addition

One of the postulates of special relativity states that the speed of light c in free space has the same value for all observers, regardless of their relative motion. But "common sense" tells us that, if we throw a ball forward at 50 ft/sec from a car moving at 80 ft/sec, the ball's speed relative to the ground is 130 ft/sec, the sum of the two speeds. Hence we would expect that a ray of light emitted in a frame of reference S' in the direction of its motion at velocity v relative to another frame S will have a speed of $c + v$ as measured in S, contradicting the above postulate. "Common sense" is no more reliable as a guide in science than it is elsewhere, and we must turn to the Lorentz transformation equations for the correct scheme of velocity addition.

Let us consider something moving relative to both S and S'. An observer in S measures its three velocity components to be

$$V_x = \frac{dx}{dt}$$

$$V_y = \frac{dy}{dt}$$

$$V_z = \frac{dz}{dt}$$

while to an observer in S' they are

$$V'_x = \frac{dx'}{dt'}$$

$$V'_y = \frac{dy'}{dt'}$$

$$V'_z = \frac{dz'}{dt'}$$

By differentiating the Lorentz transformation equations for x', y', z', and t', we obtain

$$dx' = \frac{dx - v\,dt}{\sqrt{1 - v^2/c^2}}$$

$$dy' = dy$$

$$dz' = dz$$

$$dt' = \frac{dt - \dfrac{v\,dx}{c^2}}{\sqrt{1 - v^2/c^2}}$$

and so

$$V'_x = \frac{dx'}{dt'}$$

$$= \frac{dx - v\,dt}{dt - \dfrac{v\,dx}{c^2}}$$

$$= \frac{\dfrac{dx}{dt} - v}{1 - \dfrac{v}{c^2}\dfrac{dx}{dt}}$$

2.1 $$= \frac{V_x - v}{1 - \dfrac{vV_x}{c^2}}$$ Relativistic velocity transformation

Velocity Addition   **31**

$$V'_y = \frac{dy'}{dt'}$$

$$= \frac{dy\ \sqrt{1 - v^2/c^2}}{dt - \dfrac{v\ dx}{c^2}}$$

$$= \frac{\dfrac{dy}{dt}\ \sqrt{1 - v^2/c^2}}{1 - \dfrac{v}{c^2}\dfrac{dx}{dt}}$$

2.2
$$= \frac{V_y\ \sqrt{1 - v^2/c^2}}{1 - \dfrac{vV_x}{c^2}}$$
  Relativistic velocity transformation

and, similarly,

2.3
$$V'_z = \frac{V_z\ \sqrt{1 - v^2/c^2}}{1 - \dfrac{vV_x}{c^2}}$$
  Relativistic velocity transformation

Equations 2.1 to 2.3 constitute the relativistic velocity transformation. When v is small compared with c, these equations reduce to the classical equations

$$V'_x = V_x - v$$
$$V'_y = V_y$$
$$V'_z = V_z$$

To express V_x, V_y, V_z in terms of V'_x, V'_y, V'_z, we need only replace v with $-v$ and exchange primed and unprimed quantities in Eqs. 2.1 to 2.3. Hence

2.4
$$V_x = \frac{V'_x + v}{1 + \dfrac{vV'_x}{c^2}}$$
  Inverse velocity transformation

2.5
$$V_y = \frac{V'_y\ \sqrt{1 - v^2/c^2}}{1 + \dfrac{vV'_x}{c^2}}$$
  Inverse velocity transformation

2.6
$$V_z = \frac{V'_z\ \sqrt{1 - v^2/c^2}}{1 + \dfrac{vV'_x}{c^2}}$$
  Inverse velocity transformation

If $V'_x = c$, that is, if a ray of light is emitted in the moving reference frame S' in its direction of motion relative to S, an observer in frame S will measure the velocity

$$V_x = \frac{V'_x + v}{1 + \dfrac{vV'_x}{c^2}}$$

$$= \frac{c + v}{1 + \dfrac{vc}{c^2}}$$

$$= \frac{c(c + v)}{c + v}$$

$$= c$$

Both observers determine the same value for the speed of light, as they must according to the first postulate of relativity.

The relativistic velocity transformation has other peculiar consequences. For instance, we might imagine wishing to pass a spaceship whose speed with respect to the earth is $0.9c$ at a relative speed of $0.5c$. According to conventional mechanics our required speed relative to the earth would have to be $1.4c$, more than the velocity of light. According to Eq. 2.4, however, with $V'_x = 0.5c$ and $v = 0.9c$, the necessary speed is only

$$V_x = \frac{V'_x + v}{1 + \dfrac{vV'_x}{c^2}}$$

$$= \frac{0.5c + 0.9c}{1 + \dfrac{(0.9c)(0.5c)}{c^2}}$$

$$= 0.9655c$$

which is less than c. We need go less than 10 percent faster than a spaceship traveling at $0.9c$ in order to pass it at a relative speed of $0.5c$.

2.2   The Relativity of Mass

Until now we have been considering only the purely kinematical aspects of special relativity. The dynamical consequences of relativity are at least as remarkable, including as they do the variation of mass with velocity and the equivalence of mass and energy.

We begin by considering an elastic collision (that is, a collision in which kinetic energy is conserved) between two particles A and B, as witnessed by observers in the reference frames S and S' which are in uniform relative motion. The properties of A and B are identical when determined in reference frames in which they are at rest. The frames S and S' are oriented as in Fig. 2.1, with S' moving in the $+x$ direction with respect to S at the velocity v.

Before the collision, particle A has been at rest in frame S and particle B in frame S'.

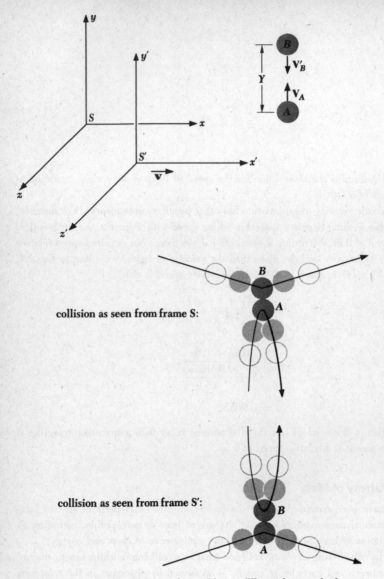

collision as seen from frame S:

collision as seen from frame S':

FIGURE 2.1   An elastic collision as observed in two different frames of reference.

Then, at the same instant, A is thrown in the $+y$ direction at the speed V_A while B is thrown in the $-y'$ direction at the speed V_B', where

2.7 $$V_A = V_B'$$

Hence the behavior of A as seen from S is exactly the same as the behavior of B as seen from S'. When the two particles collide, A rebounds in the $-y$ direction at the speed V_A, while B rebounds in the $+y'$ direction at the speed V_B. If the particles are thrown from positions Y apart, an observer in S finds that the collision occurs at $y = \frac{1}{2}Y$ and one in S' finds that it occurs at $y' = \frac{1}{2}Y$. The round-trip time T_0 for A as measured in frame S is therefore

2.8
$$T_0 = \frac{Y}{V_A}$$

and it is the same for B in S',

$$T_0 = \frac{Y}{V_B'}$$

If momentum is conserved in the S frame, it must be true that

2.9
$$m_A V_A = m_B V_B$$

where m_A and m_B are the masses of A and B, and V_A and V_B their velocities *as measured in the S frame*. In S the speed V_B is found from

2.10
$$V_B = \frac{Y}{T}$$

where T is the time required for B to make its round trip *as measured in S*. In S', however, B's trip requires the time T_0, where

2.11
$$T = \frac{T_0}{\sqrt{1 - v^2/c^2}}$$

according to our previous results. Although observers in both frames see the same event, they disagree as to the length of time the particle thrown from the other frame requires to make the collision and return.

Replacing T in Eq. 2.10 with its equivalent in terms of T_0, we have

$$V_B = \frac{Y\sqrt{1 - v^2/c^2}}{T_0}$$

From Eq. 2.8

$$V_A = \frac{Y}{T_0}$$

Inserting these expressions for V_A and V_B in Eq. 2.9, we see that momentum is conserved if

2.12
$$m_A = m_B\sqrt{1 - v^2/c^2}$$

The Relativity of Mass 35

Our original hypothesis was that A and B are identical when at rest with respect to an observer; the difference between m_A and m_B therefore means that measurements of mass, like those of space and time, depend upon the relative speed between an observer and whatever he is observing.

In the above example both A and B are moving in S. In order to obtain a formula giving the mass m of a body measured while in motion in terms of its mass m_0 when measured at rest, we need only consider a similar example in which V_A and V_B' are very small. In this case an observer in S will see B approach A with the velocity v, make a glancing collision (since $V_B' \ll v$), and then continue on. In S

$$m_A = m_0$$

and

$$m_B = m$$

and so

2.13
$$m = \frac{m_0}{\sqrt{1 - v^2/c^2}}$$
Relativistic mass

The mass of a body moving at the speed v relative to an observer is larger than its mass when at rest relative to the observer by the factor $1/\sqrt{1 - v^2/c^2}$. This mass increase is reciprocal; to an observer in S'

$$m_A = m$$

and

$$m_B = m_0$$

Measured from the earth, a spaceship in flight is shorter than its twin still on the ground and its mass is greater. To somebody on the spaceship in flight the ship on the ground also appears shorter and to have a greater mass. (The effect is, of course, unobservably small for present-day rocket speeds.) Equation 2.13 is plotted in Fig. 2.2.

Provided that momentum is defined as

2.14
$$mv = \frac{m_0 v}{\sqrt{1 - v^2/c^2}}$$

conservation of momentum is valid in special relativity just as in classical physics. However, Newton's second law of motion is correct only in the form

$$F = \frac{d}{dt}(mv)$$

2.15
$$= \frac{d}{dt}\left(\frac{m_0 v}{\sqrt{1 - v^2/c^2}}\right)$$

This is *not* equivalent to saying that

$$F = ma$$

$$= m \frac{dv}{dt}$$

even with m given by Eq. 2.13 because

$$\frac{d}{dt}(mv) = m\frac{dv}{dt} + v\frac{dm}{dt}$$

and dm/dt does not vanish if the speed of the body varies with time. The resultant force on a body is always equal to the time rate of change of its momentum.

Relativistic mass increases are significant only at speeds approaching that of light. At a speed one-tenth that of light the mass increase amounts to only 0.5 percent, but this increase is over 100 percent at a speed nine-tenths that of light. Only atomic particles such as electrons, protons, mesons, and so on have sufficiently high speeds for relativistic effects to be measurable, and in dealing with these particles the "ordinary" laws of physics cannot be used. Historically, the first confirmation of Eq. 2.13 was the discovery by Bücherer in 1908 that the ratio e/m of the electron's charge to its mass is smaller for fast electrons than for slow ones; this equation, like the others of special relativity, has been verified by so many experiments that it is now recognized as one of the basic formulas of physics.

FIGURE 2.2   The relativity of mass.

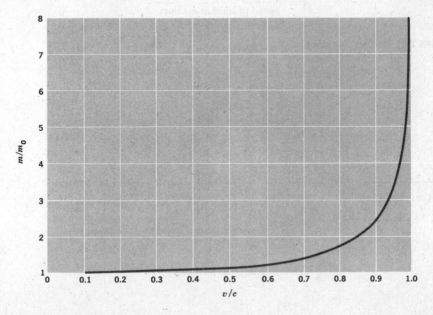

2.3 The Čerenkov Effect

The relativity of mass places a limit on the speed a body can have relative to an observer. As the body's speed v approaches the speed of light c, the ratio v^2/c^2 approaches 1 and the quantity $\sqrt{1 - v^2/c^2}$ approaches 0. Hence the mass m of the body, given by

$$m = \frac{m_0}{\sqrt{1 - v^2/c^2}}$$

approaches infinity as v approaches c. The relative speed v therefore must be less than c. An infinite force would be needed to accelerate a body to a speed at which its mass is infinite, but there are neither infinite forces nor infinite masses in the universe.

The speed of light c in special relativity is the *speed of light in free space*, 3×10^8 m/sec. In all material media, such as water, glass, or air, light travels more slowly than this, and atomic particles are capable of moving with higher speeds in such media than the speed of light *in them*, though never faster than the speed of light in free space. When an electrically charged particle moves through a substance at a speed exceeding that of light in the substance, a cone of light waves is emitted in a process roughly similar to that in which a ship produces a bow wave as it moves through the water at a speed greater than that of water waves. These light waves are known as *Čerenkov radiation*.

The production of Čerenkov radiation is shown in Fig. 2.3. The medium through which the particle is moving has an index of refraction n, which is defined as

$$n = \frac{\text{speed of light in free space}}{\text{speed of light in medium}}$$

FIGURE 2.3 The Čerenkov effect.

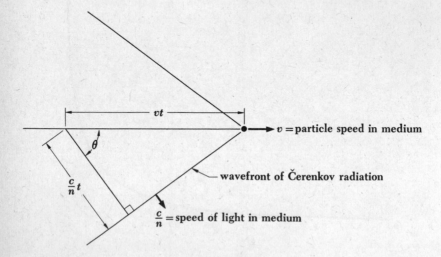

vt

$v =$ particle speed in medium

θ

$\frac{c}{n}t$

wavefront of Čerenkov radiation

$\frac{c}{n} =$ speed of light in medium

The speed of light in the medium is therefore c/n, which is less than the particle's speed v. As the particle travels through the medium, it creates disturbances along its path from which light waves spread out at the speed c/n. The envelope of these waves is a cone with the particle at its apex, where, from the figure, the angle θ between the direction of propagation of the radiation and the particle's direction is specified by

$$\cos \theta = \frac{(c/n)t}{vt}$$

2.16
$$= \frac{c}{nv}$$

When energetic atomic particles move through such media as glass or transparent plastics, the cone of Čerenkov radiation can be detected and the angle θ measured, making it possible to determine the speed of the particles. The Čerenkov radiation is visible as a bluish glow when an intense beam of such particles is involved.

2.4   Mass and Energy

The most famous relationship Einstein obtained from the postulates of special relativity concerns mass and energy. This relationship can be derived directly from the definition of the kinetic energy T of a moving body as the work done in bringing it from rest to its state of motion. That is,

$$T = \int_0^s F \, ds$$

where F is the component of the applied force in the direction of the displacement ds and s is the distance over which the force acts. Using the relativistic form of the second law of motion

$$F = \frac{d\,(mv)}{dt}$$

we have

$$T = \int_0^s \frac{d(mv)}{dt} \, ds$$
$$= \int_0^{mv} v \, d(mv)$$
$$= \int_0^v v \, d\left(\frac{m_0 v}{\sqrt{1 - v^2/c^2}}\right)$$

Integrating by parts ($\int x \, dy = xy - \int y \, dx$),

$$T = \frac{m_0 v^2}{\sqrt{1 - v^2/c^2}} - m_0 \int_0^v \frac{v\, dv}{\sqrt{1 - v^2/c^2}}$$

$$= \frac{m_0 v^2}{\sqrt{1 - v^2/c^2}} + m_0 c^2 \sqrt{1 - v^2/c^2} \Big|_0^v$$

$$= \frac{m_0 c^2}{\sqrt{1 - v^2/c^2}} - m_0 c^2$$

2.17
$$= mc^2 - m_0 c^2$$

Equation 2.17 states that the kinetic energy of a body is equal to the increase in its mass consequent upon its relative motion multiplied by the square of the speed of light.

Equation 2.17 may be rewritten

2.18
$$mc^2 = T + m_0 c^2 \qquad \text{Total energy}$$

If we interpret mc^2 as the *total energy* E of the body, it follows that, when the body is at rest and $T = 0$, it nevertheless possesses the energy $m_0 c^2$. Accordingly $m_0 c^2$ is called the *rest energy* E_0 of a body whose mass at rest is m_0. Equation 2.18 therefore becomes

2.19
$$E = E_0 + T$$

where

2.20
$$E = mc^2$$

and

2.21
$$E_0 = m_0 c^2 \qquad \text{Rest energy}$$

In addition to its kinetic, potential, electromagnetic, thermal, and other familiar guises, then, energy can manifest itself as mass. The conversion factor between the unit of mass (kg) and the unit of energy (joule) is c^2, so 1 kg of matter has an energy content of 9×10^{16} joules. Even a minute bit of matter represents a vast amount of energy, and, in fact, the conversion of matter into energy is the source of the power liberated in all of the exothermic reactions of physics and chemistry.

Since mass and energy are not independent entities, the separate conservation principles of energy and mass are properly a single one, the principle of conservation of mass energy. Mass *can* be created or destroyed, but only if an equivalent amount of energy simultaneously vanishes or comes into being, and vice versa.

When the relative speed v is small compared with c, the formula for kinetic energy must reduce to the familiar $\frac{1}{2}m_0 v^2$, which has been verified by experiment at low speeds. Let us see whether this is true. As we saw in Sec. 1.1, the binomial theorem of algebra tells us that if some quantity x is much smaller than 1,

$$(1 \pm x)^n \approx 1 \pm nx$$

The relativistic formula for kinetic energy is

$$T = mc^2 - m_0c^2$$

$$= \frac{m_0c^2}{\sqrt{1 - v^2/c^2}} - m_0c^2$$

Expanding the first term of this formula with the help of the binomial theorem, with $v^2/c^2 \ll 1$ since v is much less than c,

$$T = (1 + \tfrac{1}{2}v^2/c^2)m_0c^2 - m_0c^2$$

$$= \tfrac{1}{2}m_0v^2$$

Hence at low speeds the relativistic expression for the kinetic energy of a moving particle reduces to the classical one. The total energy of such a particle is

$$E = m_0c^2 + \tfrac{1}{2}m_0v^2$$

In the foregoing calculation relativity has once again met an important test; it has yielded exactly the same results as those of ordinary mechanics at low speeds, where we know by experience that the latter are perfectly valid. It is nevertheless important to keep in mind that, so far as is known, the correct formulation of mechanics has its basis in relativity, with classical mechanics no more than an approximation correct only under certain circumstances.

2.5   Some Relativistic Formulas

It is often convenient to express several of the relativistic formulas obtained above in forms somewhat different from their original ones. The new equations are so easy to derive that we shall simply state them without proof:

2.22
$$E = \sqrt{m_0^2c^4 + p^2c^2}$$

2.23
$$p = m_0c \sqrt{\frac{1}{1 - v^2/c^2} - 1}$$

2.24
$$T = m_0c^2 \left(\frac{1}{\sqrt{1 - v^2/c^2}} - 1 \right)$$

2.25
$$\frac{v}{c} = \sqrt{1 - \frac{1}{[1 + (T/m_0c^2)]^2}}$$

2.26
$$\frac{1}{\sqrt{1 - v^2/c^2}} = \sqrt{1 + \frac{p^2}{m_0^2c^2}}$$

2.27
$$= 1 + \frac{T}{m_0c^2}$$

These formulas are particularly useful in nuclear and elementary-particle physics, where the kinetic energies of moving particles are customarily specified, rather than their velocities. Equation 2.25, for instance, permits us to find v/c directly from T/m_0c^2, the ratio between the kinetic and rest energies of a particle. Table 2.1 is a list of the rest energies of the most stable elementary particles expressed in millions of electron volts (Mev), where 1 Mev $= 1.6 \times 10^{-13}$ joule; the electron volt is the usual energy unit in atomic and nuclear physics.

From Eq. 2.22 we see that, when $E \gg m_0c^2$,

2.28 $$E \approx pc$$

Hence, if we establish as our momentum unit the Mev/c, where, since 1 Mev $= 1$ million ev $= 10^6$ ev,

$$1 \text{ Mev}/c = \frac{10^6 \text{ ev} \times 1.6 \times 10^{-19} \text{ joule/ev}}{3 \times 10^8 \text{ m/sec}}$$

$$= 5.3 \times 10^{-22} \text{ kg-m/sec}$$

the energy and momentum of a particle will be numerically the same when $E \gg m_0c^2$.

TABLE 2.1. REST MASSES OF THE MOST STABLE ELEMENTARY PARTICLES

Particle	Symbol*	Rest mass, Mev
Photon	$\gamma(\overline{\gamma} \equiv \gamma)$	0
Neutrino	$\nu_e, \nu_\mu (\overline{\nu}_e, \overline{\nu}_\mu)$	0
Electron	$e(e^+)$	0.51
Mu meson	$\mu^-(\mu^+)$	106
Pi meson	$\pi^+(\pi^-)$	140
	$\pi^0(\overline{\pi^0} \equiv \pi^0)$	135
K meson	$K^+(K^-)$	494
	$K^0(\overline{K^0})$	498
Eta meson	$\eta^0(\overline{\eta^0} \equiv \eta^0)$	548
Proton	$p(\overline{p})$	938.2
Neutron	$n(\overline{n})$	939.5
Λ Hyperon	$\Lambda(\overline{\Lambda})$	1115
Σ Hyperon	$\Sigma^+(\overline{\Sigma^-})$	1192
	$\Sigma^0(\overline{\Sigma^0})$	1194
	$\Sigma^-(\overline{\Sigma^+})$	1197
Ξ Hyperon	$\Xi^0(\overline{\Xi^0})$	1310
	$\Xi^-(\overline{\Xi^+})$	1320
Ω Hyperon	$\Omega^-(\overline{\Omega^+})$	1676

* Antiparticle symbol in parenthesis.

2.6 Mass and Energy: Alternative Derivation

The equivalence of mass and energy can be demonstrated in a number of different ways. An interesting derivation that is somewhat different from the one given in the text, also suggested by Einstein, makes use of the basic notion that the center of mass of an isolated system (one that does not interact with its surroundings) cannot be changed by any process occurring within the system. In this derivation we imagine a closed box from one end of which a burst of electromagnetic radiation is emitted, as in Fig. 2.4. This radiation carries energy and momentum, and when the emission occurs, the box recoils in order that the total momentum of the system remain constant. When the radiation is absorbed at the opposite end of the box, its momentum cancels the momentum of the box, which then comes to rest. During the time in which the radiation was in transit, the box has moved a distance s. If the center of mass of the system is still to be in the same location in space, the radiation must have transferred mass from the end at which it was emitted to the end at which it was absorbed. We shall compute the amount of mass that must be transferred if the center of mass of the system is to remain unchanged.

The first step is to find out how much momentum is associated with a burst of electromagnetic radiation whose energy content is W. (We shall use W for energy temporarily to avoid confusion with electric-field intensity E.) Let us consider what happens when an electromagnetic wave traveling in the $+z$ direction interacts with a charge e; since all such waves interact with matter in the way to be described, the discussion is a perfectly general one. As in Fig. 2.5, the coordinate axes are chosen so that the electric field $\mathbf{E}$ is in the x direction and the magnetic field $\mathbf{B}$ in the y direction. At the instant shown in the diagram,

Figure 2.4 Radiant energy possesses inertial mass.

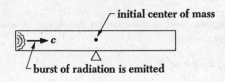

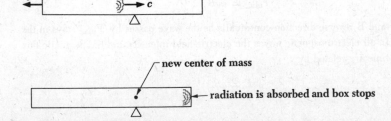

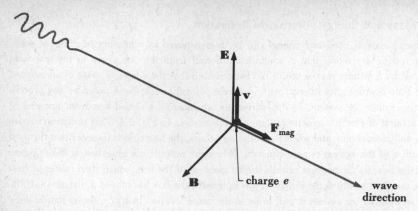

FIGURE 2.5 The magnetic field of an electromagnetic wave exerts a forward force on a charge in the path of the wave that has been set in transverse motion by the wave's electric field.

the electric field of the wave has exerted an upward electric force

$$F_{elec} = eE$$

on the charge that has given it an upward speed of v. The rate dW/dt at which the charge is absorbing energy from the wave at this instant is therefore

$$\frac{dW}{dt} = F_{elec} \times v_{charge}$$

2.29
$$= eEv$$

Now we turn to the effect of the magnetic field of the wave on the charge. A moving charge in a magnetic field is acted upon by the force

$$\mathbf{F}_{mag} = q\,\mathbf{v} \times \mathbf{B}$$

so that here, with $\mathbf{v}$ and $\mathbf{B}$ in the $+x$ and $+y$ directions respectively, $\mathbf{F}_{mag}$ is in the $+z$ direction (the direction of the wave) and has the magnitude

$$F_{mag} = evB$$

Even though $\mathbf{v}$ and $\mathbf{B}$ reverse direction continually as the wave passes by, $\mathbf{F}_{mag}$ stays in the $+z$ direction. In all electromagnetic waves the electric-field intensity and the magnetic flux density at any time are related by

$$E = cB$$

and so

$$F_{mag} = \frac{eEv}{c}$$

From Eq. 2.29, $eEv = dW/dt$, the rate at which energy is being absorbed from the wave, which means we can write

$$F_{mag} = \frac{1}{c}\frac{dW}{dt}$$

for the force in the z direction and

$$F_{mag}\, dt = \frac{1}{c}\, dW$$

for the impulse given to the charge in the time dt. The total momentum p transferred to the charge during the passage of the wave is the integral of the impulse, and so

$$p = \int F_{mag}\, dt = \frac{1}{c}\int dW$$

2.30
$$= \frac{W}{c}$$

where W is the energy transferred to the charge.

The preceding mechanism is responsible for *radiation pressure*, an example of which is the deflection of comet tails by sunlight so that they always point away from the sun.

Returning to the hypothetical experiment of Fig. 2.4, for simplicity we shall consider the sides of the box to be massless and its ends to have the mass $\frac{1}{2}M$ each. The center of mass is therefore at the center of the box, a distance $\frac{1}{2}L$ from each end. A burst of electromagnetic radiation that has the energy E carries the momentum E/c and, we suppose, has associated with it an amount of mass m. When the radiation is emitted, the box, whose mass is now $M - m$, recoils with the velocity v. From the principle of conservation of momentum,

$$p_{box} = p_{radiation}$$

$$(M - m)v = \frac{E}{c}$$

and so the recoil velocity of the box is

$$v = \frac{E}{(M - m)c} \approx \frac{E}{Mc}$$

since m is much smaller than M. The time t during which the box moves is equal to the time required by the radiation to reach the opposite end of the box, a distance L away; this means that $t = L/c$. During the time t the box is displaced to the left by $s = vt = EL/Mc^2$.

Mass and Energy: Alternative Derivation

45

After the box has stopped, the mass of its left-hand end is $\tfrac{1}{2}M - m$ and the mass of its right-hand end is $\tfrac{1}{2}M + m$ owing to the transfer of the mass m associated with the energy E of the radiation. If the center of mass is to be in the same place it was originally,

$$(\tfrac{1}{2}M - m)(\tfrac{1}{2}L + s) = (\tfrac{1}{2}M + m)(\tfrac{1}{2}L - s)$$

or

$$m = \frac{Ms}{L}$$

Inserting the value of the displacement s,

2.31
$$m = \frac{E}{c^2}$$

The mass associated with an amount of energy E is equal to E/c^2.

In the above derivation we assumed that the box is a perfectly rigid body: that the entire box starts to move when the radiation is emitted and the entire box comes to a stop when the radiation is absorbed. Actually, of course, there is no such thing as a rigid body that meets this specification; for example, the radiation, which travels with the speed of light, will arrive at the right-hand end of the box *before* that end begins to move! When the finite speed of elastic waves in the box is taken into account in a more elaborate calculation, however, the same result that $m = E/c^2$ is obtained.

2.7 General Relativity

The special theory of relativity originated in an attempt to express the laws of physics in such a way that they hold in all frames of reference moving at constant velocity with respect to one another. In fact, the statement that the laws of physics *can* be expressed in this way constitutes the first postulate of special relativity. The corresponding postulate of general relativity states that **the laws of physics may be expressed in equations having the same form in all frames of reference, regardless of their states of motion.** Thus the general theory covers accelerated as well as uniform motion, and in consequence it is able to describe gravitational phenomena. Although electromagnetism fits naturally into the framework of special relativity, its incorporation with gravity into the general theory to yield a unified field theory has not as yet been accomplished. Even the detailed description of gravity given by Einstein in 1915 in the framework of general relativity involves mathematical techniques too advanced for us to go into here.

The second postulate of general relativity states that **there is no way for an observer in a closed laboratory to distinguish between the effects produced by a gravitational field and those produced by an acceleration of the laboratory.** This postulate is known as the *principle of equivalence*, and follows from the experimental observation that the inertial mass of a body is always exactly equal to its gravitational mass. (Properly speaking, inertial mass is proportional to gravitational mass; the constant of proportionality is set equal to 1 by the appropriate choice of the gravitational constant G.) The inertial mass m_i of a body determines the acceleration a it receives from an applied force according to Newton's second law of motion,

2.32 $$F = m_i a$$

The gravitational mass m_g of a body determines the force it experiences when another body of mass M is located a distance r away according to the law of universal gravitation,

2.33 $$F_g = G \frac{M m_g}{r^2}$$

Let us consider a body near the earth's surface. If we drop this body, the force on it is given by Eq. 2.33 with $M = M_{earth}$ and $r = r_{earth}$, and its downward acceleration g is then given by Eq. 2.32 as

$$g = \frac{F_g}{m_i} = \frac{GM_{earth}}{r^2_{earth}} \frac{m_g}{m_i}$$

Since the acceleration of gravity g is observed to be the same for all bodies in free fall at the same location on the earth, we conclude that m_g/m_i is a constant. Recent experiments show that m_g/m_i is constant to at least 1 part in 10^{10}.

In the previous section we found that the inertial mass E/c^2 is associated with a light pulse of energy E. From the principle of equivalence, a gravitational mass of the same magnitude must also be associated with the light pulse, so that light ought to be subject to gravitational forces. For example, light rays that pass near the sun ought to be deflected toward it, just as the paths of the planets are deflected by the sun's gravitational field. A straightforward calculation based on the principle of equivalence predicts an angular deflection of 1.75 sec for light rays that just miss the sun; measurements made on light from stars near the sun in the sky during solar eclipses indicate that deflections of almost exactly this amount indeed occur.

Since we normally think of light rays in vacuum as traveling in "straight lines"—in fact, a straight line is usually defined as the path a light ray would take—the finding that light is bent in a gravitational field might lead us to reexamine our notions of the structure of space-time. In the previous chapter the geometry of space-time was explicitly regarded as euclidean, so that the one-dimensional motion of a particle could be described on a plot of x versus ct on a flat surface. An important result of general relativity is that space-time is non-euclidean; in a gravitational field, a plot of x versus ct for a particle should be made on a *curved* surface. According to general relativity, gravity alters the configuration of space-time near a body of matter, so that freely moving particles and light rays naturally pursue curved paths there rather than straight ones. What we think of as the force of gravity can thus be traced to a warping of the coordinate system of space-time.

Two other predictions of general relativity have also been confirmed by experiment.

General Relativity

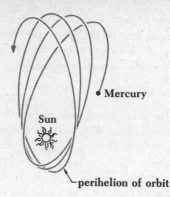

FIGURE 2.6 The advance of the perihelion of Mercury's orbit.

Mercury

Sun

perihelion of orbit

One is the advance of the perihelion of Mercury. The *perihelion* of a planet is the point in its orbit nearest the sun; in the case of Mercury, the perihelion precesses a little less than 2° per century (Fig. 2.6). All but about 40 sec per century of the perihelion advance can be accounted for in terms of orbital perturbations produced by the other planets—and Einstein showed that an advance of 43 sec per century is to be expected on the basis of general relativity.

The third prediction of general relativity is that a stationary clock runs more slowly in a strong gravitational field than in a weak one. A clock that has an interval of t_0 between ticks when it is infinitely far from all matter will have an interval between ticks of

2.34
$$t = \frac{t_0}{\sqrt{1 - \dfrac{2GM}{R}}}$$
Clock in gravitational field

when it is the distance R from a mass M. This prediction is confirmed in the gravitational red shift in spectral lines, an effect which will be examined in Chap. 3.

2.8 The Twin Paradox

We are now in a position to understand the famous relativistic phenomenon known as the twin paradox. This paradox involves two identical clocks, one of which remains on the earth while the other goes on a voyage into space at the speed v and returns a time t later. It is customary to replace actual clocks with a pair of identical male twins named A and B; this substitution is perfectly acceptable, because the processes of life—heartbeats, respiration, and so on—constitute biological clocks on the average as regular as any other.

Twin A takes off when he is 20 years old and travels at a speed of $0.99c$. To his brother B on the earth, A seems to be living more slowly, in fact at a rate only

$$\sqrt{1 - v^2/c^2} = \sqrt{1 - (0.99c)^2/c^2} = 0.14 = 14 \text{ percent}$$

as fast as B does. For every breath that A takes, B takes 7; for every meal that A eats, B eats 7; for every thought that A thinks, B thinks 7. Finally, after 70 years have elapsed by B's reckoning, A returns home, a man of only 30 while B is then 90 years old.

Where is the paradox? If we examine the situation from the point of view of twin A in the spaceship, B on the earth is in motion at $0.99c$. Therefore we might expect B to be 30 years old upon the return of the spaceship while A is 90 at this time—the precise opposite of what was concluded in the preceding paragraph.

The resolution of the paradox depends upon the fact that the spaceship is accelerated at various times in its journey: when it takes off, when it turns around, and when it finally comes to a stop. Therefore the formulas of special relativity, which hold only for frames of reference in relative motion at constant velocity, cannot be applied to the situation at all, and we must turn to general relativity. By the principle of equivalence, a large acceleration produces effects indistinguishable from those produced by a strong gravitational field—and clocks tick more slowly in strong gravitational fields. (The earth's gravitational field is small compared with the gravitational field needed to cause accelerations of the magnitude a spaceship of speed $0.99c$ must experience.) Hence A, the space traveler, is indeed younger on his return than his twin brother. Of course, A's lifespan has not been extended *to him*, since however long his 10 years on the spaceship may have seemed to his brother B, it has been only 10 years as far as he is concerned.

Problems

1. A man on the moon observes two spaceships coming toward him from opposite directions at speeds of $0.8c$ and $0.9c$ respectively. What is the relative speed of the two spaceships as measured by an observer on either one?

2. An observer moving in the $+x$ direction at a speed (in the laboratory system) of 2.9×10^8 m/sec finds the speed of an object moving in the $-x$ direction to be 2.998×10^8 m/sec. What is the speed of the object in the laboratory system?

3. A man has a mass of 100 kg on the ground. When he is in a rocket ship in flight, his mass is 101 kg as determined by an observer on the ground. What is the speed of the rocket ship?

4. How fast must an electron move in order that its mass equal the rest mass of the proton?

5. Find the speed of a 0.1-Mev electron according to classical and relativistic mechanics.

6. It is possible for the electron beam in a television picture tube to move across the screen at a speed faster than the speed of light. Why does this not contradict special relativity?

7. A beam of 0.8-Mev electrons enters a transparent material whose index of refraction is 1.4. Find the angle between the beam and the emitted Čerenkov radiation.

8. What is the minimum speed a charged particle must have if its passage through a medium whose index of refraction is 1.33 is to be accompanied by the production of Čerenkov radiation?

9. A beam of protons of unknown energy moves through a transparent material whose index of refraction is 1.8. Čerenkov radiation is emitted at an angle of 11° with the beam. Find the kinetic energy of the protons in Mev.

10. How much mass does a proton gain when it is accelerated to a kinetic energy of 500 Mev?

11. How much mass does an electron gain when it is accelerated to a kinetic energy of 500 Mev?

12. The total energy of a particle is exactly twice its rest energy. Find its speed.

13. How much energy per kilogram of rest mass is required to accelerate a rocket ship to a speed of $0.98c$?

14. An electron has an initial speed of 1.4×10^8 m/sec. How much additional energy must be imparted to it for its speed to double?

15. A certain quantity of ice at $0°C$ melts into water at $0°C$ and in so doing gains 1 kg of mass. What was its initial mass?

16. Find the mass and speed of a 2-Mev electron.

17. What is the mass of a 1-Gev electron in terms of its rest mass? ($1 \text{ Gev} = 10^3 \text{ Mev} = 10^9 \text{ ev.}$)

18. Express the energy equivalent of 1 g of mass in kilowatt hours.

19. How much work must be done in order to increase the speed of an electron from 1.2×10^8 m/sec to 2.4×10^8 m/sec?

20. A particle has a kinetic energy of 62 Mev and a momentum of 335 Mev/c. Find its mass and speed.

21. A proton has a kinetic energy of m_0c^2. Find its momentum in units of Mev/c.

22. Prove that $\frac{1}{2}mv^2$, where $m = m_0/\sqrt{1 - v^2/c^2}$, does *not* equal the kinetic energy of a particle moving at relativistic speeds.

23. Express the relativistic form of the second law of motion, $F = d(mv)/dt$, in terms of m_0, v, c, and dv/dt.

24. A particle moving in the $+x$ direction in the reference frame S has the total energy E and linear momentum p in that frame. Show that the energy E' and momentum p' of this particle as determined by an observer in the frame S' that moves in the $+x$ direction relative to S with the speed v are given by

$$E' = \frac{E - pv}{\sqrt{1 - v^2/c^2}}$$

$$p' = \frac{p - vE/c^2}{\sqrt{1 - v^2/c^2}}$$

25. Show that the period of a pendulum is proportional to the square root of the ratio between the inertial and gravitational masses of the bob.

Chapter 3
Particle Properties
of Waves

In our everyday experience there is nothing mysterious or ambiguous about the concepts of *particle* and *wave*. A stone dropped into a lake and the ripples that spread out from its point of impact apparently have in common only the ability to carry energy and momentum from one place to another. Classical physics, which mirrors the "physical reality" of our sense impressions, treats particles and waves as separate components of that reality. The mechanics of particles and the optics of waves are traditionally independent disciplines, each with its own chain of experiments and hypotheses. But the physical reality we perceive arises from phenomena that occur in the microscopic world of atoms and molecules, electrons and nuclei, and in this world there are neither particles nor waves in our sense of these terms. We regard electrons as particles because they possess charge and mass and behave according to the laws of particle mechanics in such familiar devices as television picture tubes. We shall see, however, that there is as much evidence in favor of interpreting a moving electron as a wave manifestation as there is in favor of interpreting it as a particle manifestation. We regard electromagnetic waves as waves because under suitable circumstances they exhibit diffraction, interference, and polarization. Similarly, we shall see that under other circumstances electromagnetic waves behave as though they consist of streams of particles. Together with special relativity, the wave-particle duality is central to an understanding of modern physics, and in this book there are few arguments that do not draw upon these fundamental principles.

3.1   The Photoelectric Effect

Late in the nineteenth century a series of experiments revealed that electrons are emitted from a metal surface when light of sufficiently high frequency (ultraviolet light is required for all but the alkali metals) falls upon it. This phenomenon is known as the *photoelectric effect*.

Figure 3.1 illustrates the type of apparatus that was employed in the more precise of these experiments. An evacuated tube contains two electrodes connected to an external circuit like that shown schematically, with the metal plate whose surface is to be irradiated as the anode. Some of the photoelectrons that emerge from the irradiated surface have sufficient energy to reach the cathode despite its negative polarity, and they constitute the current that is measured by the ammeter in the circuit. As the retarding potential V is increased, fewer and fewer electrons get to the cathode and the current drops. Ultimately, when V equals or exceeds a certain value V^0, of the order of a few volts, no further electrons strike the cathode and the current ceases.

The existence of the photoelectric effect ought not to be surprising; after all, light waves carry energy, and some of the energy absorbed by the metal may somehow concentrate on individual electrons and reappear as kinetic energy. When we look more closely at the data, however, we find that the photoelectric effect can hardly be interpreted so simply.

One of the features of the photoelectric effect that particularly puzzled its discoverers is that the energy distribution in the emitted electrons (called *photoelectrons*) is independent of the intensity of the light. A strong light beam yields more photoelectrons than a weak one of the same frequency, but the average electron energy is the same (Fig. 3.2). Also, within the limits of experimental accuracy (about 3×10^{-9} sec), there is no time lag between the arrival of light at a metal surface and the emission of photoelectrons. These observations cannot be understood from the electromagnetic theory of light. Let us consider violet light falling on a sodium surface in an apparatus like that of Fig. 3.1. There will be a detectable photoelectric current when 10^{-6} watt/m^2 of electromagnetic energy is absorbed by the surface (a more intense beam than this is required, of course, since sodium is a good reflector of light). Now there are about 10^{19} atoms in a layer of sodium one atom thick and $1 \ m^2$ in area, so that, if we assume that the incident light is absorbed in the 10 upper-

FIGURE 3.1 Experimental observation of the photoelectric effect.

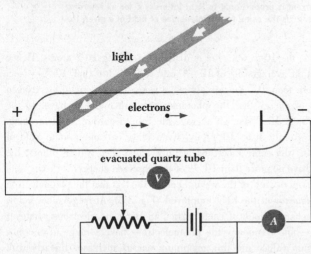

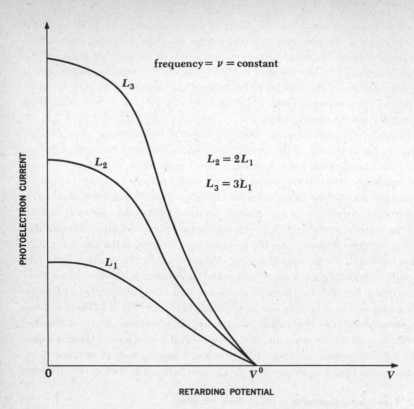

FIGURE 3.2   Photoelectron current is proportional to light intensity *L* for all retarding voltages. The extinction voltage V^0 is the same for all intensities of light of a given frequency ν.

most layers of sodium atoms, the 10^{-6} watt/m^2 is distributed among 10^{20} atoms. Hence each atom receives energy at the average rate of 10^{-26} watt, which is less than 10^{-7} ev/sec. It should therefore take more than 10^7 sec, or almost a year, for any single electron to accumulate the 1 ev or so of energy that the photoelectrons are found to have! In the maximum possible time of 3×10^{-9} sec, an average electron, according to electromagnetic theory, will have gained only 3×10^{-16} ev. Even if we call upon some kind of resonance process to explain why some electrons acquire more energy than others, the fortunate electrons will still have no more than 10^{-10} of the observed energy.

Equally odd from the point of view of the wave theory is the fact that the photoelectron energy depends upon the *frequency* of the light employed (Fig. 3.3). At frequencies below a certain critical frequency characteristic of each particular metal, no electrons whatever are emitted. Above this threshold frequency the photoelectrons have a range of energies from 0 to a certain maximum value, and this maximum energy increases linearly with

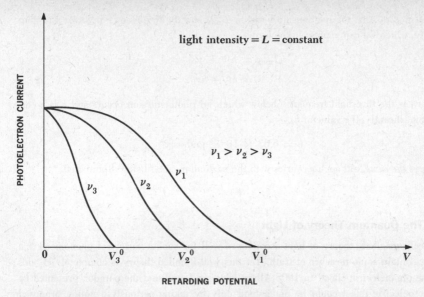

light intensity $= L =$ constant

$\nu_1 > \nu_2 > \nu_3$

ν_3 ν_2 ν_1

0 V_3^0 V_2^0 V_1^0 V

PHOTOELECTRON CURRENT

RETARDING POTENTIAL

FIGURE 3.3 The extinction voltage V^0 depends upon the frequency ν of the light. When the retarding potential is $V = 0$, the photoelectric current is the same for light of a given intensity L regardless of its frequency.

increasing frequency. High frequencies result in high maximum photoelectron energies, low frequencies in low maximum photoelectron energies. Thus a faint blue light produces electrons with more energy than those produced by a bright red light, although the latter yields a greater number of them.

Figure 3.4 is a plot of maximum photoelectron energy T_{max} versus the frequency ν of the incident light in a particular experiment. Since $T = eV$ in general, $T_{max} = eV^0$

FIGURE 3.4 The maximum photoelectron energy as a function of the frequency of the incident light.

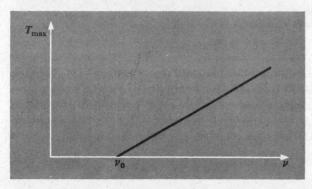

T_{max}

ν_0 ν

The Photoelectric Effect

here. It is clear that the relationship between T_{max} and the frequency ν involves a proportionality, which we can express in the form

$$T_{max} = h(\nu - \nu_0)$$

3.1
$$= h\nu - h\nu_0$$

where ν_0 is the threshold frequency below which no photoemission occurs and h is a constant. Significantly, the value of h,

$$h = 6.63 \times 10^{-34} \text{ joule-sec}$$

is *always the same*, although ν_0 varies with the particular metal being illuminated.

3.2   The Quantum Theory of Light

The electromagnetic theory of light accounts so well for such a variety of phenomena that it must contain some measure of truth. Yet this well-founded theory is completely at odds with the photoelectric effect. In 1905 Albert Einstein found that the paradox presented by the photoelectric effect could be understood only by taking seriously a notion proposed five years earlier by the German theoretical physicist Max Planck. Planck was seeking to explain the characteristics of the radiation emitted by bodies hot enough to be luminous, a problem notorious at the time for its resistance to solution. Planck was able to derive a formula for the spectrum of this radiation (that is, the relative brightness of the various colors present) as a function of the temperature of the body that was in agreement with experiment, provided he assumed that the radiation is emitted *discontinuously* as little bursts of energy (see Secs. 16.2 to 16.4 for details of this analysis). These bursts of energy are called *quanta*. Planck found that the quanta associated with a particular frequency ν of light all have the same energy and that this energy E is directly proportional to ν. That is,

3.2
$$E = h\nu$$
Quantum energy

where h, today known as *Planck's constant*, has the value

$$h = 6.63 \times 10^{-34} \text{ joule-sec}$$

While he had to assume that the electromagnetic energy radiated by a hot object emerges intermittently, Planck did not doubt that it propagates continuously through space as electromagnetic waves. Einstein proposed that light not only is emitted a quantum at a time, but also propagates as individual quanta. In terms of this hypothesis the photoelectric effect can be readily explained. The empirical formula Eq. 3.1 may be rewritten

3.3
$$h\nu = T_{max} + h\nu_0$$
Photoelectric effect

Einstein's proposal means that the three terms of Eq. 3.3 are to be interpreted as follows:

$$h\nu = \text{energy content of each quantum of the incident light}$$

$$T_{max} = \text{maximum photoelectron energy}$$

$$h\nu_0 = \text{minimum energy needed to dislodge an electron from the metal surface being illuminated}$$

There must be a minimum energy required by an electron in order to escape from a metal surface, or else electrons would pour out even in the absence of light. The energy $h\nu_0$ characteristic of a particular surface is called its *work function*. Hence Eq. 3.3 states that

$$\frac{\text{Quantum}}{\text{energy}} = \frac{\text{maximum electron}}{\text{energy}} + \frac{\text{work function}}{\text{of surface}}$$

It is easy to see why not all photoelectrons have the same energy, but emerge with all energies up to T_{max}: $h\nu_0$ is the work that must be done to take an electron through the metal surface from just beneath it, and more work is required when the electron originates deeper in the metal.

The validity of this interpretation of the photoelectric effect is confirmed by studies of *thermionic emission*. Long ago it became known that the presence of a very hot object increases the electrical conductivity of the surrounding air, and late in the nineteenth century the reason for this phenomenon was found to be the emission of electrons from such an object. Thermionic emission makes possible the operation of the vacuum tubes so widely used in electronics, where metal filaments or specially coated cathodes at high temperature supply dense streams of electrons. The emitted electrons evidently obtain their energy from the thermal agitation of the particles constituting the metal, and we should expect that the electrons must acquire a certain minimum energy in order to escape. This minimum energy can be determined for many surfaces, and it is always almost identical to the photoelectric work function for the same surfaces. In photoelectric emission, photons of light provide the energy required by an electron to escape, while in thermionic emission heat does so: in both cases the physical processes involved in the emergence of an electron from a metal surface are the same.

Let us apply Eq. 3.3 to a specific situation. The work function of potassium is 2.0 ev. When ultraviolet light of wavelength 3,500 A (1 A = 1 Angstrom unit = 10^{-10} m) falls on a potassium surface, what is the maximum energy in electron volts of the photoelectrons? From Eq. 3.3,

$$T_{max} = h\nu - h\nu_0$$

Since $h\nu_0$ is already expressed in electron volts, we need only compute the quantum energy $h\nu$ of 3,500-A light. This is

$$h\nu = \frac{hc}{\lambda}$$

$$= \frac{6.63 \times 10^{-34} \text{ joule-sec} \times 3 \times 10^8 \text{ m/sec} \times 10^{10} \text{ A/m}}{3,500 \text{ A}}$$

$$= 5.7 \times 10^{-19} \text{ joule}$$

To convert this energy from joules to electron volts, we recall that

$$1 \text{ ev} = 1.6 \times 10^{-19} \text{ joule}$$

and so

$$h\nu = \frac{5.7 \times 10^{-19} \text{ joule}}{1.6 \times 10^{-19} \text{ joule/ev}}$$

$$= 3.6 \text{ ev}$$

Hence the maximum photoelectron energy is

$$T_{\text{max}} = h\nu - h\nu_0$$

$$= 3.6 \text{ ev} - 2.0 \text{ ev}$$

$$= 1.6 \text{ ev}$$

The view that light propagates as a series of little packets of energy (sometimes called photons) is directly opposed to the wave theory of light. The latter, which provides the sole means of explaining a host of optical effects—notably diffraction and interference—is one of the most securely established of physical theories. Planck's suggestion that a hot object emits light in separate quanta was not incompatible with the propagation of light as a wave. Einstein's suggestion in 1905 that light travels through space in the form of distinct photons, however, elicited incredulity from his contemporaries. According to the wave theory, light waves spread out from a source in the way ripples spread out on the surface of a lake when a stone falls into it. The energy carried by the light, in this analogy, is distributed continuously throughout the wave pattern. According to the quantum theory, on the other hand, light spreads out from a source as a series of localized concentrations of energy, each sufficiently small to be capable of absorption by a single electron. Curiously, the quantum theory of light, which treats it strictly as a particle phenomenon, explicitly involves the light frequency ν, strictly a wave concept.

The quantum theory of light is strikingly successful in explaining the photoelectric effect. It predicts correctly that the maximum photoelectron energy should depend upon the frequency of the incident light and not upon its intensity, contrary to what the wave theory suggests, and it is able to explain why even the feeblest light can lead to the immediate emission of photoelectrons, again contrary to the wave theory. The wave theory can give no reason why there should be a threshold frequency such that, when light of lower frequency is employed, no photoelectrons are observed no matter how strong the light beam, something that follows naturally from the quantum theory.

Which theory are we to believe? A great many physical hypotheses have had to be altered or discarded when they were found to disagree with experiment, but never before have we had to devise two totally different theories to account for a single physical phenomenon. The situation here is fundamentally different from what it is, say, in the case of relativistic versus newtonian mechanics, where the latter turns out to be an approximation of the former. There is no way of deriving the quantum theory of light from the wave theory of light or vice versa.

In a specific event light exhibits *either* a wave or a particle nature, never both simultaneously. The same light beam that is diffracted by a grating can cause the emission of photoelectrons from a suitable surface, but these processes occur independently. **The wave theory of light and the quantum theory of light complement each other.** Electromagnetic waves provide the sole possible explanation for certain experiments involving light and optical phenomena, while photons provide the sole possible explanation for all of the other experiments in this field. We have no alternative to regarding light as something that manifests itself as a stream of discrete photons on occasion and as a wave train the rest of the time. The "true nature" of light is no longer a meaningful concept in terms of everyday notions, and we must accept both wave and quantum theories, apparent contradictions and all, as necessary for a complete description of light.

3.3 X Rays

The photoelectric effect provides convincing evidence that photons of light can transfer energy to electrons. Is the inverse process also possible? That is, can part or all of the kinetic energy of a moving electron be converted into a photon? As it happens, the inverse photoelectric effect not only does occur, but had been discovered (though not at all understood) prior to the theoretical work of Planck and Einstein.

In 1895 Wilhelm Roentgen made the classic observation that a highly penetrating radiation of unknown nature is produced when fast electrons impinge on matter. These *X rays* were soon found to travel in straight lines, even through electric and magnetic fields, to pass readily through opaque materials, to cause phosphorescent substances to glow, and to expose photographic plates. The faster the original electrons, the more penetrating the resulting X rays, and the greater the number of electrons, the greater the intensity of the X-ray beam.

Not long after their discovery it began to be suspected that X rays are electromagnetic waves. After all, electromagnetic theory predicts that an accelerated electric charge will radiate electromagnetic waves, and a rapidly moving electron suddenly brought to rest is certainly accelerated. Radiation produced under these circumstances is given the German name *bremsstrahlung* ("braking radiation"). The absence of any perceptible X-ray refraction in the early work could be attributed to very short wavelengths, below those in the

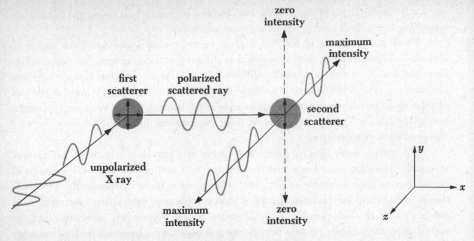

FIGURE 3.5 Barkla's experiment to demonstrate X-ray polarization.

ultraviolet range, since the refractive index of a substance decreases to unity (corresponding to straight-line propagation) with decreasing wavelength.

The wave nature of X rays was first established in 1906 by Barkla, who was able to exhibit their polarization. Barkla's experimental arrangement is sketched in Fig. 3.5. We shall analyze this experiment under the assumption that X rays are electromagnetic waves. At the left a beam of unpolarized X rays heading in the $-z$ direction is incident upon a small block of carbon. These X rays are scattered by the carbon; this means that electrons in the carbon atoms are set in vibration by the electric vectors of the X rays and then re-radiate. Because the electric vector in an electromagnetic wave is perpendicular to its direction of propagation, the initial beam of X rays contains electric vectors that lie in the xy plane only. The target electrons therefore are induced to vibrate in the xy plane. A scattered X ray that proceeds in the $+x$ direction can have an electric vector in the y direction only, and so it is plane-polarized. To demonstrate this polarization, another carbon block is placed in the path of the ray, as at the right. The electrons in this block are restricted to vibrate in the y direction and therefore reradiate X rays that propagate in the xz plane exclusively, and not at all in the y direction. The observed absence of scattered X rays outside the xz plane confirms the wave character of X rays.

In 1912 a method was devised for actually measuring the wavelengths of X rays. A diffraction experiment had been recognized as ideal, but, as we recall from physical optics, the spacing between adjacent lines on a diffraction grating must be of the same order of magnitude as the wavelength of the light for satisfactory results, and gratings cannot be ruled with the minute spacing required by X rays. In 1912, however, Max von Laue recognized that the wavelengths hypothesized for X rays were about the same order of magnitude as the spacing between adjacent atoms in crystals, which is about 1 A. He therefore proposed that crystals be used to diffract X rays, with their regular lattices acting as a kind of three-dimensional grating. Suitable experiments were performed in the next year, and the wave nature of X rays was successfully demonstrated. In these experiments wavelengths from 1.3×10^{-11} m to 4.8×10^{-11} m (0.13 A to 0.48 A) were found, 10^{-4} of those in visible light, and hence having quanta 10^4 times as energetic. We shall consider X-ray diffraction in Sec. 3.4.

For purposes of classification, electromagnetic radiations with wavelengths in the approximate interval from 10^{-11} to 10^{-8} m (0.1 to 100 A) are today considered as X rays.

Figure 3.6 is a diagram of an X-ray tube. A cathode, heated by an adjacent filament through which an electric current is passed, supplies electrons copiously by thermionic emission. The high potential difference V maintained between the cathode and a metallic target accelerates the electrons toward the latter. The face of the target is at an angle relative to the electron beam, and the X rays that emerge from the target pass through the side of the tube. The tube is evacuated to permit the electrons to get to the target unimpeded.

As we said earlier, classical electromagnetic theory predicts the production of bremsstrahlung when electrons are accelerated, thereby apparently accounting for the X rays emitted when fast electrons are stopped by the target of an X-ray tube. However, the agreement between the classical theory and the experimental data is not satisfactory in certain important respects. Figures 3.7 and 3.8 show the X-ray spectra that result when tungsten and molybdenum targets are bombarded by electrons at several different accelerating potentials. The curves exhibit two distinctive features not accountable in terms of electromagnetic theory:

FIGURE 3.6   An X-ray tube.

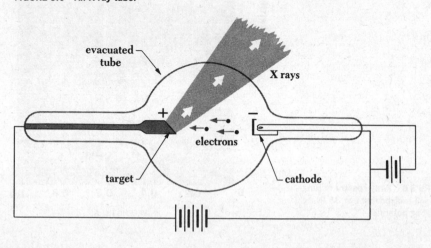

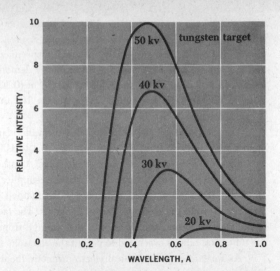

FIGURE 3.7 X-ray spectra of tungsten at various accelerating potentials.

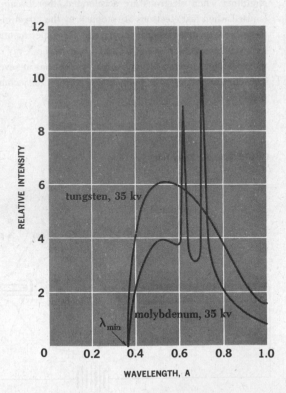

FIGURE 3.8 X-ray spectra of tungsten and molybdenum at 35 kv accelerating potential.

1. In the case of molybdenum there are pronounced intensity peaks at certain wavelengths, indicating the enhanced production of X rays. These peaks occur at various specific wavelengths for each target material and originate in rearrangements of the electron structures of the target atoms after having been disturbed by the bombarding electrons. The important thing to note at this point is the production of X rays of specific wavelengths, a decidedly nonclassical effect, in addition to the production of a continuous X-ray spectrum.

2. The X rays produced at a given accelerating potential V vary in wavelength, but none has a wavelength shorter than a certain value λ_{min}. Increasing V decreases λ_{min}. At a particular V, λ_{min} is the *same* for both the tungsten and molybdenum targets. Duane and Hunt have found that λ_{min} is inversely proportional to V; their precise relationship is

3.4
$$\lambda_{min} = \frac{1.24 \times 10^{-6} \text{ volt-m}}{V}$$
X-ray production

The second observation is readily understood in terms of the quantum theory of radiation. Most of the electrons incident upon the target lose their kinetic energy gradually in numerous collisions, their energy going simply into heat. (This is the reason that the targets in X-ray tubes are normally of high-melting-point metals, and an efficient means of cooling the target is often employed.) A few electrons, though, lose most or all of their energy in single collisions with target atoms; this is the energy that is evolved as X rays. X-ray production, then, except for the peaks mentioned in observation 1 above, represents an inverse photoelectric effect. Instead of photon energy being transformed into electron kinetic energy, electron kinetic energy is being transformed into photon energy. A short wavelength means a high frequency, and a high frequency means a high photon energy $h\nu$. It is therefore logical to interpret the short wavelength limit λ_{min} of Eq. 3.4 as corresponding to a maximum photon energy $h\nu_{max}$, where

3.5
$$h\nu_{max} = \frac{hc}{\lambda_{min}}$$

Since work functions are only several electron volts while the accelerating potentials in X-ray tubes are tens or hundreds of thousands of volts, we may assume that the kinetic energy T of the bombarding electrons is

3.6
$$T = eV$$

When the entire kinetic energy of an electron is lost to create a single photon, then

3.7
$$h\nu_{max} = T$$

Substituting Eqs. 3.5 and 3.6 into 3.7, we see that

$$h\nu_{max} = T$$

$$\frac{hc}{\lambda_{min}} = eV$$

$$\lambda_{min} = \frac{hc}{eV}$$

$$= \frac{6.63 \times 10^{-34} \text{ joule-sec} \times 3 \times 10^8 \text{ m/sec}}{1.6 \times 10^{-19} \text{ coulomb} \times V}$$

$$= \frac{1.24 \times 10^{-6}}{V} \text{ volt-m}$$

which is just the experimental relationship of Eq. 3.4. It is therefore correct to regard X-ray production as the inverse of the photoelectric effect.

A conventional X-ray machine might have an accelerating potential of 50,000 volts. To find the shortest wavelength present in its radiation, we use Eq. 3.4 with the result that

$$\lambda_{min} = \frac{1.24 \times 10^{-6} \text{ volt-m}}{5 \times 10^4 \text{ volts}}$$

$$= 2.5 \times 10^{-11} \text{ m}$$

$$= 0.25 \text{ A}$$

This wavelength corresponds to the frequency

$$\nu_{max} = \frac{c}{\lambda_{min}}$$

$$= \frac{3 \times 10^8 \text{ m/sec}}{2.5 \times 10^{-11} \text{ m}}$$

$$= 1.2 \times 10^{19} \text{ sec}^{-1}$$

3.4 X-ray Diffraction

Let us now return to the question of how X rays may be demonstrated to consist of electromagnetic waves. A crystal consists of a regular array of atoms each of which is able to scatter any electromagnetic waves that happen to strike it. The mechanism of scattering is straightforward. An atom in a constant electric field becomes polarized since its negatively charged electrons and positively charged nucleus experience forces in opposite directions; these forces are small compared with the forces holding the atom together, so the result is a distorted charge distribution equivalent to an electric dipole. In the presence of the alternating electric field of an electromagnetic wave of frequency ν, the polarization changes back and forth with the same frequency ν. An oscillating electric dipole is thus created at the expense of some of the energy of the incoming wave, whose amplitude is accordingly decreased. The oscillating dipole in turn radiates electromagnetic waves of frequency ν, and

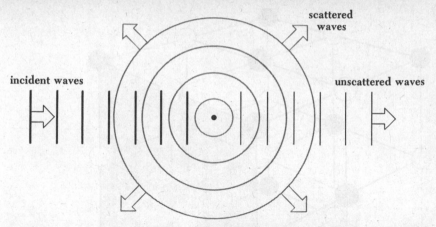

FIGURE 3.9 The scattering of electromagnetic radiation by a group of atoms. Incident plane waves are reemitted as spherical waves.

these secondary waves proceed in all directions except along the dipole axis. In an assembly of atoms exposed to unpolarized radiation, the secondary radiation is isotropic since the contributions of the individual atoms are random. In wave terminology, the secondary waves have spherical wavefronts in place of the plane wavefronts of the incoming waves (Fig. 3.9). The scattering process, then, involves an atom absorbing incident plane waves and reemitting spherical waves of the same frequency.

A monochromatic beam of X rays that falls upon a crystal is scattered in all directions within it, but, owing to the regular arrangement of the atoms, in certain directions the scattered waves constructively interfere with one another while in others they destructively interfere. The atoms in a crystal may be thought of as defining families of parallel planes, as in Fig. 3.10, with each family having a characteristic separation between its component planes. This analysis was suggested in 1913 by W. L. Bragg, in honor of whom the above planes are called *Bragg planes*. The conditions that must be fulfilled for radiation scattered by crystal atoms to undergo constructive interference may be obtained from a diagram like that in Fig. 3.11. A beam containing X rays of wavelength λ is incident upon a crystal at an angle θ with a family of Bragg planes whose spacing is a. The beam goes past atom A in the first plane and atom B in the next, and each of them scatters part of the beam in random directions. Constructive interference takes place only between those scattered rays that are parallel and whose paths differ by exactly λ, 2λ, 3λ, and so on. That is, the path difference must be $n\lambda$, where n is an integer. The only rays scattered by A and B for which this is true are those labeled I and II in Fig. 3.11. The first condition upon I and II is that their common scattering angle be equal to the angle of incidence θ of the original beam. (This condition, which is independent of wavelength, is the same as that for ordinary specular reflection in optics: angle of incidence = angle of reflection. For this reason X-ray scattering from atomic

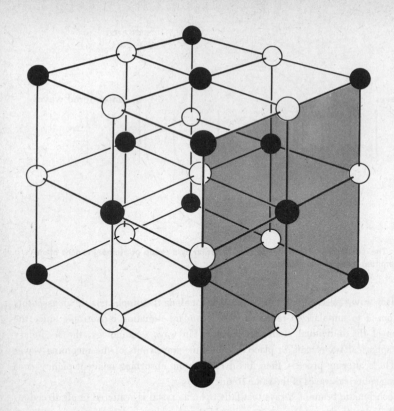

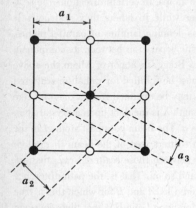

FIGURE 3.10 Structure of cubic crystal.

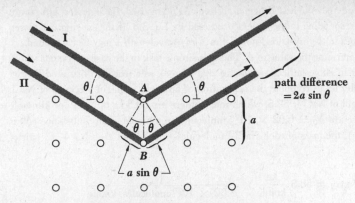

FIGURE 3.11   X-ray scattering from cubic crystal.

planes in a crystal is commonly, though incorrectly, termed *Bragg reflection*.) The second condition is that

3.8 $$2a \sin \theta = n\lambda \qquad n = 1, 2, 3, \ldots$$

since ray II must travel the distance $2a \sin \theta$ farther than ray I. The integer n is the *order* of the scattered beam.

The schematic design of an X-ray spectrometer based upon Bragg's analysis is shown in Fig. 3.12. A collimated beam of X rays falls upon a crystal at an angle θ, and a detector is placed so that it records those rays whose scattering angle is also θ. Any X rays reaching

FIGURE 3.12   X-ray spectrometer.

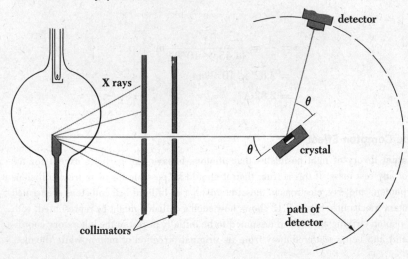

the detector therefore obey the first Bragg condition. As θ is varied, the detector will record intensity peaks corresponding to the orders predicted by Eq. 3.8. If the spacing a between adjacent Bragg planes in the crystal is known, the X-ray wavelength λ may be calculated.

How can we determine the value of a? This is a simple task in the case of crystals whose atoms are arranged in cubic lattices similar to that of rock salt (NaCl), illustrated in Fig. 3.10. As an example, let us compute the separation of adjacent atoms in a crystal of NaCl. The molecular weight of NaCl is 58.5, which means there are 58.5 kg of NaCl per kilomole (kmole). Since there are $N_0 = 6.02 \times 10^{26}$ molecules in a kmole of any substance (N_0 is Avogadro's number), the mass of each NaCl "molecule"—that is, of each Na + Cl pair of atoms—is given by

$$m_{\text{NaCl}} = 58.5 \frac{\text{kg}}{\text{kmole}} \times \frac{1}{6.02 \times 10^{26} \text{ molecules/kmole}}$$
$$= 9.72 \times 10^{-26} \text{ kg/molecule}$$

Crystalline NaCl has a density of 2.16×10^3 kg/m^3, and so, taking into account the presence of two atoms in each NaCl "molecule," the number of atoms in 1 m^3 of NaCl is

$$n = 2 \frac{\text{atoms}}{\text{molecule}} \times 2.16 \times 10^3 \frac{\text{kg}}{\text{m}^3} \times \frac{1}{9.72 \times 10^{-26} \text{ kg/molecule}}$$
$$= 4.45 \times 10^{28} \text{ atoms/m}^3$$

If a is the distance between adjacent atoms in a crystal, there are $1/a$ atoms/m along any of the crystal axes and $1/a^3$ atoms/m^3 in the entire crystal. Hence

$$\frac{1}{a^3} = n$$

and

$$a = \frac{1}{n^3} = \frac{1}{(4.45 \times 10^{28})^3} \text{ m}$$
$$= 2.82 \times 10^{-10} \text{ m}$$
$$= 2.82 \text{ A}$$

3.5 The Compton Effect

The quantum theory of light postulates that photons behave like particles except for the absence of any rest mass. If this is true, then it should be possible for us to treat collisions between photons and, say, electrons in the same manner as billiard-ball collisions are treated in elementary mechanics. Figure 3.13 shows how such a collision might be represented, with an X-ray photon striking an electron (assumed to be initially at rest in the laboratory coordinate system) and being scattered away from its original direction of motion while the electron receives an impulse and begins to move. In the collision the photon may be regarded as having lost an amount of energy that is the same as the kinetic energy T gained by the electron, though actually separate photons are involved. If the initial photon has the frequency ν associated with it, the scattered photon has the lower frequency ν', where

Loss in photon energy = gain in electron energy

3.9
$$h\nu - h\nu' = T$$

From the previous chapter we recall that

$$E = \sqrt{m_0^2 c^4 + p^2 c^2}$$

so that, since the photon has no rest mass, its total energy is

$$E = pc$$

Since

$$E = h\nu$$

for a photon, its momentum is

$$p = \frac{E}{c}$$

3.10
$$= \frac{h\nu}{c}$$

Momentum, unlike energy, is a vector quantity, incorporating direction as well as magnitude, and in the collision momentum must be conserved in each of two mutually perpendicular directions. (When more than two bodies participate in a collision, of course, momentum must be conserved in each of three mutually perpendicular directions.) The

FIGURE 3.13 The Compton effect.

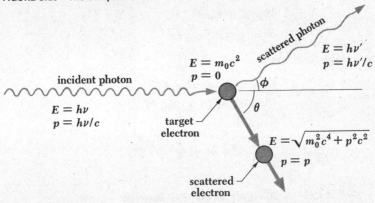

The Compton Effect

directions we choose here are that of the original photon and one perpendicular to it in the plane containing the electron and the scattered photon (Fig. 3.13). The initial photon momentum is $h\nu/c$, the scattered photon momentum is $h\nu'/c$, and the initial and final electron momenta are respectively 0 and p. In the original photon direction

$$\text{Initial momentum} = \text{final momentum}$$

3.11
$$\frac{h\nu}{c} + 0 = \frac{h\nu'}{c}\cos\phi + p\cos\theta$$

and perpendicular to this direction

$$\text{Initial momentum} = \text{final momentum}$$

3.12
$$0 = \frac{h\nu'}{c}\sin\phi - p\sin\theta$$

The angle ϕ is that between the directions of the initial and scattered photons, and θ is that between the directions of the initial photon and the recoil electron. From Eqs. 3.9, 3.11, and 3.12 we shall now obtain a formula relating the wavelength difference between initial and scattered photons with the angle ϕ between their directions, both of which are readily measurable quantities.

The first step is to multiply Eqs. 3.11 and 3.12 by c and rewrite them as

$$pc\cos\theta = h\nu - h\nu'\cos\phi$$
$$pc\sin\theta = h\nu'\sin\phi$$

By squaring each of these equations and adding the new ones together, the angle θ is eliminated, leaving

3.13
$$p^2c^2 = (h\nu)^2 - 2(h\nu)(h\nu')\cos\phi + (h\nu')^2$$

If we equate the two expressions for the total energy of a particle

$$E = T + m_0c^2$$
$$E = \sqrt{m_0^2c^4 + p^2c^2}$$

from the previous chapter, we have

$$(T + m_0c^2)^2 = m_0^2c^4 + p^2c^2$$
$$p^2c^2 = T^2 + 2m_0c^2T$$

Since

$$T = h\nu - h\nu'$$

we have

3.14
$$p^2c^2 = (h\nu)^2 - 2(h\nu)(h\nu') + (h\nu')^2 + 2m_0c^2(h\nu - h\nu')$$

Substituting this value of p^2c^2 in Eq. 3.13, we finally obtain

3.15 $$2m_0c^2(h\nu - h\nu') = 2(h\nu)(h\nu')(1 - \cos \phi)$$

It is more convenient to express the above relationship in terms of wavelength rather than frequency. Dividing Eq. 3.15 by $2h^2c^2$,

$$\frac{m_0 c}{h}\left(\frac{\nu}{c} - \frac{\nu'}{c}\right) = \frac{\nu}{c}\frac{\nu'}{c}(1 - \cos \phi)$$

and so, since $\nu/c = 1/\lambda$ and $\nu'/c = 1/\lambda'$,

$$\frac{m_0 c}{h}\left(\frac{1}{\lambda} - \frac{1}{\lambda'}\right) = \frac{(1 - \cos \phi)}{\lambda\lambda'}$$

3.16 $$\lambda' - \lambda = \frac{h}{m_0 c}(1 - \cos \phi) \qquad \text{Compton effect}$$

Equation 3.16 was derived by Arthur H. Compton in the early 1920s, and the phenomenon it describes, which he was the first to observe, is known as the *Compton effect*. It constitutes very strong evidence in support of the quantum theory of radiation.

Equation 3.16 gives the change in wavelength expected for a photon that is scattered though the angle ϕ by a particle of rest mass m_0; it is independent of the wavelength λ of the incident photon. The quantity h/m_0c is called the *Compton wavelength* of the scattering particle, which for an electron is 0.024 A (2.4×10^{-12} m). From Eq. 3.16 we note that the greatest wavelength change that can occur will take place for $\phi = 180°$, when the wavelength change will be twice the Compton wavelength h/m_0c. Because the Compton wave-

FIGURE 3.14   Experimental demonstration of the Compton effect.

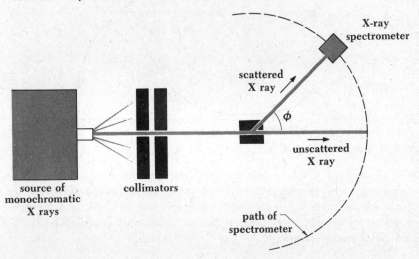

source of
monochromatic
X rays

collimators

scattered
X ray

unscattered
X ray

X-ray
spectrometer

ϕ

path of
spectrometer

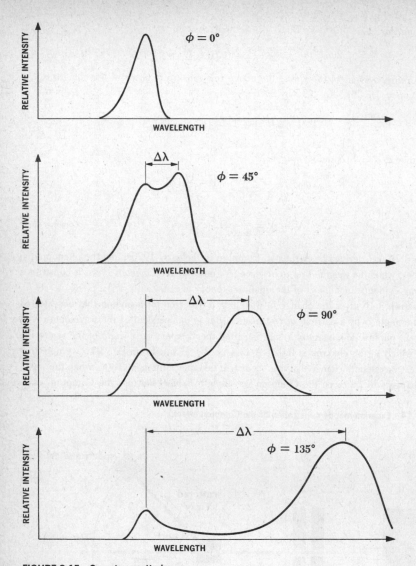

FIGURE 3.15 Compton scattering.

length of an electron is 0.024 A, while it is considerably less for other particles owing to their larger rest masses, the maximum wavelength change in the Compton effect is 0.048 A. Changes of this magnitude or less are readily observable only in X rays, since the shift in wavelength for visible light is less than 0.01 percent of the initial wavelength, while for X rays of $\lambda = 1$ A it is several percent.

The experimental demonstration of the Compton effect is straightforward. As in Fig. 3.14, a beam of X rays of a single, known wavelength is directed at a target, and the wavelengths of the scattered X rays are determined at various angles ϕ. The results, shown in Fig. 3.15, exhibit the wavelength shift predicted by Eq. 3.16, but at each angle the scattered X rays also include a substantial proportion having the initial wavelength. This is not hard to understand. In deriving Eq. 3.16, we assumed that the scattering particle is free to move at will, a reasonable assumption since many of the electrons in matter are only loosely bound to their parent atoms. Other electrons, however, are very tightly bound and, when struck by a photon, the entire atom recoils instead of the single electron. In this event the value of m_0 to use in Eq. 3.16 is that of the entire atom, usually tens of thousands of times greater than that of an electron, with the resulting Compton shift accordingly so minute as to be undetectable.

3.6 Gravitational Red Shift

Although a photon has no rest mass, it nevertheless behaves as though it possesses the inertial mass

3.17
$$m = \frac{h\nu}{c^2}$$
Photon "mass"

Does a photon possess gravitational mass as well? Since light is deflected by a gravitational field, as we learned in the previous chapter, it would seem natural to assume that photons have the same gravitational behavior as other particles.

Let us consider a photon of frequency ν emitted from the surface of a star of mass M and radius R (Fig. 3.16). The potential energy of a mass m on the star's surface is

3.18
$$V = -\frac{GMm}{R}$$

FIGURE 3.16 The frequency of a photon emitted from the surface of a star decreases as it moves away from the star.

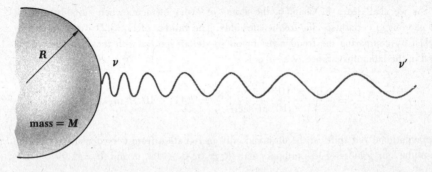

The potential energy of the photon is accordingly

$$V = - \frac{GMh\nu}{c^2 R}$$

and its total energy E, the sum of V and the quantum energy $h\nu$, is

$$E = h\nu - \frac{GMh\nu}{c^2 R}$$

3.19
$$= h\nu \left(1 - \frac{GM}{c^2 R} \right)$$

At a large distance from the star, for instance at the earth, the photon is beyond the star's gravitational field but its total energy remains the same. The photon's energy is now entirely electromagnetic, and

3.20
$$E = h\nu'$$

where ν' is the frequency of the arriving photon. (The potential energy of the photon in the earth's gravitational field is negligible compared with that in the star's field.) Hence

$$h\nu' = h\nu \left(1 - \frac{GM}{c^2 R} \right)$$

$$\frac{\nu'}{\nu} = 1 - \frac{GM}{c^2 R}$$

3.21
$$\frac{\Delta\nu}{\nu} = \frac{\nu - \nu'}{\nu} = 1 - \frac{\nu'}{\nu} = \frac{GM}{c^2 R}$$

The photon has a *lower* frequency at the earth, corresponding to its loss in energy as it leaves the field of the star. A photon in the visible region of the spectrum is thus shifted toward the red end, and this phenomenon is accordingly known as the *gravitational red shift*. It must be distinguished from the doppler red shift observed in the spectra of distant galaxies due to their apparent recession from the earth, a recession attributed to a general expansion of the universe.

As we shall learn in Chap. 6, the atoms of every element, when suitably excited, emit photons of certain specific frequencies only. The validity of Eq. 3.21 can therefore be checked by comparing the frequencies found in stellar spectra with those in spectra obtained in the laboratory. Since G/c^2 is only

$$\frac{G}{c^2} = \frac{6.67 \times 10^{-11} \text{ n-m}^2/\text{kg}^2}{(3 \times 10^8 \text{ m/sec})^2} = 7.41 \times 10^{-28} \text{ m/kg}$$

the gravitational red shift can be observed only in radiation from very dense stars. In the case of the sun, a more or less ordinary star, $R = 6.96 \times 10^8$ m and $M = 1.99 \times 10^{30}$

kg, and

$$\frac{\Delta\nu}{\nu} = \frac{GM}{c^2R} = 7.41 \times 10^{-28} \frac{\text{m}}{\text{kg}} \times \frac{1.99 \times 10^{30} \text{ kg}}{6.96 \times 10^8 \text{ m}}$$

$$= 2.12 \times 10^{-6}$$

Since $\Delta\lambda/\lambda = \Delta\nu/\nu$, the gravitational red shift in solar radiation amounts to only about 0.01 A for green light of wavelength 5,000 A and is undetectable due to the doppler broadening of the spectral lines.

However, there is a class of stars in the final stages of their evolution called *white dwarfs* that are composed of atoms whose electron structures have "collapsed," and such stars have quite enormous densities—typically $\sim$ 5 tons/in.[3] A white dwarf might have a radius of 9×10^6 m, about 0.01 that of the sun, and a mass of 1.2×10^{30} kg, about 0.6 that of the sun, so that

$$\frac{\Delta\nu}{\nu} = \frac{GM}{c^2R} \approx 7.41 \times 10^{-28} \frac{\text{m}}{\text{kg}} \times \frac{1.2 \times 10^{30} \text{ kg}}{9 \times 10^6 \text{ m}}$$

$$\approx 10^{-4}$$

Here the gravitational red shift would be $\approx$ 0.5 A for light of wavelength 5,000 A, which is measurable under favorable circumstances. In the case of the white dwarf Sirius B (the "companion of Sirius"), the predicted red shift is $\Delta\nu/\nu \approx 5.9 \times 10^{-5}$ and the observed shift is 6.6×10^{-5}; in view of the uncertainty in the M/R ratio for Sirius B, these figures would seem to confirm the attribution of gravitational mass to photons.

Recently a gravitational frequency shift has been detected in a laboratory experiment by measuring the change in frequency of gamma rays after they had "fallen" through a height h near the earth's surface. A body of mass m that falls a height h gains mgh of energy. If a falling photon of original frequency ν is taken to have the constant mass $h\nu/c^2$ (the frequency shift is so small that the change in mass may be neglected), its final energy $h\nu'$ is given by

$$h\nu' = h\nu + mgh = h\nu + \frac{h\nu gh}{c^2}$$

3.22

$$= h\nu \left(1 + \frac{gh}{c^2}\right)$$

For $h = 20$ m,

$$\frac{\Delta\nu}{\nu} = \frac{gh}{c^2} = \frac{9.8 \text{ m/sec}^2 \times 20 \text{ m}}{(3 \times 10^8 \text{ m/sec})^2}$$

$$= 2.2 \times 10^{-15}$$

A shift of this magnitude is just detectable, and the results confirm Eq. 3.22.

Gravitational Red Shift

Problems

1. The threshold wavelength for photoelectric emission in tungsten is 2,300 A. What wavelength of light must be used in order for electrons with a maximum energy of 1.5 ev to be ejected?

2. The threshold frequency for photoelectric emission in copper is 1.1×10^{15} sec^{-1}. Find the maximum energy of the photoelectrons (in joules and electron volts) when light of frequency 1.5×10^{15} sec^{-1} is directed on a copper surface.

3. The work function of sodium is 2.3 ev. What is the maximum wavelength of light that will cause photoelectrons to be emitted from sodium? What will the maximum kinetic energy of the photoelectrons be if 2,000 A light falls on a sodium surface?

4. Find the wavelength and frequency of a 100-Mev photon.

5. Find the energy of a 7,000-A photon.

6. Under favorable circumstances the human eye can detect 10^{-18} joule of electromagnetic energy. How many 6,000-A photons does this represent?

7. A 1,000-watt radio transmitter operates at a frequency of 880 kc/sec. How many photons per second does it emit?

8. How many photons per second are emitted by a 10-watt yellow lamp? (Assume the light is monochromatic with a wavelength of 6,000 A.)

9. Light from the sun arrives at the earth at the rate of about 1,400 watts/m^2 of area perpendicular to the direction of the light. (a) Find the maximum pressure (in lb/in.^{2}) this light can exert on the earth's surface. (b) Assume that sunlight consists exclusively of 6,000-A photons. How many photons per m^2 arrive at that part of the earth directly facing the sun in each second? (c) The average radius of the earth's orbit is 1.49×10^{11} m. What is the power output of the sun in watts, and how many photons per second does it emit? (d) How many photons per m^3 are there near the earth?

10. What is the wavelength of the X rays emitted when 100-kev electrons strike a target? What is their frequency?

11. An X-ray machine produces 0.1-A X rays. What accelerating voltage does it employ?

12. The distance between adjacent atomic planes in calcite is 3×10^{-8} cm. What is the smallest angle between these planes and an incident beam of 0.3-A X rays at which these X rays can be detected?

13. A potassium chloride crystal has a density of 1.98×10^3 kg/m^3. The molecular weight of KCl is 74.55. Find the distance between adjacent atoms.

14. How much energy must a photon have if it is to have the momentum of a 10-Mev proton?

15. What is the frequency of an X-ray photon whose momentum is 1.1×10^{-23} kg-m/sec?

16. Prove that it is impossible for a photon to give up all its energy and momentum to a free electron, so that the photoelectric effect can take place only when photons strike bound electrons.

17. A beam of X rays is scattered by free electrons. At 45° from the beam direction the scattered X rays have a wavelength of 0.022 A. What is the wavelength of the X rays in the direct beam?

18. An X-ray photon whose initial frequency was 1.5×10^{19} sec^{-1} emerges from a collision with an electron with a frequency of 1.2×10^{19} sec^{-1}. How much kinetic energy was imparted to the electron?

19. An X-ray photon of initial frequency 3×10^{19} sec^{-1} collides with an electron and is scattered through 90°. Find its new frequency.

20. Find the energy of an X-ray photon which can impart a maximum energy of 50 kev to an electron.

21. A monochromatic X-ray beam whose wavelength is 0.558 A is scattered through 46° Find the wavelength of the scattered beam.

22. In Sec. 3.4 the X rays scattered by a crystal were assumed to undergo no change in wavelength. Show that this assumption is reasonable by calculating the Compton wavelength of a Na atom and comparing it with the typical X-ray wavelength of 1 A.

23. As discussed in Chap. 23, certain atomic nuclei emit photons in undergoing transitions from "excited" energy states to their "ground" or normal states. These photons constitute *gamma rays*. When a nucleus emits a photon, it recoils in the opposite direction. (*a*) The $_{27}Co^{57}$ nucleus decays by K capture to $_{26}Fe^{57}$, which then emits a photon in losing 14.4 kev to reach its ground state. The mass of a $_{26}Fe^{57}$ atom is 9.5×10^{-26} kg. By how much is the photon energy reduced from the full 14.4 kev available as a result of having to share energy with the recoiling atom? (*b*) In certain crystals the atoms are so tightly bound that the entire crystal recoils when a gamma-ray photon is emitted, instead of the individual atom. This phenomenon is known as the *Mössbauer effect*. By how much is the photon energy reduced in this situation if the excited $_{26}Fe^{57}$ nucleus is part of a 1-g crystal? (c) The essentially recoil-free emission of gamma rays in situations like that of (*b*) means that it is possible to construct a source of essentially monoenergetic and hence monochromatic photons. Such a source was used in the experiment described in the last paragraph of Sec. 3.6. What is the original frequency and the change in frequency of a 14.4-kev gamma-ray photon after it has fallen 20 m near the earth's surface?

Chapter 4
Wave
Properties of
Particles

In retrospect it may seem odd that two decades passed between the discovery in 1905 of the particle properties of waves and the speculation in 1924 that the converse might also be true. It is one thing, however, to suggest a revolutionary hypothesis to explain otherwise mysterious data and quite another to advance an equally revolutionary hypothesis in the absence of a strong experimental mandate. The latter is just what Louis de Broglie did in 1924 when he proposed that matter possesses wave as well as particle characteristics. So different was the intellectual climate at the time from that prevailing at the turn of the century that de Broglie's notion received immediate and respectful attention, whereas the earlier quantum theory of light of Planck and Einstein created hardly any stir despite its striking empirical support. Although the existence of de Broglie waves was not demonstrated until 1927, the duality principle they represent provided the starting point for Schrödinger's successful development of quantum mechanics in the previous year.

4.1 De Broglie Waves

A photon of light of frequency ν has the momentum

$$p = \frac{h\nu}{c}$$

which can be expressed in terms of wavelength λ as

$$p = \frac{h}{\lambda}$$

since $\lambda\nu = c$. The wavelength of a photon is therefore specified by its momentum according to the relation

4.1
$$\lambda = \frac{h}{p}$$

Drawing upon an intuitive expectation that nature is symmetric, de Broglie asserted that Eq. 4.1 is a completely general formula that applies to material particles as well as to photons. The momentum of a particle of mass m and velocity v is

$$p = mv$$

and consequently its *de Broglie wavelength* is

4.2
$$\lambda = \frac{h}{mv}$$
De Broglie waves

The greater the particle's momentum, the shorter its wavelength. In Eq. 4.2 m is the relativistic mass

$$m = \frac{m_0}{\sqrt{1 - v^2/c^2}}$$

Equation 4.2 has been amply verified by experiments involving the diffraction of fast electrons by crystals, experiments analogous to those that showed X rays to be electromagnetic waves. Before we consider these experiments, it is appropriate to look into the question of what kind of wave phenomenon is involved in the matter waves of de Broglie. In a light wave the electromagnetic field varies in space and time, in a sound wave pressure varies in space and time; what is it whose variations constitute de Broglie waves?

4.2 Wave Function

The variable quantity characterizing de Broglie waves is called the *wave function*, denoted by the symbol Ψ (the Greek letter *psi*). **The value of the wave function associated with a moving body at the particular point** x, y, z **in space at the time** t **is related to the likelihood of finding the body there at the time.** Ψ itself, however, has no direct physical significance. There is a simple reason why Ψ cannot be interpreted in terms of an experiment. The probability P that something be somewhere at a given time can have any value between two limits: 0, corresponding to the certainty of its absence, and 1, corresponding to the certainty of its presence. (A probability of 0.2, for instance, signifies a 20 percent chance of finding the body.) But the amplitude of any wave may be negative as well as positive, and a negative probability is meaningless. Hence Ψ itself cannot be an observable quantity.

This objection does not apply to $|\Psi|^2$, the square of the absolute value of the wave function. For this and other reasons $|\Psi|^2$ is known as *probability density*. **The probability of experimentally finding the body described by the wave function** Ψ **at the point** x, y, z **at the time** t **is proportional to the value of** $|\Psi|^2$ **there at** t. A large value of $|\Psi|^2$ means the strong possibility of the body's presence, while a small value of $|\Psi|^2$ means the slight possibility of its presence. As long as $|\Psi|^2$ is not actually 0 somewhere, however, there is a definite chance, however small, of detecting it there. This interpretation was first made by Max Born in 1926.

There is a big difference between the probability of an event and the event itself. Although we shall speak of the wave function Ψ that describes a particle as being spread out in space, this does not mean that the particle itself is thus spread out. When an experiment is performed to detect electrons, for instance, a whole electron is either found at a certain time and place or it is not; there is no such thing as 20 percent of an electron. However, it is entirely possible for there to be a 20 percent chance that the electron be found at that time and place, and it is this likelihood that is specified by $|\Psi|^2$.

Alternatively, if an experiment involves a great many identical bodies all described by the same wave function Ψ, the *actual density* of bodies at x, y, z at the time t is proportional to the corresponding value of $|\Psi|^2$.

While the wavelength of the de Broglie waves associated with a moving body is given by the simple formula

$$\lambda = \frac{h}{mv}$$

determining their amplitude Ψ as a function of time and position usually presents a formidable problem. We shall discuss the calculation of Ψ in Chaps. 7 and 8 and then go on to apply the ideas developed there to the structure of the atom in Chap. 9. Until then we shall assume that we have whatever knowledge of Ψ is required by the situation at hand.

In the event that a wave function Ψ is complex, with both real and imaginary parts, the probability density is given by the product $\Psi^* \Psi$ of Ψ and its *complex conjugate* Ψ^*. The complex conjugate of any function is obtained by replacing i ($= \sqrt{-1}$) by $-i$ wherever it appears in the function. Every complex function Ψ can be written in the form

$$\Psi = A + iB$$

where A and B are real functions. The complex conjugate Ψ^* of Ψ is

$$\Psi^* = A - iB$$

and so

$$\Psi^* \Psi = A^2 - i^2 B^2 = A^2 + B^2$$

since $i^2 = -1$. Hence $\Psi^* \Psi$ is always a positive real quantity.

4.3   De Broglie Wave Velocity

With what velocity do de Broglie waves travel? Since we associate a de Broglie wave with a moving body, it is reasonable to expect that this wave travels at the same velocity v as that of the body. If we call the de Broglie wave velocity w, we may apply the usual formula

$$w = \nu\lambda$$

to determine the value of w. The wavelength λ is just the de Broglie wavelength

$$\lambda = \frac{h}{mv}$$

We shall take the frequency ν to be that specified by the quantum equation

$$E = h\nu$$

Hence

$$\nu = \frac{E}{h}$$

or, since

$$E = mc^2$$

we have

$$\nu = \frac{mc^2}{h}$$

The de Broglie wave velocity is therefore

$$w = \nu\lambda$$

$$= \frac{mc^2}{h}\frac{h}{mv}$$

4.3 $$= \frac{c^2}{v}$$

Since the particle velocity v cannot equal or exceed the velocity of light c, the de Broglie wave velocity w is always greater than c! Clearly v and w are never equal for a moving body. In order to understand this unexpected result, we shall digress briefly to consider the notions of *wave velocity* and *group velocity*. (Wave velocity is sometimes called *phase velocity*.)

Let us begin by reviewing how waves are described mathematically. For clarity we shall consider a string stretched along the x axis whose vibrations are in the y direction, as in Fig. 4.1, and are simple harmonic in character. If we choose $t = 0$ when the displacement y of the string at $x = 0$ is a maximum, its displacement at any future time t at the same place is given by the formula

4.4 $$y = A \cos 2\pi\nu t$$

where A is the amplitude of the vibrations (that is, their maximum displacement on either side of the x axis) and ν their frequency.

De Broglie Wave Velocity 81

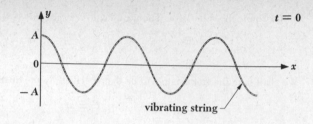

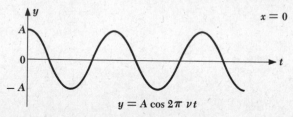

FIGURE 4.1 Wave motion.

$$y = A \cos 2\pi \nu t$$

Equation 4.4 tells us what the displacement of a single point on the string is as a function of time t. A complete description of a given wave motion in a stretched string, however, should tell us what y is at *any* point on the string at *any* time. What we want is a formula giving y as a function of both x and t. To obtain such a formula, let us imagine that we shake the string at $x = 0$ when $t = 0$, so that a harmonic wave starts to travel down the string in the $+x$ direction (Fig. 4.2). This wave has some speed w that depends upon the properties of the string. The wave travels the distance $x = wt$ in the time t; hence the time interval between the formation of the wave at $x = 0$ and its arrival at the point x is x/w. Accordingly the displacement y of the string at x at any time t is exactly the same as the value of y at $x = 0$ *at the earlier time* $t - x/w$. By simply replacing t in Eq. 4.4 with $t - x/w$, then, we have the desired formula giving y in terms of both x and t:

4.5
$$y = A \cos 2\pi\nu \left(t - \frac{x}{w} \right)$$

As a check, we note that Eq. 4.5 reduces to Eq. 4.4 at $x = 0$.

Equation 4.5 may be rewritten

$$y = A \cos 2\pi \left(\nu t - \frac{\nu x}{w} \right)$$

Since

$$w = \nu\lambda$$

we have

4.6
$$y = A \cos 2\pi \left(\nu t - \frac{x}{\lambda} \right)$$

Equation 4.6 is often more convenient to apply than Eq. 4.5.

Perhaps the most widely used description of a harmonic wave, however, is still another form of Eq. 4.5. We define the quantities *angular frequency* ω and *propagation constant k* by the formulas

4.7 $$\omega = 2\pi\nu$$ **Angular frequency**

4.8 $$k = \frac{2\pi}{\lambda}$$ **Propagation constant**

4.9 $$= \frac{\omega}{w}$$

Angular frequency gets its name from uniform circular motion, where a particle that moves around a circle ν times per second sweeps out $2\pi\nu$ rad/sec. The propagation constant (also called *angular wave number*) is equal to the number of radians corresponding to a wave train 1 m long, since there are 2π rad in one complete wave. In terms of ω and k, Eq. 4.5 becomes

4.10 $$y = A \cos(\omega t - kx)$$

In three dimensions k becomes a vector **k** normal to the wave fronts and x is replaced by the radius vector **r**. The scalar product **k · r** is then used instead of kx in Eq. 4.10.

FIGURE 4.2   Wave propagation.

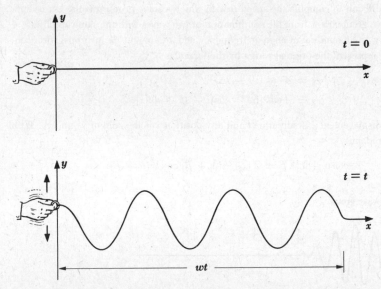

4.4 Wave and Group Velocities

The amplitude of the de Broglie waves that correspond to a moving body reflects the probability that it be found at a particular place at a particular time. It is clear that de Broglie waves cannot be represented simply by a formula resembling Eq. 4.10, which describes an indefinite series of waves all with the same amplitude A. Instead, we expect the wave representation of a moving body to correspond to a *wave packet*, or *wave group*, like that shown in Fig. 4.3, whose constituent waves have amplitudes which vary with the likelihood of detecting the body.

A familiar example of how wave groups come into being is the case of *beats*. When two sound waves of the same amplitude, but of slightly different frequencies, are produced simultaneously, the sound we hear has a frequency equal to the average of the two original frequencies and its amplitude rises and falls periodically. The amplitude fluctuations occur a number of times per second equal to the difference between the two original frequencies. If the original sounds have frequencies of, say, 440 and 442 cycles/sec, we will hear a fluctuating sound of frequency 441 cycles/sec with two loudness peaks, called beats, per second. The production of beats is illustrated in Fig. 4.4.

A way of mathematically describing a wave group, then, is in terms of a series of individual waves, differing slightly in wavelength, whose interference with one another results in the variation in amplitude that defines the group shape. If the speeds of the waves are the same, the speed with which the wave group travels is identical with the common wave speed. However, if the wave speed varies with wavelength, the different individual waves do not proceed together, and the wave group has a speed different from that of the waves that compose it.

It is not difficult to compute the speed u with which a wave group travels. Let us suppose that a wave group arises from the combination of two waves with the same amplitude A but differing by an amount $d\omega$ in angular frequency and an amount dk in propagation constant. We may represent the original waves by the formulas

$$y_1 = A \cos (\omega t - kx)$$
$$y_2 = A \cos [(\omega + d\omega)t - (k + dk)x]$$

The resultant displacement y at any time t and any position x is the sum of y_1 and y_2. With the help of the identity

$$\cos \alpha + \cos \beta = 2 \cos \tfrac{1}{2}(\alpha + \beta) \cos \tfrac{1}{2}(\alpha - \beta)$$

FIGURE 4.3   A wave group.

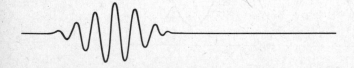

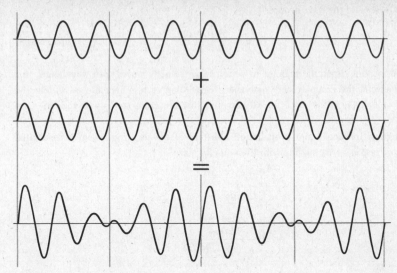

FIGURE 4.4 The production of beats.

and the relation

$$\cos(-\theta) = \cos\theta$$

we find that

$$y = y_1 + y_2$$
$$= 2A \cos \tfrac{1}{2}[(2\omega + d\omega)t - (2k + dk)x] \cos \tfrac{1}{2}(d\omega\, t - dk\, x)$$

Since $d\omega$ and dk are small compared with ω and k respectively,

$$2\omega + d\omega \approx 2\omega$$
$$2k + dk \approx 2k$$

and

4.11
$$y = 2A \cos(\omega t - kx) \cos\left(\frac{d\omega}{2}t - \frac{dk}{2}x\right)$$

Equation 4.11 represents a wave of angular frequency ω and propagation constant k that has superimposed upon it a modulation of angular frequency $\tfrac{1}{2}d\omega$ and propagation constant $\tfrac{1}{2}dk$. The effect of the modulation is to produce successive wave groups, as in Fig. 4.4. The wave velocity w is

4.12
$$w = \frac{\omega}{k}$$
Wave velocity

while the velocity u of the wave groups is

4.13
$$u = \frac{d\omega}{dk}$$

Group velocity

In general, depending upon the manner in which wave velocity varies with wavelength in a particular medium, the group velocity may be greater than or less than the wave velocity. If the wave velocity w is the same for all wavelengths, the group and wave velocities are the same.

The angular frequency and propagation constant of the de Broglie waves associated with a body of rest mass m_0 moving with the velocity v are

$$\omega = 2\pi\nu$$
$$= \frac{2\pi mc^2}{h}$$

4.14
$$= \frac{2\pi m_0 c^2}{h\sqrt{1 - v^2/c^2}}$$

and

$$k = \frac{2\pi}{\lambda}$$
$$= \frac{2\pi mv}{h}$$

4.15
$$= \frac{2\pi m_0 v}{h\sqrt{1 - v^2/c^2}}$$

Both ω and k are functions of the velocity v. The wave velocity w is, as we found earlier,

$$w = \frac{\omega}{k}$$
$$= \frac{c^2}{v}$$

which exceeds both the velocity of the body v and the velocity of light c, since $v < c$. The group velocity u of the de Broglie waves associated with the body is

$$u = \frac{d\omega}{dk}$$

4.16
$$= \frac{d\omega/dv}{dk/dv}$$

Now

86   Wave Properties of Particles

$$\frac{d\omega}{dv} = \frac{2\pi m_0}{h(1 - v^2/c^2)^{3/2}}$$

and

$$\frac{dk}{dv} = \frac{2\pi m_0}{h(1 - v^2/c^2)^{3/2}}$$

and so the group velocity is

$$u = v$$

The de Broglie wave group associated with a moving body travels with the same velocity as the body. The wave velocity w of the de Broglie waves evidently has no simple physical significance in itself.

4.5 The Diffraction of Particles

A wave manifestation having no analog in the behavior of newtonian particles is diffraction. In 1927 Davisson and Germer in the United States and G. P. Thomson in England independently confirmed de Broglie's hypothesis by demonstrating that electrons exhibit diffraction when they are scattered from crystals whose atoms are spaced appropriately. We shall consider the experiment of Davisson and Germer because its interpretation is more direct.

Davisson and Germer were studying the scattering of electrons from a solid, using an apparatus like that sketched in Fig. 4.5. The energy of the electrons in the primary beam, the angle at which they are incident upon the target, and the position of the detector can all be varied. Classical physics predicts that the scattered electrons will emerge in all directions

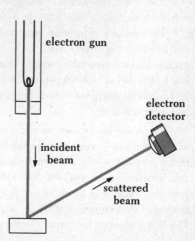

FIGURE 4.5 The Davisson-Germer experiment.

The Diffraction of Particles

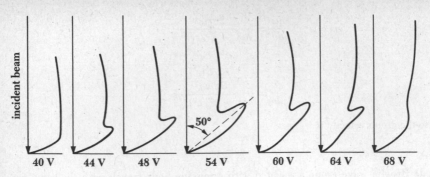

FIGURE 4.6   Results of the Davisson-Germer experiment.

with only a moderate dependence of their intensity upon scattering angle and even less upon the energy of the primary electrons. Using a block of nickel as the target, Davisson and Germer verified these predictions.

In the midst of their work there occurred an accident that allowed air to enter their apparatus and oxidize the metal surface. To reduce the oxide to pure nickel, the target was baked in a high-temperature oven. After this treatment, the target was returned to the apparatus and the measurements resumed. Now the results were very different from what had been found before the accident: instead of a continuous variation of scattered electron intensity with angle, distinct maxima and minima were observed whose positions depended upon the electron energy! Typical polar graphs of electron intensity after the accident are shown in Fig. 4.6; the method of plotting is such that the intensity at any angle is proportional to the distance of the curve at that angle from the point of scattering.

Two questions come to mind immediately: what is the reason for this new effect, and why did it not appear until after the nickel target was baked?

De Broglie's hypothesis suggested the interpretation that electron waves were being diffracted by the target, much as X rays are diffracted by Bragg reflection from crystal planes. This interpretation received support when it was realized that the effect of heating a block of nickel at high temperature is to cause the many small individual crystals of which it is normally composed to form into a single large crystal, all of whose atoms are arranged in a regular lattice.

Let us see whether we can verify that de Broglie waves are responsible for the findings of Davisson and Germer. In a particular determination, a beam of 54-ev electrons was directed perpendicularly at the nickel target, and a sharp maximum in the electron distribution occurred at an angle of 50° with the original beam. The angles of incidence and scattering relative to the family of Bragg planes shown in Fig. 4.7 will both be 65°. The spacing of the planes in this family, which can be measured by X-ray diffraction, is 0.91 A. The Bragg equation for maxima in the diffraction pattern is

$$n\lambda = 2a \sin \theta$$

Here $a = 0.91$ A and $\theta = 65°$; assuming that $n = 1$, the de Broglie wavelength λ of the diffracted electrons is

$$n\lambda = 2a \sin \theta$$
$$= 2 \times 0.91 \text{ A} \times \sin 65°$$
$$= 1.65 \text{ A}$$

Now we use de Broglie's formula

$$\lambda = \frac{h}{mv}$$

to calculate the expected wavelength of the electrons. The electron kinetic energy of 54 ev is small compared with its rest energy m_0c^2 of 5.1×10^5 ev, and so we can ignore relativistic considerations. Since

$$T = \tfrac{1}{2}mv^2$$

the electron momentum mv is

$$mv = \sqrt{2mT}$$
$$= \sqrt{2 \times 9.1 \times 10^{-31} \text{ kg} \times 54 \text{ ev} \times 1.6 \times 10^{-19} \text{ joule/ev}}$$
$$= 4.0 \times 10^{-24} \text{ kg-m/sec}$$

The electron wavelength is therefore

$$\lambda = \frac{h}{mv}$$
$$= \frac{6.63 \times 10^{-34} \text{ joule-sec}}{4.0 \times 10^{-24} \text{ kg-m/sec}}$$
$$= 1.66 \times 10^{-10} \text{ m}$$
$$= 1.66 \text{ A}$$

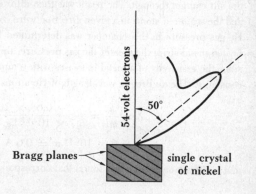

FIGURE 4.7 The diffraction of de Broglie waves by the target is responsible for the results of Davisson and Germer.

54-volt electrons

50°

Bragg planes

single crystal of nickel

The Diffraction of Particles

in excellent agreement with the observed wavelength. The Davisson-Germer experiment thus provides direct verification of de Broglie's hypothesis of the wave nature of moving bodies.

The analysis of the Davisson-Germer experiment is actually less straightforward than indicated above, since the energy of an electron increases when it enters a crystal by an amount equal to the work function of the surface. Hence the electron speeds in the experiment were greater inside the crystal and the corresponding de Broglie wavelength shorter than the corresponding values outside. An additional complication arises from interference between waves diffracted by different families of Bragg planes, which restricts the occurrence of maxima to certain combinations of electron energy and angle of incidence rather than merely to any combination that obeys the Bragg equation.

Electrons are not the only bodies whose wave behavior can be demonstrated. The diffraction of neutrons and of whole atoms when scattered by suitable crystals has been observed, and in fact neutron diffraction, like X-ray and electron diffraction, is today a widely used tool for investigating crystal structures.

The wave behavior of moving helium atoms was observed by Estermann, Frisch, and Stern in 1930. The source of the atoms was an enclosure filled with He and heated to $400°$ K, from which a beam of atoms emerged through a slit. The mean energy of the atoms was

$$\overline{\frac{1}{2}mv^2} = \frac{3}{2}kT = \frac{3}{2} \times 1.38 \times 10^{-23} \text{ joule/}°K \times 400°K$$

$$= 8.3 \times 10^{-21} \text{ joule} = 0.052 \text{ ev}$$

and their root-mean-square velocity was, since $m_{\text{He}} = 6.7 \times 10^{-27}$ kg,

$$v_{\text{rms}} = \sqrt{\overline{v^2}} = 1.6 \times 10^3 \text{ m/sec}$$

The emerging atoms had a maxwellian distribution of velocities (Chap. 15), and to secure a beam of more uniform velocity a device like that shown in Fig. 4.8 was employed. Only atoms whose velocity is such that they arrive at the second disk at exactly the same time as the slit can get through. The beam was then allowed to fall on the surface of a LiF crystal, and the scattered atoms in a given direction were collected in a small chamber. The resulting He gas pressure in this chamber was determined by measuring the electric resistance of a heated metal strip; the greater the gas pressure, the greater the loss of heat by the strip, and since the resistivity of a metal increases with temperature, the cooler the strip, the lower its resistance. The de Broglie wavelength of He atoms whose speed is 1.6×10^3 m/sec is

$$\lambda = \frac{h}{mv} = \frac{6.63 \times 10^{-34} \text{ joule·sec}}{6.7 \times 10^{-27} \text{ kg} \times 1.6 \times 10^3 \text{ m/sec}}$$

$$= 6 \times 10^{-11} \text{ m} = 0.6 \text{ A}$$

and a diffraction pattern was found that corresponded very closely to this wavelength.

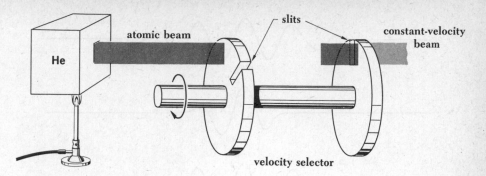

FIGURE 4.8   The production of a beam of constant-velocity helium atoms.

As in the case of electromagnetic waves, the wave and particle aspects of moving bodies can never be simultaneously observed, so that we cannot determine which is the "correct" description. All we can say is that in some respects a moving body exhibits wave properties and in other respects it exhibits particle properties. Which set of properties is most conspicuous depends upon how the de Broglie wavelength compares with the dimensions of the bodies involved: the 1.66 A wavelength of a 54-ev electron is of the same order of magnitude as the lattice spacing in a nickel crystal, but the wavelength of an automobile moving at 60 mi/hr is about 5×10^{-38} ft, far too small to manifest itself.

4.6   The Uncertainty Principle

The fact that a moving body must be regarded as a de Broglie wave group rather than as a localized entity suggests that there is a fundamental limit to the accuracy with which we can measure its particle properties. Figure 4.9a shows a de Broglie wave group: the particle may be anywhere within the wave group. If the group is very narrow, as in Fig. 4.9b, the position of the particle is readily found, but the wavelength is impossible to establish. At the other extreme, a wide group, as in Fig. 4.9c, permits a satisfactory wavelength estimate, but where is the particle located?

A straightforward argument based upon the nature of wave groups permits us to relate the inherent uncertainty Δx in a measurement of particle position with the inherent uncertainty Δp in a simultaneous measurement of its momentum. The simplest example of the formation of wave groups is that given in Sec. 4.4, where two wave trains slightly different in angular frequency ω and propagation constant k were superimposed to yield the series of groups shown in Fig. 4.4. Here let us consider the wave groups that arise when the de Broglie waves

$$\Psi_1 = A \cos (\omega t - kx)$$
$$\Psi_2 = A \cos (\omega + \Delta\omega)t - (k + \Delta k)x$$

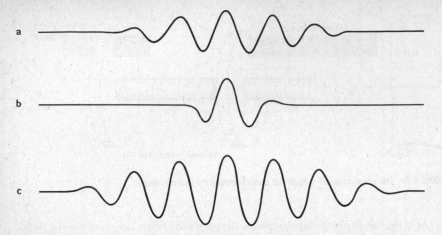

FIGURE 4.9 The width of a wave group is a measure of the uncertainty in the location of the particle it represents. The narrower the wave group, the greater the uncertainty in the wavelength.

are combined. From a calculation identical with the one used in obtaining Eq. 4.11, we find that

$$\Psi = \Psi_1 + \Psi_2$$

4.17
$$\approx 2A \cos (\omega t - kx) \cos (\tfrac{1}{2}\Delta\omega\, t - \tfrac{1}{2}\Delta k\, x)$$

which is plotted in Fig. 4.10. The width of each group is evidently equal to half the wavelength λ_m of the modulation. It is reasonable to suppose that this width is of the same order of magnitude as the inherent uncertainty Δx in the position of the group, that is,

4.18
$$\Delta x \approx \tfrac{1}{2}\lambda_m$$

The modulation wavelength is related to its propagation constant k_m by

$$\lambda_m = \frac{2\pi}{k_m}$$

From Eq. 4.17 we see that the propagation constant of the modulation is

$$k_m = \tfrac{1}{2}\Delta k$$

with the result that

$$\lambda_m = \frac{2\pi}{\tfrac{1}{2}\Delta k}$$

and

4.19

$$\Delta x = \frac{2\pi}{\Delta k}$$

Because the waves that constitute the groups are a combination of waves of propagation constant k and waves of propagation constant $k + \Delta k$, the best measurement of the propagation constant we can hope to make will still have an inherent uncertainty of Δk. This uncertainty is related to the uncertainty in the position of a wave group by Eq. 4.19, namely,

4.20

$$\Delta k = \frac{2\pi}{\Delta x}$$

The de Broglie wavelength of a particle of momentum p is

$$\lambda = \frac{h}{p}$$

The propagation constant corresponding to this wavelength is

$$k = \frac{2\pi}{\lambda}$$

$$= \frac{2\pi p}{h}$$

Hence an uncertainty Δk in the propagation constant of the de Broglie waves associated with the particle results in an uncertainty Δp in the particle's momentum according to the formula

4.21

$$\Delta p = \frac{h\, \Delta k}{2\pi}$$

Substituting the value of Δk given by Eq. 4.20, we obtain

$$\Delta p = \frac{h}{\Delta x}$$

FIGURE 4.10 Wave groups that result from the interference of wave trains having the same amplitudes but different frequencies.

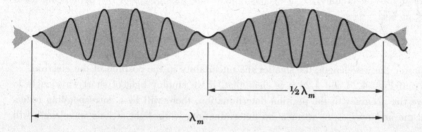

The Uncertainty Principle

or

4.22
$$\Delta x \, \Delta p \geq h$$

The sign $\geq$ is a consequence of the fact that wave groups may have different shapes. The Δx and Δp of Eq. 4.22 are thus *irreducible minima that are consequences of the wave natures of moving bodies*; any instrumental or statistical uncertainties that arise in the actual conduct of the measurement only augment the product $\Delta x \, \Delta p$.

Equation 4.22 is one form of the *uncertainty principle* first obtained by Werner Heisenberg in 1927. It states that the product of the uncertainty Δx in the position of a body at some instant and the uncertainty Δp in its momentum at the same instant is at best equal to Planck's constant h. We cannot measure simultaneously both position and momentum with perfect accuracy. If we arrange matters so that Δx is small, corresponding to the narrow wave group of Fig. 4.9b, Δp will be large. If we reduce Δp in some way, corresponding to the wide wave group of Fig. 4.9c, Δx will be large. These uncertainties are not in our apparatus but in nature.

The uncertainty principle can be derived in a variety of ways. Let us obtain it by basing our argument upon the particle nature of waves instead of upon the wave nature of particles as we did above. Suppose that we wish to measure the position and momentum of something at a certain moment. To accomplish this, we must prod it with something else that is to carry the desired information back to us; that is, we have to touch it with our finger, illuminate it with light, or interact with it in some other way. We might be examining an electron with the help of light of wavelength λ, as in Fig. 4.11. In this process photons of light strike the electron and bounce off it. Each photon possesses the momentum h/λ, and when it collides with the electron, the electron's original momentum p is changed. The precise change cannot be predicted, but it is likely to be of the same order of magnitude as the photon momentum h/λ. Hence the *act of measurement* introduces an uncertainty of

4.23
$$\Delta p = \frac{h}{\lambda}$$

in the momentum of the electron. The longer the wavelength of the light we employ in "seeing" the electron, the smaller the consequent uncertainty in its momentum.

Because light has wave properties, we cannot expect to determine the electron's position with infinite accuracy under any circumstances, but we might reasonably hope to keep the irreducible uncertainty Δx in its position to one wavelength of the light being used. That is, at best

4.24
$$\Delta x = \lambda$$

The shorter the wavelength, the smaller the uncertainty in the position of the electron.

From Eqs. 4.23 and 4.24 it is clear that, if we employ light of short wavelength to improve the accuracy of the position determination, there will be a corresponding reduction in the accuracy of the momentum determination, while light of long wavelength will

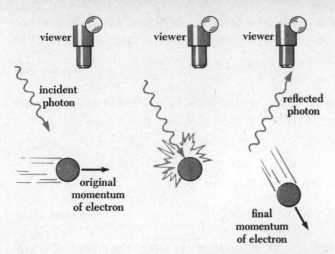

FIGURE 4.11 An electron cannot be observed without changing its momentum by an indeterminate amount.

yield an accurate momentum value but an inaccurate position value. Substituting $\lambda = \Delta x$ into Eq. 4.23,

$$\Delta p = \frac{h}{\Delta x}$$

from which we again obtain Eq. 4.22,

$$\Delta x \, \Delta p \geq h$$

Arguments like the preceding one, though superficially attractive, must as a rule be approached with caution. The above argument implies that the electron possesses a definite position and momentum at all times, and that it is the measurement process that introduces the indeterminacy in $\Delta x \, \Delta p$. On the contrary, this indeterminacy is inherent in the nature of a moving body. The justification for the many "derivations" of this kind is, first, that they show it is impossible to imagine a way around the uncertainty principle, and second, that they present a view of the principle that can be appreciated in a more familiar context than that of wave groups.

In the preceding discussions no effort was made to be precise in estimating the "uncertainty" in a measurable quantity. For instance, in obtaining Eq. 4.22 from an analysis of wave groups, we considered only two superposed wave trains. But two superposed wave trains yield a series of groups, as in Fig. 4.10, instead of a single group like one of those in Fig. 4.9. To achieve a single wave group, an infinite number of wave trains whose frequencies—and hence propagation constants k—differ only infinitesimally from one another is required. The narrower the group, the greater the range of propagation constants that is needed. Fourier analysis shows that the width Δx of a single wave group is related to the range Δk of the propagation constants of the waves that must be superposed to achieve it by

4.25
$$\Delta x \approx \frac{1}{\Delta k}$$
Single wave group

instead of the relationship $\Delta x \approx 2\pi/\Delta k$ which holds for the width of successive groups formed by only two wave trains. Hence, since Eq. 4.21

$$\Delta p = \frac{h \, \Delta k}{2\pi}$$

is still valid, we have the more realistic expression

4.26
$$\Delta x \, \Delta p \geq \frac{h}{2\pi}$$
Uncertainty principle

This form of the uncertainty principle is the proper one to use: **The product of the uncertainty Δx in the position of a body at some instant and the uncertainty Δp in its momentum at the same instant is equal to or greater than $h/2\pi$.**

The quantity $h/2\pi$ appears quite often in modern physics because, besides its connection with the uncertainty principle, $h/2\pi$ also turns out to be the basic unit of angular momentum. It is therefore customary to abbreviate $h/2\pi$ by the symbol $\hbar$:

$$\hbar = \frac{h}{2\pi} = 1.054 \times 10^{-34} \text{ joule-sec}$$

In the remainder of this book we shall use $\hbar$ in place of $h/2\pi$.

4.7 Applications of the Uncertainty Principle

Planck's constant h is so minute—only 6.63×10^{-34} joule-sec—that the limitations imposed by the uncertainty principle are significant only in the realm of the atom. On this microscopic scale, however, there are many phenomena that can be understood in terms of this principle; we shall consider several of them here.

One interesting question is whether electrons are present in atomic nuclei. As we shall learn later, in Sec. 21.5, typical nuclei are less than 10^{-14} m in radius. For an electron to be confined within such a nucleus, the uncertainty in its position may not exceed 10^{-14} m. The corresponding uncertainty in the electron's momentum is

$$\Delta p \geq \frac{\hbar}{\Delta x}$$

$$\geq \frac{1.054 \times 10^{-34} \text{ joule-sec}}{10^{-14} \text{ m}}$$

$$\geq 1.1 \times 10^{-20} \text{ kg-m/sec}$$

If this is the uncertainty in the electron's momentum, the momentum itself must be at least comparable in magnitude. An electron whose momentum is 1.1×10^{-20} kg-m/sec has a kinetic energy T many times greater than its rest energy m_0c^2, and we may accordingly use the extreme relativistic formula

$$T = pc$$

to find T. Substituting for p and c, we obtain

$$T = 1.1 \times 10^{-20} \text{ kg-m/sec} \times 3 \times 10^8 \text{ m/sec}$$
$$= 3.3 \times 10^{-12} \text{ joule}$$

Since 1 ev $= 1.6 \times 10^{-19}$ joule, the kinetic energy of the electron must be more than 20 Mev if it is to be a nuclear constituent. Experiments indicate that the electrons associated even with unstable atoms never have more than a fraction of this energy, and we conclude that electrons cannot be present within nuclei.

Let us now ask how much energy an electron needs to be confined to an atom. The hydrogen atom is about 5×10^{-11} m in radius, and therefore the uncertainty in the position of its electron may not exceed this figure. The corresponding momentum uncertainty is

$$\Delta p = 2.2 \times 10^{-24} \text{ kg-m/sec}$$

An electron whose momentum is of this order of magnitude is nonrelativistic in behavior, and its kinetic energy is

$$T = \frac{p^2}{2m}$$
$$= \frac{(2.2 \times 10^{-24} \text{ kg-m/sec})^2}{2 \times 9.1 \times 10^{-31} \text{ kg}}$$
$$= 2.7 \times 10^{-18} \text{ joule}$$

or about 17 ev. This is a wholly plausible figure.

Another form of the uncertainty principle is sometimes useful. We might wish to measure the energy E emitted sometime during the time interval Δt in an atomic process. If the energy is in the form of electromagnetic waves, the limited time available restricts the accuracy with which we can determine the frequency ν of the waves. Let us assume that the irreducible uncertainty in the number of waves we count in a wave group is one wave. Since the frequency of the waves under study is equal to the number of them we count divided by the time interval, the uncertainty $\Delta \nu$ in our frequency measurement is

$$\Delta \nu = \frac{1}{\Delta t}$$

The corresponding energy uncertainty is

Applications of the Uncertainty Principle

$$\Delta E = h \, \Delta \nu$$

and so

$$\Delta E = \frac{h}{\Delta t}$$

or

4.27 $$\Delta E \, \Delta t \geq h$$

A more realistic calculation changes Eq. 4.27 to

4.28 $$\Delta E \, \Delta t \geq \hbar$$

Equation 4.28 states that the product of the uncertainty ΔE in an energy measurement and the uncertainty Δt in the time at which the measurement was made is equal to or greater than $\hbar$.

As an example of the significance of Eq. 4.28 we can consider the radiation of light from an "excited" atom. Such an atom divests itself of its excess energy by emitting one or more photons of characteristic frequency. The average period that elapses between the excitation of an atom and the time it radiates is 10^{-8} sec. Thus the photon energy is uncertain by an amount

$$\Delta E = \frac{\hbar}{\Delta t}$$

$$= \frac{1.054 \times 10^{-34} \text{ joule-sec}}{10^{-8} \text{ sec}}$$

$$= 1.1 \times 10^{-26} \text{ joule}$$

and the frequency of the light is uncertain by

$$\Delta \nu = \frac{\Delta E}{h}$$

$$= 1.6 \times 10^{7} \text{ cycles/sec}$$

This is the irreducible limit to the accuracy with which we can determine the frequency of the radiation emitted by an atom.

4.8   The Wave-Particle Duality

Despite the abundance of experimental confirmation, many of us find it hard to appreciate how what we normally think of as a wave can also be a particle and how what we normally think of as a particle can also be a wave. The uncertainty principle provides a valuable perspective on this question which makes it possible to put such statements as those at the end of Sec. 3.2 on a more concrete basis.

Figure 4.12 shows an experimental arrangement in which light that has been diffracted by a double slit is detected on a "screen" that consists of many adjacent photoelectric cells. The photoelectric cells respond to photons, which have all the properties we associate with particles. However, when we plot the number of photons each cell counts in a certain period of time against the location of the cell, we find the characteristic pattern produced by the interference of a pair of coherent wave trains. This pattern even occurs when the light intensity is so low that, on the average, only one photon at a time is in the apparatus. The problem is, how can a photon that passes through one of the slits be affected by the presence of the other slit? In other words, how can a photon interfere with itself? This problem does not arise in the case of waves, which are spread out in space, but it would seem to have meaning in the case of photons, whose behavior suggests that they are localized in very small regions of space.

To have meaning, every question or statement in science must ultimately be reducible to an experiment. Here the relevant experiment is one that would detect which of the slits a particular photon passes through on its way to the screen. Let us imagine that we intro-

FIGURE 4.12 Hypothetical experiment to determine which slit each photon contributing to a double-slit interference pattern has passed through.

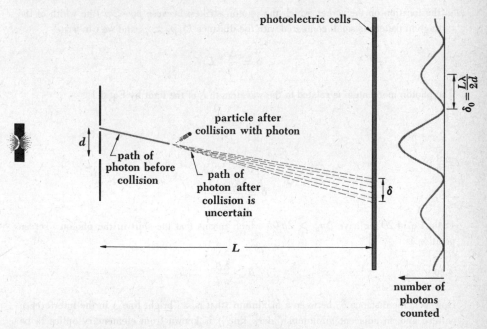

The Wave-Particle Duality

duce a cloud of small particles between the slits and the screen. A photon that passes through one of the slits strikes a particle and gives it a certain impulse which enables us to detect it (Fig. 4.12). Provided that we can establish the position of the particle with an uncertainty Δy that is less than half the space d between the slits, we can determine which slit the photon passed through. Therefore

$$\Delta y < \frac{d}{2}$$

But if we are able to limit the uncertainty in the y coordinate of the struck particle to Δy, the uncertainty Δp_y in the y component of its momentum is

4.29
$$\Delta p_y \geq \frac{\hbar}{\Delta y} > \frac{2\hbar}{d}$$

Since the collision introduces a change of Δp_y in the particle's momentum, the same change must have occurred in the photon's momentum. A change of Δp_y in the photon's momentum means a shift of

$$\delta = \frac{\Delta p_y}{p_x} L$$

in the location on the screen which the photon strikes; because $p_y \ll p$ (the width of the diffraction pattern is small compared with the distance L), $p_x \approx p$, and we can write

$$\delta = \frac{\Delta p_y}{p} L$$

The photon momentum is related to the wavelength λ of the light by Eq. 4.1,

$$p = \frac{h}{\lambda}$$

and so

$$\delta = \frac{\Delta p_y \lambda L}{h}$$

From Eq. 4.29 we have $\Delta p_y > 2\hbar/d$, which means that the shift in the photon's screen position is

4.30
$$\delta > \frac{\lambda L}{\pi d}$$

However, the distance δ_0 between a maximum (that is, a "bright line") in the interference pattern and an adjacent minimum ("dark line") is known from elementary optics to be

$$\delta_0 = \frac{\lambda L}{2d}$$

This distance is almost the same as the minimum shift involved in establishing which slit each photon passes through. What would otherwise be a pattern of alternating bright and dark lines becomes blurred owing to the interactions between the photons and the particles used to trace their paths. Thus no interference can be observed: the price of determining the exact path of each photon is the destruction of the interference pattern. If our interest is in the wave aspects of a phenomenon, they can be demonstrated; if our interest is in the particle aspects of the same phenomenon, they too can be demonstrated; but it is impossible to demonstrate *both* aspects in a simultaneous experiment. (Using photoelectric cells to detect an interference pattern is not a simultaneous experiment in this sense, because there is no way by which a photoelectric cell can determine which slit a particular photon striking it has passed through.)

The original question of how a photon can interfere with itself therefore turns out to be meaningless. It is important to be aware of the distinction between a legitimate question that cannot be answered because our existing knowledge is not sufficiently detailed or advanced to cope with it and a question whose very statement is in contradiction with experiment. Questions that seek to pry apart the elements of the wave-particle duality fall into the latter class in view of the uncertainty principle, whose own empirical validity has been thoroughly established.

Problems

1. Show that the de Broglie wavelength of a particle approaches zero faster than $1/v$ as its speed approaches the speed of light.

2. Find the de Broglie wavelength of a 15-ev proton.

3. Find the de Broglie wavelength of a 15-kev electron.

4. What is the de Broglie wavelength of an electron whose speed is 9×10^7 m/sec?

5. Find the wavelength of a 1-kg object whose speed is 1 m/sec.

6. Neutrons in equilibrium with matter at room temperature (300°K) have average energies of about ⅟₂₅ ev. (Such neutrons are often called "thermal neutrons.") Find their de Broglie wavelength.

7. Derive a formula expressing the de Broglie wavelength (in A) of an electron in terms of the potential difference V (in volts) through which it is accelerated.

8. Derive a formula for the de Broglie wavelength of a particle in terms of its kinetic energy T and its rest energy m_0c^2.

9. Assume that electromagnetic waves are a special case of de Broglie waves. Show that photons must travel with the wave velocity c and that the rest mass of the photon must be 0.

10. Obtain the de Broglie wavelength of a moving particle in the following way, which parallels de Broglie's original treatment. Consider a particle of rest mass m_0 as having a characteristic frequency of vibration of ν_0 specified by the relationship $h\nu_0 = m_0 c^2$. The particle travels with the speed v relative to an observer. With the help of special relativity show that the observer sees a progressive wave whose phase velocity is $w = c^2/v$ and whose wavelength is h/mv, where $m = m_0/\sqrt{1 - v^2/c^2}$.

11. The velocity of ocean waves is $\sqrt{g\lambda/2\pi}$, where g is the acceleration of gravity. Find the group velocity of these waves.

12. The velocity of ripples on a liquid surface is $\sqrt{2\pi/\lambda\rho}$, where S is the surface tension and ρ the density of the liquid. Find the group velocity of these waves.

13. The position and momentum of a 1-kev electron are simultaneously determined. If its position is located to within 1 A, what is the percentage of uncertainty in its momentum?

14. An electron microscope uses 40-kev electrons. Find its ultimate resolving power on the assumption that this is equal to the wavelength of the electrons.

15. Compare the uncertainties in the velocities of an electron and a proton confined in a 10-A box.

16. Wavelengths can be determined with accuracies of one part in 10^6. What is the uncertainty in the position of a 1-A X-ray photon when its wavelength is simultaneously measured?

17. In Sec. 4.7 the uncertainty principle was shown to limit the accuracy with which the frequency of atomic radiation can be determined by an amount $\Delta\nu \sim 1.6 \times 10^7$ cycles/sec. What percentage of the frequency of a 5,000-A photon is this?

18. The atoms in a solid possess a certain minimum *zero-point energy* even at $0°K$, while no such restriction holds for the molecules in an ideal gas. Use the uncertainty principle to explain these statements.

19. Verify that the uncertainty principle can be expressed in the form $\Delta L\, \Delta\theta \geq \hbar$, where ΔL is the uncertainty in the angular momentum of a body and $\Delta\theta$ is the uncertainty in its angular position.

Chapter 5
Atomic Structure

Far in the past people began to suspect that matter, despite its appearance of being continuous, possesses a definite structure on a microscopic level beyond the direct reach of our senses. This suspicion did not take on a more concrete form until a little over a century and a half ago; since then the existence of atoms and molecules, the ultimate particles of matter in its common forms, has been amply demonstrated, and their own ultimate particles, electrons, protons, and neutrons, have been identified and studied as well. In this chapter and in others to come our chief concern will be the structure of the atom, since it is this structure that is responsible for nearly all of the properties of matter that have shaped the world around us.

5.1 Atomic Models

While the scientists of the nineteenth century accepted the idea that the chemical elements consist of atoms, they knew virtually nothing about the atoms themselves. The discovery of the electron and the realization that all atoms contain electrons provided the first important insight into atomic structure. Electrons contain negative electrical charges, while atoms themselves are electrically neutral: every atom must therefore contain enough positively charged matter to balance the negative charge of its electrons. Furthermore, electrons are thousands of times lighter than whole atoms; this suggests that the positively charged constituent of atoms is what provides them with nearly all of their mass. When J. J. Thomson proposed in 1898 that atoms are uniform spheres of positively charged matter in which electrons are embedded, his hypothesis then seemed perfectly reasonable. Thomson's plum-pudding model of the atom—so called from its resemblance to that raisin-studded delicacy—is sketched in Fig. 5.1. Despite the importance of the problem, 13 years passed before a definite experimental test of the plum-pudding model was made. This experiment, as we shall see, compelled the abandonment of this apparently plausible model, leaving in its place a concept of atomic structure incomprehensible in the light of classical physics.

The most direct way to find out what is inside a plum pudding is to plunge a finger into it, a technique not very different from that used by Geiger and Marsden to find out what is inside an atom. In their classic experiment, performed in 1911 at the suggestion of Ernest Rutherford, they employed as probes the fast *alpha particles* spontaneously emitted

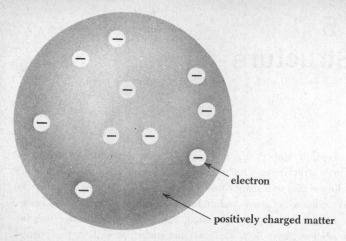

electron

positively charged matter

FIGURE 5.1 The Thomson model of the atom. This model is not in accord with experiment.

by certain radioactive elements. Alpha particles are helium atoms that have lost two electrons, leaving them with a charge of $+2e$; we shall examine their origin and properties in more detail in Chap. 23. Geiger and Marsden placed a sample of an alpha-particle-emitting substance behind a lead screen that had a small hole in it, as in Fig. 5.2, so that a narrow beam of alpha particles was produced. This beam was directed at a thin gold foil. A movable zinc sulfide screen, which gives off a visible flash of light when struck by an alpha particle, was placed on the other side of the foil. It was anticipated that most of the alpha particles would go right through the foil, while the remainder would at most suffer only slight deflections. The behavior follows from the Thomson atomic model, in which the charges within an atom are assumed to be uniformly distributed throughout its volume. If the Thomson model is correct, only weak electric forces are exerted on alpha particles passing through a thin metal foil, and their initial momenta should be enough to carry them through with only minor departures from their original paths.

What Geiger and Marsden actually found was that, while most of the alpha particles indeed emerged without deviation, some were scattered through very large angles. A few were even scattered in the backward direction. Since alpha particles are relatively heavy (over 7,000 times more massive than electrons) and those used in this experiment traveled at high speed, it was clear that strong forces had to be exerted upon them to cause such marked deflections. To explain the results, Rutherford was forced to picture an atom as being composed of a tiny *nucleus*, in which its positive charge and nearly all of its mass are concentrated, with its electrons some distance away (Fig. 5.3). Considering an atom as largely empty space, it is easy to see why most alpha particles go right through a thin foil. When an alpha particle approaches a nucleus, however, it encounters an intense electric field

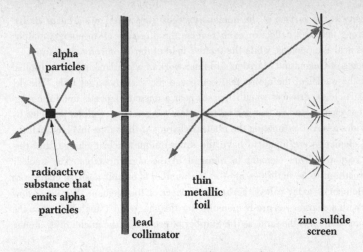

alpha
particles

radioactive
substance that
emits alpha
particles

lead
collimator

thin
metallic
foil

zinc sulfide
screen

FIGURE 5.2   The Rutherford scattering experiment.

and is likely to be scattered through a considerable angle. The atomic electrons, being so light, do not appreciably affect the motion of incident alpha particles.

Numerical estimates of electric-field intensities within the Thomson and Rutherford models emphasize the difference between them. If we assume with Thomson that the positive charge within a gold atom is spread evenly throughout its volume, and if we neglect the electrons completely, the electric-field intensity at the atom's surface (where it is a maximum) is about 10^{13} volts/m. On the other hand, if we assume with Rutherford that the positive charge within a gold atom is concentrated in a small nucleus at its center, the

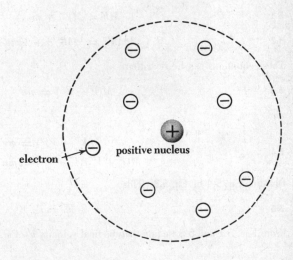

electron

positive nucleus

FIGURE 5.3   The Rutherford model of the atom.

Atomic Models

electric-field intensity at the surface of the nucleus exceeds 10^{21} volts/m—a factor of 10^8 greater. Such a strong field can deflect or even reverse the direction of an energetic alpha particle that comes near the nucleus, while the feebler field of the Thomson atom cannot.

The experiments of Geiger and Marsden and later work of a similar kind also supplied information about the nuclei of the atoms that composed the various target foils. The deflection an alpha particle experiences when it passes near a nucleus depends upon the magnitude of the nuclear charge, and so comparing the relative scattering of alpha particles by different foils provides a way of estimating the nuclear charges of the atoms involved. All the atoms of any one element were found to have the same unique nuclear charge, and this charge increased regularly from element to element in the periodic table. The nuclear charges always turned out to be multiples of $+e$; the number Z of unit positive charges in the nuclei of an element is today called the *atomic number* of the element. We know now that protons, each with a charge $+e$, are responsible for the charge on a nucleus, and so the atomic number of an element is the same as the number of protons in the nuclei of its atoms.

5.2   The Thomson Model

Before going further it is appropriate to verify in a quantitative way that the Thomson model cannot explain the occurrence of large-angle deflections of alpha particles. We shall first discuss the influence of atomic electrons on alpha-particle motion.

We begin by examining a head-on collision between a particle of mass M and initial velocity V and a particle of mass m initially at rest. (The transfer of momentum to the struck particle is a maximum for head-on collisions.) If the final velocity of M is V' and that of m is v', conservation of momentum and conservation of kinetic energy respectively require that

5.1
$$MV = MV' + mv'$$

5.2
$$\tfrac{1}{2}MV^2 = \tfrac{1}{2}MV'^2 + \tfrac{1}{2}mv'^2$$

These equations can be rewritten

5.3
$$M(V - V') = mv'$$

and

5.4
$$M(V^2 - V'^2) = mv'^2$$
$$M(V + V')(V - V') = mv'^2$$

Dividing Eq. 5.4 by Eq. 5.3 yields

5.5
$$V + V' = v'$$

From Eqs. 5.1 and 5.5 we find that the final velocity v' of m, which was initially at rest, is

5.6
$$v' = \frac{2M}{M + m} V$$

The mass of an alpha particle is about 7,000 times that of an electron, so that in the case of an alpha particle striking a stationary electron $M \gg m$ and $v' = 2V$. Thus the alpha particle gives up

5.7
$$\Delta p = mv' = 2mV$$

of its initial momentum p to the electron in a head-on collision. In such a collision the alpha particle is not deflected, of course. If the collision is a glancing one, the change in momentum Δp is less than $2mV$. To obtain an estimate of the greatest possible deflection θ of an alpha particle colliding with an electron, we can assume that the struck electron goes off at right angles to the initial path of the alpha particle (Fig. 5.4) with the momentum $2mV$. Hence

5.8
$$\theta_{electron} \approx \frac{\Delta p}{p} < \frac{2mV}{MV}$$

and, since $M \approx 7,000m$,

$$\theta_{electron} < 3 \times 10^{-4} \text{ rad}$$
$$< 0.02°$$

Now we turn to the effect a positively charged cloud the size of an atom has on an incident alpha particle. Outside the cloud the electric-field intensity a distance r from the center has the magnitude

$$E_{outside} = \frac{1}{4\pi\varepsilon_0} \frac{Q}{r^2}$$

where Q is the total charge. Inside the cloud the electric-field intensity has the magnitude

$$E_{inside} = \frac{1}{4\pi\varepsilon_0} \frac{Qr}{R^3}$$

where R is the radius of the cloud, which in the Thomson model is the radius of the entire

FIGURE 5.4 The deflection of an alpha particle by an atomic electron.

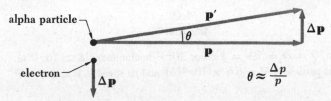

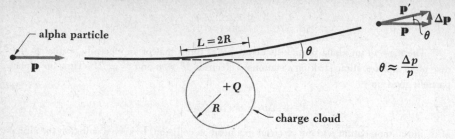

FIGURE 5.5 The deflection of an alpha particle by a charge cloud.

atom. The strongest field is evidently at the surface of the cloud where $r = R$, and the greatest deflection of an alpha particle will occur when it just grazes the surface. At the surface the force exerted on an alpha particle of charge $2e$ is

$$F_{max} = 2eE = \frac{2eQ}{4\pi\varepsilon_0 R^2}$$

Because the force drops off sharply with distance, to obtain an approximate result we need only consider the interaction that occurs over a short distance L near the charge cloud. We shall assume that $L \approx 2R$ (Fig. 5.5) and that the entire force F_{max} acts during the time interval $\Delta t = L/V \approx 2R/V$ that the alpha particle is near the charge cloud. The transverse impulse given to the alpha particle is therefore

$$F_{max}\,\Delta t \approx \frac{4eQ}{4\pi\varepsilon_0 RV}$$

and the resulting change in the alpha particle's momentum $p = MV$ is

$$\Delta p = F_{max}\,\Delta t$$

$$\approx \frac{4eQ}{4\pi\varepsilon_0 RV}$$

Hence the maximum angle of deflection is specified by

5.9

$$\theta_{\text{Thomson atom}} < \frac{\Delta p}{p}$$

$$< \frac{4eQ}{4\pi\varepsilon_0 RMV^2}$$

In the case of a gold atom, $Q = Ze = 79e = 1.26 \times 10^{-17}$ coulomb and $R \approx 10^{-10}$ m, while the mass of an alpha particle is $M = 6.6 \times 10^{-27}$ kg and its speed is $V \approx 2 \times 10^7$ m/sec. Accordingly

$$\theta_{\text{Thomson atom}} < \frac{1}{4\pi\varepsilon_0} \times \frac{4eQ}{RMV^2}$$

$$< 9 \times 10^9 \frac{\text{n-m}^2}{\text{coulomb}^2} \times \frac{4 \times 1.6 \times 10^{-19} \text{ coulomb} \times 1.26 \times 10^{-17} \text{ coulomb}}{10^{-10} \text{ m} \times 6.6 \times 10^{-27} \text{ kg} \times (2 \times 10^7 \text{ m/sec})^2}$$

$$< 3 \times 10^{-4} \text{ rad}$$

$$< 0.02°$$

Even though the deflection of an incident alpha particle by either an atomic electron or a positive charge cloud the size of an atom is minute, might not a succession of such deflections produce an appreciable total scattering angle? To determine the effect of a series of N random deflections, the statistical methods used to study random-walk problems are required. The results show that, if the average individual scattering angle has the magnitude $\bar{\theta}$, the average total deflection $\bar{\Theta}$ after N encounters will be

5.10
$$\bar{\Theta} = \sqrt{N}\,\bar{\theta}$$

A typical foil used in Geiger and Marsden's experiments was about 10^4 atomic layers thick, so on the basis of the Thomson model and assuming that one scattering event averaging $0.01°$ (either by an electron or by the positive charge cloud) occurs per atomic layer traversed, $\bar{\Theta} \approx 1°$. In fact, the average total deflection found by Geiger and Marsden was about $1°$. However, as was mentioned earlier, a definite proportion of alpha particles was found to be deflected through angles much greater than $1°$; for example, about 1 in every 8,000 were scattered through $90°$ or more. This is the really crucial result of the experiments. One out of 8,000 may not seem a very impressive number, but the probability $P(\geq \Theta)$ that a total deflection of more than some specific angle Θ occurs when the average deflection is $\bar{\Theta}$ is

5.11
$$P(\geq \Theta) = e^{-(\Theta/\bar{\Theta})^2}$$

The likelihood on the basis of the Thomson model that an alpha particle be scattered through $90°$ or more is therefore

$$P(\geq 90°) = e^{-(90°/1°)^2} = e^{-90^2} = e^{-8100}$$
$$= 10^{-3500}$$

Thus only 1 out of every 10^{3500} alpha particles should be scattered through $90°$ or more—a convincing disagreement with experiment. The Thomson model of the atom must therefore be discarded.

5.3 Alpha-particle Scattering

Rutherford arrived at a formula, describing the scattering of alpha particles by thin foils on the basis of his atomic model, that agreed with the experimental results. The derivation

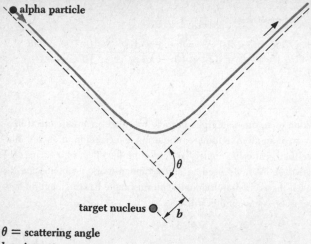

alpha particle

θ

target nucleus ● b

θ = scattering angle
b = impact parameter

FIGURE 5.6   Rutherford scattering.

of this formula both illustrates the application of fundamental physical laws in a novel setting and introduces certain notions, such as that of the *cross section* for an interaction, that
are important in many other aspects of modern physics.

Rutherford began by assuming that the alpha particle and the nucleus it interacts with
are both small enough to be considered as point masses and charges; that the electrostatic
repulsive force between alpha particle and nucleus (which are both positively charged) is
the only one acting; and that the nucleus is so massive compared with the alpha particle that
it does not move during their interaction. Owing to the variation of the electrostatic force
with $1/r^2$, where r is the instantaneous separation between alpha particle and nucleus, the
alpha particle's path is a hyperbola with the nucleus at the outer focus (Fig. 5.6). The *impact
parameter b* is the minimum distance to which the alpha particle would approach the nucleus
if there were no force between them, and the *scattering angle θ* is the angle between the
asymptotic direction of approach of the alpha particle and the asymptotic direction in which
it recedes. Our first task is to find a relationship between b and θ.

As a result of the impulse $\int \mathbf{F}\, dt$ given it by the nucleus, the momentum of the alpha
particle changes by Δp from the initial value $\mathbf{p}_1$ to the final value $\mathbf{p}_2$. That is,

$$\Delta p = p_2 - p_1$$

5.12
$$= \int \mathbf{F}\, dt$$

Because the nucleus remains stationary during the passage of the alpha particle, by hypothesis, the alpha-particle kinetic energy remains constant; hence the *magnitude* of its momen-

tum also remains constant, and

$$p_1 = p_2 = mv$$

Here v is the alpha-particle velocity far from the nucleus. From Fig. 5.7 we see that, according to the law of sines,

$$\frac{\Delta p}{\sin \theta} = \frac{mv}{\sin \dfrac{(\pi - \theta)}{2}}$$

Since

$$\sin \frac{1}{2}(\pi - \theta) = \cos \frac{\theta}{2}$$

and

$$\sin \theta = 2 \sin \frac{\theta}{2} \cos \frac{\theta}{2}$$

we have for the momentum change

5.13
$$\Delta p = 2\, mv \sin \frac{\theta}{2}$$

Because the impulse $\int \mathbf{F}\, dt$ is in the same direction as the momentum change $\Delta \mathbf{p}$, its magnitude is

5.14
$$\int F \cos \phi\, dt$$

FIGURE 5.7 Geometrical relationships in Rutherford scattering.

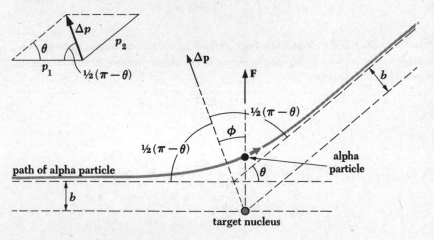

path of alpha particle

target nucleus

where ϕ is the instantaneous angle between $\mathbf{F}$ and $\Delta \mathbf{p}$ along the path of the alpha particle. Inserting Eqs. 5.13 and 5.14 in Eq. 5.12,

$$2 \, mv \sin \frac{\theta}{2} = \int_0^\infty F \cos \phi \, dt$$

To change the variable on the right-hand side from t to ϕ, we note that the limits of integration will change to $-\frac{1}{2}(\pi - \theta)$ and $+\frac{1}{2}(\pi - \theta)$, corresponding to ϕ at $t = 0$ and $t = \infty$ respectively, and so

5.15
$$2 \, mv \sin \frac{\theta}{2} = \int_{-(\pi-\theta)/2}^{+(\pi-\theta)/2} F \cos \phi \, \frac{dt}{d\phi} \, d\phi$$

The quantity $d\phi/dt$ is just the angular velocity ω of the alpha particle about the nucleus (this is evident from Fig. 5.7). The electrostatic force exerted by the nucleus on the alpha particles acts along the radius vector joining them, and so there is no torque on the alpha particle and its angular momentum $m\omega r^2$ is constant. Hence

$$m\omega r^2 = \text{constant}$$

$$= mr^2 \frac{d\phi}{dt}$$

$$= mvb$$

from which we see that

$$\frac{dt}{d\phi} = \frac{r^2}{vb}$$

Substituting this expression for $dt/d\phi$ in Eq. 5.15,

5.16
$$2 \, mv^2b \sin \frac{\theta}{2} = \int_{-(\pi-\theta)/2}^{+(\pi-\theta)/2} Fr^2 \cos \phi \, d\phi$$

As we recall, F is the electrostatic force exerted by the nucleus on the alpha particle. The charge on the nucleus is Ze, corresponding to the atomic number Z, and that on the alpha particle is $2e$. Therefore

$$F = \frac{1}{4\pi\varepsilon_0} \frac{2Ze^2}{r^2}$$

and

$$\frac{4\pi\varepsilon_0 mv^2 b}{Ze^2} \sin \frac{\theta}{2} = \int_{-(\pi-\theta)/2}^{+(\pi-\theta)/2} \cos \phi \, d\phi$$

$$= 2 \cos \frac{\theta}{2}$$

The scattering angle θ is related to the impact parameter b by the equation

$$\cot \frac{\theta}{2} = \frac{2\pi\varepsilon_0 m v^2}{Z e^2} b$$

It is more convenient to specify the alpha-particle energy T instead of its mass and velocity separately; with this substitution,

5.17
$$\cot \frac{\theta}{2} = \frac{4\pi\varepsilon_0 T}{Z e^2} b$$

Figure 5.8 is a schematic representation of Eq. 5.17; the rapid decrease in θ as b increases is evident. A very near miss is required for a substantial deflection.

5.4   The Rutherford Scattering Formula

Equation 5.17 cannot be directly confronted with experiment since there is no way of measuring the impact parameter corresponding to a particular observed scattering angle. An indirect strategy is required. Our first step is to note that all alpha particles approaching a target nucleus with an impact parameter from 0 to b will be scattered through an angle of θ or more, where θ is given in terms of b by Eq. 5.17. This means that an alpha particle that is initially directed anywhere within the area πb^2 around a nucleus will be scattered through θ or more (Fig. 5.8); the area πb^2 is accordingly called the *cross section* for the interaction. The general symbol for cross section is σ, and so here

5.18
$$\sigma = \pi b^2$$

We must keep in mind that the incident alpha particle is actually scattered before it reaches the immediate vicinity of the nucleus and hence does not necessarily pass within a distance b of it.

FIGURE 5.8   **The scattering angle decreases with increasing impact parameter.**

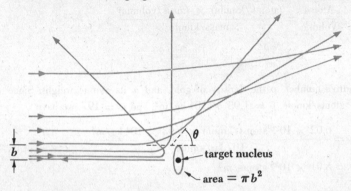

Now we consider a foil of thickness t that contains n atoms per unit volume. The number of target nuclei per unit area is nt, and an alpha-particle beam incident upon an area A therefore encounters ntA nuclei. The aggregate cross section for scatterings of θ or more is the number of target nuclei ntA multiplied by the cross section σ for such scattering per nucleus, or $ntA\sigma$. Hence the fraction f of incident alpha particles scattered by θ or more is the ratio between the aggregate cross section $ntA\sigma$ for such scattering and the total target area A. That is,

$$ f = \frac{\text{alpha particles scattered by } \theta \text{ or more}}{\text{incident alpha particles}} $$

$$ = \frac{\text{aggregate cross section}}{\text{target area}} $$

$$ = \frac{ntA\sigma}{A} $$

$$ = nt\pi b^2 $$

Substituting for b from Eq. 5.17,

5.19
$$ f = \pi nt \left(\frac{Ze^2}{4\pi\varepsilon_0 T} \right)^2 \cot^2 \frac{\theta}{2} $$

In the above calculation it was assumed that the foil is sufficiently thin so that the cross sections of adjacent nuclei do not overlap and that a scattered alpha particle receives its entire deflection from a single encounter with a nucleus.

Let us use Eq. 5.19 to determine what fraction of a beam of 7.7-Mev alpha particles is scattered through angles of more than $45°$ when incident upon a gold foil 3×10^{-7} m thick. (These values are typical of the alpha-particle energies and foil thicknesses used by Geiger and Marsden; for comparison, a human hair is about 10^{-4} m in diameter.) We begin by finding n, the number of gold atoms per unit volume in the foil, from the relationship

$$ \frac{\text{Atoms}}{\text{Volume}} = \frac{(\text{atoms/kmole}) \times (\text{mass/volume})}{\text{mass/kmole}} $$

$$ n = \frac{N_0 \rho}{w} $$

where N_0 is Avogadro's number, ρ the density of gold, and w its atomic weight. Since $N_0 = 6.02 \times 10^{26}$ atoms/kmole, $\rho = 1.93 \times 10^4$ kg/m³, and $w = 197$, we have

$$ n = \frac{6.02 \times 10^{26} \text{ atoms/kmole} \times 1.93 \times 10^4 \text{ kg/m}^3}{197 \text{ kg/kmole}} $$

$$ = 5.91 \times 10^{28} \text{ atoms/m}^3 $$

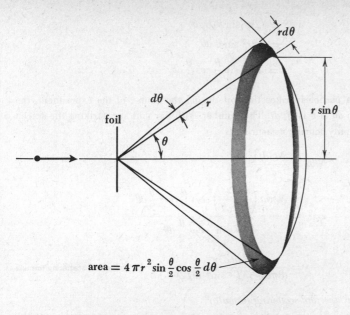

$$\text{area} = 4\pi r^2 \sin\frac{\theta}{2} \cos\frac{\theta}{2}\, d\theta$$

**FIGURE 5.9   In the Rutherford experiment, particles are detected that have
been scattered between θ and $\theta + d\theta$.**

The atomic number Z of gold is 79, a kinetic energy of 7.7 Mev is equal to 1.23×10^{-12} joule, and $\theta = 45°$; from these figures we find that

$$f = 7 \times 10^{-5}$$

of the incident alpha particles are scattered through 45° or more—only 0.007 percent! A foil this thin is quite transparent to alpha particles.

In an actual experiment, a detector measures alpha particles scattered between θ and $\theta + d\theta$, as in Fig. 5.9. The fraction of incident alpha particles so scattered is found by differentiating Eq. 5.19 with respect to θ, an operation that yields

5.20
$$df = -\pi nt \left(\frac{Ze^2}{4\pi\varepsilon_0 T}\right)^2 \cot\frac{\theta}{2}\csc^2\frac{\theta}{2}\, d\theta$$

(The minus sign expresses the fact that f decreases with increasing θ.) In the experiment, a fluorescent screen was placed a distance r from the foil, and the scattered alpha particles were detected by means of the scintillations they caused. Those alpha particles scattered between θ and $\theta + d\theta$ reach a zone of a sphere of radius r whose width is $r\,d\theta$. The zone radius itself is $r \sin\theta$, and so the area dS of the screen struck by these particles is

$$dS = (2\pi r \sin \theta)(r\,d\theta)$$
$$= 2\pi r^2 \sin \theta \, d\theta$$
$$= 4\pi r^2 \sin \frac{\theta}{2} \cos \frac{\theta}{2} \, d\theta$$

If a total of N_i alpha particles strikes the foil during the course of the experiment, the number scattered into $d\theta$ at θ is $N_i \, df$. The number $N(\theta)$ per unit area striking the screen at θ, which is the quantity actually measured, is

$$N(\theta) = \frac{N_i \, |df|}{dS}$$

$$= \frac{N_i \pi n t \left(\dfrac{Ze^2}{4\pi\varepsilon_0 T}\right)^2 \cot \dfrac{\theta}{2} \csc^2 \dfrac{\theta}{2} \, d\theta}{4\pi r^2 \sin \dfrac{\theta}{2} \cos \dfrac{\theta}{2} \, d\theta}$$

5.21 $$= \frac{N_i n t Z^2 e^4}{(8\pi\varepsilon_0)^2 r^2 T^2 \sin^4 (\theta/2)}$$ **Rutherford scattering formula**

Equation 5.21 is the *Rutherford scattering formula*.

According to Eq. 5.21, the number of alpha particles per unit area arriving at the fluorescent screen a distance r from the scattering foil should be directly proportional to the thickness t of the foil, the number of foil atoms per unit volume n, and the square of the atomic number Z of the foil atoms, and it should be inversely proportional to the square of the kinetic energy T of the alpha particles and to $\sin^4 (\theta/2)$, where θ is the scattering angle. These predictions agreed with the measurements of Geiger and Marsden mentioned earlier, which led Rutherford to conclude that his assumptions, chief among them the hypothesis of the nuclear atom, were correct. Rutherford is therefore credited with the "discovery" of the nucleus. Figure 5.10 shows how $N(\theta)$ varies with θ.

5.5 Nuclear Dimensions

When we say that the experimental data on the scattering of alpha particles by thin foils verifies our assumption that atomic nuclei are point particles, what is really meant is that their dimensions are insignificant compared with the minimum distance to which the incident alpha particles approach the nuclei. Rutherford scattering therefore permits us to determine an upper limit to nuclear dimensions. Let us compute the distance of closest approach r_0 of the most energetic alpha particles employed in the early experiments. An alpha particle will have its smallest r_0 when its impact parameter is $b = 0$, corresponding to a head-on approach followed by a 180° scattering. At the instant of closest approach the initial kinetic energy T of the particle is entirely converted to electrostatic potential energy, and so at that instant

5.22
$$T = \frac{1}{4\pi\varepsilon_0} \frac{2Ze^2}{r_0}$$

since the charge of the alpha particle is $2e$ and that of the nucleus Ze. Hence

$$r_0 = \frac{2Ze^2}{4\pi\varepsilon_0 T}$$

The maximum T found in alpha particles of natural origin is 7.7 Mev, which is

$$7.7 \times 10^6 \text{ ev} \times 1.6 \times 10^{-19} \text{ joule/ev} = 1.2 \times 10^{-12} \text{ joule}$$

FIGURE 5.10   Rutherford scattering.

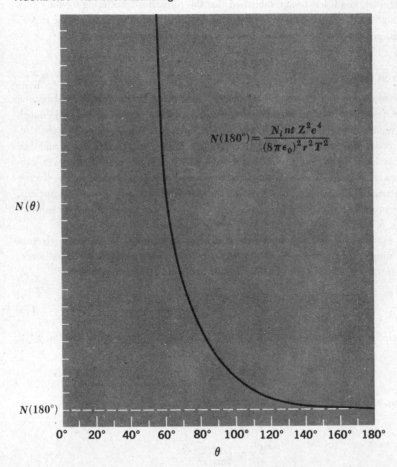

$$N(180°) = \frac{N_i nt\, Z^2 e^4}{(8\pi\epsilon_0)^2 r^2 T^2}$$

$N(\theta)$

$N(180°)$

0°   20°   40°   60°   80°   100°   120°   140°   160°   180°

θ

Since $1/4\pi\varepsilon_0 = 9 \times 10^9$ n-m²/coulomb²

$$r_0 = \frac{2 \times 9 \times 10^9 \text{ n-m}^2/\text{coulomb}^2 \times (1.6 \times 10^{-19} \text{ coulomb})^2 Z}{1.2 \times 10^{-12} \text{ joule}}$$

$$= 3.8 \times 10^{-16} Z \text{ m}$$

The atomic number of gold, a typical foil material, is $Z = 79$, so that

$$r_0(\text{Au}) = 3.0 \times 10^{-14} \text{ m}$$

The radius of the gold nucleus is therefore less than 3.0×10^{-14} m, well under 1/10,000 the radius of the atom as a whole.

In more recent years particles of much higher energies than 7.7 Mev have been artificially accelerated, and it has been established that the Rutherford scattering formula does indeed eventually fail to agree with experiment, as expected. We shall discuss these experiments and the information they provide on actual nuclear dimensions in Chap. 21.

5.6 Electron Orbits

The Rutherford model of the atom, so convincingly confirmed by experiment, postulates a tiny, massive, positively charged nucleus surrounded at a relatively great distance by enough electrons to render the atom, as a whole, electrically neutral. Thomson visualized the electrons in his model atom as embedded in the positively charged matter that fills it, and thus not necessarily being in motion. The electrons in Rutherford's model atom, however, cannot be stationary, because there is nothing that can keep them in place against the electrostatic force attracting them to the nucleus. If the electrons are in motion around the nucleus, however, dynamically stable orbits (comparable with those of the planets about the sun) are possible (Fig. 5.11).

Let us examine the classical dynamics of the hydrogen atom, whose single electron makes it the simplest of all atoms. We shall assume a circular electron orbit for convenience, though it might as reasonably be assumed elliptical in shape. The centripetal force

$$F_c = \frac{mv^2}{r}$$

holding the electron in an orbit r from the nucleus is provided by the electrostatic force

$$F_e = \frac{1}{4\pi\varepsilon_0} \frac{e^2}{r^2}$$

between them, and the condition for orbit stability is

$$F_c = F_e$$

5.23
$$\frac{mv^2}{r} = \frac{1}{4\pi\varepsilon_0} \frac{e^2}{r^2}$$

The electron velocity v is therefore related to its orbit radius r by the formula

5.24
$$v = \frac{e}{\sqrt{4\pi\varepsilon_0 mr}}$$

The total energy E of the electron in a hydrogen atom is the sum of its kinetic energy

$$T = \tfrac{1}{2}mv^2$$

and its potential energy

$$V = -\frac{e^2}{4\pi\varepsilon_0 r}$$

(The minus sign signifies that the force on the electron is in the $-r$ direction.) Hence

$$E = T + V$$
$$= \frac{mv^2}{2} - \frac{e^2}{4\pi\varepsilon_0 r}$$

Substituting for v from Eq. 5.24,

$$E = \frac{e^2}{8\pi\varepsilon_0 r} - \frac{e^2}{4\pi\varepsilon_0 r}$$

5.25
$$= -\frac{e^2}{8\pi\varepsilon_0 r}$$

The total energy of an atomic electron is negative; this is necessary if it is to be bound to

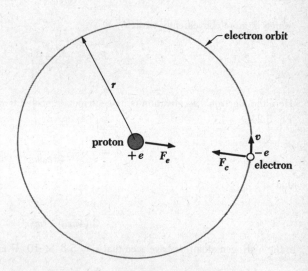

FIGURE 5.11 Force balance in the hydrogen atom.

Electron Orbits

the nucleus. If E were greater than zero, the electron would have too much energy to remain in a closed orbit about the nucleus.

Experiments indicate that 13.6 ev is required to separate a hydrogen atom into a proton and an electron; that is, its binding energy E is -13.6 ev. Since 13.6 ev $= 2.2 \times 10^{-18}$ joule, we can find the orbital radius of the electron in a hydrogen atom from Eq. 5.25:

$$r = -\frac{e^2}{8\pi\varepsilon_0 E}$$

$$= -\frac{(1.6 \times 10^{-19} \text{ coulomb})^2}{8\pi \times 8.85 \times 10^{-12} \text{ farad/m} \times (-2.2 \times 10^{-18} \text{ joule})}$$

$$= 5.3 \times 10^{-11} \text{ m}$$

An atomic radius of this order of magnitude agrees with estimates made in other ways.

5.7 Failure of Classical Physics

The analysis presented in the preceding section is a straightforward application of Newton's laws of motion and Coulomb's law of electric force—both pillars of classical physics—and is in accord with the experimental observation that atoms are stable. However, this picture is *not* in accord with electromagnetic theory—another pillar of classical physics—which states that accelerated electric charges radiate energy in the form of electromagnetic waves. The rate at which a charge e with the acceleration a radiates energy is

5.26
$$P = \frac{e^2 a^2}{6\pi\varepsilon_0 c^3}$$

which is more conveniently written

$$P = \frac{2}{3}\frac{e^2 a^2}{4\pi\varepsilon_0 c^3}$$

Here the electron's acceleration is its centripetal acceleration v^2/r, so that, with v given by Eq. 5.24,

$$a = \frac{v^2}{r} = \frac{e^2}{4\pi\varepsilon_0 m r^2}$$

and

$$P = \frac{2}{3}\frac{e^6}{(4\pi\varepsilon_0)^3 c^3 m^2 r^4}$$

In the hydrogen atom, we have seen that $r = 5.3 \times 10^{-11}$ m. Hence

$$P = \frac{2}{3} \left(9 \times 10^9 \frac{\text{n-m}^2}{\text{coulomb}^2} \right)^3 \times \frac{(1.6 \times 10^{-19} \text{ coulomb})^6}{(3 \times 10^8 \text{ m/sec})^3 (9.1 \times 10^{-31} \text{ kg})^2 (5.3 \times 10^{-11} \text{ m})^4}$$

$$= 4.6 \times 10^{-9} \text{ joule/sec}$$

$$= 2.9 \times 10^{10} \text{ ev/sec}$$

which is an extremely rapid rate of energy loss for such a system. And this is only the initial rate. As the electron loses energy, E becomes more and more negative. From Eq. 5.25 this means that r decreases, and since P is proportional to $1/r^4$, the electron radiates energy faster and faster as it spirals in toward the nucleus (Fig. 5.12). Only about 10^{-16} sec would be required for a "stable" hydrogen atom to collapse.

Whenever they have been directly tested, the predictions of electromagnetic theory have always agreed with experiment, yet atoms do not collapse. This contradiction can mean only one thing: The laws of physics that are valid in the macroscopic world do not hold true in the microscopic world of the atom.

The reason for the failure of classical physics to yield a meaningful analysis of atomic structure is that it approaches nature exclusively in terms of the abstract concepts of "pure" particles and "pure" waves. As we learned in the two preceding chapters, particles and waves have many properties in common, though the smallness of Planck's constant renders the wave-particle duality imperceptible in the macroscopic world. The validity of classical physics decreases as the scale of the phenomena under study decreases, and full allowance must be made for the particle behavior of waves and the wave behavior of particles if the

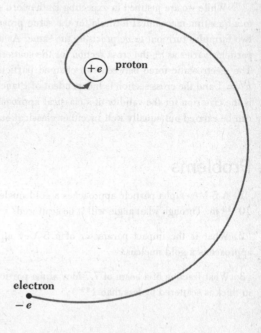

FIGURE 5.12 An atomic electron should, classically, spiral rapidly into the nucleus as it radiates energy due to its acceleration.

Failure of Classical Physics

atom is to be understood. In the next chapter we shall see how the Bohr atomic model, which combines classical and modern notions, accomplishes part of the latter task. Not until we consider the atom from the point of view of quantum mechanics, which makes no compromise with intuitive notions acquired in our daily lives, will we find a really successful theory of the atom.

An interesting question arises at this point. In our derivation of the Rutherford scattering formula we made use of the same laws of physics that proved such dismal failures when applied to atomic stability. Is it not therefore possible, even likely, that the formula is not correct, and that the atom in reality does not resemble the Rutherford model of a small central nucleus surrounded by distant electrons? This question is not a trivial one. To verify that a classical calculation ought to be at least approximately correct, we note that the de Broglie wavelength of an alpha particle whose speed is 2×10^7 m/sec is

$$\lambda = \frac{h}{mv} = \frac{6.63 \times 10^{-34} \text{ joule-sec}}{6.6 \times 10^{-27} \text{ kg} \times 2 \times 10^7 \text{ m/sec}}$$

$$= 5 \times 10^{-15} \text{ m}$$

As we saw in Sec. 5.5, the closest an alpha particle with this wavelength ever gets to a gold nucleus is 3×10^{-14} m, which is 6 de Broglie wavelengths, and so it is reasonable to regard the alpha particle as a classical particle in the interaction. We are therefore correct in thinking of the atom in terms of Rutherford's model, though the dynamics of the atomic electrons—which is another matter entirely—requires a nonclassical approach.

While we are justified in expecting Rutherford's classical formula to be at least similar to a quantum-mechanical formula for the same process, it is a curious coincidence that the two formulas turn out to be precisely the same. As a general rule, if the force between two particles varies as r^n, the cross section for the scattering of one by the other varies as h^{4+2n}. The electrostatic force between two charged particles is proportional to r^{-2}, so $h^{4+2n} = h^0 = 1$ and the cross section is independent of Planck's constant h. Since independence of h is the criterion for the validity of a classical approach, the analysis of Rutherford scattering can be carried out equally well by either classical or quantum-mechanical methods.

Problems

1. A 5-Mev alpha particle approaches a gold nucleus with an impact parameter of 2.6×10^{-13} m. Through what angle will it be scattered?

2. What is the impact parameter of a 5-Mev alpha particle scattered by $10°$ when it approaches a gold nucleus?

3. What fraction of a beam of 7.7-Mev alpha particles incident upon a gold foil 3×10^{-7} m thick is scattered by less than $1°$?

4. What fraction of a beam of 7.7-Mev alpha particles incident upon a gold foil 3×10^{-7} m thick is scattered by $90°$ or more?

5. Show that twice as many alpha particles are scattered by a foil through angles between 60 and $90°$ as are scattered through angles of $90°$ or more.

6. A beam of 8.3-Mev alpha particles is directed at an aluminum foil. It is found that the Rutherford scattering formula ceases to be obeyed at scattering angles exceeding about $60°$. If the alpha particle is assumed to have a radius of 2×10^{-15} m, find the radius of the aluminum nucleus.

7. Find the distance of closest approach of 1-Mev protons incident upon gold nuclei.

8. Find the distance of closest approach of 8-Mev protons incident upon gold nuclei.

9. The derivation of the Rutherford scattering formula was made nonrelativistically. Justify this approximation by computing the mass ratio between an 8-Mev alpha particle and an alpha particle at rest.

10. Find the frequency of rotation of the electron in the classical model of the hydrogen atom. In what region of the spectrum are electromagnetic waves of this frequency?

11. The electric-field intensity at a distance r from the center of a uniformly charged sphere of radius R and total charge Q is $Qr/4\pi\varepsilon_0 R^3$ when $r < R$. Such a sphere corresponds to the Thomson model of the atom. Show that an electron in this sphere executes simple harmonic motion about its center and derive a formula for the frequency of this motion. Evaluate the frequency of the electron oscillations for the case of the hydrogen atom and compare it with the frequencies of the spectral lines of hydrogen (see Fig. 6.6).

Chapter 6
Bohr Model
of the
Atom

The first theory of the hydrogen atom to succeed in accounting for the more conspicuous aspects of its behavior was presented by Niels Bohr in 1913. Bohr applied quantum ideas to atomic structure to obtain a model which, despite its serious inadequacies and subsequent replacement by a quantum-mechanical description of greater accuracy and usefulness, nevertheless persists as the mental picture many scientists have of the atom. While it is not the general policy of this book to go deeply into hypotheses that have had to be discarded, we shall discuss Bohr's theory of the hydrogen atom because it provides a valuable transition to the more abstract quantum theory of the atom. For this reason our account of the Bohr theory differs somewhat from the original one given by Bohr, though all of the results are identical.

6.1 Atomic Spectra

The ability of the Bohr theory to explain the origin of spectral lines is among its most spectacular accomplishments, and so it is appropriate to preface our exposition of the theory itself with a look at atomic spectra.

We have already mentioned that heated solids emit radiation in which all wavelengths are present, though with different intensities. We shall learn in Chap. 16 that the observed features of this radiation can be explained on the basis of the quantum theory of light independently of the details of the radiation process itself or of the nature of the solid. From this fact it follows that, when a solid is heated to incandescence, we are witnessing the collective behavior of a great many interacting atoms rather than the characteristic behavior of the individual atoms of a particular element.

At the other extreme, the atoms or molecules in a rarefied gas are so far apart on the average that their only mutual interactions occur during occasional collisions. Under these circumstances we would expect any emitted radiation to be characteristic of the individual atoms or molecules present, an expectation that is realized experimentally. When an atomic gas or vapor at somewhat less than atmospheric pressure is suitably "excited," usually by

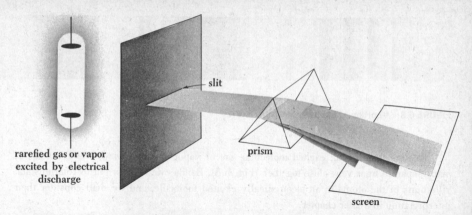

FIGURE 6.1 An idealized spectrometer.

slit

prism

screen

rarefied gas or vapor
excited by electrical
discharge

the passage of an electric current through it, the emitted radiation has a spectrum which contains certain discrete wavelengths only. An idealized laboratory arrangement for observing such atomic spectra is sketched in Fig. 6.1. Figure 6.2 shows the atomic spectra of several elements; they are called *emission line spectra*. Every element displays a unique line spectrum when a sample of it in the vapor phase is excited; spectroscopy is therefore a useful tool for analyzing the composition of an unknown substance.

FIGURE 6.2 Portions of the emission spectra of hydrogen, helium, and mercury.

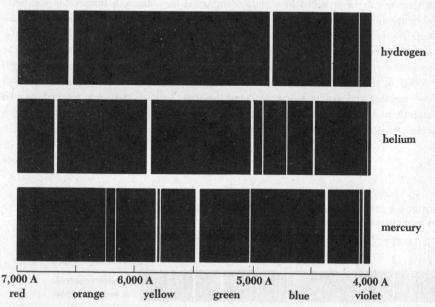

hydrogen

helium

mercury

| 7,000 A | | 6,000 A | | 5,000 A | | 4,000 A |
| red | orange | yellow | green | | blue | violet |

←——— λ

FIGURE 6.3   A portion of the band spectrum of PN.

The spectrum of an excited molecular gas or vapor contains *bands* which consist of many separate lines very close together (Fig. 6.3). Bands owe their origin to rotations and vibrations of the atoms in an electronically excited molecule, and we shall consider their interpretation in a later chapter.

When white light is passed through a gas, it is found to absorb light of certain of the wavelengths present in its emission spectrum. The resulting *absorption line spectrum* consists of a bright background crossed by dark lines corresponding to the missing wavelengths (Fig. 6.4); emission spectra consist of bright lines on a dark background. The dark Fraunhofer lines in the solar spectrum occur because the luminous part of the sun, which radiates almost exactly according to theoretical predictions for any object heated to 5800°K, is surrounded by an envelope of cooler gas which absorbs light of certain wavelengths only.

In the latter part of the nineteenth century it was discovered that the wavelengths present in atomic spectra fall into definite sets called *spectral series*. The wavelengths in each series can be specified by a simple empirical formula, with remarkable similarity among the formulas for the various series that comprise the complete spectrum of an element. The first such spectral series was found by J. J. Balmer in 1885 in the course of a study of the visible part of the hydrogen spectrum. Figure 6.5 shows the *Balmer series*. The line with the longest wavelength, 6,563 A, is designated H_α, the next, whose wavelength is 4,863 A, is designated H_β, and so on. As the wavelength decreases, the lines are found closer together and weaker in intensity until the *series limit* at 3,646 A is reached, beyond which there are no further separate lines but only a faint continuous spectrum. Balmer's formula for the wavelengths of this series is

6.1
$$\frac{1}{\lambda} = R\left(\frac{1}{2^2} - \frac{1}{n^2}\right) \qquad n = 3, 4, 5, \ldots \qquad \text{Balmer}$$

FIGURE 6.4   The dark lines in the absorption spectrum of an element correspond to certain of the bright lines in its emission spectrum.

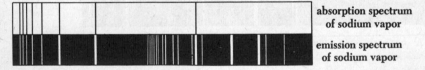

absorption spectrum
    of sodium vapor

emission spectrum
    of sodium vapor

The quantity R, known as the *Rydberg number*, has the value

$$R = 1.097 \times 10^7 \, \text{m}^{-1}$$
$$= 1.097 \times 10^{-3} \, \text{A}^{-1}$$

The H_α line corresponds to $n = 3$, the H_β line to $n = 4$, and so on. The series limit corresponds to $n = \infty$, so that it occurs at a wavelength of $4/R$, in agreement with experiment.

The Balmer series contains only those wavelengths in the visible portion of the hydrogen spectrum. The spectral lines of hydrogen in the ultraviolet and infrared regions fall into several other series. In the ultraviolet the *Lyman series* contains the wavelengths specified by the formula

6.2
$$\frac{1}{\lambda} = R \left(\frac{1}{1^2} - \frac{1}{n^2} \right) \qquad n = 2, 3, 4, \ldots$$
Lyman

In the infrared, three spectral series have been found whose component lines have the wavelengths specified by the formulas

6.3
$$\frac{1}{\lambda} = R \left(\frac{1}{3^2} - \frac{1}{n^2} \right) \qquad n = 4, 5, 6, \ldots$$
Paschen

6.4
$$\frac{1}{\lambda} = R \left(\frac{1}{4^2} - \frac{1}{n^2} \right) \qquad n = 5, 6, 7, \ldots$$
Brackett

6.5
$$\frac{1}{\lambda} = R \left(\frac{1}{5^2} - \frac{1}{n^2} \right) \qquad n = 6, 7, 8, \ldots$$
Pfund

The above spectral series of hydrogen are plotted in terms of wavelength in Fig. 6.6; the Brackett series evidently overlaps the Paschen and Pfund series. The value of R is the same in Eqs. 6.1 to 6.5.

The existence of such remarkable regularities in the hydrogen spectrum, together

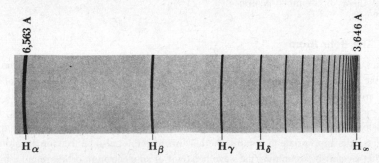

FIGURE 6.5   The Balmer series of hydrogen.

Atomic Spectra

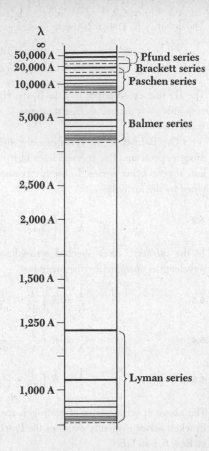

λ
∞

50,000 A — Pfund series
20,000 A — Brackett series
10,000 A — Paschen series

5,000 A — Balmer series

2,500 A —

2,000 A —

1,500 A —

1,250 A —

Lyman series

1,000 A —

FIGURE 6.6   The spectral series of hydrogen.

with similar regularities in the spectra of more complex elements, poses a definitive test for any theory of atomic structure.

6.2   The Bohr Atom

We saw in the previous chapter that the principles of classical physics are incompatible with the observed stability of the hydrogen atom. The electron in this atom is obliged to whirl around the nucleus to keep from being pulled into it and yet must radiate electromagnetic energy continuously. Because other apparently paradoxical phenomena, like the photoelectric effect and the diffraction of electrons, find explanation in terms of quantum concepts, it is appropriate to inquire whether this might not also be true for the atom.

Let us start by examining the wave behavior of an electron in orbit around a hydrogen nucleus. The de Broglie wavelength of this electron is

$$\lambda = \frac{h}{mv}$$

where the electron speed v is that given by Eq. 5.24:

$$v = \frac{e}{\sqrt{4\pi\varepsilon_0 mr}}$$

Hence

6.6
$$\lambda = \frac{h}{e}\sqrt{\frac{4\pi\varepsilon_0 r}{m}}$$

By substituting 5.3×10^{-11} m for the radius r of the electron orbit, we find the electron wavelength to be

$$\lambda = \frac{6.63 \times 10^{-34} \text{ joule-sec}}{1.6 \times 10^{-19} \text{ coulomb}}\sqrt{\frac{4\pi \times 8.85 \times 10^{-12} \text{ farad/m} \times 5.3 \times 10^{-11} \text{ m}}{9.1 \times 10^{-31} \text{ kg}}}$$

$$= 33 \times 10^{-11} \text{ m}$$

This wavelength is exactly the same as the circumference of the electron orbit,

$$2\pi r = 33 \times 10^{-11} \text{ m}$$

The orbit of the electron in a hydrogen atom corresponds to one complete electron wave joined on itself (Fig. 6.7).

The fact that the electron orbit in a hydrogen atom is one electron wavelength in circumference provides the clue we need to construct a theory of the atom. If we consider the vibrations of a wire loop (Fig. 6.8), we find that their wavelengths always fit an integral number of times into the loop's circumference so that each wave joins smoothly with the next. If the wire were perfectly elastic, these vibrations would continue indefinitely. Why are these the only vibrations possible in a wire loop? If a fractional number of wavelengths is placed around the loop, as in Fig. 6.9, destructive interference will occur as the waves travel around the loop, and the vibrations will die out rapidly. By considering the behavior of electron waves in the hydrogen atom as analogous to the vibrations of a wire loop, then, we may postulate that **an electron can circle a nucleus indefinitely without radiating energy provided that its orbit contains an integral number of de Broglie wavelengths.**

This postulate is the decisive one in our understanding of the atom. It combines both the particle and wave characters of the electron into a single statement, since the electron wavelength is computed from the orbital speed required to balance the electrostatic attraction of the nucleus. While we can never observe these antithetical characters simultaneously, they are inseparable in nature.

It is a simple matter to express the condition that an electron orbit contain an integral number of de Broglie wavelengths. The circumference of a circular orbit of radius r is $2\pi r$,

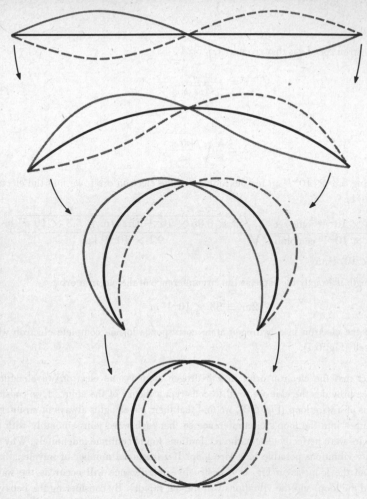

——— electron path
——— de Broglie electron wave

FIGURE 6.7 The orbit of the electron in a hydrogen atom corresponds to a complete electron de Broglie wave joined on itself.

and so we may write the condition for orbit stability as

6.7 $$n\lambda = 2\pi r_n \qquad n = 1, 2, 3, \ldots$$

where r_n designates the radius of the orbit that contains n wavelengths. The integer n is called the *quantum number* of the orbit. Substituting for λ the electron wavelength given by Eq. 6.6 yields

$$\frac{nh}{e}\sqrt{\frac{4\pi\varepsilon_0 r_n}{m}} = 2\pi r_n$$

and so the stable electron orbits are those whose radii are given by

6.8
$$r_n = \frac{n^2 h^2 \varepsilon_0}{\pi m e^2} \qquad n = 1, 2, 3, \ldots$$

The radius of the innermost orbit is customarily called the *Bohr radius* of the hydrogen atom and is denoted by the symbol a_0:

FIGURE 6.8   The vibrations of a wire loop.

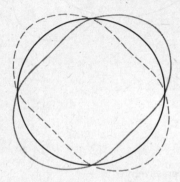

circumference = 2 wavelengths

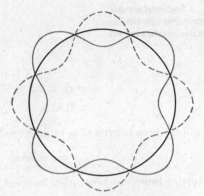

circumference = 4 wavelengths

circumference = 8 wavelengths

The Bohr Atom

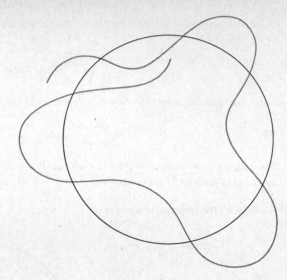

FIGURE 6.9 A fractional number
of wavelengths cannot persist because destructive interference will
occur.

$$a_0 = r_1 = 5.3 \times 10^{-11} \text{ m}$$
$$= 0.53 \text{ A}$$

The other radii are given in terms of a_0 by the formula

$$r_n = n^2 a_0$$

so that the spacing between adjacent orbits increases progressively.

6.3 Energy Levels and Spectra

The various permitted orbits involve different electron energies. The electron energy E_n is given in terms of the orbit radius r_n by Eq. 5.25 as

$$E_n = -\frac{e^2}{8\pi\varepsilon_0 r_n}$$

Substituting for r_n from Eq. 6.8, we see that

6.9
$$E_n = -\frac{me^4}{8\varepsilon_0^2 h^2}\left(\frac{1}{n^2}\right) \qquad n = 1, 2, 3, \ldots \qquad$$ **Energy levels**

The energies specified by Eq. 6.9 are called the *energy levels* of the hydrogen atom and are plotted in Fig. 6.10. These levels are all negative, signifying that the electron does not have enough energy to escape from the atom. The lowest energy level E_1 is called the *ground state* of the atom, and the higher levels E_2, E_3, E_4, . . . are called *excited states*. As the quantum number n increases, the corresponding energy E_n approaches closer and closer to 0; in the limit of $n = \infty$, $E_\infty = 0$ and the electron is no longer bound to the nucleus to form an atom. (A positive energy for a nucleus-electron combination means that the electron is not bound to the nucleus and has no quantum conditions to fulfill; such a combination does not constitute an atom, of course.)

The ground-state energy E_1 of the hydrogen atom is a convenient energy unit to use in discussing various aspects of atomic and molecular physics. This energy unit is called the *rydberg* (ry), and its numerical value is

$$1 \text{ ry} = \frac{me^4}{8\varepsilon_0^2 h^2} = 2.17 \times 10^{-18} \text{ joule}$$

$$= 13.6 \text{ ev}$$

It is now necessary for us to confront directly the equations we have developed with

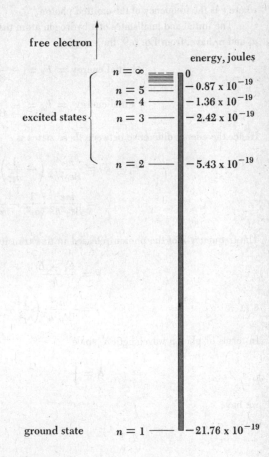

FIGURE 6.10 Energy levels of the hydrogen atom.

experiment. An especially striking experimental result is that atoms exhibit line spectra in both emission and absorption; do these spectra follow from our atomic model?

The presence of definite, discrete energy levels in the hydrogen atom suggests a connection with line spectra. Let us tentatively assert that, when an electron in an excited state drops to a lower state, the lost energy is emitted as a single photon of light. According to our model, electrons cannot exist in an atom except in certain specific energy levels. The jump of an electron from one level to another, with the difference in energy between the levels being given off all at once in a photon rather than in some more gradual manner, fits in well with this model. If the quantum number of the initial (higher energy) state is n_i and the quantum number of the final (lower energy) state is n_f, we are asserting that

6.10

$$\text{Initial energy} - \text{final energy} = \text{photon energy}$$
$$E_i - E_f = h\nu$$

where ν is the frequency of the emitted photon.

The initial and final states of a hydrogen atom that correspond to the quantum numbers n_i and n_f have, from Eq. 6.9, the energies

$$\text{Initial energy} = E_i = - \frac{me^4}{8\varepsilon_0{}^2 h^2}\left(\frac{1}{n_i{}^2}\right)$$

$$\text{Final energy} = E_f = - \frac{me^4}{8\varepsilon_0{}^2 h^2}\left(\frac{1}{n_f{}^2}\right)$$

Hence the energy difference between these states is

$$E_i - E_f = \frac{me^4}{8\varepsilon_0{}^2 h^2}\left(-\frac{1}{n_i{}^2}\right) - \left(-\frac{1}{n_f{}^2}\right)$$

$$= \frac{me^4}{8\varepsilon_0{}^2 h^2}\left(\frac{1}{n_f{}^2} - \frac{1}{n_i{}^2}\right)$$

The frequency ν of the photon released in this transition is

$$\nu = \frac{E_i - E_f}{h}$$

6.11

$$= \frac{me^4}{8\varepsilon_0{}^2 h^3}\left(\frac{1}{n_f{}^2} - \frac{1}{n_i{}^2}\right)$$

In terms of photon wavelength λ, since

$$\lambda = \frac{c}{\nu}$$

we have

$$\frac{1}{\lambda} = \frac{\nu}{c}$$

6.12
$$= \frac{me^4}{8\varepsilon_0^2 ch^3} \left(\frac{1}{n_f^2} - \frac{1}{n_i^2} \right)$$

Equation 6.12 states that the radiation emitted by excited hydrogen atoms should contain certain wavelengths only. These wavelengths, furthermore, fall into definite sequences that depend upon the quantum number n_f of the final energy level of the electron. Since the initial quantum number n_i must always be greater than the final quantum number n_f in each case, in order that there be an excess of energy to be given off as a photon, the calculated formulas for the first five series are

6.13 $\qquad n_f = 1$: $\qquad \dfrac{1}{\lambda} = \dfrac{me^4}{8\varepsilon_0^2 ch^3} \left(\dfrac{1}{1^2} - \dfrac{1}{n^2} \right) \qquad n = 2, 3, 4, \dots$

6.14 $\qquad n_f = 2$: $\qquad \dfrac{1}{\lambda} = \dfrac{me^4}{8\varepsilon_0^2 ch^3} \left(\dfrac{1}{2^2} - \dfrac{1}{n^2} \right) \qquad n = 3, 4, 5, \dots$

6.15 $\qquad n_f = 3$: $\qquad \dfrac{1}{\lambda} = \dfrac{me^4}{8\varepsilon_0^2 ch^3} \left(\dfrac{1}{3^2} - \dfrac{1}{n^2} \right) \qquad n = 4, 5, 6, \dots$

6.16 $\qquad n_f = 4$: $\qquad \dfrac{1}{\lambda} = \dfrac{me^4}{8\varepsilon_0^2 ch^3} \left(\dfrac{1}{4^2} - \dfrac{1}{n^2} \right) \qquad n = 5, 6, 7, \dots$

6.17 $\qquad n_f = 5$: $\qquad \dfrac{1}{\lambda} = \dfrac{me^4}{8\varepsilon_0^2 ch^3} \left(\dfrac{1}{5^2} - \dfrac{1}{n^2} \right) \qquad n = 6, 7, 8, \dots$

These sequences are identical in form with the empirical spectral series discussed earlier. The Lyman series, Eq. 6.2, corresponds to $n_f = 1$; the Balmer series, Eq. 6.1, corresponds to $n_f = 2$; the Paschen series, Eq. 6.3, corresponds to $n_f = 3$; the Brackett series, Eq. 6.4, corresponds to $n_f = 4$; and the Pfund series, Eq. 6.5, corresponds to $n_f = 5$.

We still cannot consider our assertion that the line spectrum of hydrogen originates in electron transitions from high to low energy states as proved, however. The final step is to compare the value of the constant term in Eqs. 6.13 to 6.17 with that of the Rydberg number R of the empirical equations 6.1 to 6.5. The value of this constant term is

$$\frac{me^4}{8\varepsilon_0^2 ch^3}$$

$$= \frac{9.1 \times 10^{-31} \text{ kg} \times (1.6 \times 10^{-19} \text{ coulomb})^4}{8 \times (8.85 \times 10^{-12} \text{ farad/m})^2 \times 3 \times 10^8 \text{ m/sec} \times (6.63 \times 10^{-34} \text{ joule-sec})^3}$$

$$= 1.097 \times 10^7 \text{ m}^{-1}$$

which is indeed the same as R! This theory of the hydrogen atom, which is essentially that

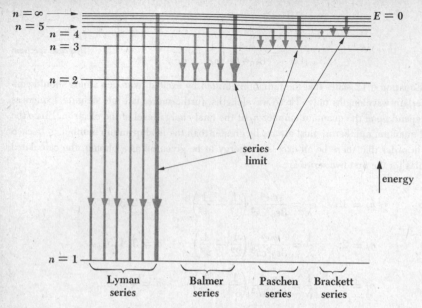

$n = \infty$ $E = 0$

$n = 5$
$n = 4$
$n = 3$

$n = 2$

series limit

energy

$n = 1$

Lyman series   Balmer series   Paschen series   Brackett series

FIGURE 6.11 Spectral lines originate in transitions between energy levels.

developed by Bohr in 1913, therefore agrees both qualitatively and quantitatively with experiment. Figure 6.11 shows schematically how spectral lines are related to atomic energy levels.

6.4 Atomic Excitation

There are two principal mechanisms that can excite an atom to an energy level above its ground state, thereby enabling it to radiate. One mechanism is a collision with another particle during which part of their joint kinetic energy is absorbed by the atom. An atom excited in this way will return to its ground state in an average of 10^{-8} sec by emitting one or more photons. To produce an electric discharge in a rarefied gas, an electric field is established which accelerates electrons and atomic ions until their kinetic energies are sufficient to excite atoms they happen to collide with. Neon signs and mercury-vapor lamps are familiar examples of how a strong electric field applied between electrodes in a gas-filled tube leads to the emission of the characteristic spectral radiation of that gas, which happens to be reddish light in the case of neon and bluish light in the case of mercury vapor.

A different excitation mechanism is involved when an atom absorbs a photon of light whose energy is just the right amount to raise the atom to a higher energy level. For example, a photon of wavelength 1,217 A is emitted when a hydrogen atom in the $n = 2$ state drops to the $n = 1$ state; the absorption of a photon of wavelength 1,217 A by a hydrogen atom initially in the $n = 1$ state will therefore bring it up to the $n = 2$ state. This process explains the origin of absorption spectra. When white light, which contains all wavelengths, is passed through hydrogen gas, photons of those wavelengths that correspond to transitions between energy levels are absorbed. The resulting excited hydrogen atoms reradiate their excitation energy almost at once, but these photons come off in random directions with only a few in the same direction as the original beam of white light. The dark lines in an absorption spectrum are therefore never completely black, but only appear so by contrast with the bright background. We expect the absorption spectrum of any element to be identical with its emission spectrum, then, which agrees with observation.

6.5 The Franck-Hertz Experiment

Atomic spectra are not the only means of investigating the presence of discrete energy levels within atoms. A series of experiments based on the first of the excitation mechanisms of the previous section was performed by Franck and Hertz starting in 1914. These experiments provided a direct demonstration that atomic energy levels do indeed exist and, furthermore, that these levels are the same as those suggested by observations of line spectra. Franck and Hertz bombarded the vapors of various elements with electrons of known energy, using an apparatus like that shown in Fig. 6.12. A small potential difference V_0 is maintained between the grid and collecting plate, so that only electrons having energies greater than a certain minimum contribute to the current i through the galvanometer. As the accelerating potential V is increased, more and more electrons arrive at the plate and i rises (Fig. 6.13). If kinetic energy is conserved in a collision between an electron and one of the atoms in the vapor, the electron merely bounces off in a direction different from its original one. Because an atom is so much heavier than an electron, the latter loses almost no kinetic energy in the process. After a certain critical electron energy is reached, however, the plate current drops abruptly. The interpretation of this effect is that an electron colliding with one of the atoms gives up some or all of its kinetic energy in exciting the atom to an energy level above its ground state. Such a collision is called *inelastic*, in contrast to an *elastic* collision in which kinetic energy is conserved. The critical electron energy corresponds to the excitation energy of the atom. Then, as the accelerating potential V is raised further, the plate current again increases, since the electrons now have sufficient energy left after experiencing an inelastic collision to reach the plate. Eventually another sharp drop in plate current i occurs, which is interpreted as arising from the excitation of a higher energy level. As Fig. 6.13 indicates, a series of critical potentials for a particular atomic species is obtained in this way. (The highest potentials, of course, result from several inelastic collisions and are multiples of the lower ones.)

To check the interpretation of critical potentials as being due to discrete atomic energy levels, Franck and Hertz observed the emission spectra of vapors during electron bombard-

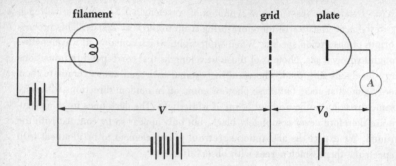

FIGURE 6.12 Apparatus for the Franck-Hertz experiment.

ment. In the case of mercury vapor, for example, they found that a minimum electron energy of 4.9 ev was required to excite the 2,536-A spectral line of mercury—and a photon of 2,536-A light has an energy of just 4.9 ev! The Franck-Hertz experiments were performed shortly after Bohr announced his theory of the hydrogen atom, and they provided independent confirmation of his basic ideas.

6.6 The Correspondence Principle

The principles of quantum physics, so different from those of classical physics in the microscopic world that lies beyond the reach of our senses, must nevertheless yield results

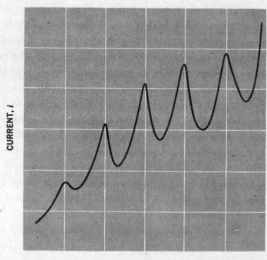

FIGURE 6.13 Results of the Franck-Hertz experiment, showing critical potentials.

ACCELERATING POTENTIAL, V

identical with those of classical physics in the domain where experiment indicates the latter to be valid. We have already seen that this fundamental requirement is satisfied by the theory of relativity, the quantum theory of radiation, and the wave theory of matter; we shall now show that it is satisfied also by Bohr's theory of the atom.

According to electromagnetic theory, an electron moving in a circular orbit radiates electromagnetic waves whose frequencies are equal to its frequency of revolution and to harmonics (that is, integral multiples) of that frequency. In a hydrogen atom the electron's speed is

$$v = \frac{e}{\sqrt{4\pi\varepsilon_0 mr}}$$

according to Eq. 5.24, where r is the radius of its orbit. Hence the frequency of revolution f of the electron is

$$f = \frac{\text{electron speed}}{\text{orbit circumference}}$$

$$= \frac{v}{2\pi r}$$

$$= \frac{e}{2\pi \sqrt{4\pi\varepsilon_0 mr^3}}$$

The radius r_n of a stable orbit is given in terms of its quantum number n by Eq. 6.8 as

$$r_n = \frac{n^2 h^2 \varepsilon_0}{\pi m e^2}$$

and so the frequency of revolution is

6.18
$$f = \frac{me^4}{8\varepsilon_0^2 h^3} \left(\frac{2}{n^3}\right)$$

Under what circumstances should the Bohr atom behave classically? If the electron orbit is so large that we might expect to be able to measure it directly, quantum effects should be entirely inconspicuous. An orbit 1 cm across, for example, meets this specification; its quantum number is very close to $n = 10,000$, and while hydrogen atoms so grotesquely large do not actually occur because their energies would be only infinitesimally below the ionization energy, they are not prohibited in theory. What does the Bohr theory predict that such an atom will radiate? According to Eq. 6.11, a hydrogen atom dropping from the n_ith energy level to the n_fth energy level emits a photon whose frequency is

$$v = \frac{me^4}{8\varepsilon_0^2 h^3} \left(\frac{1}{n_f^2} - \frac{1}{n_i^2}\right)$$

Let us write n for the initial quantum number n_i and $n - p$ (where $p = 1, 2, 3, \ldots$) for the final quantum number n_f. With this substitution,

$$\nu = \frac{me^4}{8\varepsilon_0^2 h^3}\left[\frac{1}{(n-p)^2} - \frac{1}{n^2}\right]$$

$$= \frac{me^4}{8\varepsilon_0^2 h^3}\left[\frac{2np - p^2}{n^2(n-p)^2}\right]$$

Now, when n_i and n_f are both very large, n is much greater than p, and

$$2np - p^2 \approx 2np$$

$$(n-p)^2 \approx n^2$$

so that

6.19
$$\nu = \frac{me^4}{8\varepsilon_0^2 h^3}\left(\frac{2p}{n^3}\right)$$

When $p = 1$, the frequency ν of the radiation is exactly the same as the frequency of rotation f of the orbital electron given in Eq. 6.18! Harmonics of this frequency are radiated when $p = 2, 3, 4, \ldots$. Hence both quantum and classical pictures of the hydrogen atom make identical predictions in the limit of very large quantum numbers. When $n = 2$, Eq. 6.18 predicts a radiation frequency that differs from that given by Eq. 6.11 by almost 300 percent, while when $n = 10,000$, the discrepancy is only about 0.01 percent.

The requirement that quantum physics give the same results as classical physics in the limit of large quantum numbers was called by Bohr the *correspondence principle*. It has played an important role in the development of the quantum theory of matter.

6.7 Nuclear Motion and Reduced Mass

In developing the theory of the hydrogen atom, we assumed that its nucleus (a single proton) remains stationary while the orbital electron revolves around it. This is not an unreasonable assumption: the proton mass is 1,836 times greater than the electron mass, and the electrostatic force

$$F_e = \frac{1}{4\pi\varepsilon_0}\frac{e^2}{r^2}$$

that each exerts on the other has the same magnitude for both. It is nevertheless of interest to know precisely what effect nuclear motion has on the behavior of the hydrogen atom.

We begin by expressing in a different way the condition that a stable orbit in an atom consist of an integral number of electron de Broglie wavelengths. This condition states that

$$n\lambda = 2\pi r$$

and, since the de Broglie wavelength λ is given by

$$\lambda = \frac{h}{mv}$$

we may equivalently write

6.20
$$mvr = \frac{nh}{2\pi} = n\hbar$$

The quantity mvr we recognize as the *angular momentum* of the electron in its orbit. Hence an alternate expression of Bohr's first postulate is that the angular momentum of a hydrogen atom must be an integral multiple of $\hbar$. In fact, the quantization of angular momentum in units of $\hbar$ was the starting point of Bohr's original work, since the hypothesis of de Broglie waves had not been proposed as yet. (We shall see in Chap. 9 that the quantization rule expressed in Eq. 6.20 holds only for the component of the angular momentum of a system in a particular direction, while the magnitude of the angular momentum itself is quantized in a somewhat different way.) In terms of the angular velocity ω of the electron, the quantization rule of Eq. 6.20 is stated

6.21
$$m\omega r^2 = n\hbar$$

since

$$\omega = \frac{v}{r}$$

If the nucleus has a mass M that is not infinite, both it and its orbital electron revolve around a common *center of mass*. The center of mass of a body is its balance point; if we think of the nucleus and electron in a hydrogen atom as being at opposite ends of a massless rod r long, in effect constituting a lopsided dumbbell (Fig. 6.14), the center of mass, which

FIGURE 6.14 Both the electron and nucleus of a hydrogen atom revolve around a common center of mass.

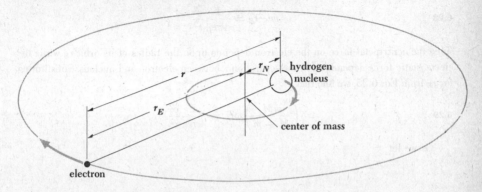

is r_N distant from the nucleus and r_E distant from the electron, may be found from the requirement that

6.22
$$mr_E = Mr_N$$

where

6.23
$$r = r_E + r_N$$

The total angular momentum of the hydrogen atom is the sum of the angular momentum of the electron and that of the nucleus:

Total angular momentum = electron angular momentum + nuclear angular momentum
$$= m\omega r_E{}^2 + M\omega r_N{}^2$$

The angular velocity ω is the same for both particles. According to Bohr's first postulate, the total angular momentum of the atom must be an integral multiple of $\hbar$, and so

6.24
$$m\omega r_E{}^2 + M\omega r_N{}^2 = n\hbar$$

From Eqs. 6.22 and 6.23 we find that

6.25
$$r_E = \left(\frac{M}{M+m}\right)r$$

6.26
$$r_N = \left(\frac{m}{M+m}\right)r$$

and so Eq. 6.24 becomes

6.27
$$\left(\frac{mM}{M+m}\right)\omega r^2 = n\hbar$$

The condition for force balance, Eq. 5.23, must also be generalized to take into account nuclear motion. Its proper expression is

6.28
$$m\omega^2 r_E = \frac{1}{4\pi\varepsilon_0}\frac{e^2}{r^2}$$

since the centripetal force on the electron depends upon the radius of its orbit r_E while the electrostatic force depends upon the separation r between electron and nucleus. Substituting for r_E from Eq. 6.25, we find that

6.29
$$\left(\frac{mM}{M+m}\right)\omega^2 r = \frac{1}{4\pi\varepsilon_0}\frac{e^2}{r^2}$$

If we let

6.30
$$m' = \left(\frac{mM}{M + m}\right)$$

we see that Eqs. 6.27 and 6.29 become respectively

6.31
$$m'\omega r^2 = n\hbar$$

and

6.32
$$m'\omega^2 r = \frac{1}{4\pi\varepsilon_0}\frac{e^2}{r^2}$$

From Eqs. 6.31 and 6.32 we can proceed as we did earlier in this chapter to obtain an expression for the energy levels of the hydrogen atom. The result is

6.33
$$E_n = -\frac{m'e^4}{8\varepsilon_0^2 h^2}\left(\frac{1}{n^2}\right)$$

which is exactly the same as Eq. 6.9 except that m' replaces the electron mass m. The quantity m' is known as the *reduced mass* of the electron because its value is less than m. Owing to motion of the nucleus, all of the energy levels of hydrogen are changed by the fraction

$$\frac{m'}{m} = \frac{M}{M + m}$$
$$= \frac{1,836}{1,837}$$
$$= 0.99945$$

an increase of 0.055 percent since the energies E_n, being smaller in absolute value, are therefore less negative. The use of Eq. 6.33 in place of Eq. 6.9 removes a small but definite discrepancy between the predicted wavelengths of the various spectral lines of hydrogen and the actual experimentally determined wavelengths. The value of the Rydberg number R to eight significant figures without correcting for nuclear motion is 1.0973731×10^7 m^{-1}; the correction lowers it to 1.0967758×10^7 m^{-1}.

The notion of reduced mass played an important part in the discovery of *deuterium*, an isotope of hydrogen whose atomic mass is almost exactly double that of ordinary hydrogen owing to the presence of a neutron as well as a proton in the nucleus. Because of the greater nuclear mass, the spectral lines of deuterium are all shifted slightly to wavelengths shorter than those of ordinary hydrogen. The H$_\alpha$ line of deuterium, for example, has a wavelength of 6,561 A, while that of hydrogen is 6,563 A: a small but definite difference, sufficient for the identification of deuterium.

6.8 Hydrogenic Atoms

The Bohr theory can be applied to ions containing single electrons as well as to hydrogen atoms. That is, He$^+$ (singly ionized helium), Li^{++} (doubly ionized lithium), and so on behave just like hydrogen except for the effects of greater nuclear charge and mass. They are accordingly called *hydrogenic atoms*. We have already seen how to take nuclear mass into account through the substitution of the reduced mass m' for the electron mass m. The nucleus of a hydrogenic atom of atomic number Z carries the charge $+Ze$, and so the electrostatic force it exerts on the orbital electron is

6.34
$$F_e = \frac{1}{4\pi\varepsilon_0} \frac{Ze^2}{r^2}$$

instead of simply $e^2/4\pi\varepsilon_0 r^2$ as in the case of hydrogen, where $Z = 1$. Consequently the formula for the energy levels of a hydrogenic atom is

6.35
$$E_n = -\frac{m'Z^2e^4}{8\varepsilon_0^2 h^2}\left(\frac{1}{n^2}\right)$$
Hydrogenic atoms

These levels are different by a factor of Z^2 from those of the hydrogen atom. Figure 6.15 shows the energy levels of He$^+$ and H together with some possible transitions in each; evidently certain spectral lines in He$^+$ should correspond closely to lines found in H. Since the general formula for the spectral lines of a hydrogenic atom of atomic number Z is

6.36
$$\frac{1}{\lambda} = \frac{m'Z^2e^4}{8\varepsilon_0^2 ch^3}\left(\frac{1}{n_f^2} - \frac{1}{n_i^2}\right)$$

a transition from the $n = 4$ state of He$^+$ to its $n = 2$ state involves the emission of a photon whose wavelength is nearly the same as that emitted in a transition from the $n = 2$ state of H to its $n = 1$ state. The small difference between the wavelengths arises from the difference between the nuclear masses of He$^+$ and H, which affects the value of the reduced mass m'.

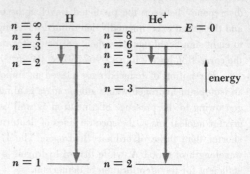

FIGURE 6.15 Energy levels in hydrogen and singly ionized helium.

Problems

1. Determine the value of the longest wavelength found in the Paschen series.

2. Find the wavelength of the spectral line corresponding to the transition in hydrogen from the $n = 6$ state to the $n = 3$ state.

3. Find the wavelength of the photon emitted when a hydrogen atom goes from the $n = 10$ state to its ground state.

4. How much energy is required to remove an electron in the $n = 2$ state from a hydrogen atom?

5. A beam of electrons bombards a sample of hydrogen. Through what potential difference must the electrons have been accelerated if the first line of the Balmer series is to be emitted?

6. A μ^- meson ($m = 207\, m_e$) can be captured by a proton to form a "mesic atom." Find the radius of the first Bohr orbit of such an atom.

7. A μ^- meson is in the $n = 2$ state of a titanium atom. Find the energy radiated when the mesic atom drops to its ground state.

8. Find the recoil speed of a hydrogen atom after it emits a photon in going from the $n = 4$ state to the $n = 1$ state.

9. How many revolutions does an electron in the $n = 2$ state of a hydrogen atom make before dropping to the $n = 1$ state? (The average lifetime of an excited state is about 10^{-8} sec.)

10. The average lifetime of an excited atomic state is 10^{-8} sec. If the wavelength of the spectral line associated with the decay of this state is 4,000 A, find the width of the line.

11. At what temperature will the average molecular kinetic energy in gaseous hydrogen equal the binding energy of a hydrogen atom?

12. An electron joins a bare helium nucleus to form a He^+ ion. Find the wavelength of the photon emitted in this process if the electron is assumed to have had no kinetic energy when it combined with the nucleus.

13. A mixture of ordinary hydrogen and tritium, a hydrogen isotope whose nucleus is approximately three times more massive than ordinary hydrogen, is excited and its spectrum observed. How far apart in wavelength will the H_α lines of the two kinds of hydrogen be?

14. A positronium atom is a system consisting of a positron (positive electron) and an electron. (a) Compare the wavelength of the photon emitted in the $n = 3 \rightarrow n = 2$ transition in

positronium with that of the H_α line. (*b*) Compare the ionization energy in positronium with that in hydrogen.

15. Use the uncertainty principle to determine the ground-state radius r_1 of the hydrogen atom in the following way. First find a formula for the electron kinetic energy in terms of the momentum an electron must have if confined to a region of linear dimension r_1. Add this kinetic energy to the electrostatic potential energy of an electron the distance r_1 from a proton, and differentiate with respect to r_1 the resulting expression for the total electron energy E to find the value of r_1 for which E is a minimum. Compare the result with that given by Eq. 6.8 with $n = 1$.

Chapter 7
Schrödinger's
Equation

The Bohr theory of the atom, discussed in the previous chapter, is able to account for certain experimental data in a convincing manner, but it has a number of severe limitations. While the Bohr theory correctly predicts the spectral series of hydrogen, hydrogen isotopes, and hydrogenic atoms, it is incapable of being extended to treat the spectra of complex atoms having two or more electrons each; it can give no explanation of why certain spectral lines are more intense than others (that is, why certain transitions between energy levels have greater probabilities of occurrence); and it cannot account for the observation that many spectral lines actually consist of several separate lines whose wavelengths differ slightly. And, perhaps most important, it does not permit us to obtain what a really successful theory of the atom should make possible: an understanding of how individual atoms interact with one another to endow macroscopic aggregates of matter with the physical and chemical properties we observe.

These objections to the Bohr theory are not put forward in an unfriendly way, for the theory was one of those seminal achievements that transform scientific thought, but rather to emphasize that an approach to atomic phenomena of greater generality is required. Such an approach was developed in 1925–1926 by Erwin Schrödinger, Werner Heisenberg, and others under the apt name of *quantum mechanics*. By the early 1930s the application of quantum mechanics to problems involving nuclei, atoms, molecules, and matter in the solid state made it possible to understand a vast body of otherwise-puzzling data and—a vital attribute of any theory—led to predictions of remarkable accuracy.

7.1 Quantum Mechanics

The fundamental difference between newtonian mechanics and quantum mechanics lies in what it is that they describe. Newtonian mechanics is concerned with the motion of a particle under the influence of applied forces, and it takes for granted that such quantities as the particle's position, mass, velocity, and acceleration can be measured. This assumption is, of course, completely valid in our everyday experience, and newtonian mechanics provides the "correct" explanation for the behavior of moving bodies in the sense that the values it predicts for observable magnitudes agree with the measured values of those magnitudes.

Quantum mechanics, too, consists of relationships between observable magnitudes, but the uncertainty principle radically alters the definition of "observable magnitude" in the atomic realm. According to the uncertainty principle, the position and momentum of a particle cannot be accurately measured at the same time, while in newtonian mechanics both are assumed to have definite, ascertainable values at every instant. The quantities whose relationships quantum mechanics explores are *probabilities*. Instead of asserting, for example, that the radius of the electron's orbit in a ground-state hydrogen atom is always exactly 5.3×10^{-11} m, quantum mechanics states that this is the *most probable* radius; if we conduct a suitable experiment, most trials will yield a different value, either larger or smaller, but the value most likely to be found will be 5.3×10^{-11} m.

At first glance quantum mechanics seems a poor substitute for newtonian mechanics, but closer inspection reveals a striking fact: *Newtonian mechanics is nothing but an approximate version of quantum mechanics.* The certainties proclaimed by newtonian mechanics are illusory, and their agreement with experiment is a consequence of the fact that macroscopic bodies consist of so many individual atoms that departures from average behavior are unnoticeable. Instead of two sets of physical principles, one for the macroscopic universe and one for the microscopic universe, there is only a single set, and quantum mechanics represents our best effort to date in formulating it.

7.2 The Wave Function

As mentioned in Chap. 4, the quantity with which quantum mechanics is concerned is the *wave function* Ψ of a particle. While Ψ itself has no physical interpretation, the square of its absolute magnitude $|\Psi|^2$ (or $\Psi^*\Psi$ if Ψ is complex) evaluated at a particular point at a particular time is proportional to the probability of experimentally finding the particle there at that time. The problem of quantum mechanics is to determine Ψ for a particle when its freedom of motion is limited by the action of external forces.

Even before we consider the actual calculation of Ψ, we can establish certain requirements it must always fulfill. For one thing, since $|\Psi|^2$ evaluated at a point is proportional to the probability P of finding the particle described by Ψ at that point, the integral of $|\Psi|^2$ over all space must be finite—the particle is *somewhere*, after all. If

$$\int_{-\infty}^{\infty} |\Psi|^2 \, dV$$

is 0, the particle does not exist, and if it is ∞, the particle is everywhere simultaneously; $|\Psi|^2$ cannot be negative or complex because of the way it is defined, and so the only possibility left is that its integral be a finite quantity if Ψ is to describe properly a real body.

It is usually convenient to have $|\Psi|^2$ be *equal* to the probability P of finding the particle described by Ψ, rather than merely be proportional to P. If $|\Psi|^2$ is to equal P, then it must be true that

7.1
$$\int_{-\infty}^{\infty} |\Psi|^2 \, dV = 1$$
Normalization

since

$$\int_{-\infty}^{\infty} P \, dV = 1$$

is the mathematical statement that the particle exists somewhere at all times. A wave function that obeys Eq. 7.1 is said to be *normalized*. Every acceptable wave function can be normalized by multiplying by an appropriate constant; in Chap. 8 we shall see exactly how this is done.

Besides being normalizable, Ψ must be single-valued, since P can have only one value at a particular place and time. A further condition that Ψ must obey is that it and its partial derivatives $\partial \Psi / \partial x$, $\partial \Psi / \partial y$, $\partial \Psi / \partial z$ be continuous everywhere.

Schrödinger's equation, which is the fundamental equation of quantum mechanics in the same sense that the second law of motion is the fundamental equation of newtonian mechanics, is a wave equation in the variable Ψ. It is appropriate for us to review the nature and solution of a simpler wave equation, that governing wave propagation along a stretched string, before we tackle Schrödinger's equation itself.

7.3 The Wave Equation

Let us consider a stretched string lying on the x axis whose displacements take place in the xy plane. The tension in the string is T and its mass per unit length is μ. To obtain the differential equation governing the propagation of waves in this string, we apply the second law of motion, $\mathbf{F} = m\mathbf{a}$, to the small segment Δl of the string shown in Fig. 7.1.

Our discussion will be limited to small displacements of the string which do not involve abrupt changes in its shape. A number of approximations are valid for such displacements:

1. The tension in the string has the same magnitude T everywhere.
2. The angles θ_1 and θ_2 are sufficiently small so that

7.2
$$\cos \theta_1 \approx \cos \theta_2 \approx 1$$

7.3
$$\sin \theta_1 \approx \tan \theta_1$$

7.4
$$\sin \theta_2 \approx \tan \theta_2$$

3. The length Δl of the displaced segment is very nearly the same as its normal length Δx. Hence $\Delta l \approx \Delta x$ and the mass of the segment is

$$\Delta m \approx \mu \, \Delta x$$

From Fig. 7.1 we see that

$$F_{1x} = -T \cos \theta_1 \qquad F_{2x} = T \cos \theta_2$$
$$F_{1y} = -T \sin \theta_1 \qquad F_{2y} = T \sin \theta_2$$

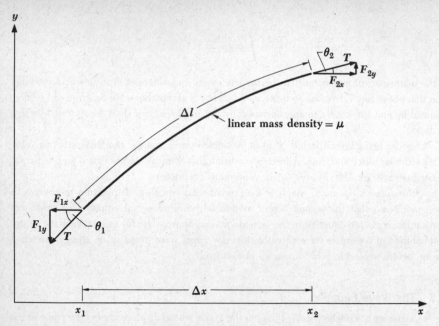

FIGURE 7.1 Segment of a stretched string displaced from its normal position along the x axis as a wave travels past. The displacement is greatly exaggerated here.

In view of Eq. 7.2, the resultant force on the string segment in the x direction is

$$F_x = F_{1x} + F_{2x} = T \left(\cos \theta_2 - \cos \theta_1 \right) = 0$$

Hence we need consider only forces in the y direction. The resultant force in the y direction is

$$F_y = F_{1y} + F_{2y} = T \left(\sin \theta_2 - \sin \theta_1 \right)$$
$$= T \left(\tan \theta_2 - \tan \theta_1 \right)$$

Since $\tan \theta = dy/dx$,

$$F_y = T \left[\left(\frac{dy}{dx} \right)_{x_2} - \left(\frac{dy}{dx} \right)_{x_1} \right]$$

which we can write as

$$F_y = T \Delta \left(\frac{dy}{dx} \right)$$

The quantity $\Delta(dy/dx)$ represents the change in dy/dx over the distance Δx between x_1 and x_2.

Applying the second law of motion to the string segment yields

$$F_y = \Delta m\, a_y$$

$$T\Delta\left(\frac{dy}{dx}\right) = \mu\,\Delta x\,\frac{d^2y}{dt^2}$$

7.5
$$\frac{\Delta}{\Delta x}\left(\frac{dy}{dx}\right) = \frac{\mu}{T}\frac{d^2y}{dt^2}$$

The quantity

$$\frac{\Delta}{\Delta x}\left(\frac{dy}{dx}\right)$$

is the rate of change of dy/dx with distance x. As $\Delta x \to 0$,

$$\frac{\Delta}{\Delta x}\left(\frac{dy}{dx}\right) \to \frac{d^2y}{dx^2}$$

so that in the limit of an infinitely small segment of the string Eq. 7.5 becomes

7.6
$$\frac{d^2y}{dx^2} = \frac{\mu}{T}\frac{d^2y}{dt^2}$$

This expression is not quite correct, however, since y is a function of both t and x and d^2y/dx^2 therefore varies with time t while d^2y/dt^2 varies with position x. In Eq. 7.6 we are referring to differentiations made at a specific time and place, which means that the total derivatives must be replaced by the partial derivatives.

$$\frac{\partial^2 y}{\partial x^2} = \left(\frac{d^2y}{dx^2}\right)_{t\,=\,\text{constant}}$$

$$\frac{\partial^2 y}{\partial t^2} = \left(\frac{d^2y}{dt^2}\right)_{x\,=\,\text{constant}}$$

The resulting partial differential equation

7.7
$$\frac{\partial^2 y}{\partial x^2} = \frac{\mu}{T}\frac{\partial^2 y}{\partial t^2}$$

Wave equation for stretched string

is the *wave equation* for a stretched string.

Solutions to the wave equation may be of many kinds, which reflects the variety of waves that can occur—a single traveling pulse, a train of waves of constant amplitude and

The Wave Equation

wavelength, a train of superposed waves of constant amplitude and wavelength, a train of superposed waves of different amplitudes and wavelengths, a standing wave in a string fastened at both ends, and so on. All solutions must be of the form

7.8
$$y = F\left(t \pm \sqrt{\frac{\mu}{T}}\,x\right)$$

where F is any function that can be differentiated. The solutions $F(t - \sqrt{\mu/T}\,x)$ represent waves traveling in the $+x$ direction, and the solutions $F(t + \sqrt{\mu/T}\,x)$ represent waves traveling in the $-x$ direction.

Comparing Eq. 7.8 with Eq. 4.5 shows that the quantity $\sqrt{T/\mu}$ is equal to the wave speed, for which we will use the symbol v from now on. With

7.9
$$\sqrt{\frac{T}{\mu}} = v$$

the wave equation for a stretched string becomes

7.10
$$\frac{\partial^2 y}{\partial x^2} = \frac{1}{v^2}\frac{\partial^2 y}{\partial t^2} \qquad\qquad \textbf{Wave equation}$$

This form of the wave equation holds for waves in *any* medium in which the wave speed v is independent of the precise character of the waves, that is, when v is the same regardless of the particular shape, frequency, and wavelength of the waves. Waves in a stretched string, sound waves in air, and light waves in vacuum are examples of actual waves that obey Eq. 7.10; in each case the speed v depends only upon the properties of the medium.

Our interest here is in the wave equivalent of a "free particle," namely a particle that is not under the influence of any forces and therefore pursues a straight path at constant speed. This equivalent corresponds to the general solution of Eq. 7.10 for undamped (that is, constant amplitude A), monochromatic (constant angular frequency ω) harmonic waves in the $+x$ direction,

7.11
$$y = Ae^{-i\omega(t-x/v)}$$

In this formula y is a complex quantity, with both real and imaginary parts. Because

7.12
$$e^{-i\theta} = \cos\theta - i\sin\theta$$

Eq. 7.11 can be written in the form

7.13
$$y = A\cos\omega\left(t - \frac{x}{v}\right) - iA\sin\omega\left(t - \frac{x}{v}\right)$$

Only the real part of Eq. 7.13 has significance in the case of waves in a stretched string,

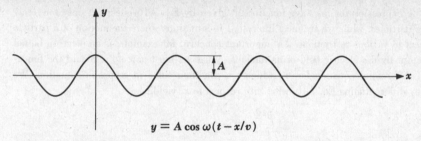

$$y = A \cos \omega(t - x/v)$$

FIGURE 7.2 Waves in the xy plane traveling in the +x direction along a stretched string lying on the x axis.

where y represents the displacement of the string from its normal position (Fig. 7.2); in this case the imaginary part is discarded as irrelevant.

7.4 Schrödinger's Equation: Time-dependent Form

In quantum mechanics the wave function Ψ corresponds to the displacement y of wave motion in a string. However, Ψ, unlike y, is not itself a measurable quantity and may therefore be complex. For this reason we shall assume that Ψ is specified in the x direction by

7.14
$$\Psi = Ae^{-i\omega(t-x/v)}$$

When we replace ω in the above formula by $2\pi\nu$ and v by $\lambda\nu$, we obtain

7.15
$$\Psi = Ae^{-2\pi i(\nu t - x/\lambda)}$$

which is convenient since we already know what ν and λ are in terms of the total energy E and momentum p of the particle being described by Ψ. Since

$$E = h\nu = 2\pi\hbar\nu$$

and

$$\lambda = \frac{h}{p} = \frac{2\pi\hbar}{p}$$

we have

7.16
$$\Psi = Ae^{-(i/\hbar)(Et-px)}$$

Equation 7.16 is a mathematical description of the wave equivalent of an unrestricted particle of total energy E and momentum p moving in the $+x$ direction, just as Eq. 7.11 is a mathematical description of a harmonic displacement wave moving freely along a stretched string.

The expression for the wave function Ψ given by Eq. 7.16 is correct only for freely moving particles, while we are most interested in situations where the motion of a particle is subject to various restrictions. An important concern, for example, is an electron bound to an atom by the electric field of its nucleus. What we must now do is obtain the fundamental differential equation for Ψ, which we can then solve in a specific situation. We begin by differentiating Eq. 7.16 twice with respect to x, yielding

7.17
$$\frac{\partial^2 \Psi}{\partial x^2} = -\frac{p^2}{\hbar^2} \Psi$$

and once with respect to t, yielding

7.18
$$\frac{\partial \Psi}{\partial t} = -\frac{iE}{\hbar} \Psi$$

At speeds small compared with that of light, the total energy E of a particle is the sum of its kinetic energy $p^2/2m$ and its potential energy V, where V is in general a function of position x and time t:

7.19
$$E = \frac{p^2}{2m} + V$$

Multiplying both sides of this equation by the wave function Ψ,

7.20
$$E\Psi = \frac{p^2\Psi}{2m} + V\Psi$$

From Eqs. 7.17 and 7.18 we see that

7.21
$$E\Psi = -\frac{\hbar}{i}\frac{\partial \Psi}{\partial t}$$

and

7.22
$$p^2\Psi = -\hbar^2 \frac{\partial^2 \Psi}{\partial x^2}$$

Substituting these expressions for $E\Psi$ and $p^2\Psi$ into Eq. 7.20, we obtain

7.23
$$\frac{\hbar}{i}\frac{\partial \Psi}{\partial t} = \frac{\hbar^2}{2m}\frac{\partial^2 \Psi}{\partial x^2} - V\Psi$$

Time-dependent Schrödinger's equation in one dimension

Equation 7.23 is the *time-dependent form of Schrödinger's equation*. In three dimensions the time-dependent form of Schrödinger's equation is

7.24
$$\frac{\hbar}{i}\frac{\partial \Psi}{\partial t} = \frac{\hbar^2}{2m}\left(\frac{\partial^2 \Psi}{\partial x^2} + \frac{\partial^2 \Psi}{\partial y^2} + \frac{\partial^2 \Psi}{\partial z^2}\right) - V\Psi$$

where the particle's potential energy V is some function of x, y, z, and t. Any restrictions that may be present on the particle's motion will affect the potential-energy function V. Once V is known, Schrödinger's equation may be solved for the wave function Ψ of the particle, from which its probability density $|\Psi^2|$ may be determined for a specified x, y, z, t.

The quantity

7.25
$$\frac{\partial^2 \Psi}{\partial x^2} + \frac{\partial^2 \Psi}{\partial y^2} + \frac{\partial^2 \Psi}{\partial z^2}$$

is often abbreviated $\nabla^2 \Psi$, where ∇^2, the *laplacian operator*, is defined by

7.26
$$\nabla^2 \equiv \frac{\partial^2}{\partial x^2} + \frac{\partial^2}{\partial y^2} + \frac{\partial^2}{\partial z^2}$$

Laplacian operator in
cartesian coordinates

An operator is a mathematical instruction that tells us what operation to carry out on the quantity that follows it. When the laplacian operator ∇^2 appears in an equation, we know immediately that we are to take the second partial derivative of whatever follows it, here the wave function Ψ, with respect to each coordinate. An advantage of using $\nabla^2 \Psi$ to represent Eq. 7.25 is that $\nabla^2 \Psi$ does not specifically involve cartesian coordinates, so that Schrödinger's equation in the form

7.27
$$\frac{\hbar}{i} \frac{\partial \Psi}{\partial t} = \frac{\hbar^2}{2m} \nabla^2 \Psi - V \Psi$$

Time-dependent
Schrödinger's equation
in three dimensions

holds in all coordinate systems provided that ∇^2 is suitably defined in each system. Often the proper choice of a coordinate system facilitates solving a differential equation; for example, we shall find that Schrödinger's equation for the hydrogen atom is most easily solved for Ψ using spherical polar coordinates, while in the case of the hydrogen molecular ion H_2^+ (a stable system composed of two protons and an electron) prolate spheroidal coordinates are appropriate.

Let us return to the manner in which Schrödinger's equation was obtained starting from the wave function of a freely moving particle. The extension of Schrödinger's equation from the special case of an unrestricted particle (potential energy V = constant) to the general case of a particle subject to arbitrary forces that vary in time and space [$V = V(x,y,z,t)$] is entirely plausible, but there is no a priori way to *prove* that this extension is correct. All we can do is to postulate Schrödinger's equation, solve it for a variety of physical situations, and compare the results of the calculations with the results of experiments. If they agree, the postulate embodied in Schrödinger's equation is valid; if they disagree, the postulate must be discarded and some other approach would have to be explored. In other words, Schrödinger's equation cannot be derived from "first principles," but represents a first principle itself.

In practice, Schrödinger's equation has turned out to be completely accurate in predicting the results of experiments. To be sure, we must keep in mind that Eq. 7.27 can be used only for nonrelativistic problems, and a more elaborate formulation is required when

Schrödinger's Equation: Time-dependent Form

particle speeds comparable with that of light are involved. Because it is in accord with experiment within its range of applicability, we are entitled to regard Schrödinger's equation as representing a successful postulate concerning certain aspects of the physical world. But for all its success, this equation remains a postulate in the same sense as the postulates of special relativity or the laws of thermodynamics: None of these can be derived from some other principle, and each is a fundamental generalization neither more nor less valid than the empirical data it is based upon. It is worth noting in this connection that Schrödinger's equation does not represent an increase in the number of postulates required to describe the workings of the physical world, because Newton's second law of motion, regarded in classical mechanics as a postulate, can be derived from Schrödinger's equation provided that the quantities it relates are understood to be averages rather than definite values.

7.5 Probability Current

The quantity $|\Psi|^2 = \Psi^*\Psi$ has been spoken of as the probability density of the particle described by the wave function Ψ. This interpretation can be justified by considering a particle moving in the x direction. If $|\Psi|^2$ is indeed the probability density of the particle, then it should be conserved as the particle travels from place to place; if $|\Psi|^2$ is *not* conserved, then

$$\int_{-\infty}^{\infty} |\Psi|^2 \, dV = 1$$

is not correct at all times, and $|\Psi|^2$ cannot represent the probability density of a real particle.

Let us consider the motion of the particle between x_1 and x_2 in Fig. 7.3. At x_1 the flux of probability, or *probability current*, has some value S_1 and at x_2 it has some value S_2. For probability density $|\Psi|^2$ to be conserved, any difference between S_1 and S_2 must be equal to the rate at which the total probability

FIGURE 7.3 The rate of change of the total probability within a region must be equal to the difference between the inward flow S_1 and the outward flow S_2 of probability density.

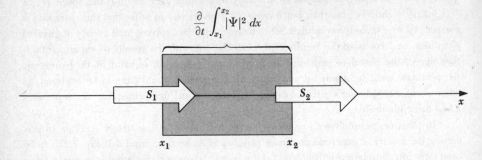

$$\int_{x_1}^{x_2} |\Psi|^2 \, dx$$

in the region between x_1 and x_2 is changing with time. That is, the difference between the inward and outward flows of probability must be the same as the rate of change of the probability within the region:

7.28
$$\frac{\partial}{\partial t} \int_{x_1}^{x_2} |\Psi|^2 \, dx = S_1 - S_2 \qquad \text{**Probability conservation**}$$

What we shall now do is to show that Eq. 7.28 is actually obeyed.

We start with the time-dependent Schrödinger's equation

7.29
$$i\hbar \frac{\partial \Psi}{\partial t} = -\frac{\hbar^2}{2m} \frac{\partial^2 \Psi}{\partial x^2} + V\Psi$$

and its complex conjugate, which is obtained by replacing i by $-i$ and Ψ by Ψ^*:

7.30
$$-i\hbar \frac{\partial \Psi^*}{\partial t} = -\frac{\hbar^2}{2m} \frac{\partial^2 \Psi^*}{\partial x^2} + V\Psi^*$$

The first step is to multiply Eq. 7.29 by Ψ^* and Eq. 7.30 by Ψ, which yields

7.31
$$i\hbar \Psi^* \frac{\partial \Psi}{\partial t} = -\frac{\hbar^2}{2m} \left(\Psi^* \frac{\partial^2 \Psi}{\partial x^2} \right) + \Psi^* V\Psi$$

7.32
$$-i\hbar \, \Psi \frac{\partial \Psi^*}{\partial t} = -\frac{\hbar^2}{2m} \left(\Psi \frac{\partial^2 \Psi^*}{\partial x^2} \right) + \Psi V\Psi^*$$

Now we subtract Eq. 7.32 from Eq. 7.31 to obtain

$$i\hbar \left(\Psi^* \frac{\partial \Psi}{\partial t} + \Psi \frac{\partial \Psi^*}{\partial t} \right) = -\frac{\hbar^2}{2m} \left(\Psi^* \frac{\partial^2 \Psi}{\partial x^2} - \Psi \frac{\partial^2 \Psi^*}{\partial x^2} \right)$$

which simplifies to

$$i\hbar \frac{\partial}{\partial t} (\Psi^* \Psi) = -\frac{\hbar^2}{2m} \frac{\partial}{\partial x} \left(\Psi^* \frac{\partial \Psi}{\partial x} - \Psi \frac{\partial \Psi^*}{\partial x} \right)$$

Multiplying through by $-i/\hbar$ and integrating both sides of the equation from x_1 to x_2,

$$\frac{\partial}{\partial t} \int_{x_1}^{x_2} \Psi^* \Psi \, dx = \frac{i\hbar}{2m} \int_{x_1}^{x_2} \frac{\partial}{\partial x} \left(\Psi^* \frac{\partial \Psi}{\partial x} - \Psi \frac{\partial \Psi^*}{\partial x} \right) dx$$

7.33
$$= \frac{i\hbar}{2m} \left[\Psi^* \frac{\partial \Psi}{\partial x} - \Psi \frac{\partial \Psi^*}{\partial x} \right]_{x_1}^{x_2}$$

Equation 7.33 is the same as Eq. 7.28 for conservation of probability provided that we define probability current S as

7.34
$$S = -\frac{i\hbar}{2m}\left(\Psi^* \frac{\partial \Psi}{\partial x} - \Psi \frac{\partial \Psi^*}{\partial x}\right)$$
Probability current

The minus sign is required in order that the right-hand side of Eq. 7.33 equal $S_1 - S_2$ to agree with Eq. 7.28.

The preceding definition of probability current can be made plausible by considering a freely moving particle of energy E and momentum p. The wave function of such a particle is, by Eq. 7.16,

$$\Psi = Ae^{-(i/\hbar)(Et-px)}$$

and its complex conjugate is

$$\Psi^* = A^*e^{(i/\hbar)(Et-px)}$$

Hence

$$\frac{\partial \Psi}{\partial x} = \frac{ip}{\hbar}Ae^{-(i/\hbar)(Et-px)} = \frac{ip}{\hbar}\Psi$$

$$\frac{\partial \Psi^*}{\partial x} = -\frac{ip}{\hbar}Ae^{*(i/\hbar)(Et-px)} = -\frac{ip}{\hbar}\Psi^*$$

and so, from Eq. 7.34, we have for the probability current

$$S = -\frac{i\hbar}{2m}\left(\frac{ip}{\hbar}\Psi^*\Psi + \frac{ip}{\hbar}\Psi^*\Psi\right)$$

$$= \frac{p}{m}\Psi^*\Psi$$

7.35
$$= |\Psi|^2 v$$

since $p/m = v$, the speed of the particle. The probability current for a free particle is simply the product of its probability density and its speed. Equation 7.28 for the conservation of probability in this case can be written

7.36
$$\frac{\partial}{\partial t}\int_{x_1}^{x_2}|\Psi|^2\,dx = |\Psi|^2_{x_1}v_1 - |\Psi|^2_{x_2}v_2$$

This is a familiar type of equation; a conservation theorem of exactly the same form but with mass density ρ in place of probability density $|\Psi|^2$ is one of the basic equations of fluid mechanics.

It is appropriate to ask why Eq. 7.35 cannot be used to define probability current in general, instead of Eq. 7.34. The answer is that Eq. 7.35 was obtained by using the wave function for a free particle, that is, for a particle whose potential energy $V(x,t)$ is constant.

In such a situation no forces act on the particle, its speed v is the same at x_1 and x_2, and both sides of Eq. 7.36 are equal to 0 no matter where x_1 and x_2 are chosen. When $V(x,t)$ varies with x and t, the particle energy E and the particle momentum p are both functions of x and t, and their partial derivatives must be considered in evaluating $\partial\Psi/\partial x$ and $\partial\Psi^*/\partial x$. Of course, if $V(x,t)$ varies only slowly with x and t, the derivatives of E and p can be neglected in evaluating $\partial\Psi/\partial x$ and $\partial\Psi^*/\partial x$ and Eq. 7.35 will then be a valid approximation; but Eq. 7.34 is correct under all circumstances.

7.6   Expectation Values

Once Schrödinger's equation has been solved for a particle in a given physical situation, the resulting wave function $\Psi(x,y,z,t)$ contains all the information about the particle that is permitted by the uncertainty principle. Except for those variables that happen to be quantized in some cases, this information is in the form of probabilities and not specific numbers. As an example, let us calculate the *expectation value* $\overline{x}$ of the position of a particle confined to the x axis that is described by the wave function $\Psi(x,t)$. This is the value of x we would obtain if we determined experimentally the positions of a great many particles described by the same wave function at some instant t and then averaged the results.

To make the procedure clear, we shall first answer a slightly different question: What is the average position $\overline{x}$ of a number of particles distributed along the x axis in such a way that there are N_1 particles at x_1, N_2 particles at x_2, and so on? The average position in this case is the same as the center of mass of the distribution, and so

$$\overline{x} = \frac{N_1x_1 + N_2x_2 + N_3x_3 + \cdots}{N_1 + N_2 + N_3 + \cdots}$$

$$= \frac{\Sigma N_ix_i}{\Sigma N_i}$$

When we are dealing with a single particle, we must replace the number N_i of particles at x_i by the probability P_i that the particle be found in an interval dx at x_i. This probability is

$$P_i = |\Psi_i|^2\, dx$$

where Ψ_i is the particle wave function evaluated at $x = x_i$. Making this substitution and changing the summations to integrals, we see that the expectation value of the position of the single particle is

7.37
$$\overline{x} = \frac{\displaystyle\int_{-\infty}^{\infty} x|\Psi|^2\, dx}{\displaystyle\int_{-\infty}^{\infty} |\Psi|^2\, dx}$$

If Ψ is a normalized wave function, the denominator of Eq. 7.37 is equal to the probability that the particle exist somewhere between $x = -\infty$ and $x = \infty$, and therefore has the value 1. Hence

7.38
$$\bar{x} = \int_{-\infty}^{\infty} x|\Psi|^2 \, dx$$

This formula states that $\bar{x}$ is located at the center of mass (so to speak) of $|\Psi|^2$; if $|\Psi|^2$ is plotted versus x on a graph and the area enclosed by the curve and the x axis is cut out, the balance point will be at $\bar{x}$.

Equation 7.38 is customarily written in the equivalent form

7.39
$$\bar{x} = \int_{-\infty}^{\infty} \Psi^* x \Psi \, dx$$

for reasons that will be apparent in the next section.

The same procedure as that followed above can be used to obtain the expectation value $\overline{G(x)}$ of any quantity [for instance, potential energy $V(x)$] that is a function of the position x of a particle described by a wave function Ψ. The result is

7.40
$$\overline{G(x)} = \int_{-\infty}^{\infty} \Psi^* \, G(x) \, \Psi \, dx \qquad \text{Expectation value}$$

This formula holds even if $G(x)$ varies with time, because $\overline{G(x)}$ in any event must be evaluated at a particular time t since Ψ is itself a function of t.

7.7 Operators

In the previous section we saw how an expectation value can be obtained for any quantity that is a function of the position x of a particle represented by a wave function Ψ. Thus we can find expectation values at any time t for x itself and for $V(x)$, the potential energy of the particle, both of which are part of a complete description of the state of the particle. Other dynamical quantities, such as the particle's momentum p and total energy E, cannot be treated in quite the same manner, however. The expectation values of p and E would seem to be given by

$$\bar{p} = \int_{-\infty}^{\infty} \Psi^* p \Psi \, dx$$

$$\bar{E} = \int_{-\infty}^{\infty} \Psi^* E \Psi \, dx$$

These formulas are perfectly straightforward until we realize that, because $\Psi = \Psi(x,t)$, we must express p and E as functions of x and t in order to carry out the integrations. But the uncertainty principle implies that no such functions as $p(x,t)$ and $E(x,t)$ can exist, once x and t are specified, the relationships

$$\Delta p\,\Delta x \geq \hbar$$

$$\Delta E\,\Delta t \geq \hbar$$

mean that we cannot, in principle, determine p and E exactly. (Or, if p and E are specified, as in the case of a stationary state such as that represented by an atomic energy level, x and t cannot then be determined exactly.)

In classical physics no such limitation occurs, because the uncertainty principle can be neglected in the macroscopic world. When we apply the second law of motion to the motion of a body subject to various forces, we expect to get $p(x,t)$ and $E(x,t)$ from the solution as well as $x(t)$; to solve a problem in classical mechanics means in essence to ascertain the entire future course of the body's motion. In quantum physics, on the other hand, all we get directly by applying Schrödinger's equation to the motion of a particle is the wave function Ψ, and the future course of the particle's motion—like its initial state—is a matter of probabilities instead of definite figures.

A suggestion as to the proper way to evaluate $\overline{p}$ and $\overline{E}$ may be obtained by differentiating the free-particle wave function

$$\Psi = Ae^{-(i/\hbar)(Et-px)}$$

with respect to x and to t. We find that

$$\frac{\partial \Psi}{\partial x} = \frac{i}{\hbar}p\Psi$$

$$\frac{\partial \Psi}{\partial t} = -\frac{i}{\hbar}E\Psi$$

which can be written in the suggestive forms

7.41
$$p\Psi = \frac{\hbar}{i}\frac{\partial}{\partial x}\Psi$$

7.42
$$E\Psi = i\hbar\frac{\partial}{\partial t}\Psi$$

Evidently the dynamical quantity p in some sense corresponds to the differential operator $(\hbar/i)\,\partial/\partial x$ and the dynamical quantity E similarly corresponds to the differential operator $i\hbar\,\partial/\partial t$. (As stated earlier, an operator tells us what operation to carry out on the quantity that follows it. The operator $i\hbar\,\partial/\partial t$ instructs us to take the partial derivative of what comes after it with respect to t and multiply the result by $i\hbar$.)

It is customary to denote operators by sans-serif boldface letters, so that **p** is the operator that corresponds to momentum p and E is the operator that corresponds to total energy E. From Eqs. 7.41 and 7.42 these operators are

7.43
$$\mathsf{p} = \frac{\hbar}{i} \frac{\partial}{\partial x}$$
Momentum operator

7.44
$$\mathsf{E} = i\hbar \frac{\partial}{\partial t}$$
Total energy operator

Though we have only shown that the correspondences expressed in Eqs. 7.43 and 7.44 hold for free particles, they are entirely general results whose validity is the same as that of Schrödinger's equation. To support this statement, we can replace the equation

$$E = T + V$$

for the total energy of a particle with the operator equation

7.45
$$\mathsf{E} = \mathsf{T} + V$$

Since the kinetic energy T is given in terms of momentum p by

$$T = \frac{p^2}{2m}$$

we have

$$\mathsf{T} = \frac{\mathsf{p}^2}{2m} = \frac{1}{2m} \left(\frac{\hbar}{i} \frac{\partial}{\partial x} \right)^2$$

7.46
$$= -\frac{\hbar^2}{2m} \frac{\partial^2}{\partial x^2}$$
Kinetic energy operator

Equation 7.45 therefore reads

7.47
$$i\hbar \frac{\partial}{\partial t} = -\frac{\hbar^2}{2m} \frac{\partial^2}{\partial x^2} + V$$

Now we multiply the identity $\Psi = \Psi$ by Eq. 7.47 and obtain

$$i\hbar \frac{\partial \Psi}{\partial t} = -\frac{\hbar^2}{2m} \frac{\partial^2 \Psi}{\partial x^2} + V\Psi$$

which is Schrödinger's equation. Postulating Eqs. 7.43 and 7.44 is equivalent to postulating Schrödinger's equation.

Because p and E can be replaced by their corresponding operators in an equation, we can use these operators to obtain expectation values for p and E. Thus the expectation value for p is

$$\overline{p} = \int_{-\infty}^{\infty} \Psi^* \mathsf{p} \Psi \, dx$$

$$= \int_{-\infty}^{\infty} \Psi^* \left(\frac{\hbar}{i} \frac{\partial}{\partial x} \right) \Psi \, dx$$

7.48
$$= \frac{\hbar}{i} \int_{-\infty}^{\infty} \Psi^* \frac{\partial \Psi}{\partial x} \, dx$$

and the expectation value for E is

$$\overline{E} = \int_{-\infty}^{\infty} \Psi^* E \Psi \, dx$$

$$= \int_{-\infty}^{\infty} \Psi^* \left(i\hbar \frac{\partial}{\partial t} \right) \Psi \, dx$$

7.49
$$= i\hbar \int_{-\infty}^{\infty} \Psi^* \frac{\partial \Psi}{\partial t} \, dx$$

Both Eqs. 7.48 and 7.49 can be evaluated for any acceptable wave function $\Psi(x,t)$.

It is now clear why it is necessary to express expectation values involving operators in the form

$$\overline{p} = \int_{-\infty}^{\infty} \Psi^* \mathsf{p} \Psi \, dx$$

The other alternatives are

$$\int_{-\infty}^{\infty} \mathsf{p} \Psi^* \Psi \, dx = \frac{\hbar}{i} \int_{-\infty}^{\infty} \frac{\partial}{\partial x} (\Psi^* \Psi) \, dx$$

$$= \frac{\hbar}{i} \left[\Psi^* \Psi \right]_{-\infty}^{\infty} = 0$$

since Ψ^* and Ψ must be 0 at $x = \pm\infty$; and

$$\int_{-\infty}^{\infty} \Psi^* \Psi \, \mathsf{p} \, dx = \frac{\hbar}{i} \int_{-\infty}^{\infty} \Psi^* \Psi \frac{\partial}{\partial x} \, dx$$

which makes no sense. In the case of algebraic quantities such as x and $V(x)$, the order of factors in the integrand is unimportant, but when differential operators are involved, the correct order of factors must be observed.

Every observable quantity G characteristic of a physical system may be represented by a suitable quantum-mechanical operator G. To obtain this operator, it is only necessary to express G in terms of x and p and then replace p by $(\hbar/i) \, \partial/\partial x$. If the wave function Ψ of the system is known, the expectation value of $G(x,p)$ is

7.50
$$\overline{G(x,p)} = \int_{-\infty}^{\infty} \Psi^* G \Psi \, dx$$
Expectation value of an operator

This result substantiates the statement made earlier that from Ψ can be obtained all the information about a system that is permitted by the uncertainty principle.

7.8 Schrödinger's Equation: Steady-state Form

In a great many situations the potential energy of a particle does not depend upon time explicitly; the forces that act upon it, and hence V, vary with the position of the particle only. When this is true, Schrödinger's equation may be simplified by removing all reference to t. We note that the one-dimensional wave function Ψ of an unrestricted particle may be written

$$\Psi = Ae^{-(i/\hbar)(Et-px)}$$
$$= Ae^{-(iE/\hbar)t}e^{+(ip/\hbar)x}$$
7.51
$$= \psi e^{-(iE/\hbar)t}$$

That is, Ψ is the product of a time-dependent function $e^{-(iE/\hbar)t}$ and a position-dependent function ψ. As it happens, the time variations of *all* functions of particles acted upon by stationary forces have the same form as that of an unrestricted particle. Substituting the Ψ of Eq. 7.51 into the time-dependent form of Schrödinger's equation, we find that

$$- E\psi e^{-(iE/\hbar)t} = \frac{\hbar^2}{2m} e^{-(iE/\hbar)t} \frac{\partial^2 \psi}{\partial x^2} - V\psi e^{-(iE/\hbar)t}$$

and so, dividing through by the common exponential factor,

7.52
$$\frac{\partial^2 \psi}{\partial x^2} + \frac{2m}{\hbar^2}(E - V)\psi = 0$$
Steady-state Schrödinger's equation in one dimension

Equation 7.52 is the *steady-state form of Schrödinger's equation*. In three dimensions it is

7.53
$$\nabla^2 \psi + \frac{2m}{\hbar^2}(E - V)\psi = 0$$
Steady-state Schrödinger's equation in three dimensions

The steady-state form of Schrödinger's equation can be obtained in another way. Let us express the total energy $E = T + V$ of a particle in terms of the operators E, T, and V. Because the potential energy V does not vary with time, the total energy E is constant, and the operator E is simply a number:

$$E = E$$

The potential energy operator is, as before,

$$V = V$$

since $V = V(x)$, while the kinetic energy operator T has been found (Eq. 7.46) to be

$$T = \frac{p^2}{2m} = -\frac{\hbar^2}{2m}\frac{\partial^2}{\partial x^2}$$

Hence the operator equation that corresponds to $E = T + V$ when E is constant is

7.54
$$E = -\frac{\hbar^2}{2m}\frac{\partial^2}{\partial x^2} + V$$

We now multiply $\psi = \psi$ by this operator equation and obtain

7.55
$$E\psi = -\frac{\hbar^2}{2m}\frac{\partial^2\psi}{\partial x^2} + V\psi$$

which is the same as Eq. 7.52.

7.9 Eigenvalues and Eigenfunctions

The total energy operator

7.56
$$H = -\frac{\hbar^2}{2m}\frac{\partial^2}{\partial x^2} + V \qquad \text{Hamiltonian operator}$$

is called the *hamiltonian operator* because it is reminiscent of the hamiltonian function in advanced classical mechanics, which is an expression for the total energy of a system in terms of coordinates and momenta only. Evidently the steady-state Schrödinger's equation can be written simply as

7.57
$$E\psi = H\psi$$

In general, Schrödinger's steady-state equation can be solved only for certain values of the energy E. What is meant by this statement has nothing to do with any mathematical difficulties that may be present, but is something much more fundamental. To "solve" Schrödinger's equation for a given system means to obtain a wave function ψ that not only obeys the equation and whatever boundary conditions there are, but also fulfills the requirements for an acceptable wave function—namely, that it and its derivatives be continuous, finite, and single-valued. If there is no such wave function, the system cannot exist in a steady state. Thus energy quantization appears in wave mechanics as a natural element of the theory, and energy quantization in the physical world is revealed as a universal phenomenon characteristic of *all* stable systems.

A familiar and quite close analogy to the manner in which energy quantization occurs in solutions of Schrödinger's equation is with standing waves in a stretched string of length L

that is fixed at both ends. Here, instead of a single wave propagating indefinitely in one direction, waves are traveling in both the $+x$ and $-x$ directions simultaneously subject to the condition that the displacement y always be zero at both ends of the string. An acceptable function $y(x,t)$ for the displacement must, with its derivatives, obey the same requirements of continuity, finiteness, and single-valuedness as Ψ and, in addition, must be real since y represents a directly measurable quantity. The only solutions of the wave equation

$$\frac{\partial^2 y}{\partial x^2} = \frac{1}{v^2}\frac{\partial^2 y}{\partial t^2}$$

that are in accord with these various limitations are those in which the wavelengths are given by

$$\lambda_n = \frac{2L}{n+1} \qquad n = 0, 1, 2, 3, \ldots$$

as shown in Fig. 7.4. It is the *combination* of the wave equation and the restrictions placed

FIGURE 7.4 Standing waves in a stretched string fastened at both ends.

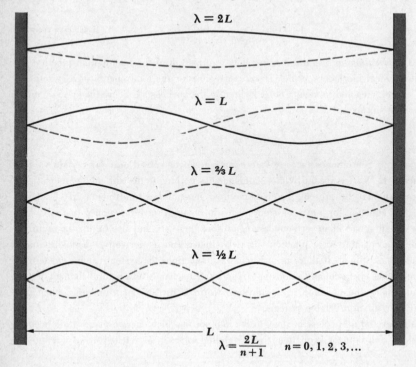

$$\lambda = \frac{2L}{n+1} \qquad n = 0, 1, 2, 3, \ldots$$

on the nature of its solution that leads us to conclude that $y(x,t)$ can exist only for certain wavelengths λ_n.

The values of energy E_n for which Schrödinger's steady-state equation can be solved are called *eigenvalues* and the corresponding wave functions ψ_n are called *eigenfunctions*. (These terms come from the German *Eigenwert*, meaning "proper or characteristic value," and *Eigenfunktion*, or "proper or characteristic function.") The discrete energy levels of the hydrogen atom

$$E_n = -\frac{me^4}{32\pi^2\varepsilon_0^2\hbar^2}\left(\frac{1}{n^2}\right) \qquad n = 1, 2, 3, \ldots$$

are an example of a set of eigenvalues; we shall see in Chap. 9 why these particular values of E are the only ones that yield acceptable wave functions for the electron in the hydrogen atom.

Eigenvalues are not restricted solely to energy. Schrödinger's steady-state equation can be written

7.58 $$E_n\psi_n = \mathsf{H}\psi_n$$

so we can say that the various E_n are the eigenvalues of the hamiltonian operator H. This kind of association between eigenvalues and quantum-mechanical operators is quite general. The condition that a certain dynamical variable G be restricted to the discrete values G_n— in other words, that G be quantized—is that the wave functions ψ_n of the system be such that

7.59 $$\mathsf{G}\psi_n = G_n\psi_n \qquad\qquad \text{Eigenvalue equation}$$

where G is the operator that corresponds to G and each G_n is a real number. When Eq. 7.59 holds for the wave functions of a system, it is a fundamental postulate (in fact, *the* fundamental postulate) of quantum mechanics that any measurement of G can only yield one of the values G_n. If measurements of G are made on a number of identical systems all in states described by the particular eigenfunction ψ_k, each measurement will yield the single value G_k.

An important example of a dynamical variable other than total energy that is found to be quantized in stable systems is angular momentum L. In the case of the hydrogen atom, we shall find that the eigenvalues of the magnitude of the total angular momentum are specified by

$$L_l = \sqrt{l(l+1)}\,\hbar \qquad l = 0, 1, 2, \ldots, (n-1)$$

Of course, a dynamical variable G may not obey Eq. 7.59 and therefore not be quantized. In this case measurements of G made on a number of identical systems will not yield a unique result but instead a spread of values whose average is the expectation value

$$G = \int_{-\infty}^{\infty} \psi\,\mathsf{G}\,\psi\,dx$$

Eigenvalues and Eigenfunctions 167

In the hydrogen atom, the electron's position is not quantized, for instance, so that we must think of the electron as being present in the vicinity of the nucleus with a certain probability $|\psi(x,y,z)|^2$ per unit volume but with no predictable position or even orbit in the classical sense. This probabilistic statement does not conflict with the fact that experiments performed on hydrogen atoms always show that it contains one whole electron, not 27 percent of an electron in a certain region and 73 percent elsewhere; the probability is one of *finding* the electron, and although this probability is smeared out in space, the electron itself is not.

Problems

1. Verify that all solutions of the wave equation

$$\frac{\partial^2 y}{\partial x^2} = \frac{1}{v^2}\frac{\partial^2 y}{\partial t^2}$$

must be of the form $y = F(t \pm x/v)$ as asserted in Sec. 7.3.

2. Show that the dimensions of $|\Psi|^2$ are m^{-3}, so that it properly describes probability density.

3. If $\Psi_1(x,t)$ and $\Psi_2(x,t)$ are both solutions of Schrödinger's equation for a given potential $V(x)$, show that the linear combination

$$\Psi = a_1\Psi_1 + a_2\Psi_2$$

in which a_1 and a_2 are arbitrary constants is also a solution. (This result is in accord with the empirical observation of the interference of de Broglie waves, for instance in the Davisson-Germer experiment discussed in Chap. 4.)

4. Show that Schrödinger's equation, which is complex, can be replaced by a pair of real equations by substituting

$$\Psi(x,t) = \Psi_1(x,t) + i\Psi_2(x,t)$$

(in which Ψ_1 and Ψ_2 are real) and separating the result into real and imaginary parts. Is there anything to be gained by doing this?

5. The angular momentum about the z axis of a particle moving in the x, y plane is $L_z = xp_y - yp_x$. Verify that

$$L_z = \frac{\hbar}{i}\frac{\partial}{\partial\phi}$$

where ϕ is the angle between the x axis and the radius vector to the particle from the origin.

6. Show that the expectation values $\overline{px}$ and $\overline{xp}$ are related by

$$\overline{px} - \overline{xp} = \frac{\hbar}{i}$$

This result is described by saying that p and x do not *commute* and it is intimately related to the uncertainty principle statement that $\Delta p \, \Delta x \geq \hbar$.

7. If the expectation value of the product of p and x is required, the proper formula to use is $(\overline{xp} + \overline{px})/2$. Show that this formula always yields a real number while neither $\overline{xp}$ nor $\overline{px}$ does.

8. Schrödinger's time-independent equation for a free particle restricted to the x axis is

$$\frac{d^2\psi}{dx^2} + \frac{2mE}{\hbar^2}\,\psi = 0$$

(a) Show that

$$\psi = Ae^{i\sqrt{2mE}\,x/\hbar} + Be^{-i\sqrt{2mE}\,x/\hbar}$$

where A and B are constants is a solution of this equation. (In fact, it is the general solution.) (b) By calculating the momentum of the particle, show that $A^2/(A^2 + B^2)$ is the probability that the particle moves in the $+x$ direction and that $B^2/(A^2 + B^2)$ is the probability that it moves in the $-x$ direction. (c) Show that the magnitude of the particle's momentum is the same regardless of its direction of motion.

Chapter 8
Applications of
Quantum
Mechanics

To solve Schrödinger's equation, even in its simpler steady-state form, usually requires sophisticated mathematical techniques. For this reason the study of quantum mechanics has traditionally been reserved for advanced students who have the required proficiency in mathematics. However, since quantum mechanics is the theoretical structure whose results are closest to experimental reality, we must explore its methods and applications if we are to achieve any understanding of modern physics. As we shall see, even a relatively limited mathematical background is sufficient for us to follow the trains of thought that have led quantum mechanics to its greatest achievements.

8.1 The Particle in a Box: Energy Quantization

Our first problem using Schrödinger's equation is that of a particle bouncing back and forth between the walls of a box (Fig. 8.1). Our interest in this problem is threefold: to see how Schrödinger's equation is solved when the motion of a particle is subject to restrictions; to learn the characteristic properties of solutions of this equation, such as the limitation of particle energy to certain specific values only; and to compare the predictions of quantum mechanics with those of newtonian mechanics.

We may specify the particle's motion by saying that it is restricted to traveling along the x axis between $x = 0$ and $x = L$ by infinitely hard walls. A particle does not lose energy when it collides with such walls, so that its total energy stays constant. From the formal point of view of quantum mechanics, the potential energy V of the particle is infinite on both sides of the box, while V is a constant—say 0 for convenience—on the inside. Since the particle cannot have an infinite amount of energy, it cannot exist outside the box, and so its wave function ψ is 0 for $x \leq 0$ and $x \geq L$. Our task is to find what ψ is within the box, namely, between $x = 0$ and $x = L$.

Within the box Schrödinger's equation becomes

8.1
$$\frac{d^2\psi}{dx^2} + \frac{2m}{\hbar^2}\,E\psi = 0$$

since $V = 0$ there. (The total derivative $d^2\psi/dx^2$ is used instead of the partial derivative $\partial^2\psi/\partial x^2$ because ψ is a function of x only in this problem.) Equation 8.1 has the two possible solutions

8.2
$$\psi = A \sin \sqrt{\frac{2mE}{\hbar^2}}\, x$$

8.3
$$\psi = B \cos \sqrt{\frac{2mE}{\hbar^2}}\, x$$

which we can verify by substitution back into Eq. 8.1; their sum is also a solution. A and B are constants to be evaluated. These solutions are subject to the important boundary condition that $\psi = 0$ for $x = 0$ and for $x = L$. Since $\cos 0° = 1$, the second solution cannot describe the particle because it does not permit ψ to be 0 at $x = 0$. Hence we conclude that $B = 0$. Since $\sin 0° = 0$, the first solution always yields $\psi = 0$ at $x = 0$, as required, but ψ will be 0 at $x = L$ only when

8.4
$$\sqrt{\frac{2mE}{\hbar^2}}\,L = \pi, 2\pi, 3\pi, \ldots$$
$$= n\pi \qquad n = 1, 2, 3, \ldots$$

This result comes about because the sines of the angles $\pi, 2\pi, 3\pi, \ldots$ are all 0.

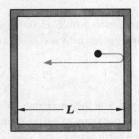

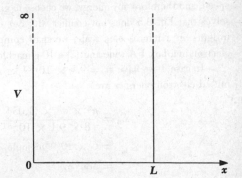

FIGURE 8.1 A particle confined to a box of width L.

The Particle in a Box: Energy Quantization

From Eq. 8.4 it is clear that the energy of the particle can have only certain values, which are the eigenvalues mentioned in the previous chapter. These eigenvalues, constituting the *energy levels* of the system, are

8.5
$$E_n = \frac{n^2 \pi^2 \hbar^2}{2mL^2} \qquad n = 1, 2, 3, \ldots$$

Particle in a box

The integer n corresponding to the energy level E_n is called its *quantum number*. A particle confined to a box cannot have an arbitrary energy: the fact of its confinement leads to restrictions on its wave function that permit it to have only those energies specified by Eq. 8.5.

It is significant that the electron cannot have zero energy; if it did, the electron wave function ψ would have to be zero everywhere in the box, and this means that the electron cannot be present there. The exclusion of $E = 0$ as a possible value for the energy of a trapped electron, like the limitation of E to a discrete set of definite values, is a quantum-mechanical result that has no counterpart in classical mechanics, where all energies, including zero, are presumed possible.

The uncertainty principle provides confirmation that $E = 0$ is not admissible. Because the particle is trapped in the box, the uncertainty in its position is $\Delta x = L$, the width of the box. The uncertainty in its momentum must therefore be

$$\Delta p \geq \frac{\hbar}{L}$$

which is not compatible with $E = 0$. We note that the momentum corresponding to E_1 is, since the particle energy here is entirely kinetic,

$$p_1 = \pm \sqrt{2mE_1} = \pm \frac{\pi \hbar}{L}$$

which is in accord with the uncertainty principle.

Why are we not aware of energy quantization in our own experience? Surely a marble rolling back and forth between the sides of a level box with a smooth floor can have any speed, and therefore any energy, we choose to give it, including zero. In order to assure ourselves that Eq. 8.5 does not conflict with our direct observations while providing unique insights on a microscopic scale, we shall compute the permitted energy levels of (1) an electron in a box 1 A wide and (2) a 10-g marble in a box 10 cm wide.

In case 1 we have $m = 9.1 \times 10^{-31}$ kg and $L = 1$ A $= 10^{-10}$ m, so that the permitted electron energies are

$$E_n = \frac{n^2 \times \pi^2 \times (1.054 \times 10^{-34} \text{ joule-sec})^2}{8 \times 9.1 \times 10^{-31} \text{ kg} \times (10^{-10} \text{ m})^2}$$
$$= 6 \times 10^{-18} n^2 \text{ joule}$$
$$= 38 n^2 \text{ ev}$$

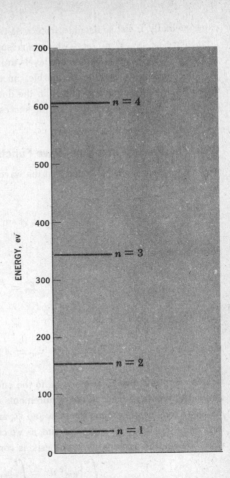

FIGURE 8.2 Energy levels of an electron confined to a box 1 A wide.

The minimum energy the electron can have is 38 ev, corresponding to $n = 1$. The sequence of energy levels continues with $E_2 = 152$ ev, $E_3 = 342$ ev, $E_4 = 608$ ev, and so on (Fig. 8.2). These energy levels are sufficiently far apart to make the quantization of electron energy in such a box conspicuous if such a box actually did exist.

In case 2 we have $m = 10$ g $= 10^{-2}$ kg and $L = 10$ cm $= 10^{-1}$ m, so that the permitted marble energies are

$$E_n = \frac{n^2 \times \pi^2 \times (1.054 \times 10^{-34} \text{ joule-sec})^2}{8 \times 10^{-2} \text{ kg} \times (10^{-1} \text{ m})^2}$$

$$= 5.5 \times 10^{-64} n^2 \text{ joule}$$

The minimum energy the marble can have is 5.5×10^{-64} joule, corresponding to $n = 1$. A marble with this kinetic energy has a speed of only 3.3×10^{-31} m/sec and is therefore

The Particle in a Box: Energy Quantization

experimentally indistinguishable from a stationary marble. A reasonable speed a marble might have is, say, ⅓ m/sec—which corresponds to the energy level of quantum number $n = 10^{30}$! The permissible energy levels are so very close together, then, that there is no way of determining whether the marble can take on only those energies predicted by Eq. 8.5 or any energy whatever. Hence in the domain of everyday experience quantum effects are imperceptible; this accounts for the success in this domain of newtonian mechanics.

8.2   The Particle in a Box: Wave Functions

In the previous section we found that the wave function of a particle in a box whose energy is E is

$$\psi = A \sin \sqrt{\frac{2mE}{\hbar^2}} \, x$$

Since the possible energies are

$$E_n = \frac{n^2 \pi^2 \hbar^2}{2mL^2}$$

substituting E_n for E yields

8.6
$$\psi_n = A \sin \frac{n\pi x}{L}$$

for the eigenfunctions corresponding to the energy eigenvalues E_n. It is easy to verify that these eigenfunctions meet all the requirements we have discussed: for each quantum number n, ψ_n is a single-valued function of x, and ψ_n and $\partial \psi_n / \partial x$ are continuous. Furthermore, the integral of $|\psi_n|^2$ over all space is finite, as we can see by integrating $|\psi_n|^2 \, dx$ from $x = 0$ to $x = L$ (since the particle, by hypothesis, is confined within these limits):

$$\int_{-\infty}^{\infty} |\psi_n|^2 \, dx = \int_0^L |\psi_n|^2 \, dx$$

8.7
$$= A^2 \int_0^L \sin^2 \left(\frac{n\pi x}{L} \right) dx$$

$$= A^2 \frac{L}{2}$$

To normalize ψ we must assign a value to A such that $|\psi_n|^2$ is *equal* to the probability P of finding the particle at x, rather than merely proportional to P. If $|\psi_n|^2$ is to equal P, then it must be true that

8.8
$$\int_{-\infty}^{\infty} |\psi_n|^2 \, dx = 1$$

since

$$\int_{-\infty}^{\infty} P\, dx = 1$$

is the mathematical way of stating that the particle exists somewhere at all times. Comparing Eqs. 8.7 and 8.8, we see that the wave functions of a particle in a box are normalized if

8.9
$$A = \sqrt{\frac{2}{L}}$$

The normalized wave functions of the particle are therefore

8.10
$$\psi_n = \sqrt{\frac{2}{L}} \sin \frac{n\pi x}{L}$$

The normalized wave functions ψ_1, ψ_2, and ψ_3 together with the probability densities $|\psi_1|^2$, $|\psi_2|^2$, and $|\psi_3|^2$ are plotted in Fig. 8.3. While ψ_n may be negative as well as positive, $|\psi_n|^2$ is always positive and, since ψ_n is normalized, its value at a given x is equal to the probability P of finding the particle there. In every case $|\psi_n|^2 = 0$ at $x = 0$ and $x = L$, the

FIGURE 8.3 Wave functions and probability densities of a particle confined to a box with rigid walls.

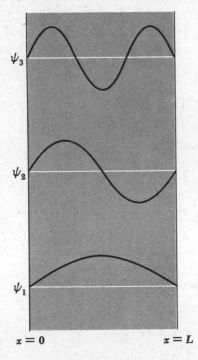
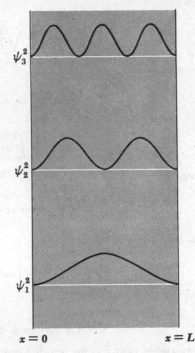

The Particle in a Box: Wave Functions

boundaries of the box. At a particular point in the box the probability of the particle being present may be very different for different quantum numbers. For instance, $|\psi_1|^2$ has its maximum value of $\frac{1}{2}L$ in the middle of the box, while $|\psi_2|^2 = 0$ there: a particle in the lowest energy level of $n = 1$ is most likely to be in the middle of the box, while a particle in the next higher state of $n = 2$ is *never* there! Classical physics, of course, predicts the same probability for the particle being anywhere in the box.

The wave functions shown in Fig. 8.3 resemble the possible vibrations of a string fixed at both ends, such as those of the stretched string of Fig. 7.4. This is a consequence of the fact that waves in a stretched string and the wave representing a moving particle obey wave equations of the same form, so that, when identical restrictions are placed upon each kind of wave, the solutions are identical.

8.3 The Particle in a Box: Momentum Quantization

As an exercise, let us calculate the expectation value $\bar{p}$ of the momentum of a particle trapped in a one-dimensional box. Since

$$\psi_n^* = \psi_n = \sqrt{\frac{2}{L}} \sin \frac{n\pi x}{L}$$

and

$$\frac{d\psi}{dx} = \sqrt{\frac{2}{L}} \frac{n\pi}{L} \cos \frac{n\pi x}{L}$$

we have

$$\bar{p} = \int_{-\infty}^{\infty} \psi^* p \psi \, dx = \int_{-\infty}^{\infty} \psi^* \left(\frac{\hbar}{i} \frac{d}{dx} \right) \psi \, dx$$

$$= \frac{\hbar}{i} \frac{2}{L} \frac{n\pi}{L} \int_0^L \sin \frac{n\pi x}{L} \cos \frac{n\pi x}{L} \, dx$$

We note that

$$\int \sin ax \cos ax \, dx = \frac{1}{2a} \sin^2 ax$$

so that here, with $a = n\pi/L$,

$$\bar{p} = \frac{\hbar}{iL} \left[\sin^2 \frac{n\pi x}{L} \right]_0^L = 0$$

since

$$\sin^2 0 = \sin^2 n\pi = 0 \qquad n = 1, 2, 3, \ldots$$

The expectation value $\bar{p}$ of the particle's momentum is 0.

At first glance this conclusion seems strange. After all, $E = p^2/2m$, and so we would anticipate that

8.11
$$p_n = \pm \sqrt{2mE_n} = \pm \frac{n\pi\hbar}{L}$$

The $\pm$ sign provides the explanation: The particle is moving back and forth, and so its *average* momentum for any value of n is

$$p_{average} = \frac{(+n\pi\hbar/L) + (-n\pi\hbar/L)}{2} = 0$$

in agreement with the expectation value.

According to Eq. 8.11, there should be two momentum eigenfunctions for every energy eigenfunction, corresponding to the two possible directions of motion. The general procedure for finding the eigenvalues of a quantum-mechanical operator, here p, is to start from the eigenvalue equation

8.12
$$p\,\psi_n = p_n\,\psi_n$$

where each p_n is a real number. This equation holds only when the wave functions ψ_n are eigenfunctions of the momentum operator p. Since

$$p = \frac{\hbar}{i}\frac{d}{dx}$$

here, we can see at once that the energy eigenfunctions

$$\psi_n = \sqrt{\frac{2}{L}}\sin\frac{n\pi x}{L}$$

are not also momentum eigenfunctions, because

$$\frac{\hbar}{i}\frac{d}{dx}\left(\sqrt{\frac{2}{L}}\sin\frac{n\pi x}{L}\right) = \frac{\hbar}{i}\frac{n\pi}{L}\sqrt{\frac{2}{L}}\cos\frac{n\pi x}{L} \neq p_n\,\psi_n$$

To find the correct momentum eigenfunctions, we note that

$$\sin\theta = \frac{e^{i\theta} - e^{-i\theta}}{2i} = \frac{1}{2i}e^{i\theta} - \frac{1}{2i}e^{-i\theta}$$

Hence each energy eigenfunction can be expressed as a linear combination of the two wave functions

8.13
$$\psi_n^+ = \frac{1}{2i}\sqrt{\frac{2}{L}}\,e^{in\pi x/L}$$

8.14
$$\psi_n^- = \frac{1}{2i}\sqrt{\frac{2}{L}}\,e^{-in\pi x/L}$$

Inserting the first of these wave functions in the eigenvalue equation 8.12 we have

The Particle in a Box: Momentum Quantization

$$\mathsf{p}\psi_n{}^+ = p_n{}^+\,\psi_n{}^+$$

$$\frac{\hbar}{i}\frac{d}{dx}\,\psi_n{}^+ = \frac{\hbar}{i}\frac{1}{2i}\sqrt{\frac{2}{L}}\,\frac{in\pi}{L}\,e^{in\pi x/L} = \frac{n\pi\hbar}{L}\,\psi_n{}^+ = p_n{}^+\,\psi_n{}^+$$

so that

$$p_n{}^+ = +\,\frac{n\pi\hbar}{L}$$

Similarly the wave function $\psi_n{}^-$ yields the momentum eigenfunctions

$$p_n{}^- = -\,\frac{n\pi\hbar}{L}$$

We conclude that $\psi_n{}^+$ and $\psi_n{}^-$ are indeed the momentum eigenfunctions for a particle in a box, and that Eq. 8.11 correctly states the corresponding momentum eigenvalues.

8.4 The Particle in a Nonrigid Box

It is interesting to solve the problem of the particle in a box when the walls of the box are no longer assumed to be infinitely rigid. In this case the potential energy V outside the box is a finite quantity; the corresponding situation in the case of a vibrating string would involve the nonrigid support of the string at each end, so that the ends can move slightly. This problem is more difficult to treat, and we shall simply present the result here. (We shall take another look at a particle in a nonrigid box when we examine the theory of the deuteron in Chap. 22.) The first few wave functions for a particle in such a box are shown in Fig. 8.4. The wave functions ψ_n now do *not* equal zero outside the box. Even though the particle's energy is smaller than the value of V outside the box, there is still a definite probability that it be found outside it! In other words, even though the particle does not have enough energy to break through the walls of the box according to "common sense," it may nevertheless somehow penetrate them. This peculiar situation is readily understandable in terms of the uncertainty principle. Because the uncertainty Δp in a particle's momentum is related to the uncertainty Δx in its position by the formula

$$\Delta p\,\Delta x \geq \hbar$$

an infinite uncertainty in particle momentum outside the box is the price of definitely establishing that the particle is never there. A particle requires an infinite amount of energy if its momentum is to have an infinite uncertainty, implying that $V = \infty$ outside the box. If V instead has a finite value outside the box, then, there is some probability—not necessarily great, but not zero either—that the particle will "leak" out. As we shall see in Chap. 23, the quantum-mechanical prediction that particles always have some chance of escaping from confinement (since potential energies are never infinite in the real world, our original rigid-walled box has no physical counterpart) exactly fits the observed behavior of those radio-active nuclei that emit alpha particles.

When the confining box has nonrigid walls, the particle wave function ψ_n does not equal zero at the walls. The particle wavelengths that can fit into the box are therefore somewhat longer than in the case of the box with rigid walls, corresponding to lower particle momenta and hence to lower energy levels.

The condition that the potential energy V outside the box be finite has another consequence: it is now possible for a particle to have an energy E that exceeds V. Such a particle is not trapped inside the box, since it always has enough energy to penetrate its walls, and its energy is not quantized but may have any value above V. However, the particle's kinetic energy outside the box, $E - V$, is always less than its kinetic energy inside, which is just E since $V = 0$ in the box according to our original specification. Less energy means longer wavelength, and so ψ has a longer wavelength outside the box than inside.

In the optics of light waves, it is readily observed that when a light wave reaches a region where its wavelength changes (that is, a region of different index of refraction), reflection as well as transmission occurs. This is the reason we see our reflections in shop windows. The effect is common to all types of waves, and it may be shown mathematically to follow from the requirement that the wave variable (electric-field intensity E in the case of electromagnetic waves, pressure p in the case of sound waves, wave height h in the case of water waves, etc.) and its first derivative be continuous at the boundary where the wavelength change takes place. Exactly the same considerations apply to the wave function ψ representing a moving particle. The wave function of a particle encountering a region in which it has a different potential energy, as we saw above, decreases in wavelength if V decreases and increases in wavelength if V increases. In either situation some reflection occurs at the boundaries between the regions. What does "some" reflection mean when we are discussing the motion of a single particle? Since ψ is related to the probability of finding the particle in a particular place, the partial reflection of ψ means that there is a chance that the particle will be reflected.

FIGURE 8.4 Wave functions and probability densities of a particle confined to a box with nonrigid walls.

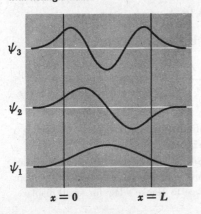 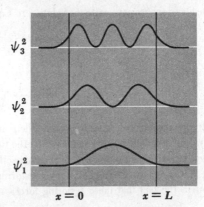

ψ_3 ψ_3^2

ψ_2 ψ_2^2

ψ_1 ψ_1^2

$x = 0$ $x = L$ $x = 0$ $x = L$

The Particle in a Nonrigid Box

What we have been saying, then, is that particles with enough energy to penetrate a wall nevertheless stand some chance of bouncing off instead. This prediction complements the "leaking" out of particles trapped in the box despite the fact that they have insufficient energy to penetrate its walls. Both of these predictions are unique with quantum mechanics and do not correspond to any behavior expected in classical physics. Their confirmation in numerous atomic and nuclear experiments (see Secs. 22.1 and 23.4) supports the validity of the quantum-mechanical approach.

8.5 The Harmonic Oscillator

Harmonic motion occurs when a system of some kind vibrates about an equilibrium configuration. The system may be an object supported by a spring or floating in a liquid, a diatomic molecule, an atom in a crystal lattice—there are countless examples in both the macroscopic and microscopic realms. The condition for harmonic motion to occur is the presence of a restoring force that acts to return the system to its equilibrium configuration when it is disturbed; the inertia of the masses involved causes them to overshoot equilibrium, and the system oscillates indefinitely if no dissipative processes are also present.

In the special case of simple harmonic motion, the restoring force F on a particle of mass m is linear; that is, F is proportional to the particle's displacement from its equilibrium position, so that

8.15
$$F = -kx$$

This relationship is customarily called Hooke's law. According to the second law of motion, $\mathbf{F} = m\mathbf{a}$, and so here

$$-kx = m \frac{d^2x}{dt^2}$$

8.16
$$\frac{d^2x}{dt^2} + \frac{k}{m}x = 0$$

There are various ways to write the solution to Eq. 8.16. A convenient one is

8.17
$$x = A \cos (2\pi\nu t + \phi)$$

where

8.18
$$\nu = \frac{1}{2\pi} \sqrt{\frac{k}{m}}$$

is the frequency of the oscillations, A is their amplitude, and ϕ, the phase constant, is a quantity that depends upon the value of x at the time $t = 0$.

The importance of the simple harmonic oscillator in both classical and modern physics lies not in the strict adherence of actual restoring forces to Hooke's law, which is seldom true, but in the fact that these restoring forces reduce to Hooke's law for small displacements x. To appreciate this point we note that any force which is a function of x can be expressed in a Maclaurin's series about the equilibrium position $x = 0$ as

$$F(x) = F_{x=0} + \left(\frac{dF}{dx}\right)_{x=0} x + \frac{1}{2}\left(\frac{d^2F}{dx^2}\right)_{x=0} x^2 + \frac{1}{6}\left(\frac{d^3F}{dx^3}\right)_{x=0} x^3 + \cdots$$

Since $x = 0$ is the equilibrium position, $F_{x=0} = 0$, and since for small x the values of x^2, x^3, ... are very small compared with x, the third and higher terms of the series can be neglected. The only term of significance when x is small is therefore the second one. Hence

$$F(x) = \left(\frac{dF}{dx}\right)_{x=0} x$$

which is Hooke's law when $(dF/dx)_{x=0}$ is negative, as of course it is for any restoring force. The conclusion, then, is that *all* oscillations are simple harmonic in character when their amplitudes are sufficiently small.

The potential energy function $V(x)$ that corresponds to a Hooke's law force may be found by calculating the work needed to bring a particle from $x = 0$ to $x = x$ against such a force. The result is

8.19
$$V(x) = -\int_0^x F(x)\,dx = k \int_0^x x\,dx = \tfrac{1}{2}kx^2$$

and is plotted in Fig. 8.5. If the energy of the oscillator is E, the particle vibrates back and forth between $x = -A$ and $x = +A$, where E and A are related by $E = \tfrac{1}{2}kA^2$.

FIGURE 8.5 The potential energy of a harmonic oscillator is proportional to x^2, where x is the displacement from the equilibrium position. The amplitude A of the motion is determined by the total energy E of the oscillator, which classically can have any value.

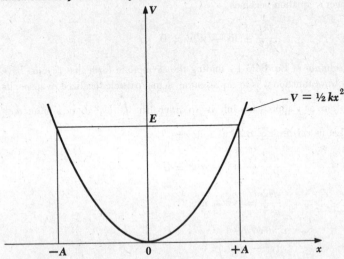

Even before we make a detailed calculation we can anticipate three quantum-mechanical modifications to this classical picture. First, there will not be a continuous spectrum of allowed energies but a discrete spectrum consisting of certain specific values only. Second, the lowest allowed energy will not be $E = 0$ but will be some definite minimum $E = E_0$. Third, there will be a certain probability that the particle can "penetrate" the potential well it is in and go beyond the limits of $-A$ and $+A$.

Schrödinger's equation for the harmonic oscillator is, with $V = \frac{1}{2}kx^2$,

8.20
$$\frac{d^2\psi}{dx^2} + \frac{2m}{\hbar^2}\left(E - \frac{1}{2}kx^2\right)\psi = 0$$

It is convenient to simplify Eq. 8.20 by introducing the dimensionless quantities

8.21
$$y = \left(\frac{1}{\hbar}\sqrt{km}\right)^{1/2} x$$
$$= \sqrt{\frac{2\pi m\nu}{\hbar}}\, x$$

and

8.22
$$\alpha = \frac{2E}{\hbar}\sqrt{\frac{m}{k}}$$
$$= \frac{2E}{h\nu}$$

where ν is the classical frequency of the oscillation given by Eq. 8.18. In making these substitutions, what we have essentially done is change the units in which x and E are expressed from meters and joules, respectively, to appropriate dimensionless units. In terms of y and α Schrödinger's equation becomes

8.23
$$\frac{d^2\psi}{dy^2} + (\alpha - y^2)\psi = 0$$

We begin the solution of Eq. 8.23 by finding the asymptotic form that ψ must have as $y \to \pm\infty$. If any wave function ψ is to represent an actual particle localized in space, its value must approach zero as y approaches infinity in order that $\int_{\infty}^{\infty} |\psi|^2\, dy$ be a finite, nonvanishing quantity. Let us rewrite Eq. 8.23 as follows:

$$\frac{d^2\psi}{dy^2} - (y^2 - \alpha)\psi = 0$$
$$\frac{d^2\psi}{dy^2} = (y^2 - \alpha)\psi$$
$$\frac{d^2\psi/dy^2}{(y^2 - \alpha)\psi} = 1$$

As $y \to \infty$, $y^2 \gg \alpha$ and we have

8.24
$$\lim_{y \to \infty} \frac{d^2\psi/dy^2}{y^2\psi} = 1$$

A function ψ_∞ that satisfies Eq. 8.24 is

8.25
$$\psi_\infty = e^{-y^2/2}$$

since

$$\lim_{y \to \infty} \frac{d^2\psi_\infty}{dy^2} = \lim_{y \to \infty} (y^2 - 1)e^{-y^2/2} = y^2 e^{-y^2/2}$$

Equation 8.25 is the required asymptotic form of ψ.

We are now able to write

$$\psi = f(y)\psi_\infty$$
8.26
$$= f(y)e^{-y^2/2}$$

where $f(y)$ is a function of y that remains to be found. By inserting the ψ of Eq. 8.26 in Eq. 8.23 we obtain

8.27
$$\frac{d^2f}{dy^2} - 2y\frac{df}{dy} + (\alpha - 1)f = 0$$

which is the differential equation that f obeys.

The standard procedure for solving differential equations like Eq. 8.27 is to assume that $f(y)$ can be expanded in a power series in y, namely

$$f(y) = A_0 + A_1y + A_2y^2 + A_3y^3 + \cdots$$

8.28
$$= \sum_{n=0}^{\infty} A_n y^n$$

and then to determine the values of the coefficients A_n. Differentiating f yields

$$\frac{df}{dy} = A_1 + 2A_2y + 3A_3y^2 + \cdots$$

$$= \sum_{n=1}^{\infty} nA_n y^{n-1}$$

By multiplying this equation by y we obtain

$$y\frac{df}{dy} = A_1y + 2A_2y^2 + 3A_3y^3 + \cdots$$

8.29
$$= \sum_{n=0}^{\infty} nA_n y^n$$

The Harmonic Oscillator

The second derivative of f with respect to y is

$$\frac{d^2 f}{dy^2} = 1 \cdot 2 \, A_2 + 2 \cdot 3 \, A_3 y + 3 \cdot 4 \, A_4 y^2 + \cdots$$

$$= \sum_{n=2}^{\infty} n(n-1) A_n y^{n-2}$$

which is equal to

8.30
$$\frac{d^2 f}{dy^2} = \sum_{n=0}^{\infty} (n+2)(n+1) A_{n+2} y^n$$

(That the latter two series are indeed equal can be verified by working out the first terms of each.) We now substitute Eqs. 8.28 to 8.30 in Eq. 8.27 to obtain

8.31
$$\sum_{n=0}^{\infty} [(n+2)(n+1) A_{n+2} - (2n+1-\alpha) A_n] y^n = 0$$

In order for this equation to hold for all values of y, the quantity in brackets must be zero for all values of n. Hence we have the condition that

$$(n+2)(n+1) A_{n+2} = (2n+1-\alpha) A_n$$

and so

8.32
$$A_{n+2} = \frac{2n+1-\alpha}{(n+2)(n+1)} A_n$$

This *recursion formula* enables us to find the coefficients $A_2, A_3, A_4, \ldots$ in terms of A_0 and A_1. (Since Eq. 8.27 is a second-order differential equation, its solution has two arbitrary constants, which are A_0 and A_1 here.) Starting from A_0 we obtain the sequence of coefficients $A_2, A_4, A_6 \ldots$, and starting from A_1 we obtain the other sequence $A_3, A_5, A_7, \ldots$.

8.6   The Harmonic Oscillator: Energy Levels

It is necessary for us to inquire into the behavior of

$$\psi = f(y) e^{-y^2/2}$$

as $y \to \infty$; only if $\psi \to 0$ as $y \to \infty$ can ψ be a physically acceptable wave function. Because $f(y)$ is multiplied by $e^{-y^2/2}$, ψ will meet this requirement provided that

$$\lim_{y \to \infty} f(y) < e^{y^2/2}$$

(As we shall see, it is unnecessary for us to specify just how much smaller f must be in the limit than $e^{y^2/2}$.)

A suitable way to compare the asymptotic behaviors of $f(y)$ and $e^{y^2/2}$ is to express the latter in a power series (f is already in the form of a power series) and to examine the ratio between successive coefficients of each series as $n \to \infty$. From the recursion formula of Eq. 8.32 we can tell by inspection that

$$\lim_{n\to\infty} \frac{A_{n+2}}{A_n} = \frac{2}{n}$$

Since

$$e^z = 1 + z + \frac{z^2}{2!} + \frac{z^3}{3!} + \cdots$$

we can express $e^{y^2/2}$ in a power series as

$$e^{y^2/2} = 1 + \frac{y^2}{2} + \frac{y^4}{2^2 \cdot 2!} + \frac{y^6}{2^3 \cdot 3!} + \cdots$$

$$= \sum_{n=0,2,4,\ldots}^{\infty} \frac{1}{2^{n/2}\left(\dfrac{n}{2}\right)!} y^n$$

$$= \sum_{n=0,2,4,\ldots}^{\infty} B_n y^n$$

The ratio between successive coefficients of y^n here is

$$\frac{B_{n+2}}{B_n} = \frac{2^{n/2}\left(\dfrac{n}{2}\right)!}{2^{(n+2)/2}\left(\dfrac{n+2}{2}\right)!} = \frac{2^{n/2}\left(\dfrac{n}{2}\right)!}{2\cdot 2^{n/2}\left(\dfrac{n}{2}+1\right)\left(\dfrac{n}{2}\right)!}$$

$$= \frac{1}{2\left(\dfrac{n}{2}+1\right)} = \frac{1}{n+2}$$

In the limit of $n \to \infty$ this ratio becomes

$$\lim_{n\to\infty} \frac{B_{n+2}}{B_n} = \frac{1}{n}$$

Thus successive coefficients in the power series for f decrease *less* rapidly than those in the power series for $e^{y^2/2}$ instead of *more* rapidly, which means that $f(y)e^{-y^2/2}$ does not vanish as $y \to \infty$.

There is a simple way out of this dilemma. If the series representing f terminates at a certain value of n, so that all the coefficients A_n are zero for values of n higher than this one, ψ will go to zero as $y \to \infty$ because of the $e^{-y^2/2}$ factor. In other words, if f is a polynomial with a finite number of terms instead of an infinite series, it is acceptable. From the recursion formula

$$A_{n+2} = \frac{2n + 1 - \alpha}{(n + 2)(n + 1)} A_n$$

it is clear that if

8.33
$$\alpha = 2n + 1$$

for any value of n, then $A_{n+2} = A_{n+4} = A_{n+6} = \cdots = 0$, which is what we want.

(Equation 8.33 takes care of only one sequence of coefficients, either the sequence of even n starting with A_0 or the sequence of odd n starting with A_1. If n is even, it must be true that $A_1 = 0$ and only even powers of y appear in the polynomial, while if n is odd, it must be true that $A_0 = 0$ and only odd powers of y appear. We shall see the result in the next section, where the polynomial is tabulated for various values of n.)

The condition that $\alpha = 2n + 1$ is a necessary and sufficient condition for the wave equation 8.23 to have solutions that meet the various requirements that ψ must fulfill. From Eq. 8.22, the definition of α, we have

$$\alpha_n = \frac{2E}{h\nu} = 2n + 1$$

or

8.34
$$E_n = \left(n + \frac{1}{2}\right)h\nu \qquad n = 0, 1, 2, \ldots \qquad \text{Energy levels of harmonic oscillator}$$

The energy of a harmonic oscillator is thus quantized in steps of $h\nu$, where ν is the classical frequency of oscillation and h is Planck's constant. The energy levels here are evenly spaced (Fig. 8.6), unlike the energy levels of a particle in a box whose spacing diverges. We note that, when $n = 0$,

8.35
$$E_0 = \tfrac{1}{2}h\nu \qquad \text{Zero-point energy}$$

which is the lowest value the energy of the oscillator can have. This value is called the *zero-point energy* because a harmonic oscillator in equilibrium with its surroundings would approach an energy of $E = E_0$ and not $E = 0$ as the temperature approaches $0°K$.

8.7 The Harmonic Oscillator: Wave Functions

For each choice of the parameter α_n there is a different wave function ψ_n. Each function consists of a polynomial $H_n(y)$ (called a *Hermite polynomial*) in either odd or even powers

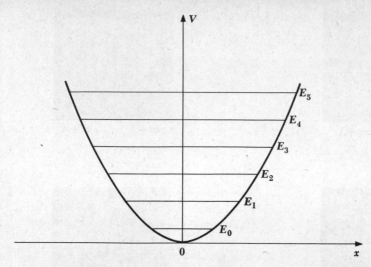

FIGURE 8.6 Energy levels of a harmonic oscillator according to quantum mechanics.

of y, the exponential factor $e^{-y^2/2}$, and a numerical coefficient which is needed for ψ_n to meet the normalization condition

$$\int_{-\infty}^{\infty} |\psi_n|^2 \, dy = 1 \qquad n = 0, 1, 2, \dots$$

The general formula for the nth wave function is

8.36
$$\psi_n = \left(\frac{2m\nu}{\hbar}\right)^{1/4} (2^n n!)^{-1/2} H_n(y) e^{-y^2/2}$$

The first six Hermite polynomials $H_n(y)$ are listed in Table 8.1, and the corresponding wave functions ψ_n are plotted in Fig. 8.7. In each case the range to which a particle oscillat-

TABLE 8.1. SOME HERMITE POLYNOMIALS

n	$H_n(y)$	α_n	E_n
0	1	1	$\frac{1}{2}h\nu$
1	$2y$	3	$\frac{3}{2}h\nu$
2	$4y^2 - 2$	5	$\frac{5}{2}h\nu$
3	$8y^3 - 12y$	7	$\frac{7}{2}h\nu$
4	$16y^4 - 48y^2 + 12$	9	$\frac{9}{2}h\nu$
5	$32y^5 - 160y^3 + 120y$	11	$\frac{11}{2}h\nu$

The Harmonic Oscillator: Wave Functions

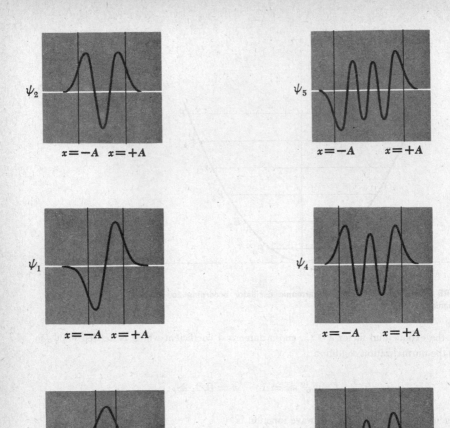

FIGURE 8.7 The first six harmonic-oscillator wave functions. The vertical lines show the limits $-A$ and $+A$ between which a classical oscillator with the same energy would vibrate.

ing classically with the same total energy E_n would be confined is indicated; evidently the particle is able to penetrate into classically forbidden regions—in other words, to exceed the amplitude A determined by the energy—with an exponentially decreasing probability, just as in the situation of a particle in a box with nonrigid walls.

It is interesting and instructive to compare the probability densities of a classical harmonic oscillator and a quantum-mechanical harmonic oscillator of the same energy. The upper graph of Fig. 8.8 shows this density for the classical oscillator: The probability P of

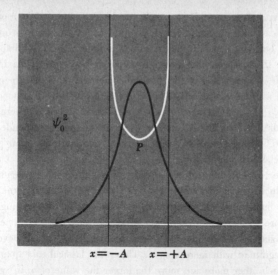

ψ_0^2

P

$x=-A$ $x=+A$

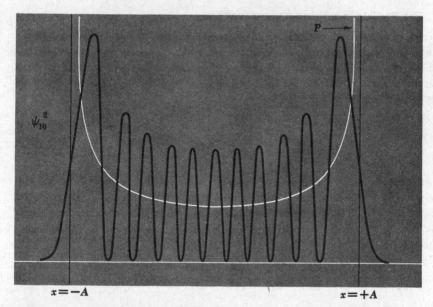

ψ_{10}^2

P

$x=-A$ $x=+A$

FIGURE 8.8 Probability densities for the $n = 0$ and $n = 10$ states of a quantum-mechanical harmonic oscillator. The probability densities for classical harmonic oscillators with the same energies are shown in white.

finding the particle at a given position is greatest at the end-points of its motion, where it moves slowly, and least near the equilibrium position ($x = 0$), where it moves rapidly. Exactly the opposite behavior is manifested by a quantum-mechanical oscillator in its lowest energy state of $n = 0$. As shown, the probability density $|\psi_0|^2$ has its maximum value at $x = 0$ and drops off on either side of this position. However, this disagreement becomes less and less marked with increasing n: The lower graph of Fig. 8.8 corresponds to $n = 10$, and it is clear that $|\psi_{10}|^2$ when averaged over x has approximately the general character of the classical probability P. This is another example of the correspondence principle mentioned in Sec. 6.6: In the limit of large quantum numbers, quantum physics yields the same results as classical physics.

It might be objected that, although $|\psi_{10}|^2$ does indeed approach P when smoothed out, nevertheless $|\psi_{10}|^2$ fluctuates rapidly with x whereas P does not. However, this objection has meaning only if the fluctuations are observable, and the smaller the spacing of the peaks and hollows, the more strongly the uncertainty principle prevents their detection without altering the physical state of the oscillator. The exponential "tails" of $|\psi|^2$ beyond $x = \pm A$ also decrease in magnitude with increasing n. Thus the classical and quantum pictures begin to resemble each other more and more the larger the value of n, in agreement with the correspondence principle, although they are radically different for small n.

8.8   The Particle in a Three-dimensional Box

Although a number of interesting physical problems are limited to one dimension, others—which include the fundamental problem of atomic structure—involve three dimensions. By examining the case of a particle confined to a three-dimensional box, we shall be introduced to the manner in which Schrödinger's equation in three dimensions is solved and to the general character of the solutions.

In three dimensions, the steady-state form of Schrödinger's equation is

8.37
$$\frac{\partial^2 \psi}{\partial x^2} + \frac{\partial^2 \psi}{\partial y^2} + \frac{\partial^2 \psi}{\partial z^2} + \frac{2m}{\hbar^2}(E - V)\,\psi = 0$$

The simplest box for our purposes is an extension of the one-dimensional box of Fig. 8.1, namely a cube L long on each side with infinitely hard walls which are parallel to the coordinate axes (Fig. 8.9). Such a box corresponds to a potential energy function $V(x,y,z)$ whose value is $V = 0$ inside the box and $V = \infty$ outside it. Inside the box Schrödinger's equation is

8.38
$$\frac{\partial^2 \psi}{\partial x^2} + \frac{\partial^2 \psi}{\partial y^2} + \frac{\partial^2 \psi}{\partial z^2} + \frac{2m}{\hbar^2} E\psi = 0$$

which we must solve for the wave function ψ subject to the boundary condition that $\psi = 0$ at the sides of the box.

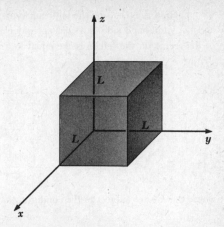

FIGURE 8.9 A cubical box.

Equation 8.38 involves all three coordinates x, y, z. In order to obtain a solution, we must first separate it into three independent equations, each of which involves only one coordinate. The procedure is to assume that the wave function $\psi(x,y,z)$ is really the product of three different functions: $\psi_x(x)$, which depends upon x alone; $\psi_y(y)$, which depends upon y alone; and $\psi_z(z)$, which depends upon z alone. That is,

8.39
$$\psi(x,y,z) = \psi_x(x)\,\psi_y(y)\,\psi_z(z)$$

This assumption is reasonable, since all it suggests is that the variation of ψ with each coordinate is independent of its variation with the others. Taking partial derivatives of $\psi = \psi_x\psi_y\psi_z$,

$$\frac{\partial^2\psi}{\partial x^2} = \psi_y\,\psi_z\,\frac{d^2\psi_x}{dx^2}$$

$$\frac{\partial^2\psi}{\partial y^2} = \psi_x\,\psi_z\,\frac{d^2\psi_y}{dy^2}$$

$$\frac{\partial^2\psi}{\partial z^2} = \psi_x\,\psi_y\,\frac{d^2\psi_z}{dz^2}$$

Now we substitute these partial derivatives together with $\psi = \psi_x\psi_y\psi_z$ into Eq. 8.38 and obtain

$$\psi_y\psi_z\frac{d^2\psi_x}{dx^2} + \psi_x\psi_z\frac{d^2\psi_y}{dy^2} + \psi_x\psi_y\frac{d^2\psi_z}{dz^2} + \frac{2mE}{\hbar^2}\psi_x\psi_y\psi_z = 0$$

Dividing through by $\psi_x\psi_y\psi_z$ and rearranging terms yields

8.40
$$\frac{1}{\psi_x}\frac{d^2\psi_x}{dz^2} + \frac{1}{\psi_y}\frac{d^2\psi_y}{dy^2} + \frac{1}{\psi_z}\frac{d^2\psi_z}{dz^2} = -\frac{2mE}{\hbar^2}$$

The Particle in a Three-dimensional Box

Each of the terms on the left-hand side of Eq. 8.40 is a function of a different variable, and the right-hand side is a constant independent of the values of x, y, z. Therefore each of the terms on the left must be equal to a separate constant, which we can represent by writing

8.41
$$\frac{1}{\psi_x} \frac{d^2 \psi_x}{dx^2} = -k_x^2$$

8.42
$$\frac{1}{\psi_y} \frac{d^2 \psi_y}{dy^2} = -k_y^2$$

8.43
$$\frac{1}{\psi_z} \frac{d^2 \psi_z}{dz^2} = -k_z^2$$

where the k's are subject to the condition that

8.44
$$k_x^2 + k_y^2 + k_z^2 = \frac{2mE}{\hbar^2}$$

Equations 8.41 to 8.43 are ordinary differential equations of exactly the same form as Eq. 8.1, and as in the case of Eq. 8.1 they can have sine and cosine solutions. The boundary conditions on ψ state that $\psi = 0$ at the sides of the box, that is, when $x = 0$ or L, $y = 0$ or L, and $z = 0$ or L. Only the sine function is 0 at the origin as required, so this is the proper function to use in the solutions. We therefore have for the complete wave function

8.45
$$\psi = \psi_x \psi_y \psi_z = A \sin k_x x \sin k_y y \sin k_z z$$

where A is a normalization constant whose value must be such that $\int |\psi|^2 \, dV = 1$.

We have assured that $\psi = 0$ at the origin by choosing the sine functions, and now we must determine the values of k_x, k_y, and k_z that will result in $\psi = 0$ at x, y, $z = L$. These values are

$$k_x L = n_x \pi \qquad n_x = 1, 2, 3, \ldots$$
$$k_y L = n_y \pi \qquad n_y = 1, 2, 3, \ldots$$
$$k_z L = n_z \pi \qquad n_z = 1, 2, 3, \ldots$$

The wave functions of a particle in a cubical box are accordingly given by the formula

8.46
$$\psi = A \sin \frac{n_x \pi x}{L} \sin \frac{n_y \pi y}{L} \sin \frac{n_z \pi z}{L} \qquad \begin{array}{l} n_x = 1, 2, 3, \ldots \\ n_y = 1, 2, 3, \ldots \\ n_z = 1, 2, 3, \ldots \end{array}$$

and the possible energies of the particle are

8.47
$$E = (n_x^2 + n_y^2 + n_z^2) \frac{\pi^2 \hbar^2}{2mL^2}$$

Three quantum numbers, one corresponding to each coordinate, are required here to specify each state of the particle, instead of the single quantum number required in the case of a particle in a one-dimensional box. This is a general characteristic of three-dimensional problems, and we will encounter it again when we take up the theory of the hydrogen atom in the next chapter.

An interesting aspect of Eq. 8.47 is that a single energy level may be associated with more than one quantum state. For example, the confined particle has an energy of $9\pi^2\hbar^2/2mL^2$ in each of the following three quantum states:

$$
\begin{Bmatrix} n_x = 2 \\ n_y = 2 \\ n_z = 1 \end{Bmatrix}
\qquad
\begin{Bmatrix} n_x = 1 \\ n_y = 2 \\ n_z = 2 \end{Bmatrix}
\qquad
\begin{Bmatrix} n_x = 2 \\ n_y = 1 \\ n_z = 2 \end{Bmatrix}
$$

An energy level of this kind is said to be *degenerate*. The above energy level is threefold degenerate because the three different wave functions

$$
\psi_{221} = A \sin \frac{2\pi x}{L} \sin \frac{2\pi y}{L} \sin \frac{\pi z}{L}
$$

$$
\psi_{122} = A \sin \frac{\pi x}{L} \sin \frac{2\pi y}{L} \sin \frac{2\pi z}{L}
$$

$$
\psi_{212} = A \sin \frac{2\pi x}{L} \sin \frac{\pi y}{L} \sin \frac{2\pi z}{L}
$$

all describe states with that energy. The states are physically different in other respects: the momentum eigenvalues of the various states are not the same, for instance.

Problems

1. Find the lowest energy of a neutron confined to a box 10^{-14} m across. (The size of a nucleus is of this order of magnitude.)

2. According to the correspondence principle, quantum theory should give the same results as classical physics in the limit of large quantum numbers. Show that, as $n \rightarrow \infty$, the probability of finding a particle trapped in a box between x and $x + dx$ is independent of x, which is the classical expectation.

3. Find the energies of the five lowest energy levels of a particle in a cubical box. Which of these levels is degenerate?

4. Find the value of the normalization constant A in Eq. 8.46 for the wave function of a particle in a cubical box when $n_x = 1$, $n_y = 2$, $n_z = 3$.

5. An electron is confined in a cubical box 1 A long on each edge. At what temperature would the average energy of the molecules of an ideal gas be the same as the lowest energy this electron can possess?

6. Use the fact that $\alpha \geq 0$ (since $E \geq 0$) to show that the coefficients A_n of Eq. 8.28 are all zero for negative values of n.

7. Show that the first three harmonic-oscillator wave functions are normalized solutions of Schrödinger's equation.

8. Find the zero-point energy in electron volts of a pendulum whose period is 1 sec.

9. Show that the expectation values $\overline{T}$ and $\overline{V}$ of the kinetic and potential energies of a harmonic oscillator are given by $\overline{T} = \overline{V} = E_0/2$ when it is in the $n = 0$ state. (This is true for all states of a harmonic oscillator, in fact.) How does this result compare with the classical values of T and V?

10. An important property of the eigenfunctions of an operator is that they are *orthogonal* to one another, which means that

$$\int_{-\infty}^{\infty} \psi_n \psi_m \, dV = 0 \qquad n \neq m$$

Verify this relationship for the eigenfunctions of a particle in a one-dimensional box with the help of the relationship $\sin \theta = (e^{i\theta} - e^{-i\theta})/2i$.

11. Show that the harmonic-oscillator eigenfunctions are orthogonal.

12. Consider a particle of mass m trapped in a two-dimensional box L long and W wide. Starting from Schrödinger's equation, show that the permitted energies of the particle are given by

$$E = \frac{h^2}{8m} \left(\frac{a^2}{L^2} + \frac{b^2}{W^2} \right)$$

where a and b are positive integers.

Chapter 9
Quantum Theory
of the
Hydrogen
Atom

The quantum-mechanical theory of the atom, which was developed shortly after the formulation of quantum mechanics itself, represents an epochal contribution to our knowledge of the physical universe. Besides revolutionizing our approach to atomic phenomena, this theory has made it possible for us to understand such related matters as how atoms interact with one another to form stable molecules, the origin of the periodic table of the elements, and why solids are endowed with their characteristic electrical, magnetic, and mechanical properties, all topics we shall explore in later chapters. For the moment we shall concentrate on the quantum theory of the hydrogen atom and how its formal mathematical results may be interpreted in terms of familiar concepts.

9.1 Schrödinger's Equation for the Hydrogen Atom

A hydrogen atom consists of a proton, a particle of electric charge $+e$, and an electron, a particle of charge $-e$ which is 1,836 times lighter than the proton. For the sake of convenience we shall consider the proton to be stationary, with the electron moving about in its vicinity but prevented from escaping because of the proton's electric field. (As in the Bohr theory, the correction for proton motion is simply a matter of replacing the electron mass m by the reduced mass m'.) Schrödinger's equation for the electron in three dimensions, which is what we must use for the hydrogen atom, is

$$\nabla^2\psi + \frac{2m}{\hbar^2}(E - V)\psi = 0$$

which in cartesian coordinates is

9.1
$$\frac{\partial^2\psi}{\partial x^2} + \frac{\partial^2\psi}{\partial y^2} + \frac{\partial^2\psi}{\partial z^2} + \frac{8\pi^2 m}{h^2}(E - V)\psi = 0$$

The potential energy V here is the electrostatic potential energy

9.2
$$V = -\frac{e^2}{4\pi\varepsilon_0 r}$$

of a charge $-e$ when it is the distance r from another charge $+e$. Since V is a function of r rather than of x, y, z, we cannot substitute Eq. 9.2 directly into Eq. 9.1. There are two alternatives: we can express V in terms of the cartesian coordinates x, y, z by replacing r by $\sqrt{x^2 + y^2 + z^2}$, or we can express Schrödinger's equation in terms of the spherical polar coordinates r, θ, ϕ defined in Fig. 9.1. As it happens, owing to the symmetry of the physical situation, doing the latter makes the problem considerably easier to solve.

The spherical polar coordinates r, θ, ϕ of the point P shown in Fig. 9.1 have the following interpretations:

$r =$ length of radius vector from origin O to point P

$\quad = \sqrt{x^2 + y^2 + z^2}$

$\theta =$ angle between radius vector and $+z$ axis

$\quad =$ zenith angle

$\quad = \cos^{-1} \dfrac{z}{\sqrt{x^2 + y^2 + z^2}}$ **Spherical polar coordinates**

$\phi =$ angle between the projection of the radius vector in the xy plane and the $+x$ axis, measured in the direction shown

$\quad =$ azimuth angle

$\quad = \tan^{-1} \dfrac{y}{x}$

On the surface of a sphere whose center is at O, lines of constant zenith angle θ are like

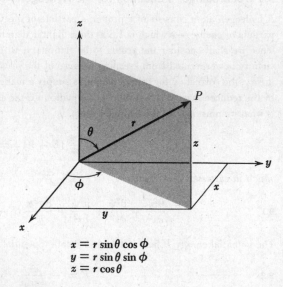

$$x = r \sin\theta \cos\phi$$
$$y = r \sin\theta \sin\phi$$
$$z = r \cos\theta$$

FIGURE 9.1 Spherical polar coordinates.

parallels of latitude on a globe (but we note that the value of θ of a point is *not* the same as its latitude; $\theta = 90°$ at the equator, for instance, but the latitude of the equator is $0°$), and lines of constant azimuth angle ϕ are like meridians of longitude (here the definitions coincide if the axis of the globe is taken as the $+z$ axis and the $+x$ axis is at $\phi = 0°$).

In spherical polar coordinates Schrödinger's equation becomes

9.3 $$\frac{1}{r^2}\frac{\partial}{\partial r}\left(r^2\frac{\partial\psi}{\partial r}\right) + \frac{1}{r^2\sin\theta}\frac{\partial}{\partial\theta}\left(\sin\theta\frac{\partial\psi}{\partial\theta}\right) + \frac{1}{r^2\sin^2\theta}\frac{\partial^2\psi}{\partial\phi^2} + \frac{2m}{\hbar^2}(E - V)\psi = 0$$

Substituting Eq. 9.2 for the potential energy V and multiplying the entire equation by $r^2\sin^2\theta$, we obtain

9.4 $$\sin^2\theta\frac{\partial}{\partial r}\left(r^2\frac{\partial\psi}{\partial r}\right) + \sin\theta\frac{\partial}{\partial\theta}\left(\sin\theta\frac{\partial\psi}{\partial\theta}\right) + \frac{\partial^2\psi}{\partial\phi^2}$$

$$+ \frac{2mr^2\sin^2\theta}{\hbar^2}\left(\frac{e^2}{4\pi\varepsilon_0 r} + E\right)\psi = 0 \qquad \text{Hydrogen atom}$$

Equation 9.4 is the partial differential equation for the wave function ψ of the electron in a hydrogen atom. Together with the various conditions ψ must obey, as discussed in Chap. 7 (for instance, that ψ have just one value at each point r, θ, ϕ), this equation completely specifies the behavior of the electron. In order to see just what this behavior is, we must solve Eq. 9.4 for ψ.

9.2 Separation of Variables

The virtue of writing Schrödinger's equation in spherical polar coordinates for the problem of the hydrogen atom is that in this form it may be readily separated into three independent equations, each involving only a single coordinate. We follow essentially the same procedure we used in Chap. 8 in treating the problem of a particle in a three-dimensional box. Here we look for solutions in which the wave function $\psi(r, \theta, \phi)$ has the form of a product of three different functions: $R(r)$, which depends upon r alone; $\Theta(\theta)$, which depends upon θ alone; and $\Phi(\phi)$, which depends upon ϕ alone. That is, we assume that

9.5 $$\psi(r,\theta,\phi) = R(r)\Theta(\theta)\Phi(\phi) \qquad \text{Hydrogen atom wave function}$$

The function $R(r)$ describes how the wave function ψ of the electron varies along a radius vector from the nucleus, with θ and ϕ constant. The function $\Theta(\theta)$ describes how ψ varies with zenith angle θ along a meridian on a sphere centered at the nucleus, with r and ϕ constant. The function $\Phi(\phi)$ describes how ψ varies with azimuth angle ϕ along a parallel on a sphere centered at the nucleus, with r and θ constant.

From Eq. 9.5, which we may write more simply as

$$\psi = R\Theta\Phi$$

we see that

$$\frac{\partial \psi}{\partial r} = \Theta \Phi \frac{\partial R}{\partial r}$$

$$\frac{\partial \psi}{\partial \theta} = R \Phi \frac{\partial \Theta}{\partial \theta}$$

$$\frac{\partial^2 \psi}{\partial \phi^2} = R \Theta \frac{\partial^2 \Phi}{\partial \phi^2}$$

Hence, when we substitute $R\Theta\Phi$ for ψ in Schrödinger's equation for the hydrogen atom and divide the entire equation by $R\Theta\Phi$, we find that

9.6
$$\frac{\sin^2 \theta}{R} \frac{\partial}{\partial r}\left(r^2 \frac{\partial R}{\partial r}\right) + \frac{\sin \theta}{\Theta} \frac{\partial}{\partial \theta}\left(\sin \theta \frac{\partial \Theta}{\partial \theta}\right)$$
$$+ \frac{1}{\Phi} \frac{\partial^2 \Phi}{\partial \phi^2} + \frac{2mr^2 \sin^2 \theta}{\hbar^2}\left(\frac{e^2}{4\pi\varepsilon_0 r} + E\right) = 0$$

The third term of Eq. 9.6 is a function of azimuth angle ϕ only, while the other terms are functions of R and θ only. Let us rearrange Eq. 9.6 to read

9.7
$$\frac{\sin^2 \theta}{R} \frac{\partial}{\partial r}\left(r^2 \frac{\partial R}{\partial r}\right) + \frac{\sin \theta}{\Theta} \frac{\partial}{\partial \theta}\left(\sin \theta \frac{\partial \Theta}{\partial \theta}\right)$$
$$+ \frac{2mr^2 \sin^2 \theta}{\hbar^2}\left(\frac{e^2}{4\pi\varepsilon_0 r} + E\right) = -\frac{1}{\Phi} \frac{\partial^2 \Phi}{\partial \phi^2}$$

This equation can be correct only if both sides of it are equal to the same constant, since they are functions of *different* variables. As we shall see, it is convenient to call this constant m_l^2. The differential equation for the function Φ is therefore

9.8
$$-\frac{1}{\Phi} \frac{d^2 \Phi}{d\phi^2} = m_l^2$$

When we substitute m_l^2 for the right-hand side of Eq. 9.7, divide the entire equation by $\sin^2 \theta$, and rearrange the various terms, we find that

9.9
$$\frac{1}{R} \frac{\partial}{\partial r}\left(r^2 \frac{\partial R}{\partial r}\right) + \frac{2mr^2}{\hbar^2}\left(\frac{e^2}{4\pi\varepsilon_0 r} + E\right)$$
$$= \frac{m_l^2}{\sin^2 \theta} - \frac{1}{\Theta \sin \theta} \frac{\partial}{\partial \theta}\left(\sin \theta \frac{\partial \Theta}{\partial \theta}\right)$$

Again we have an equation in which different variables appear on each side, requiring that both sides be equal to the same constant. This constant we shall call $l(l + 1)$, once more for reasons that will be apparent later. The equations for the functions Θ and R are therefore

9.10
$$\frac{m_l^2}{\sin^2\theta} - \frac{1}{\Theta\sin\theta}\frac{d}{d\theta}\left(\sin\theta\frac{d\Theta}{d\theta}\right) = l(l+1)$$

9.11
$$\frac{1}{R}\frac{d}{dr}\left(r^2\frac{dR}{dr}\right) + \frac{2mr^2}{\hbar^2}\left(\frac{e^2}{4\pi\varepsilon_0 r} + E\right) = l(l+1)$$

Equations 9.8, 9.10, and 9.11 are usually written

9.12
$$\frac{d^2\Phi}{d\phi^2} + m_l^2\Phi = 0$$

9.13
$$\frac{1}{\sin\theta}\frac{d}{d\theta}\left(\sin\theta\frac{d\Theta}{d\theta}\right) + \left[l(l+1) - \frac{m_l^2}{\sin^2\theta}\right]\Theta = 0$$

9.14
$$\frac{1}{r^2}\frac{d}{dr}\left(r^2\frac{dR}{dr}\right) + \left[\frac{2m}{\hbar^2}\left(\frac{e^2}{4\pi\varepsilon_0 r} + E\right) - \frac{l(l+1)}{r^2}\right]R = 0$$

Each of these is an ordinary differential equation for a single function of a single variable. We have therefore accomplished our task of simplifying Schrödinger's equation for the hydrogen atom, which began as a partial differential equation for a function ψ of three variables.

9.3   Quantum Numbers

The first of the above equations, Eq. 9.12, is readily solved, with the result

9.15
$$\Phi(\phi) = Ae^{im_l\phi}$$

where A is the constant of integration. We have already stated that one of the conditions a wave function—and hence Φ, which is a component of the complete wave function ψ—must obey is that it have a single value at a given point in space. From Fig. 9.2 it is evident that ϕ and $\phi + 2\pi$ both identify the same meridian plane. Hence it must be true that $\Phi(\phi) = \Phi(\phi + 2\pi)$, or

$$Ae^{im_l\phi} = Ae^{im_l(\phi + 2\pi)}$$

which can only happen when m_l is 0 or a positive or negative integer ($\pm 1, \pm 2, \pm 3, \ldots$). The constant m_l is known as the *magnetic quantum number* of the hydrogen atom.

The differential equation 9.13 for $\Theta(\theta)$ has a rather complicated solution in terms of polynomials called the *associated Legendre functions*. For our present purpose, the important thing about these functions is that they exist only when the constant l is an integer equal to or greater than $|m_l|$, the absolute value of m_l. This requirement can be expressed as a condition on m_l in the form

$$m_l = 0, \pm 1, \pm 2, \ldots, \pm l$$

The constant l is known as the *orbital quantum number*.

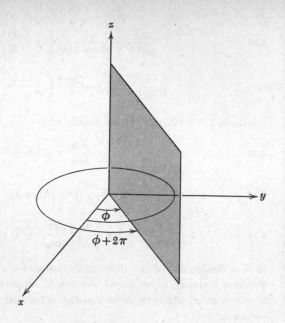

FIGURE 9.2 The angles ϕ and $\phi +$ 2π both identify the same meridian plane.

The solution of the final equation, Eq. 9.14, for the radial part $R(r)$ of the hydrogen-atom wave function ψ is also complicated, being in terms of polynomials called the *associated Laguerre functions*. Equation 9.14 can be solved only when E is positive or has one of the negative values E_n (signifying that the electron is bound to the atom) specified by

9.16
$$E_n = - \frac{me^4}{32\pi^2 \varepsilon_0^2 \hbar^2} \left(\frac{1}{n^2} \right)$$

where n is an integer. We recognize that this is precisely the same formula for the energy levels of the hydrogen atom that Bohr obtained.

Another condition that must be obeyed in order to solve Eq. 9.14 is that n, known as the *total quantum number*, must be equal to or greater than $l + 1$. This requirement may be expressed as a condition on l in the form

$$l = 0, 1, 2, \ldots, (n - 1)$$

Hence we may tabulate the three quantum numbers n, l, and m together with their permissible values as follows:

9.17

$$n = 1, 2, 3, \ldots \qquad \text{Total quantum number}$$

$$l = 0, 1, 2, \ldots, (n - 1) \qquad \text{Orbital quantum number}$$

$$m_l = 0, \pm 1, \pm 2, \ldots, \pm l \qquad \text{Magnetic quantum number}$$

The theory of a particle in a box considered in the previous chapter showed that three quantum numbers arise naturally in the case of three-dimensional motion, and so it is not surprising that three quantum numbers should also be needed to describe the three-dimensional motion of the electron in a hydrogen atom. It is worth noting again how inevitably quantum numbers appear in quantum-mechanical theories of particles trapped in a particular region of space.

To exhibit the dependence of R, Θ, and Φ upon the quantum numbers n, l, m, we may write for the electron wave function

9.18
$$\psi = R_{nl}\Theta_{lm_l}\Phi_{m_l}$$

The wave functions R, Θ, and Φ together with ψ are given in Table 9.1 for $n = 1, 2,$ and 3.

9.4   Total Quantum Number

It is interesting to consider the interpretation of the hydrogen-atom quantum numbers in terms of the classical model of the atom. This model, as we saw in Chap. 5, corresponds exactly to planetary motion in the solar system except that the inverse-square force holding the electron to the nucleus is electrical rather than gravitational. Two quantities are *conserved*—that is, maintain a constant value at all times—in planetary motion, as Newton was able to show from Kepler's three empirical laws. These are the scalar *total energy* and the vector *angular momentum* of each planet. Classically the total energy can have any value whatever, but it must, of course, be negative if the planet is to be trapped permanently in the solar system. In the quantum-mechanical theory of the hydrogen atom the electron energy is also a constant, but while it may have any positive value whatever, the *only* negative values it can have are specified by the formula

(9.16)
$$E_n = -\frac{me^4}{32\pi\varepsilon_0{}^2\hbar^2}\left(\frac{1}{n^2}\right)$$

The theory of planetary motion can also be worked out from Schrödinger's equation, and it yields an energy restriction identical in form to Eq. 9.16. However, the total quantum number n for any of the planets turns out to be so immense that the separation of permitted energy levels is far too small to be observable. For this reason classical physics provides an adequate description of planetary motion but fails within the atom.

The quantization of electron energy in the hydrogen atom is therefore described by the total quantum number n.

9.5   Orbital Quantum Number

The interpretation of the orbital quantum number l is a bit less obvious. Let us examine the differential equation for the radial part $R(r)$ of the wave function ψ:

(9.14)
$$\frac{1}{r^2}\frac{d}{dr}\left(r^2\frac{dR}{dr}\right) + \left[\frac{2m}{\hbar^2}\left(\frac{e^2}{4\pi\varepsilon r} + E\right) - \frac{l(l+1)}{r^2}\right]R = 0$$

Table 9.1. Normalized wave functions of the hydrogen atom for n = 1, 2, and 3

The quantity $a_0 = \hbar^2/me^2 = 0.53$ Å is equal to the radius of the innermost Bohr orbit

n	l	m_l	$\Phi(\phi)$	$\Theta(\theta)$	$R(r)$	$\psi\,(r,\theta,\phi)$
1	0	0	$\dfrac{1}{\sqrt{2\pi}}$	$\dfrac{1}{\sqrt{2}}$	$\dfrac{2}{a_0^{3/2}}\,e^{-r/a_0}$	$\dfrac{1}{\sqrt{\pi}\,a_0^{3/2}}\,e^{-r/a_0}$
2	0	0	$\dfrac{1}{\sqrt{2\pi}}$	$\dfrac{1}{\sqrt{2}}$	$\dfrac{1}{2\sqrt{2}\,a_0^{3/2}}\left(2-\dfrac{r}{a_0}\right)e^{-r/2a_0}$	$\dfrac{1}{4\sqrt{2\pi}\,a_0^{3/2}}\left(2-\dfrac{r}{a_0}\right)e^{-r/2a_0}$
2	1	0	$\dfrac{1}{\sqrt{2\pi}}$	$\dfrac{\sqrt{6}}{2}\cos\theta$	$\dfrac{1}{2\sqrt{6}\,a_0^{3/2}}\dfrac{r}{a_0}\,e^{-r/2a_0}$	$\dfrac{1}{4\sqrt{2\pi}\,a_0^{3/2}}\dfrac{r}{a_0}\,e^{-r/2a_0}\cos\theta$
2	1	±1	$\dfrac{1}{\sqrt{2\pi}}e^{\pm i\phi}$	$\dfrac{\sqrt{3}}{2}\sin\theta$	$\dfrac{1}{2\sqrt{6}\,a_0^{3/2}}\dfrac{r}{a_0}\,e^{-r/2a_0}$	$\dfrac{1}{8\sqrt{\pi}\,a_0^{3/2}}\dfrac{r}{a_0}\,e^{-r/2a_0}\sin\theta\,e^{\pm i\phi}$
3	0	0	$\dfrac{1}{\sqrt{2\pi}}$	$\dfrac{1}{\sqrt{2}}$	$\dfrac{2}{81\sqrt{3}\,a_0^{3/2}}\left(27-18\dfrac{r}{a_0}+2\dfrac{r^2}{a_0^2}\right)e^{-r/3a_0}$	$\dfrac{1}{81\sqrt{3\pi}\,a_0^{3/2}}\left(27-18\dfrac{r}{a_0}+2\dfrac{r^2}{a_0^2}\right)e^{-r/3a_0}$
3	1	0	$\dfrac{1}{\sqrt{2\pi}}$	$\dfrac{\sqrt{6}}{2}\cos\theta$	$\dfrac{4}{81\sqrt{6}\,a_0^{3/2}}\left(6-\dfrac{r}{a_0}\right)\dfrac{r}{a_0}\,e^{-r/3a_0}$	$\dfrac{\sqrt{2}}{81\sqrt{\pi}\,a_0^{3/2}}\left(6-\dfrac{r}{a_0}\right)\dfrac{r}{a_0}\,e^{-r/3a_0}\cos\theta$
3	1	±1	$\dfrac{1}{\sqrt{2\pi}}e^{\pm i\phi}$	$\dfrac{\sqrt{3}}{2}\sin\theta$	$\dfrac{4}{81\sqrt{6}\,a_0^{3/2}}\left(6-\dfrac{r}{a_0}\right)\dfrac{r}{a_0}\,e^{-r/3a_0}$	$\dfrac{1}{81\sqrt{\pi}\,a_0^{3/2}}\left(6-\dfrac{r}{a_0}\right)\dfrac{r}{a_0}\,e^{-r/3a_0}\sin\theta\,e^{\pm i\phi}$
3	2	0	$\dfrac{1}{\sqrt{2\pi}}$	$\dfrac{\sqrt{10}}{4}(3\cos^2\theta-1)$	$\dfrac{4}{81\sqrt{30}\,a_0^{3/2}}\dfrac{r^2}{a_0^2}\,e^{-r/3a_0}$	$\dfrac{1}{81\sqrt{6\pi}\,a_0^{3/2}}\dfrac{r^2}{a_0^2}\,e^{-r/3a_0}(3\cos^2\theta-1)$
3	2	±1	$\dfrac{1}{\sqrt{2\pi}}e^{\pm i\phi}$	$\dfrac{\sqrt{15}}{2}\sin\theta\cos\theta$	$\dfrac{4}{81\sqrt{30}\,a_0^{3/2}}\dfrac{r^2}{a_0^2}\,e^{-r/3a_0}$	$\dfrac{1}{81\sqrt{\pi}\,a_0^{3/2}}\dfrac{r^2}{a_0^2}\,e^{-r/3a_0}\sin\theta\cos\theta\,e^{\pm i\phi}$
3	2	±2	$\dfrac{1}{\sqrt{2\pi}}e^{\pm 2i\phi}$	$\dfrac{\sqrt{15}}{4}\sin^2\theta$	$\dfrac{4}{81\sqrt{30}\,a_0^{3/2}}\dfrac{r^2}{a_0^2}\,e^{-r/3a_0}$	$\dfrac{1}{162\sqrt{\pi}\,a_0^{3/2}}\dfrac{r^2}{a_0^2}\,e^{-r/3a_0}\sin^2\theta\,e^{\pm 2i\phi}$

This equation is solely concerned with the radial aspect of the electron's motion, that is, its motion toward or away from the nucleus; yet we notice the presence of E, the total electron energy, in it. The total energy E includes the electron's kinetic energy of orbital motion, which should have nothing to do with its radial motion. This contradiction may be removed by the following argument. The kinetic energy T of the electron has two parts, T_{radial} due to its motion toward or away from the nucleus, and $T_{orbital}$ due to its motion around the nucleus. The potential energy V of the electron is the electrostatic energy

$$V = - \frac{e^2}{4\pi\varepsilon_0 r}$$

Hence the total energy of the electron is

$$E = T_{radial} + T_{orbital} + V$$
$$= T_{radial} + T_{orbital} - \frac{e^2}{4\pi\varepsilon_0 r}$$

Inserting this expression for E in Eq. 9.14 we obtain, after a slight rearrangement,

9.19 $$\frac{1}{r^2} \frac{d}{dr}\left(r^2 \frac{dR}{dr}\right) + \frac{2m}{\hbar^2}\left[T_{radial} + T_{orbital} - \frac{\hbar^2 l(l+1)}{2mr^2}\right]R = 0$$

If the last two terms in the square brackets of this equation cancel each other out, we shall have what we want: a differential equation for $R(r)$ that involves functions of the radius vector r exclusively. We therefore require that

9.20 $$T_{orbital} = \frac{\hbar^2 l(l+1)}{2mr^2}$$

The orbital kinetic energy of the electron is

$$T_{orbital} = \tfrac{1}{2}mv^2_{orbital}$$

Since the angular momentum L of the electron is

$$L = mv_{orbital}r$$

we may write for the orbital kinetic energy

$$T_{orbital} = \frac{L^2}{2mr^2}$$

Hence, from Eq. 9.20,

$$\frac{L^2}{2mr^2} = \frac{\hbar^2 l(l+1)}{2mr^2}$$

or

9.21 $$L = \sqrt{l(l+1)}\,\hbar \qquad \text{Electron angular momentum}$$

Orbital Quantum Number

Our interpretation of this result is that, since the orbital quantum number l is restricted to the values

$$l = 0, 1, 2, \ldots, (n - 1)$$

the electron can have only those particular angular momenta L specified by Eq. 9.21. Like total energy E, *angular momentum is both conserved and quantized*. The quantity

$$\hbar = h/2\pi = 1.054 \times 10^{-34} \text{ joule-sec}$$

is thus the natural unit of angular momentum.

In macroscopic planetary motion, once again, the quantum number describing angular momentum is so large that the separation into discrete angular-momentum states cannot be experimentally observed. For example, an electron (or, for that matter, any other body) whose orbital quantum number is 2 has the angular momentum

$$L = \sqrt{2(2 + 1)}\ \hbar$$
$$= \sqrt{6}\ \hbar$$
$$= 2.6 \times 10^{-34} \text{ joule-sec}$$

By contrast the orbital angular momentum of the earth is 2.7×10^{40} joule-sec!

It is customary to specify angular-momentum states by a letter, with s corresponding to $l = 0$, p to $l = 1$, and so on according to the following scheme:

$$l = 0\ \ 1\ \ 2\ \ 3\ \ 4\ \ 5\ \ 6\ \ldots$$
$$s\ \ p\ \ d\ \ f\ \ g\ \ h\ \ i\ \ldots$$

Angular-momentum states

This peculiar code originated in the empirical classification of spectra into series called sharp, principal, diffuse, and fundamental which occurred before the theory of the atom was developed. Thus an s state is one with no angular momentum, a p state has the angular momentum $\sqrt{2}\ \hbar$, etc. The combination of the total quantum number with the letter that represents orbital angular momentum provides a convenient and widely used notation for atomic states. In this notation a state in which $n = 2$, $l = 0$ is a $2s$ state, for example, and one in which $n = 4$, $l = 2$ is a $4d$ state. Table 9.2 gives the designations of atomic states in hydrogen through $n = 6$, $l = 5$.

9.6 Magnetic Quantum Number

The orbital quantum number l determines the *magnitude* of the electron's angular momentum. Angular momentum, however, like linear momentum, is a vector quantity, and so to describe it completely requires that its *direction* be specified as well as its magnitude. (The vector **L**, we recall, is perpendicular to the plane in which the rotational motion takes place, and its sense is given by the right-hand rule: when the fingers of the right hand point in the

TABLE 9.2. THE SYMBOLIC DESIGNATION OF ATOMIC STATES IN HYDROGEN

	s $l=0$	p $l=1$	d $l=2$	f $l=3$	g $l=4$	h $l=5$
$n = 1$	$1s$					
$n = 2$	$2s$	$2p$				
$n = 3$	$3s$	$3p$	$3d$			
$n = 4$	$4s$	$4p$	$4d$	$4f$		
$n = 5$	$5s$	$5p$	$5d$	$5f$	$5g$	
$n = 6$	$6s$	$6p$	$6d$	$6f$	$6g$	$6h$

direction of the motion, the thumb is in the direction of **L**. This rule is illustrated in Fig. 9.3.)

What possible significance can a direction in space have for a hydrogen atom? The answer becomes clear when we reflect that an electron revolving about a nucleus is a minute current loop and has a magnetic field like that of a magnetic dipole. In an external magnetic field **B**, a magnetic dipole has an amount of potential energy V_m that depends upon both the magnitude μ of its magnetic moment and the orientation of this moment with respect to the field (Fig. 9.4). The torque τ on a magnetic dipole in a magnetic field of flux density **B** is

$$\tau = \mu B \sin \theta$$

where θ is the angle between μ and **B**. The torque is a maximum when the dipole is per-

FIGURE 9.3 The right-hand rule for angular momentum.

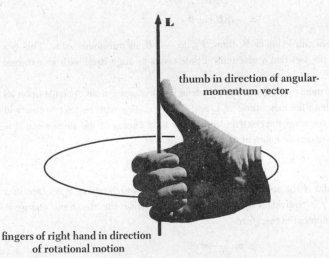

L

thumb in direction of angular-momentum vector

fingers of right hand in direction
of rotational motion

Magnetic Quantum Number

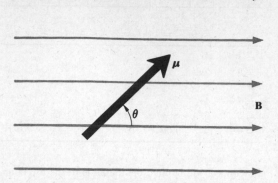

FIGURE 9.4 A magnetic dipole of moment μ at the angle θ relative to a magnetic field B.

pendicular to the field, and zero when it is parallel or antiparallel to it. To calculate the potential energy V_m, we must first establish a reference configuration at which V_m is zero by definition. (Since only *changes* in potential energy are ever experimentally observed, the choice of a reference configuration is arbitrary.) It is convenient to set $V_m = 0$ when $\theta = 90°$, that is, when μ is perpendicular to **B**. The potential energy at any other orientation of μ is equal to the external work that must be done to rotate the dipole from $\theta_0 = 90°$ to the angle θ characterizing that orientation. Hence

$$V_m = \int_{90°}^{\theta} \tau \, d\theta$$

$$= \mu B \int_{90°}^{\theta} \sin \theta \, d\theta$$

9.22
$$= -\mu B \cos \theta$$

When μ points in the same direction as **B**, then, $V_m = -\mu B$, its minimum value. This is a natural consequence of the fact that a magnetic dipole tends to align itself with an external magnetic field.

Since the magnetic moment of the orbital electron in a hydrogen atom depends upon its angular momentum **L**, both the magnitude of **L** and its orientation with respect to the field determine the extent of the magnetic contribution to the total energy of the atom when it is in a magnetic field. The magnetic moment of a current loop is

$$\mu = iA$$

where i is the current and A the area it encloses. An electron which makes ν rev/sec in a circular orbit of radius r is equivalent to a current of $-e\nu$ (since the electronic charge is $-e$), and its magnetic moment is therefore

$$\mu = -e\nu\pi r^2$$

The linear speed v of the electron is $2\pi \nu r$, and so its angular momentum is

$$L = mvr$$
$$= 2\pi m\nu r^2$$

Comparing the formulas for magnetic moment μ and angular momentum L shows that

9.23 $$\boldsymbol{\mu} = -\left(\frac{e}{2m}\right)\mathbf{L}$$ **Electron magnetic moment**

for an orbital electron. The quantity $(-e/2m)$, which involves the charge and mass of the electron only, is called its *gyromagnetic ratio*. The minus sign means that $\boldsymbol{\mu}$ is in the opposite direction to $\mathbf{L}$. While the above expression for the magnetic moment of an orbital electron has been obtained by a classical calculation, quantum mechanics yields the same result. The magnetic potential energy of an atom in a magnetic field is therefore

9.24 $$V_m = \left(\frac{e}{2m}\right) LB \cos \theta$$

a function of both B and θ.

The magnitude L of the electron's orbital angular momentum $\mathbf{L}$ we already know to be quantized, being given as a function of the orbital quantum number l by the formula

$$L = \sqrt{l(l+1)}\, \hbar$$

It is therefore not surprising to learn that the *direction* of $\mathbf{L}$ is also quantized with respect to an external magnetic field. This fact is often referred to as *space quantization*. The magnetic quantum number m_l specifies the direction of $\mathbf{L}$ by determining the component of $\mathbf{L}$ in the field direction. If we let the magnetic-field direction be parallel to the z axis, the component of $\mathbf{L}$ in this direction is

9.25 $$L_z = m_l \hbar$$ **Space quantization**

The possible values of m_l for a given value of l range from $+l$ through 0 to $-l$, so that the number of possible orientations of the angular-momentum vector $\mathbf{L}$ in a magnetic field is $2l + 1$. When $l = 0$, L_z can have only the single value of 0, which means that $\mathbf{L}$ is perpendicular to $\mathbf{B}$ (and $V_m = 0$); when $l = 1$, L_z may be $\hbar$, 0, or $-\hbar$; when $l = 2$, L_z may be $2\hbar$, $\hbar$, 0, $-\hbar$, or $-2\hbar$; and so on. We note that $\mathbf{L}$ can never be aligned exactly parallel or antiparallel to $\mathbf{B}$, since L_z is always smaller than the magnitude $\sqrt{l(l+1)}\, \hbar$ of the total angular momentum.

The space quantization of the orbital angular momentum of the hydrogen atom is shown in Fig. 9.5. We must regard an atom characterized by a certain value of m_l as standing ready to assume a certain orientation of its angular momentum $\mathbf{L}$ relative to an external magnetic field in the event it finds itself in such a field.

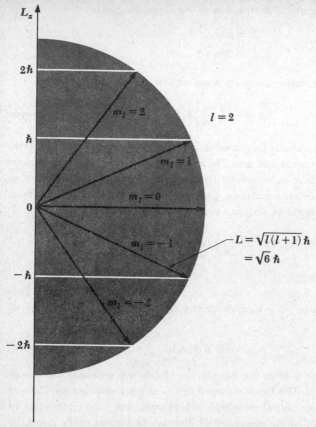

FIGURE 9.5 Space quantization of orbital angular momentum.

In the absence of an external magnetic field, the direction of the z axis is entirely arbitrary. Hence it must be true that the component of **L** in *any* direction we choose is $m_l\hbar$; the significance of an external magnetic field is that it provides an experimentally meaningful reference direction with whose help L_z can be measured.

Why is only one component of **L** quantized? The answer is closely related to the fact that **L** can never point in any specific direction z but instead traces out a cone in space such that its projection L_z is $m_l\hbar$. The reason for the latter phenomenon is the uncertainty principle: if **L** were fixed in space, so that L_x and L_y as well as L_z had definite values, the electron would be confined to a definite plane. For instance, if **L** were in the z direction, the electron would have to be in the xy plane at all times (Fig. 9.6a). This can occur only if the electron's momentum component p_z in the z direction is infinitely uncertain, which

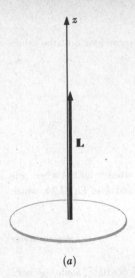

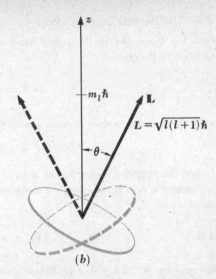

(a)   (b)

FIGURE 9.6 The uncertainty principle prohibits the angular-momentum vector L from having a definite direction in space.

of course is impossible if it is to be part of a hydrogen atom. However, since in reality only *one* component L_z of L together with its magnitude L have definite values and $|L| > |L_z|$, the electron is not limited to a single plane (Fig. 9.6b), and there is a built-in uncertainty, as it were, in the electron's z coordinate. The direction of L is constantly changing, as in Fig. 9.7, and so the average values of L_x and L_y are 0, although L_z always has the specific value $m_l\hbar$.

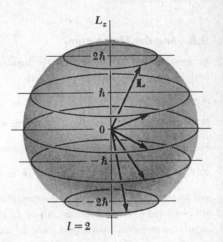

FIGURE 9.7 The angular-momentum vector L precesses constantly about the z axis.

9.7 The Normal Zeeman Effect

From Fig. 9.5 we see that the angle θ between $\mathbf{L}$ and the z direction can have only the values specified by the relation

$$\cos \theta = \frac{m_l}{\sqrt{l(l+1)}}$$

while the permitted values of L are specified by

$$L = \sqrt{l(l+1)}\,\hbar$$

To find the magnetic energy that an atom of magnetic quantum number m_l has when it is in a magnetic field $\mathbf{B}$, we insert the above expressions for $\cos \theta$ and L in Eq. 9.24, which yields

9.26
$$V_m = m_l \left(\frac{e\hbar}{2m}\right) B$$
Magnetic energy

The quantity $e\hbar/2m$ is called the *Bohr magneton*; its value is 9.27×10^{-24} joule/(weber/ m^2). In a magnetic field, then, the energy of a particular atomic state depends upon the value of m_l as well as upon that of n. A state of total quantum number n breaks up into several substates when the atom is in a magnetic field, and their energies are slightly more or slightly less than the energy of the state in the absence of the field. This phenomenon leads to a "splitting" of individual spectral lines into separate lines when atoms radiate in a magnetic field, with the spacing of the lines dependent upon the magnitude of the field. The splitting of spectral lines by a magnetic field is called the *Zeeman effect* after the Dutch physicist Zeeman, who first observed it in 1896. The Zeeman effect is a vivid confirmation of space quantization.

9.8 Angular Momentum

In Secs. 9.5 and 9.6 we saw that the angular momentum $\mathbf{L}$ of a hydrogen atom is quantized both in magnitude and direction, with

(9.21)
$$L = \sqrt{l(l+1)}\,\hbar$$

(9.25)
$$L_z = m_l\hbar$$

For our present purposes, it is more convenient to write Eq. 9.21 in the equivalent form

9.27
$$L^2 = l(l+1)\,\hbar^2$$

and to regard L^2, the square of the magnitude of $\mathbf{L}$, as being quantized. Because L_z and L^2 are restricted to certain specific values, L_z and L^2 must be eigenvalues of the corresponding angular-momentum operators L_z and L^2 and the hydrogen atom wave functions must be eigenfunctions of these operators. Let us see whether this is true.

To obtain L_z and L^2 we follow the prescription of Sec. 7.7: first we express L_z and L^2

in terms of the coordinates x, y, z and the linear-momentum components p_x, p_y, p_z, and then replace the latter by the equivalent differential operators

$$p_x = \frac{\hbar}{i} \frac{\partial}{\partial x}$$

$$p_y = \frac{\hbar}{i} \frac{\partial}{\partial y}$$

$$p_z = \frac{\hbar}{i} \frac{\partial}{\partial z}$$

The angular momentum $\mathbf{L}$ of a particle at the position $\mathbf{r}$ whose linear momentum is $\mathbf{p}$ is defined by the vector formula

$$\mathbf{L} = \mathbf{r} \times \mathbf{p}$$

(See Fig. 9.8.) The cartesian components of $\mathbf{L}$ are, from the definition of the cross product,

FIGURE 9.8 The angular momentum L of a particle whose position vector is r and whose linear momentum is p is perpendicular to the plane containing r and p and has the magnitude rp sin θ.

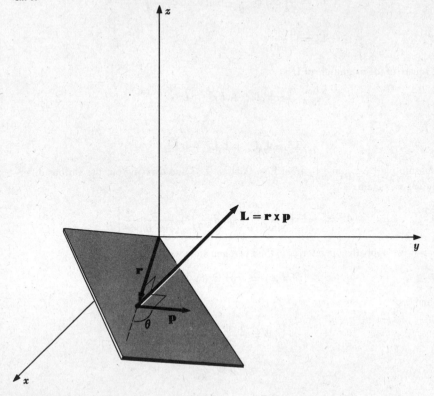

$$L_x = yp_z - zp_y$$
$$L_y = zp_x - xp_z$$
$$L_z = xp_y - yp_x$$

and so the corresponding operators are

9.28
$$\mathsf{L_x} = \frac{\hbar}{i}\left(y\frac{\partial}{\partial z} - z\frac{\partial}{\partial y}\right)$$

9.29
$$\mathsf{L_y} = \frac{\hbar}{i}\left(z\frac{\partial}{\partial x} - x\frac{\partial}{\partial z}\right)$$

9.30
$$\mathsf{L_z} = \frac{\hbar}{i}\left(x\frac{\partial}{\partial y} - y\frac{\partial}{\partial x}\right)$$

In spherical polar coordinates these operators become

9.31
$$\mathsf{L_x} = \frac{\hbar}{i}\left(-\sin\phi\,\frac{\partial}{\partial\theta} - \cot\theta\cos\phi\,\frac{\partial}{\partial\phi}\right)$$

9.32
$$\mathsf{L_y} = \frac{\hbar}{i}\left(\cos\phi\,\frac{\partial}{\partial\theta} - \cot\theta\sin\phi\,\frac{\partial}{\partial\phi}\right)$$

9.33
$$\mathsf{L_z} = \frac{\hbar}{i}\frac{\partial}{\partial\phi}$$

The square of the magnitude of **L** is

$$L^2 = L_x{}^2 + L_y{}^2 + L_z{}^2$$

and so

$$\mathsf{L}^2 = \mathsf{L_xL_x} + \mathsf{L_yL_y} + \mathsf{L_zL_z}$$

By substituting $\mathsf{L_x}$, $\mathsf{L_y}$, and $\mathsf{L_z}$ from Eqs. 9.31 to 9.33 and carrying out the various differentiations we obtain

9.34
$$\mathsf{L}^2 = -\hbar^2\left[\frac{1}{\sin\theta}\frac{\partial}{\partial\theta}\left(\sin\theta\,\frac{\partial}{\partial\theta}\right) + \frac{1}{\sin^2\theta}\frac{\partial^2}{\partial\phi^2}\right]$$

Now we apply the operator $\mathsf{L_z}$ to the hydrogen atom wave function

9.35
$$\psi(r,\theta,\phi) = R(r)\,\Theta\,(\theta)\,\Phi\,(\phi)$$

We find that

$$\mathsf{L_z}\psi = \frac{\hbar}{i}\frac{\partial\psi}{\partial\phi}$$

9.36
$$= \frac{\hbar}{i}R\,\Theta\,\frac{d\Phi}{d\phi}$$

According to Eq. 9.15 the function Φ (ϕ) is given by

(9.15)
$$\Phi = Ae^{im_l\phi}$$

and so

$$\frac{d\Phi}{d\phi} = im_l Ae^{im\phi} = im_l\Phi$$

Hence Eq. 9.36 becomes

$$L_z\psi = \frac{\hbar}{i} im_l R\Theta\Phi$$

9.37
$$= m_l\hbar\,\psi$$

This result means that the only possible values L_z can have in a hydrogen atom are those specified by $m_l\hbar$. Equation 9.37 thus confirms the identification made earlier of the magnetic quantum number m_l with space quantization of angular momentum.

Next we apply the operator L^2 to the wave function $\psi = R\Theta\Phi$. We have

$$L^2\psi = -\hbar^2\left[\frac{1}{\sin\theta}\frac{\partial}{\partial\theta}\left(\sin\theta\frac{\partial}{\partial\theta}\right) + \frac{1}{\sin^2\theta}\frac{\partial^2}{\partial\phi^2}\right]R\Theta\Phi$$

$$= -\hbar^2 R\left[\frac{\Phi}{\sin\theta}\frac{d}{d\theta}\left(\sin\theta\frac{d\Theta}{d\theta}\right) + \frac{\Theta}{\sin^2\theta}\frac{d^2\Phi}{d\phi^2}\right]$$

Differentiating

$$\Phi = Ae^{im_l\phi}$$

twice with respect to ϕ yields

$$\frac{d^2\Phi}{d\phi^2} = i^2 m_l^2\Phi = -m_l^2\Phi$$

and so

9.38
$$L^2\psi = -\hbar^2 R\Phi\left[\frac{1}{\sin\theta}\frac{d}{d\theta}\left(\sin\theta\frac{d\Theta}{d\theta}\right) - \frac{m_l^2}{\sin^2\theta}\Theta\right]$$

The function $\Theta(\theta)$ is a solution of Eq. 9.13, which is

(9.13)
$$\frac{1}{\sin\theta}\frac{d}{d\theta}\left(\sin\theta\frac{d\Theta}{d\theta}\right) + \left[l(l+1) - \frac{m_l^2}{\sin^2\theta}\right]\Theta = 0$$

If we rewrite Eq. 9.13 in the form

$$\frac{1}{\sin\theta}\frac{d}{d\theta}\left(\sin\theta\frac{d\Theta}{d\theta}\right) - \frac{m_l^2}{\sin^2\theta}\Theta = -l(l+1)\Theta$$

Angular Momentum

213

we see that the quantity in brackets in Eq. 9.38 is equal to $-l(l + 1)\Theta$. Hence

$$L^2\psi = l(l + 1)\hbar^2 R\Theta\Phi$$

9.39 $$= l(l + 1)\hbar^2\psi$$

The only possible values L^2 can have in a hydrogen atom are those specified by $l(l + 1)\hbar^2$, a conclusion we arrived at through a somewhat different route in Sec. 9.5.

The wave functions ψ_{nlm_l} of the hydrogen atom thus satisfy three eigenvalue equations:

9.40 $$\mathsf{H}\psi = E_n\psi$$

9.41 $$\mathsf{L}^2\psi = l(l + 1)\hbar^2\psi$$

9.42 $$\mathsf{L}_z\psi = m_l\hbar\psi$$

It is worth noting that these equations are satisfied by the wave functions of *all* systems for which Schrödinger's equation in spherical polar coordinates can be separated into independent equations for each coordinate. That is, if the substitution of $R(r)\Theta(\theta)\Phi(\phi)$ for $\psi(r,\theta,\phi)$ yields one equation for $R(r)$, another for $\Theta(\theta)$, and still another for $\Phi(\phi)$, then ψ is an eigenfunction of H, of L^2, and of L_z, and E, L^2, and L_z are all quantized. The eigenvalues of L^2 and L_z are given by Eqs. 9.41 and 9.42; the eigenvalues of H depend upon the specific potential energy function V involved.

Under what circumstances is Schrödinger's equation in spherical polar coordinates, namely

(9.3) $$\frac{1}{r^2}\frac{\partial}{\partial r}\left(r^2\frac{\partial\psi}{\partial r}\right) + \frac{1}{r^2\sin^2\theta}\frac{\partial}{\partial\theta}\left(\sin\theta\frac{\partial\psi}{\partial\theta}\right) + \frac{1}{r^2\sin^2\theta}\frac{\partial^2\psi}{\partial\phi^2} + \frac{2m}{\hbar^2}(E - V)\psi = 0$$

separable in the above way? A review of Sec. 9.2 shows that when the potential energy function V depends upon r alone—that is, when a particle is in a spherically symmetric force field—the substitution of $R\Theta\Phi$ for ψ will always result in separate equations for R, Θ, and Φ. In Chap. 8 we saw how energy and linear-momentum quantization arise naturally in a variety of situations; now it is evident that angular momentum, too, is quantized in many cases of physical interest. We shall encounter further examples of angular-momentum quantization in later chapters.

9.9 Electron Probability Density

In Bohr's model of the hydrogen atom the electron is visualized as revolving around the nucleus in a circular path. This model is pictured in a spherical polar coordinate system in Fig. 9.9. We see it implies that, if a suitable experiment were performed, the electron would always be found a distance of $r = n^2a_0$ (where n is the quantum number of the orbit and $a_0 = 0.53$ A is the radius of the innermost orbit) from the nucleus and in the equatorial plane $\theta = 90°$, while its azimuth angle ϕ changes with time.

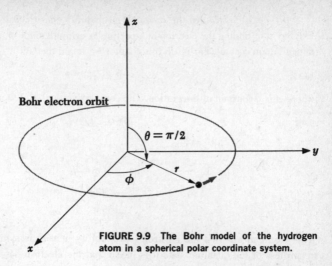

z

Bohr electron orbit

$\theta = \pi/2$

y

ϕ

r

x

FIGURE 9.9 The Bohr model of the hydrogen atom in a spherical polar coordinate system.

The quantum theory of the hydrogen atom modifies the straightforward prediction of the Bohr model in two ways. First, no definite values for r, θ, or ϕ can be given, but only the relative probabilities for finding the electron in volume elements at various locations. This imprecision is, of course, a consequence of the wave nature of the electron. Second, we cannot even think of the electron as moving around the nucleus in any conventional sense since the probability density $|\psi|^2$ is independent of time and may vary considerably from place to place.

The electron wave function ψ in a hydrogen atom is given by

$$\psi = R\Theta\Phi$$

where

$$R = R_{nl}(r)$$

describes how ψ varies with r when the orbital and total quantum numbers have the values n and l;

$$\Theta = \Theta_{lm_l}(\theta)$$

describes how ψ varies with θ when the magnetic and orbital quantum numbers have the values l and m_l; and

$$\Phi = \Phi_{m_l}(\phi)$$

describes how ψ varies with ϕ when the magnetic quantum number is m_l. The probability density $|\psi|^2$ may therefore be written

$$|\psi|^2 = |R|^2|\Theta|^2|\Phi|^2$$

Electron Probability Density

It is easy to show that the azimuthal probability density $|\Phi|^2$, which is a measure of the likelihood of finding the electron at a particular azimuth angle ϕ, is a constant that does not depend upon ϕ at all. Earlier in this chapter we found that

(9.15)
$$\Phi = Ae^{im_l\phi}$$

where A is a constant of integration. The complex conjugate of Φ is

9.43
$$\Phi^* = Ae^{-im_l\phi}$$

and so

$$|\Phi|^2 = \Phi\Phi^*$$
$$= A^2 e^{(im_l\phi - im_l\phi)}$$
$$= A^2$$

This result means that the electron's probability density is symmetrical about the z axis regardless of the quantum state it is in, so that the electron has the same chance of being found at one angle ϕ as at another.

To evaluate A, we call upon the fact that the integral of $|\Phi|^2$ over all angles must equal 1, since the electron must exist somewhere. Hence

$$\int_0^{2\pi} |\Phi|^2 d\phi = A^2 \int_0^{2\pi} d\phi$$
$$= 2\pi A^2$$
$$= 1$$

and

$$A = \frac{1}{\sqrt{2\pi}}$$

The normalized azimuthal function is therefore

9.44
$$\Phi = \frac{1}{\sqrt{2\pi}} e^{im_l\phi}$$

The radial part R of the wave function, in contrast to Φ, not only varies with r but does so in a different way for each combination of quantum numbers n and l (Table 9.1). Figure 9.10 contains graphs of R versus r for 1s, 2s, 2p, 3s, 3p, and 3d states of the hydrogen atom. Evidently R is a maximum at $r = 0$—that is, at the nucleus itself—for all s states, while it is zero at $r = 0$ for states that possess angular momentum.

The *probability density* of the electron at a distance r from the nucleus is proportional to $|R|^2$, but the *actual probability* P of finding it there is proportional to $|R|^2 \, dV$, where dV is an infinitesimal volume element between r and $r + dr$. (We recognize the parallel with mass density; water has a *mass density* of 1 g/cm³, but its *mass* depends upon the specific

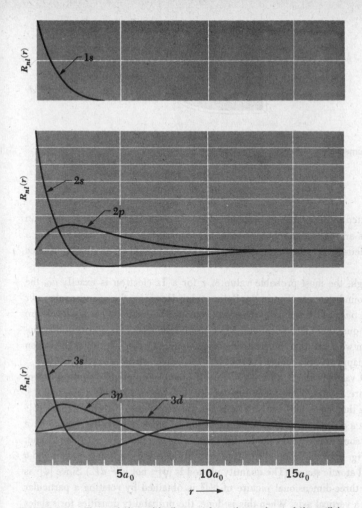

FIGURE 9.10 The variation with distance from the nucleus of the radial part of the electron wave function in hydrogen for various quantum states. The quantity $a_0 = \hbar^2/me^2 = 0.53$ A is the radius of the first Bohr orbit.

volume of water in question.) Now dV is the volume of a spherical shell whose inner radius is r and whose outer radius is $r + dr$, as in Fig. 9.11, so that

9.45
$$dV = 4\pi r^2\, dr$$

If R is a normalized function, then, the actual numerical probability P of finding the electron in a hydrogen atom at a distance between r and $r + dr$ from the nucleus is

Electron Probability Density

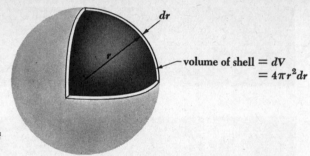

FIGURE 9.11 The volume of a spherical shell.

9.46
$$P = 4\pi r^2 |R_{nl}|^2 \, dr$$

Equation 9.46 is plotted in Fig. 9.12 for the same states whose radial functions R appear in Fig. 9.10; the curves are quite different as a rule. It is interesting to note that P is not a maximum at the nucleus for s states, as R itself is, but has its maximum at a finite distance from it.

Curiously enough, the most probable value of r for a $1s$ electron is exactly a_0, the orbital radius of a ground-state electron in the Bohr model. However, the *average* value of r for a $1s$ electron is $1.5a_0$, which seems puzzling at first sight because the energy levels are the same in both the quantum-mechanical and Bohr atomic models. This apparent discrepancy is removed when we recall that the electron energy depends upon $1/r$ rather than upon r directly, and, as it happens, the average value of $1/r$ for a $1s$ electron is exactly $1/a_0$.

The function Θ varies with zenith angle θ for all quantum numbers l and m_l except $l = m_l = 0$, which are s states. The probability density $|\Theta|^2$ for an s state is a constant ($\frac{1}{2}$, in fact), which means that, since $|\Phi|^2$ is also a constant, the electron probability density $|\psi|^2$ has the same value at a given r in all directions. Electrons in other states, however, do have angular preferences, sometimes quite complicated ones. This can be seen in Table 9.1 and, more strikingly, in Fig. 9.13, in which electron probability densities as functions of r and θ are shown for several atomic states. (The quantity plotted is $|\psi|^2$, not $|\psi|^2 \, dV$.) Since $|\psi|^2$ is independent of ϕ, a three-dimensional picture of $|\psi|^2$ is obtained by rotating a particular representation about a vertical axis. When this is done, the probability densities for s states are evidently spherically symmetric, while the others are not. The pronounced lobe patterns characteristic of many of the states turn out to be significant in chemistry since these patterns help determine the manner in which adjacent atoms in a molecule interact; we shall refer to this notion once more in Chap. 13.

A study of Fig. 9.13 also reveals quantum-mechanical states with a remarkable resemblance to those of the Bohr model. The electron probability-density distribution for a $2p$ state with $m_l = \pm 1$, for instance, is like a doughnut in the equatorial plane centered at the nucleus, and calculation shows the most probable distance of the electron from the nucleus to be $4r_0$—precisely the radius of the Bohr orbit for the same total quantum number. Similar correspondences exist for $3d$ states with $m_l = \pm 2$, $4f$ states with $m_l = \pm 3$, and so on: in

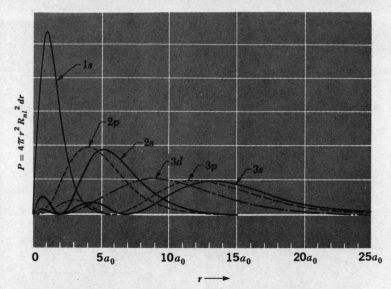

FIGURE 9.12 The probability of finding the electron in a hydrogen atom at a distance between r and $r + dr$ from the nucleus for the quantum states of Fig. 9.10.

every case the highest angular momentum possible for that energy level, and in every case the angular-momentum vector as near the z axis as possible so that the probability density is as close as possible to the equatorial plane. Thus the Bohr model predicts the most probable location of the electron in *one* of the several possible states in each energy level.

Problems

1. Prove that Eqs. 9.1 and 9.3 are equivalent.

2. Show that

$$\Theta_{20}(\theta) = \frac{\sqrt{10}}{4} (3 \cos^2 \theta - 1)$$

is a solution of Eq. 9.13 and that it is normalized.

3. Show that

$$R_{10}(r) = \frac{2}{a_0{}^{3/2}} e^{-r/a_0}$$

is a solution of Eq. 9.14 and that it is normalized.

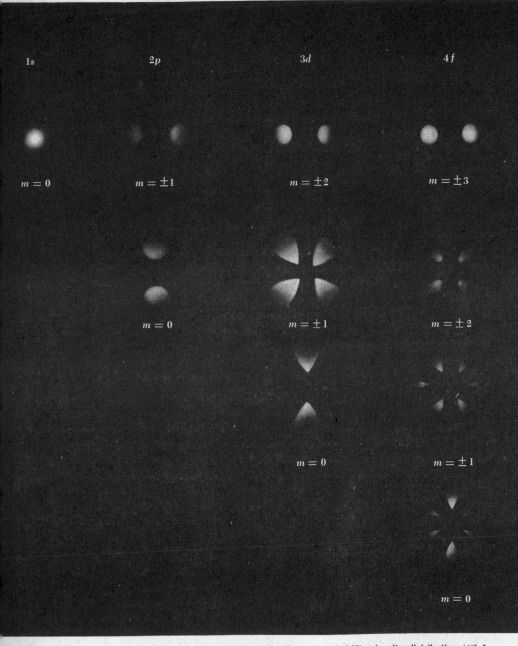

FIGURE 9.13 Photographic representation of the electron probability density distribution $|\psi|^2$ for several energy states. These may be regarded as sectional views of the distributions in a plane

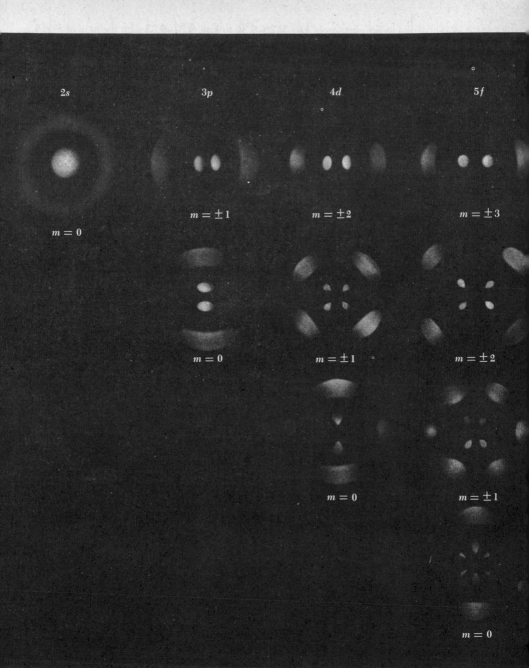

containing the polar axis, which is vertical and in the plane of the paper. The scale varies from figure to figure.

4. Verify the transformation of Eqs. 9.28–9.30 to Eqs. 9.31–9.33 with the help of Fig. 9.1. Note that

$$\frac{\partial}{\partial x} = \frac{\partial r}{\partial x}\frac{\partial}{\partial r} + \frac{\partial \theta}{\partial x}\frac{\partial}{\partial \theta} + \frac{\partial \phi}{\partial x}\frac{\partial}{\partial \phi}$$

Similar relationships hold for $\partial/\partial y$ and $\partial/\partial x$.

5. Verify Eq. 9.34.

6. Show that for the hydrogen atom

$$L_x\psi \neq k_1\psi$$
$$L_y\psi \neq k_2\psi$$

where k_1 and k_2 are constants. What is the significance of this result?

7. The probability of finding an atomic electron whose radial wave function is $R(r)$ outside a sphere of radius r_0 centered on the nucleus is

$$\int_{r_0}^{\infty} |R(r)|^2 4\pi r^2 \, dr$$

The wave function $R_{10}(r)$ of Prob. 3 corresponds to the ground state of a hydrogen atom, and a_0 there is the radius of the Bohr orbit corresponding to that state. (a) Calculate the probability of finding a ground-state electron in a hydrogen atom at a distance greater than a_0 from the nucleus. (b) When the electron in a ground-state hydrogen atom is $2a_0$ from the nucleus, all its energy is potential energy. According to classical physics, the electron therefore cannot ever exceed the distance $2a_0$ from the nucleus. Find the probability that $r > 2a_0$ for the electron in a ground-state hydrogen atom.

8. *Unsöld's theorem* states that, for any value of the orbital quantum number l, the probability densities summed over all possible states from $m_l = -l$ to $m_l = +l$ yield a constant independent of angles θ or ϕ; that is,

$$\sum_{m_l=-l}^{+l} |\Theta|^2|\Phi|^2 = \text{constant}$$

This theorem means that every closed-subshell atom or ion (Sec. 10.4) has a spherically symmetric distribution of electric charge. Verify Unsöld's theorem for $l = 0$, $l = 1$, and $l = 2$ with the help of Table 9.1.

Chapter 10
Many-electron
Atoms

Despite the accuracy with which the quantum theory accounts for certain of the properties of the hydrogen atom, and despite the elegance and essential simplicity of this theory, it cannot approach a complete description of this atom or of other atoms without the further hypothesis of electron spin and the exclusion principle associated with it. In this chapter we shall be introduced to the role of electron spin in atomic phenomena and to the reason why the exclusion principle is the key to understanding the structures of complex atomic systems.

10.1 Electron Spin

Let us begin by citing two of the most conspicuous shortcomings of the theory developed in the preceding chapter. The first is the experimental fact that many spectral lines actually consist of two separate lines that are very close together. An example of this *fine structure* is the first line of the Balmer series of hydrogen, which arises from transitions between the $n = 3$ and $n = 2$ levels in hydrogen atoms. Here the theoretical prediction is for a single line of wavelength 6,563 A, while in reality there are two lines 1.4 A apart—a small effect, but a conspicuous failure for the theory.

The second major discrepancy between the simple quantum theory of the atom and the experimental data occurs in the Zeeman effect, which we briefly mentioned in Sec. 9.7. There we saw that a hydrogen atom of magnetic quantum number m_l has the magnetic energy

10.1
$$V_m = m_l \frac{e\hbar}{2m} B$$

when it is located in a magnetic field of flux density **B**. Now m_l can have the $2l + 1$ values of $+l$ through 0 to $-l$, so a state of given orbital quantum number l is split into $2l + 1$ substates differing in energy by $(e\hbar/2m)B$ when the atom is in a magnetic field. However, because changes in m_l are restricted to $\Delta m_l = 0, \pm 1$ (see Chap. 11), a given spectral line that arises from a transition between two states of different l is split into only three components, as shown in Fig. 10.1. The *normal Zeeman effect*, then, consists of the splitting of a spectral line of frequency ν_0 into three components whose frequencies are

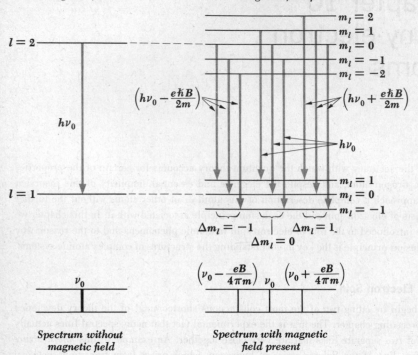

FIGURE 10.1 The normal Zeeman effect.

$$\nu_1 = \nu_0 - \frac{e\hbar}{2m}\frac{B}{h} = \nu_0 - \frac{e}{4\pi m}B$$

10.2 $\nu_2 = \nu_0$ **Normal Zeeman effect**

$$\nu_3 = \nu_0 + \frac{e\hbar}{2m}\frac{B}{h} = \nu_0 + \frac{e}{4\pi m}B$$

While the normal Zeeman effect is indeed observed in the spectra of a few elements under certain circumstances, more often it is not: four, six, or even more components may appear, and even when three components are present their spacing may not agree with Eq. 10.2. Several anomalous Zeeman patterns are shown in Fig. 10.2 together with the predictions of Eq. 10.2.

In an effort to account for both fine structure in spectral lines and the anomalous Zeeman effect, S. A. Goudsmit and G. E. Uhlenbeck proposed in 1925 that **the electron possesses an intrinsic angular momentum independent of any orbital angular momentum it might have and, associated with this angular momentum, a certain magnetic**

moment. What Goudsmit and Uhlenbeck had in mind was a classical picture of an electron as a charged sphere spinning on its axis. The rotation involves angular momentum, and because the electron is negatively charged, it has a magnetic moment $\boldsymbol{\mu}_s$ opposite in direction to its angular-momentum vector $\mathbf{L}_s$. The notion of *electron spin* proved to be successful in explaining not only fine structure and the anomalous Zeeman effect but a wide variety of other atomic effects as well. Of course, the idea that electrons are spinning charged spheres is hardly in accord with quantum mechanics, but in 1928 Dirac was able to show on the basis of a relativistic quantum-theoretical treatment that particles having the charge and mass of the electron must have just the intrinsic angular momentum and magnetic moment attributed to them by Goudsmit and Uhlenbeck.

The quantum number s is used to describe the spin angular momentum of the electron. The only value s can have is $s = \frac{1}{2}$; this restriction follows from Dirac's theory and, as we shall see below, may also be obtained empirically from spectral data. The magnitude S of the angular momentum due to electron spin is given in terms of the spin quantum number s by the formula

$$S = \sqrt{s(s+1)}\,\hbar$$

10.3
$$= \frac{\sqrt{3}}{2}\,\hbar$$

which is the same formula as that giving the magnitude L of the orbital angular momentum in terms of the orbital quantum number l:

$$L = \sqrt{l(l+1)}\,\hbar$$

FIGURE 10.2 The normal and anomalous Zeeman effects in various spectral lines.

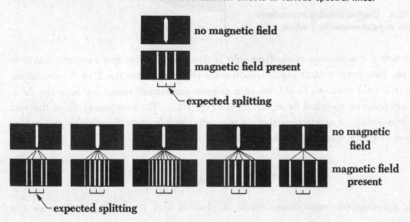

Electron Spin

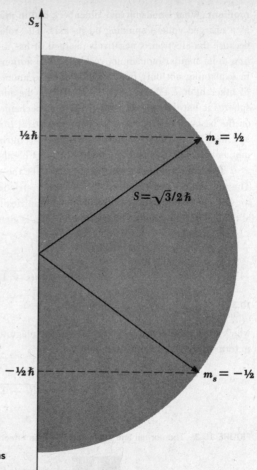

FIGURE 10.3 The two possible orientations
of the spin angular-momentum vector.

The space quantization of electron spin is described by the spin magnetic quantum number m_s. Just as the orbital angular-momentum vector can have the $2l + 1$ orientations in a magnetic field from $+l$ to $-l$, the spin angular-momentum vector can have the $2s + 1 = 2$ orientations specified by $m_s = \pm\frac{1}{2}$ (Fig. 10.3). The component S_z of the spin angular momentum of an electron along a magnetic field in the z direction is determined by the spin magnetic quantum number, so that

$$S_z = m_s\hbar$$

10.4
$$= \pm\frac{1}{2}\hbar$$

The gyromagnetic ratio characteristic of electron spin is almost exactly twice that characteristic of electron orbital motion. Thus, taking this ratio as equal to 2, the spin magnetic moment μ_s of an electron is related to its spin angular momentum S by

10.5
$$\mu_s = \frac{e}{m} \mathbf{S}$$

The possible components of μ_s along any axis, say the z axis, are therefore limited to

10.6
$$\mu_{sz} = \pm \frac{e\hbar}{2m}$$

We recognize the quantity $(e\hbar/2m)$ as the Bohr magneton.

Space quantization was first explicitly demonstrated by O. Stern and W. Gerlach in 1921. They directed a beam of neutral silver atoms from an oven through a set of collimating slits into an inhomogeneous magnetic field, as shown in Fig. 10.4. A photographic plate recorded the configuration of the beam after its passage through the field. In its normal state, the entire magnetic moment of a silver atom is due to the spin of one of its electrons. In a uniform magnetic field, such a dipole would merely experience a torque tending to align it with the field. In an inhomogeneous field, however, each "pole" of the dipole is

FIGURE 10.4 The Stern-Gerlach experiment.

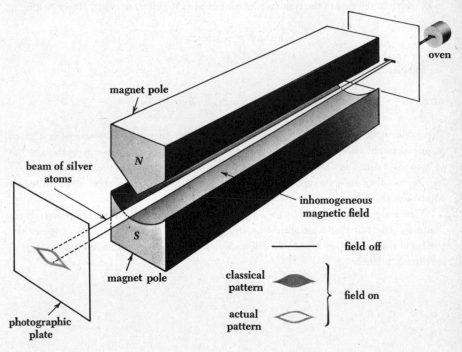

subject to a force of different magnitude, and as a result there is a resultant force on the dipole that varies with its orientation relative to the field. Classically, all orientations should be present in a beam of atoms, which would result merely in a broad trace on the photographic plate instead of the thin line formed in the absence of any magnetic field. Stern and Gerlach found, however, that the initial beam split into two distinct parts, corresponding to the two opposite spin orientations in the magnetic field that are permitted by space quantization.

10.2   Spin-orbit Coupling

The fine-structure doubling of spectral lines may be explained on the basis of a magnetic interaction between the spin and orbital angular momenta of atomic electrons. This spin-orbit coupling can be understood in terms of a straightforward classical model. An electron revolving about a proton finds itself in a magnetic field because, in its own frame of reference, the proton is circling about *it*. This magnetic field then acts upon the electron's own spin magnetic moment to produce a kind of internal Zeeman effect. The energy V_m of a magnetic dipole of moment μ in a magnetic field of flux density **B** is, in general,

10.7
$$V_m = -\mu B \cos \theta$$

where θ is the angle between μ and **B**. The quantity $\mu \cos \theta$ is the component of μ parallel to **B**, which in the case of the spin magnetic moment of the electron is μ_{sz}. Hence, letting

$$\mu \cos \theta = \mu_{sz} = \pm \frac{e\hbar}{2m}$$

we find that

10.8
$$V_m = \pm \frac{e\hbar}{2m} B$$

Depending upon the orientation of its spin vector, the energy of an electron in a given atomic quantum state will be higher or lower by $(e\hbar/2m)B$ than its energy in the absence of spin-orbit coupling. The result is the splitting of every quantum state (except s states) into two separate substates and, consequently, the splitting of every spectral line into two component lines.

The assignment of $s = \frac{1}{2}$ is the only one that conforms to the observed fine-structure doubling. The fact that what should be single states are in fact twin states imposes the condition that the $2s + 1$ possible orientations of the spin angular-momentum vector S must total 2. Hence

$$2s + 1 = 2$$
$$s = \frac{1}{2}$$

To check whether the observed fine structure in spectral lines corresponds to the energy shifts predicted by Eq. 10.8, we must compute the magnitude B of the magnetic field experienced by an atomic electron. An estimate is easy to obtain. A circular wire loop of radius r that carries the current i has a magnetic field of flux density

$$B = \frac{\mu_0 i}{2r}$$

at its center. An orbital electron, say in a hydrogen atom, "sees" itself circled f times each second by a proton of charge $+e$, for a resulting flux density of

$$B = \frac{\mu_0 f e}{2r}$$

In the ground-state Bohr atom $f \approx 7 \times 10^{15}$ and $r \approx 5 \times 10^{-11}$ m, so that

$$B \approx 14 \text{ webers/m}^2$$

which is a very strong magnetic field. The value of the Bohr magneton is

$$\frac{e\hbar}{2m} = 9.27 \times 10^{-24} \text{ joule/(weber/m}^2)$$

Hence the magnetic energy V_m of such an electron is, from Eq. 10.8,

$$V_m = \frac{e\hbar}{2m} B$$
$$\approx 9.27 \times 10^{-24} \text{ joule/(wober/m}^2) \times 14 \text{ webers/m}^2$$
$$\approx 1.30 \times 10^{-22} \text{ joule}$$

The wavelength shift corresponding to such a change in energy is about 2 A for a spectral line of unperturbed wavelength 6,563 A, slightly more than the observed splitting of the line originating in the $n = 3 \rightarrow n = 2$ transition. However, the flux density of the magnetic field at orbits of higher order is less than for ground-state orbits, which accounts for the discrepancy.

10.3 The Exclusion Principle

In the normal configuration of a hydrogen atom, the electron is in its lowest quantum state. What are the normal configurations of more complex atoms? Are all 92 electrons of a uranium atom in the the same quantum state, to be envisioned perhaps as circling the nucleus crowded together in a single Bohr orbit? Many lines of evidence make this hypothesis unlikely. One example is the great difference in chemical behavior exhibited by certain elements whose atomic structures differ by just one electron: for instance, the elements having atomic numbers 9, 10, and 11 are respectively the halogen gas fluorine, the inert gas neon, and the alkali metal sodium. Since the electronic structure of an atom controls its interactions with other atoms, it is hard to understand why the chemical properties of the elements should change so abruptly with a small change in atomic number if all of the electrons in an atom exist together in the same quantum state.

In 1925 Wolfgang Pauli discovered the fundamental principle that governs the electronic configurations of atoms having more than one electron. His *exclusion principle* states that **no two electrons in an atom can exist in the same quantum state.** Each electron in an atom must have a different set of the quantum numbers n, l, m_l, m_s. Pauli was led to this conclusion from a study of atomic spectra. It is possible to determine the various states of an atom from its spectrum, and the quantum numbers of these states can be inferred. In the spectra of every element but hydrogen a number of lines are *missing* that correspond to transitions to and from states having certain combinations of quantum numbers. Thus no transitions are observed in helium to or from the ground-state configuration in which the spins of both electrons are in the same direction to give a total spin of 1, although transitions *are* observed to and from the other ground-state configuration, in which the spins are in opposite directions to give a total spin of 0. In the absent state the quantum numbers of *both* electrons would be $n = 1$, $l = 0$, $m_l = 0$, $m_s = \frac{1}{2}$, while in the state known to exist one of the electrons has $m_s = \frac{1}{2}$ and the other $m_s = -\frac{1}{2}$. Pauli showed that every unobserved atomic state involves two or more electrons with identical quantum numbers, and the exclusion principle is a statement of this empirical finding.

Before we explore the role of the exclusion principle in determining atomic structures, let us look into its quantum-mechanical implications. We saw in the previous chapter that the complete wave function ψ of the electron in a hydrogen atom can be expressed as the product of three separate wave functions, each describing that part of ψ which is a function of one of the three coordinates r, θ, ϕ. It is possible to show in an analogous way that the complete wave function $\psi(1, 2, 3, \ldots, n)$ of a system of n particles in the same force field can be expressed as the product of the wave functions $\psi(1)$, $\psi(2)$, $\psi(3)$, $\ldots$, $\psi(n)$ of the individual particles. That is,

10.9 $$\psi(1, 2, 3, \ldots, n) = \psi(1)\,\psi(2)\,\psi(3)\ldots\psi(n)$$

We shall use this result to investigate the kinds of wave functions that can be used to describe a system of two identical particles.

Let us suppose that one of the particles is in quantum state a and the other in state b. Because the particles are identical, it should make no difference in the probability density $|\psi|^2$ of the system if the particles are exchanged, with the one in state a replacing the one in state b and vice versa. Symbolically, we require that

10.10 $$|\psi|^2(1,2) = |\psi|^2(2,1)$$

Hence the wave function $\psi(2,1)$, representing the exchanged particles, can be given by either

10.11 $$\psi(2,1) = \psi(1,2)$$ Symmetric

or

10.12 $$\psi(2,1) = -\psi(1,2)$$ Antisymmetric

and still fulfill Eq. 10.10. The wave function of the system is not itself a measurable quantity, and so it can be altered in sign by the exchange of the particles. Wave functions unaffected by an exchange of particles are said to be *symmetric*, while those reversing sign upon such an exchange are said to be *antisymmetric*.

If particle 1 is in state a and particle 2 is in state b, the wave function of the system is, according to Eq. 10.9,

10.13a $$\psi_I = \psi_a(1)\,\psi_b(2)$$

while if particle 2 is in state a and particle 1 is in state b, the wave function is

10.13b $$\psi_{II} = \psi_a(2)\,\psi_b(1)$$

Because the two particles are in fact indistinguishable, we have no way of knowing at any moment whether ψ_I or ψ_{II} describes the system. The likelihood that ψ_I is correct at any moment is the same as the likelihood that ψ_{II} is correct. Equivalently, we can say that the system spends half the time in the configuration whose wave function is ψ_I and the other half in the configuration whose wave function is ψ_{II}. Therefore a linear combination of ψ_I and ψ_{II} is the proper description of the system. There are two such combinations possible, the symmetric one

10.14 $$\psi_S = \frac{1}{\sqrt{2}}\,[\psi_a(1)\,\psi_b(2) + \psi_a(2)\,\psi_b(1)]$$

and the antisymmetric one

10.15 $$\psi_A = \frac{1}{\sqrt{2}}\,[\psi_a(1)\,\psi_b(2) - \psi_a(2)\,\psi_b(1)]$$

The factor $\sqrt{2}$ is required to normalize ψ_S and ψ_A. Exchanging particles 1 and 2 leaves ψ_S unaffected, while it reverses the sign of ψ_A. Both ψ_S and ψ_A obey Eq. 10.10.

There are a number of important distinctions between the behavior of particles in systems whose wave functions are symmetric and that of particles in systems whose wave functions are antisymmetric. The most obvious is that, in the former case, both particles 1 and 2 can simultaneously exist in the same state, with $a = b$, while in the latter case, if we set $a = b$, we find that $\psi_A = 0$: the two particles *cannot* be in the same quantum state. Comparing this quantum-mechanical statement with Pauli's empirical exclusion principle, according to which no two electrons in an atom can be in the same quantum state, we conclude that systems of electrons are described by wave functions that reverse sign upon the exchange of any pair of them.

The Exclusion Principle

The results of various experiments show that *all* particles which have a spin of ½ have wave functions that are antisymmetric to an exchange of any pair of them. Such particles, which include protons and neutrons as well as electrons, obey the exclusion principle when they are in the same system; that is, when they move in a common force field, each member of the system must be in a different quantum state. Particles of spin ½ are often referred to as *Fermi particles* or *fermions* because, as we shall learn in Chap. 16, the behavior of aggregates of them is governed by a statistical distribution law discovered by Fermi and Dirac.

Particles whose spins are 0 or an integer have wave functions that are symmetric to an exchange of any pair of them. These particles do not obey the exclusion principle. Particles of 0 or integral spin are often referred to as *Bose particles* or *bosons* because the statistical distribution law that describes aggregates of them was discovered by Bose and Einstein. Photons, alpha particles, and helium atoms are Bose particles.

There are other important consequences of the symmetry or antisymmetry of particle wave functions besides that expressed in the exclusion principle. It is these consequences that make it useful to classify particles according to the nature of their wave functions rather than simply according to whether or not they obey the exclusion principle.

10.4 Electron Configurations

Two basic rules determine the electronic structures of many-electron atoms:

1. A system of particles is stable when its total energy is a minimum.
2. Only one electron can exist in any particular quantum state in an atom.

Before we apply these rules to actual atoms, let us examine the variation of electron energy with quantum state.

While the various electrons in a complex atom certainly interact directly with one another, much about atomic structure can be understood by simply considering each electron as though it exists in a constant mean force field. For a given electron this field is approximately the electric field of the nuclear charge Ze decreased by the partial shielding of those other electrons that are closer to the nucleus. All of the electrons that have the same total quantum number n are, on the average, roughly the same distance from the nucleus. These electrons therefore interact with virtually the same electric field and have similar energies. It is conventional to speak of such electrons as occupying the same atomic *shell*. Shells are denoted by capital letters according to the following scheme:

$$n = 1 \quad 2 \quad 3 \quad 4 \quad 5 \ldots \qquad \text{\textbf{Atomic shells}}$$
$$K \quad L \quad M \quad N \quad O \ldots$$

The energy of an electron in a particular shell also depends to a certain extent upon its orbital quantum number l, though this dependence is not so great as that upon n. In a complex atom the degree to which the full nuclear charge is shielded from a given electron by intervening shells of other electrons varies with its probability-density distribution. When l is large, the distribution has roughly circular contour lines, while when l is small, the contour lines are elliptical. An electron of small l therefore is more likely to be found near the nucleus (where it is poorly shielded by the other electrons) than one of higher l (see Fig. 9.12), which results in a lower total energy (that is, higher binding energy) for it. The electrons in each shell accordingly increase in energy with increasing l. This effect is illustrated in Fig. 10.5, which is a plot of the binding energies of various atomic electrons as a function of atomic number.

FIGURE 10.5 The binding energies of atomic electrons in Ry. (1 Ry = 1 Rydberg = 13.6 ev = ground-state energy of H atom.)

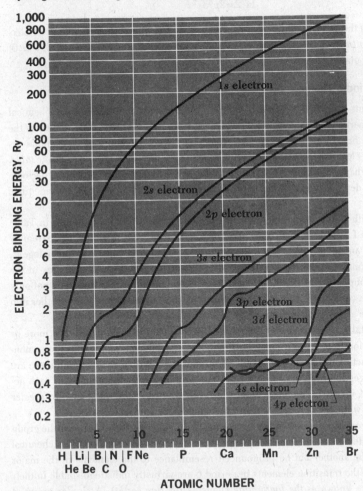

Electrons that share a certain value of l in a shell are said to occupy the same *subshell*. All of the electrons in a subshell have almost identical energies, since the dependence of electron energy upon m_l and m_s is comparatively minor.

The occupancy of the various subshells in an atom is usually expressed with the help of the notation introduced in the previous chapter for the various quantum states of the hydrogen atom. As indicated in Table 9.2, each subshell is identified by its total quantum number n followed by the letter corresponding to its orbital quantum number l. A superscript after the letter indicates the number of electrons in that subshell. For example, the electron configuration of sodium is written

$$1s^2 2s^2 2p^6 3s^1$$

which means that the $1s$ ($n = 1$, $l = 0$) and $2s$ ($n = 2$, $l = 0$) subshells contain two electrons each, the $2p$ ($n = 2$, $l = 1$) subshell contains six electrons, and the $3s$ ($n = 3$, $l = 0$) subshell contains one electron.

10.5 The Periodic Table

When the elements are listed in order of atomic number, elements with similar chemical and physical properties recur at regular intervals. This empirical observation, known as the *periodic law*, was first formulated by Dmitri Mendeleev about a century ago. A tabular arrangement of the elements exhibiting this recurrence of properties is called a *periodic table*. Table 10.1 is perhaps the simplest form of periodic table; though more elaborate periodic tables have been devised to exhibit the periodic law in finer detail, Table 10.1 is adequate for our purposes.

Elements with similar properties form the *groups* shown as vertical columns in Table 10.1. Thus group I consists of hydrogen plus the alkali metals, all of which are extremely active chemically and all of which have valences of $+1$. Group VII consists of the halogens, volatile, active nonmetals that have valences of -1 and form diatomic molecules in the gaseous state. Group VIII consists of the inert gases, elements so inactive that they not only almost never form compounds with other elements, but their atoms do not join together into molecules like the atoms of other gases.

The horizontal rows in Table 10.1 are called *periods*. Across each period is a more or less steady transition from an active metal through less active metals and weakly active nonmetals to highly active nonmetals and finally to an inert gas. Within each column there are also regular changes in properties, but they are far less conspicuous than those in each period. For example, increasing atomic number in the alkali metals is accompanied by greater chemical activity, while the reverse is true in the halogens.

A series of *transition elements* appears in each period after the third between the group II and group III elements. The transition elements are metals with a considerable chemical resemblance to one another but no pronounced resemblance to the elements in the major groups. Fifteen of the transition elements in period 6 are virtually indistinguishable in their properties, and are known as the *lanthanide* elements (or *rare earths*). A similar group of closely related metals, the *actinide* elements, is found in period 7.

TABLE 10.1 THE PERIODIC TABLE OF THE ELEMENTS The number above the symbol of each element is its atomic number, and that below is its atomic weight. The elements whose atomic weights are given in parentheses do not occur in nature, but have been prepared artificially in nuclear reactions. The atomic weight in such a case is the mass number of the most long-lived radioactive isotope of the element.

Period	Group I	Group II												Group III	Group IV	Group V	Group VI	Group VII	Group VIII
1	1 H 1.00																		2 He 4.00
2	3 Li 6.94	4 Be 9.01												5 B 10.81	6 C 12.01	7 N 14.01	8 O 16.00	9 F 19.00	10 Ne 20.18
3	11 Na 22.99	12 Mg 24.31												13 Al 26.98	14 Si 28.09	15 P 30.98	16 S 32.07	17 Cl 35.46	18 Ar 39.94
4	19 K 39.10	20 Ca 40.08	21 Sc 44.96	22 Ti 47.90	23 V 50.94	24 Cr 52.00	25 Mn 54.94	26 Fe 55.85	27 Co 58.93	28 Ni 58.71	29 Cu 63.54	30 Zn 65.37		31 Ga 69.72	32 Ge 72.59	33 As 74.92	34 Se 78.96	35 Br 79.91	36 Kr 83.8
5	37 Rb 85.47	38 Sr 87.66	39 Y 88.91	40 Zr 91.22	41 Nb 92.91	42 Mo 95.94	43 Tc (99)	44 Ru 101.1	45 Rh 102.91	46 Pd 106.4	47 Ag 107.87	48 Cd 112.40		49 In 114.82	50 Sn 118.69	51 Sb 121.75	52 Te 127.60	53 I 126.90	54 Xe 131.30
6	55 Cs 132.91	56 Ba 137.34	57–71 *	72 Hf 178.49	73 Ta 180.95	74 W 183.85	75 Re 186.2	76 Os 190.2	77 Ir 192.2	78 Pt 195.09	79 Au 197.0	80 Hg 200.59		81 Tl 204.37	82 Pb 207.19	83 Bi 208.98	84 Po (210)	85 At (210)	86 Rn 222
7	87 Fr (223)	88 Ra 226.05	89–103 **																

*Rare earths

57 La 138.91	58 Ce 140.12	59 Pr 140.91	60 Nd 144.24	61 Pm (145)	62 Sm 150.35	63 Eu 152.0	64 Gd 157.25	65 Tb 158.92	66 Dy 162.50	67 Ho 164.92	68 Er 167.26	69 Tm 168.93	70 Yb 173.04	71 Lu 174.97

**Actinides

89 Ac 227	90 Th 232.04	91 Pa 231	92 U 238.03	93 Np (237)	94 Pu (242)	95 Am (243)	96 Cm (247)	97 Bk (249)	98 Cf (251)	99 Es (254)	100 Fm (253)	101 Md (256)	102 No (254)	103 Lw (257)

The notion of electron shells and subshells fits perfectly into the pattern of the periodic table, which is just a mirror of the atomic structures of the elements. Let us see how this pattern arises.

The exclusion principle places definite limits on the number of electrons that can occupy a given subshell. A subshell is characterized by a certain total quantum number n and orbital quantum number l, where

$$l = 0, 1, 2, \ldots, (n - 1)$$

There are $2l + 1$ different values of the magnetic quantum number m_l for any l, since

$$m_l = 0, \pm 1, \pm 2, \ldots, \pm l$$

and two possible values of the spin magnetic quantum number m_s ($+\frac{1}{2}$ and $-\frac{1}{2}$) for any m_l. Hence each subshell can contain a maximum of $2(2l + 1)$ electrons and each shell a maximum of

$$\sum_{l=0}^{l=n-1} 2(2l + 1) = 2[1 + 3 + 5 + \cdots + 2(n - 1) + 1]$$

$$= 2[1 + 3 + 5 + \cdots + 2n - 1]$$

The quantity in the brackets contains n terms whose average value is $\frac{1}{2}[1 + (2n - 1)]$, so that

$$[1 + 3 + 5 + \cdots + 2n - 1] = n \times \frac{1}{2}[1 + (2n - 1)]$$

and the maximum number of electrons in the nth shell is

$$2 \times \frac{n}{2}[1 + (2n - 1)] = 2n^2$$

An atomic shell or subshell that contains its full quota of electrons is said to be *closed*. A closed s subshell ($l = 0$) holds two electrons, a closed p subshell ($l = 1$) six electrons, a closed d subshell ($l = 2$) ten electrons, and so on.

The total orbital and spin angular momenta of the electrons in a closed subshell are zero, and their effective charge distributions are perfectly symmetrical (see Problem 8 of Chap. 9). The electrons in a closed shell are all very tightly bound, since the positive nuclear charge is large relative to the negative charge of the inner shielding electrons (Fig. 10.6). Since an atom containing only closed shells has no dipole moment, it does not attract other electrons, and its electrons cannot be readily detached. Such atoms we expect to be passive chemically, like the inert gases—and the inert gases all turn out to have closed-shell electron configurations or their equivalents.

Those atoms with but a single electron in their outermost shells tend to lose this electron, which is relatively far from the nucleus and is shielded by the inner electrons from all but an effective nuclear charge of $+e$. Hydrogen and the alkali metals are in this category and accordingly have valences of $+1$. Atoms whose outer shells lack a single electron of

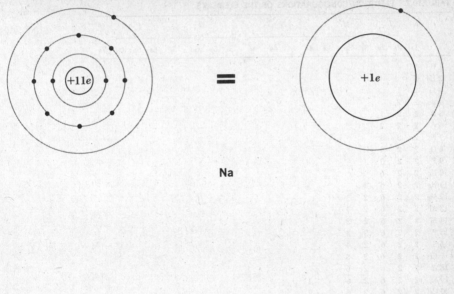

Na

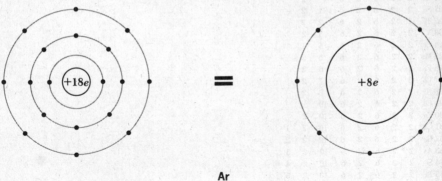

Ar

FIGURE 10.6 Electron shielding in sodium and argon. Each outer electron in an Ar atom is acted upon by an effective nuclear charge 8 times greater than that acting upon the outer electron in a Na atom, even though the outer electrons in both cases are in the M (n = 3) shell.

being closed tend to acquire such an electron through the attraction of the imperfectly shielded strong nuclear charge, which accounts for the chemical behavior of the halogens. In this manner the similarities of the members of the various groups of the periodic table may be accounted for.

Table 10.2 shows the electron configurations of the elements. The origin of the transi-

TABLE 10.2. ELECTRON CONFIGURATIONS OF THE ELEMENTS

	K	L		M			N				O				P			Q
	1s	2s	2p	3s	3p	3d	4s	4p	4d	4f	5s	5p	5d	5f	6s	6p	6d	7s
1 H	1																	
2 He	2																	
3 Li	2	1																
4 Be	2	2																
5 B	2	2	1															
6 C	2	2	2															
7 N	2	2	3															
8 O	2	2	4															
9 F	2	2	5															
10 Ne	2	2	6															
11 Na	2	2	6	1														
12 Mg	2	2	6	2														
13 Al	2	2	6	2	1													
14 Si	2	2	6	2	2													
15 P	2	2	6	2	3													
16 S	2	2	6	2	4													
17 Cl	2	2	6	2	5													
18 A	2	2	6	2	6													
19 K	2	2	6	2	6		1											
20 Ca	2	2	6	2	6		2											
21 Sc	2	2	6	2	6	1	2											
22 Ti	2	2	6	2	6	2	2											
23 V	2	2	6	2	6	3	2											
24 Cr	2	2	6	2	6	5	1											
25 Mn	2	2	6	2	6	5	2											
26 Fe	2	2	6	2	6	6	2											
27 Co	2	2	6	2	6	7	2											
28 Ni	2	2	6	2	6	8	2											
29 Cu	2	2	6	2	6	10	1											
30 Zn	2	2	6	2	6	10	2											
31 Ga	2	2	6	2	6	10	2	1										
32 Ge	2	2	6	2	6	10	2	2										
33 As	2	2	6	2	6	10	2	3										
34 Se	2	2	6	2	6	10	2	4										
35 Br	2	2	6	2	6	10	2	5										
36 Kr	2	2	6	2	6	10	2	6										
37 Rb	2	2	6	2	6	10	2	6			1							
38 Sr	2	2	6	2	6	10	2	6			2							
39 Y	2	2	6	2	6	10	2	6	1		2							
40 Zr	2	2	6	2	6	10	2	6	2		2							
41 Nb	2	2	6	2	6	10	2	6	4		1							
42 Mo	2	2	6	2	6	10	2	6	5		1							
43 Te	2	2	6	2	6	10	2	6	5		2							
44 Ru	2	2	6	2	6	10	2	6	7		1							
45 Rh	2	2	6	2	6	10	2	6	8		1							
46 Pd	2	2	6	2	6	10	2	6	10		1							
47 Ag	2	2	6	2	6	10	2	6	10		1							
48 Cd	2	2	6	2	6	10	2	6	10		2							
49 In	2	2	6	2	6	10	2	6	10		2	1						
50 Sn	2	2	6	2	6	10	2	6	10		2	2						
51 Sb	2	2	6	2	6	10	2	6	10		2	3						
52 Te	2	2	6	2	6	10	2	6	10		2	4						

TABLE 10.2. ELECTRON CONFIGURATIONS OF THE ELEMENTS (*Continued*)

	K	L		M			N				O				P			Q
	1s	2s	2p	3s	3p	3d	4s	4p	4d	4f	5s	5p	5d	5f	6s	6p	6d	7s
53 I	2	2	6	2	6	10	2	6	10		2	5						
54 Xe	2	2	6	2	6	10	2	6	10		2	6						
55 Cs	2	2	6	2	6	10	2	6	10		2	6			1			
56 Ba	2	2	6	2	6	10	2	6	10		2	6			2			
57 La	2	2	6	2	6	10	2	6	10		2	6	1		2			
58 Ce	2	2	6	2	6	10	2	6	10	2	2	6			2			
59 Pr	2	2	6	2	6	10	2	6	10	3	2	6			2			
60 Nd	2	2	6	2	6	10	2	6	10	4	2	6			2			
61 Pm	2	2	6	2	6	10	2	6	10	5	2	6			2			
62 Sm	2	2	6	2	6	10	2	6	10	6	2	6			2			
63 Eu	2	2	6	2	6	10	2	6	10	7	2	6			2			
64 Gd	2	2	6	2	6	10	2	6	10	7	2	6	1		2			
65 Tb	2	2	6	2	6	10	2	6	10	9	2	6			2			
66 Dy	2	2	6	2	6	10	2	6	10	10	2	6			2			
67 Ho	2	2	6	2	6	10	2	6	10	11	2	6			2			
68 Er	2	2	6	2	6	10	2	6	10	12	2	6			2			
69 Tm	2	2	6	2	6	10	2	6	10	13	2	6			2			
70 Yb	2	2	6	2	6	10	2	6	10	14	2	6			2			
71 Lu	2	2	6	2	6	10	2	6	10	14	2	6	1		2			
72 Hf	2	2	6	2	6	10	2	6	10	14	2	6	2		2			
73 Ta	2	2	6	2	6	10	2	6	10	14	2	6	3		2			
74 W	2	2	6	2	6	10	2	6	10	14	2	6	4		2			
75 Re	2	2	6	2	6	10	2	6	10	14	2	6	5		2			
76 Os	2	2	6	2	6	10	2	6	10	14	2	6	6		2			
77 Ir	2	2	6	2	6	10	2	6	10	14	2	6	7		2			
78 Pt	2	2	6	2	6	10	2	6	10	14	2	6	9		1			
79 Au	2	2	6	2	6	10	2	6	10	14	2	6	10		1			
80 Hg	2	2	6	2	6	10	2	6	10	14	2	6	10		2			
81 Tl	2	2	6	2	6	10	2	6	10	14	2	6	10		2	1		
82 Pb	2	2	6	2	6	10	2	6	10	14	2	6	10		2	2		
83 Bi	2	2	6	2	6	10	2	6	10	14	2	6	10		2	3		
84 Po	2	2	6	2	6	10	2	6	10	14	2	6	10		2	4		
85 At	2	2	6	2	6	10	2	6	10	14	2	6	10		2	5		
86 Rn	2	2	6	2	6	10	2	6	10	14	2	6	10		2	6		
87 Fr	2	2	6	2	6	10	2	6	10	14	2	6	10		2	6		1
88 Ra	2	2	6	2	6	10	2	6	10	14	2	6	10		2	6		2
89 Ac	2	2	6	2	6	10	2	6	10	14	2	6	10		2	6	1	2
90 Th	2	2	6	2	6	10	2	6	10	14	2	6	10		2	6	2	2
91 Pa	2	2	6	2	6	10	2	6	10	14	2	6	10	2	2	6	1	2
92 U	2	2	6	2	6	10	2	6	10	14	2	6	10	3	2	6	1	2
93 Np	2	2	6	2	6	10	2	6	10	14	2	6	10	4	2	6	1	2
94 Pu	2	2	6	2	6	10	2	6	10	14	2	6	10	5	2	6	1	2
95 Am	2	2	6	2	6	10	2	6	10	14	2	6	10	6	2	6	1	2
96 Cm	2	2	6	2	6	10	2	6	10	14	2	6	10	7	2	6	1	2
97 Bk	2	2	6	2	6	10	2	6	10	14	2	6	10	8	2	6	1	2
98 Cf	2	2	6	2	6	10	2	6	10	14	2	6	10	10	2	6		2
99 E	2	2	6	2	6	10	2	6	10	14	2	6	10	11	2	6		2
100 Fm	2	2	6	2	6	10	2	6	10	14	2	6	10	12	2	6		2
101 Md	2	2	6	2	6	10	2	6	10	14	2	6	10	13	2	6		2
102 No	2	2	6	2	6	10	2	6	10	14	2	6	10	14	2	6		2
103 Lw	2	2	6	2	6	10	2	6	10	14	2	6	10	14	2	6	1	2

The Periodic Table

tion elements evidently lies in the tighter binding of s electrons than d or f electrons in complex atoms, discussed in the previous section. The first element to exhibit this effect is potassium, whose outermost electron is in a $4s$ instead of a $3d$ substate. The difference in binding energy between $3d$ and $4s$ electrons is not very great, as can be seen in the configurations of chromium and copper. In both of these elements an additional $3d$ electron is present at the expense of a vacancy in the $4s$ subshell. In this connection another glance at Fig. 10.5 will be instructive.

The order in which electron subshells are filled in atoms is

$$1s, 2s, 2p, 3s, 3p, 4s, 3d, 4p, 5s, 4d, 5p, 6s, 4f, 5d, 6p, 7s, 6d$$

as we can see from Table 10.2 and Fig. 10.7. The remarkable similarities in chemical behavior among the lanthanides and actinides are easy to understand on the basis of this sequence. All of the lanthanides have the same $5s^2 5p^6 6s^2$ configurations but have incomplete $4f$ subshells. The addition of $4f$ electrons has virtually no effect on the chemical properties of the lanthanide elements, which are determined by the outer electrons. Similarly, all of the actinides have $6s^2 6p^6 7s^2$ configurations, and differ only in the numbers of their $5f$ and $6d$ electrons.

These irregularities in the binding energies of atomic electrons are also responsible for the lack of completely full outer shells in the heavier inert gases. Helium ($Z = 2$) and

FIGURE 10.7 The sequence of quantum states in an atom. Not to scale.

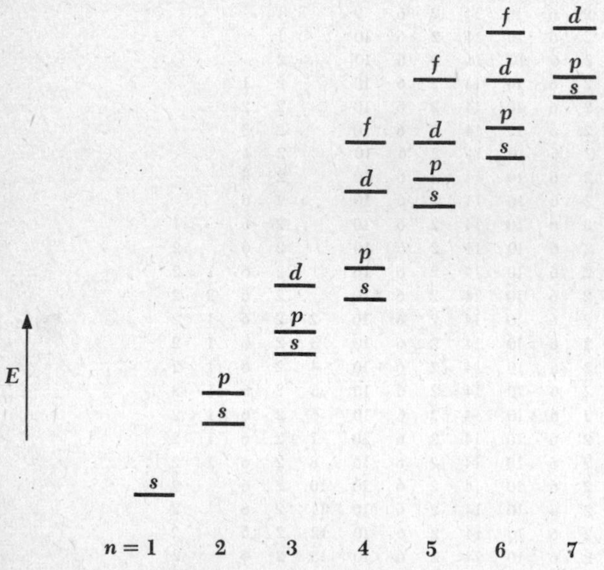

neon ($Z = 10$) contain closed K and L shells respectively, but argon ($Z = 18$) has only 8 electrons in its M shell, corresponding to closed 3s and 3p subshells. The reason the 3d subshell is not filled next is simply that 4s electrons have higher binding energies than 3d electrons, as we have said, and so the 4s subshell is filled first in potassium and calcium. As the 3d subshell is filled in successively heavier transition elements, there are still one or two outer 4s electrons that make possible chemical activity. Not until krypton ($Z = 36$) is another inert gas reached, and here a similarly incomplete outer shell occurs with only the 4s and 4p subshells filled. Following krypton is rubidium ($Z = 37$), which skips both the 4d and 4f subshells to have a 5s electron. The next inert gas is xenon ($Z = 54$), which has filled 4d, 5s, and 5p subshells, but now even the inner 4f subshell is empty as well as the 5d and 5f subshells. The same pattern recurs with the remainder of the inert gases.

While we have sketched the origins of only a few of the chemical and physical properties of the elements in terms of their electron configurations, many more can be quantitatively understood by similar reasoning.

10.6 Hund's Rule

In general, the electrons in an atom remain unpaired—that is, have parallel spins—whenever possible. This principle is called *Hund's rule*. The ferromagnetism of iron, cobalt, and nickel is a consequence of Hund's rule: their 3d subshells are only partially occupied, and the electrons in these subshells do not pair off to permit their spin magnetic moments to cancel out. In iron, for instance, five of the six 3d electrons have parallel spins, so that each iron atom has a large resultant magnetic moment. We shall examine other consequences of Hund's rule in Chap. 13 in connection with molecular bonding.

The origin of Hund's rule lies in the mutual repulsion of atomic electrons. Because of this repulsion, the farther apart the electrons in an atom are, the lower the energy of the atom. Electrons in the same subshell with the same spin must have different m_l values and accordingly are described by wave functions whose spatial distributions are different. Electrons with parallel spins are therefore more separated in space than if they paired off, and this arrangement, having less energy, is the more stable one.

10.7 Total Angular Momentum

Each electron in an atom has a certain orbital angular momentum **L** and a certain spin angular momentum **S**, both of which contribute to the total angular momentum **J** of the atom. Like all angular momenta, **J** is quantized, with a magnitude given by

10.16
$$J = \sqrt{J(J + 1)}\, \hbar$$
Total atomic angular momentum

and a component J_z in the z direction given by

10.17
$$J_z = M_J \hbar$$
z component of total atomic angular momentum

Total Angular Momentum   **241**

where J and M_J are the quantum numbers governing $\mathbf{J}$ and J_z. Our task in the remainder of this chapter is to look into the properties of $\mathbf{J}$. We shall do this in terms of the semi-classical *vector model* of the atom, which provides a more intuitively accessible framework for understanding angular-momentum considerations than does a purely quantum-mechanical approach.

Let us first consider an atom whose total angular momentum is provided by a single electron. Atoms of the elements in group I of the periodic table—hydrogen, lithium, sodium, and so on—are of this kind since they have single electrons outside closed inner shells (except for hydrogen, which has no inner electrons) and the exclusion principle assures that the total angular momentum and magnetic moment of a closed shell are zero. Also in this category are the ions He^+, Be^+, Mg^+, B^{++}, Al^{++}, and so on.

The magnitude L of the orbital angular momentum $\mathbf{L}$ of an atomic electron is determined by its orbital quantum number l according to the formula

10.18
$$L = \sqrt{l(l + 1)}\; \hbar$$

while the component L_z of $\mathbf{L}$ along the z axis is determined by the magnetic quantum number m_l according to the formula

10.19
$$L_z = m_l \hbar$$

Similarly the magnitude S of the spin angular momentum $\mathbf{S}$ is determined by the spin quantum number s (which has the sole value $+\frac{1}{2}$) according to the formula

10.20
$$S = \sqrt{s(s + 1)}\; \hbar$$

while the component S_z of $\mathbf{S}$ along the z axis is determined by the magnetic spin quantum number m_s according to the formula

10.21
$$S_z = m_s \hbar$$

Because $\mathbf{L}$ and $\mathbf{S}$ are vectors, they must be added vectorially to yield the total angular momentum $\mathbf{J}$:

10.22
$$\mathbf{J} = \mathbf{L} + \mathbf{S}$$

It is customary to use the symbols j and m_j for the quantum numbers that describe J and J_z for a single electron, so that

10.23
$$J = \sqrt{j(j + 1)}\; \hbar$$

10.24
$$J_z = m_j \hbar$$

To obtain the relationships among the various angular-momentum quantum numbers, it is simplest to start with the z components of the vectors $\mathbf{J}$, $\mathbf{L}$, and $\mathbf{S}$. Since J_z, L_z, and S_z are scalar quantities,

$$J_z = L_z \pm S_z$$
$$m_j\hbar = m_l\hbar \pm m_s\hbar$$

and

10.25 $$m_j = m_l \pm m_s$$

The possible values of m_l range from $+l$ through 0 to $-l$, and those of m_s are $\pm s$. The quantum number l is always an integer or 0 while $s = \frac{1}{2}$, and as a result m_j must be half-integral. The possible values of m_j also range from $+j$ through 0 to $-j$ in integral steps, and so, for any value of l,

10.26 $$j = l \pm s$$

Like m_j, j is always half-integral.

Because of the simultaneous quantization of **J**, **L**, and **S** they can have only certain specific relative orientations. This is a general conclusion; in the case of a one-electron atom, there are only two relative orientations possible. One of these corresponds to $j = l + s$, so that $J > L$, and the other to $j = l - s$, so that $J < L$. Figure 10.8 shows the two ways in which **L** and **S** can combine to form **J** when $l = 1$. Evidently the orbital and spin angular-momentum vectors can never be exactly parallel or antiparallel to each other or to the total angular-momentum vector.

FIGURE 10.8 The two ways in which **L** and **S** can be added to form **J** when $l = 1, s = \frac{1}{2}$.

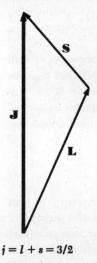

$j = l + s = 3/2$

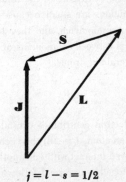

$j = l - s = 1/2$

Total Angular Momentum

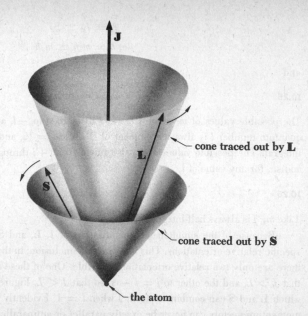

cone traced out by **L**

cone traced out by **S**

the atom

FIGURE 10.9 The orbital and spin angular-momentum vectors L and S precess about J according to the vector model of the atom.

The angular momenta **L** and **S** interact magnetically, as we saw in Sec. 10.2, and as a result exert torques on each other. If there is no external magnetic field, the total angular momentum **J** is conserved in magnitude and direction, and the effect of the internal torques can only be the precession of **L** and **S** around the direction of their resultant **J** (Fig. 10.9). However, if there is an external magnetic field **B** present, then **J** precesses about the direction of **B** while **L** and **S** continue precessing about **J**, as in Fig. 10.10. The possibility of different orientations of **J** relative to **B** is what gives rise to the anomalous Zeeman effect, since the different relative orientations involve slightly different energies.

Atomic nuclei also have intrinsic angular momenta and magnetic moments, as we shall see in Chap. 22, and these contribute to the total atomic angular momenta and magnetic moments. These contributions are small because nuclear magnetic moments are $\sim 10^{-3}$ the magnitude of electronic moments, and they lead to the *hyperfine structure* of spectral lines with typical spacings between components of $\sim 10^{-2}$ A as compared with typical fine-structure spacings of several angstroms.

10.8 LS Coupling

When more than one electron contributes orbital and spin angular momenta to the total angular momentum **J** of an atom, **J** is still the vector sum of these individual momenta. Because the electrons involved interact with one another, the manner in which their individual momenta L_i and S_i add together to form **J** follows certain definite patterns depending upon the circumstances. The usual pattern for all but the heaviest atoms is that the orbital angular momenta L_i of the various electrons are coupled together electrostatically into a single re-sultant L and the spin angular momenta S_i are coupled together independently into another single resultant S; we shall examine the reasons for this behavior later in this section. The momenta L and S then interact magnetically via the spin-orbit effect to form a total angular momentum J. This scheme, called *LS coupling*, may be summarized as follows:

10.27
$$L = \Sigma L_i$$
$$S = \Sigma S_i$$
$$J = L + S$$
LS coupling

As usual, L, S, J, L_z, S_z, and J_z are quantized, with the respective quantum numbers being L, S, J, M_L, M_S, and M_J. Hence

10.28 $$L = \sqrt{L(L + 1)}\,\hbar$$

10.29 $$L_z = M_L\hbar$$

10.30 $$S = \sqrt{S(S + 1)}\,\hbar$$

10.31 $$S_z = M_S\hbar$$

10.32 $$J = \sqrt{J(J + 1)}\,\hbar$$

10.33 $$J_z = M_J\hbar$$

Both L and M_L are always integers or 0, while the other quantum numbers are half-integral if an odd number of electrons is involved and integral or 0 if an even number of electrons is involved.

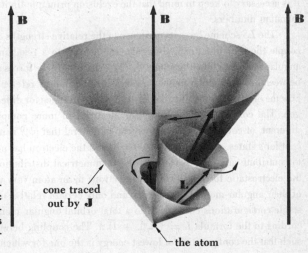

FIGURE 10.10 In the presence of an external magnetic field B, the total angular-momentum vector J precesses about B according to the vector model of the atom.

cone traced out by **J**

L

the atom

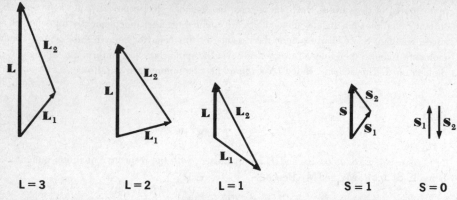

L = 3   L = 2   L = 1   S = 1   S = 0

FIGURE 10.11 When $l_1 = 1$, $s_1 = \frac{1}{2}$, and $l_2 = 2$, $s_2 = \frac{1}{2}$, there are three ways in which L_1 and L_2 can combine to form L and two ways in which S_1 and S_2 can combine to form S.

As an example, let us consider two electrons, one with $l_1 = 1$ and the other with $l_2 = 2$. There are three ways in which L_1 and L_2 can be combined into a single vector L that is quantized according to Eq. 10.28, as shown in Fig. 10.10. These correspond to L = 1, 2, and 3 since all values of L are possible from $l_1 + l_2$ to $l_1 - l_2$. The spin quantum number s is always $+\frac{1}{2}$, so there are two possibilities for the sum $S_1 + S_2$, as in Fig. 10.11, that correspond to S = 0 and S = 1. We note that L_1 and L_2 can never be exactly parallel to L, nor S_1 and S_2 to S, except when the vector sum is 0. The quantum number J can have all values between L + S and L − S, which in this case means that J can be 0, 1, 2, 3, or 4. In working out the possible values of L, S, and J for a many-electron atom, it is necessary to keep in mind that the exclusion principle limits the possible sets of electron quantum numbers.

The *LS* scheme owes its existence to the relative strengths of the electrostatic forces that couple the individual orbital angular momenta into a resultant L and the individual spin angular momenta into a resultant S. The origins of these forces are interesting. The coupling between orbital angular momenta can be understood by reference to Fig. 9.13, which shows how the electron probability density $|\psi|^2$ varies in space for different quantum states in hydrogen. The corresponding patterns for electrons in more complex atoms will be somewhat different, of course, but it remains true in general that $|\psi|^2$ is not spherically symmetric except for s states. (In the latter case $l = 0$ and the electron has no orbital angular momentum to contribute anyway.) Because of the asymmetrical distributions of their charge densities, the electrostatic forces between the electrons in an atom vary with the relative orientations of their angular-momentum vectors, and only certain relative orientations are stable. These stable configurations correspond to a total orbital angular momentum that is quantized according to the formula $L = \sqrt{L(L + 1)}\ \hbar$. The coupling between the various L_i is usually such that the configuration of lowest energy is the one for which L is a maximum. This effect is easy to understand if we imagine two electrons in the same Bohr orbit. Because the electrons repel each other electrostatically, they tend to revolve around the nucleus in the same direction, which maximizes L. If they revolved in opposite directions to minimize L, the electrons would pass each other more frequently, leading to a higher energy for the system.

The origin of the strong coupling between electron spins is harder to visualize because it is a purely quantum-mechanical effect with no classical analog. (The direct interaction between the intrinsic electron magnetic moments, it should be noted, is insignificant and not responsible for the coupling between electron spin angular momenta.) The basic idea is that the complete wave function $\psi(1, 2, \ldots, n)$ of a system of n electrons is the product of a wave function $u(1, 2, \ldots, n)$ that describes the coordinates of the electrons and a spin function $s(1, 2, \ldots, n)$ that describes the orientations of their spins. As we saw in Sec. 10.3, the complete function $\psi(1, 2, \ldots, n)$ must be antisymmetric, which means that $u(1, 2, \ldots, n)$ is not independent of $s(1, 2, \ldots, n)$. A change in the relative orientations of the electron spin angular-momentum vectors must therefore be accompanied by a change in the electronic configuration of the atom, which means a change in the atom's electrostatic potential energy. To go from one total spin angular momentum S to a different one involves altering the structure of the atom, and therefore a strong electrostatic force, besides altering the directions of the spin angular momenta $S_1, S_2, \ldots, S_n$, which requires only a weak magnetic force. This situation is what is described when it is said that the spin momenta S_i are strongly coupled together electrostatically. The S_i always combine into a ground-state configuration in which S is a maximum. This is an example of Hund's rule; as mentioned earlier, electrons with parallel spins have different m_l values and are described by different wave functions, which means that there is a greater average separation in space of the electrons and accordingly a lower total energy.

Although we shall not try to justify this conclusion, the combination of L and S that makes J a *minimum* results in the lowest energy.

10.9   jj Coupling

The electrostatic forces that couple the L_i into a single vector L and the S_i into another vector S are stronger than the magnetic spin-orbit forces that couple L and S to form J in light atoms, and dominate the situation even when a moderate external magnetic field is applied. (In the latter case the precession of J around B is accordingly slower than the precession of L and S around J.) However, in heavy atoms the nuclear charge becomes great enough to produce spin-orbit interactions comparable in magnitude to the electrostatic ones between the L_i and between the S_i, and the LS coupling scheme begins to break down. A similar breakdown occurs in strong external magnetic fields (typically ~ 10 webers/m²), which produces the Paschen-Back effect in atomic spectra. In the limit of the failure of LS coupling, the total angular momenta J_i of the individual electrons add together directly to form the angular momentum J of the entire atom, a situation referred to as *jj coupling* since each J_i is described by a quantum number j in the manner described in Sec. 10.7. Hence

$$J_i = L_i + S_i$$
$$J = \Sigma J_i$$

In Sec. 9.5 we saw that individual orbital angular-momentum states are customarily described by a lowercase letter, with s corresponding to $l = 0$, p to $l = 1$, d to $l = 2$, and so on. A similar scheme using capital letters is used to designate the entire electronic state of an atom according to its total orbital angular-momentum quantum number L as follows:

$$L = 0 \quad 1 \quad 2 \quad 3 \quad 4 \quad 5 \quad 6 \ldots$$
$$S \quad P \quad D \quad F \quad G \quad H \quad I \ldots$$

A superscript number before the letter (^2P for instance) is used to indicate the *multiplicity* of the state, which is the number of different possible orientations of L and S and hence the number of different possible values of J. The multiplicity is equal to $2S + 1$ in the usual situation where $L > S$, since J ranges from $L + S$ through 0 to $L - S$. Thus when $S = 0$, the multiplicity is 1 (a *singlet* state) and $J = L$; when $S = \frac{1}{2}$, the multiplicity is 2 (a *doublet* state) and $J = L \pm \frac{1}{2}$; when $S = 1$, the multiplicity is 3 (a *triplet* state) and $J = L + 1, L$, or $L - 1$; and so on. (In a configuration in which $S > L$, the multiplicity is given by $2L + 1$.) The total angular-momentum quantum number J is used as a subscript after the letter, so that a $^2P_{3/2}$ state (read as "doublet P three-halves") refers to an electronic configuration in which $S = \frac{1}{2}$, $L = 1$, and $J = \frac{3}{2}$. For historical reasons, these designations are called *term symbols*.

In the event that the angular momentum of the atom arises from a single outer electron, the total quantum number n of this electron can be used as a prefix: thus the ground state of the sodium atom is described by $3^2S_{1/2}$, since its electronic configuration has an electron with $n = 3$, $l = 0$, and $s = \frac{1}{2}$ (and hence $j = \frac{1}{2}$) outside closed $n = 1$ and $n = 2$ shells. For consistency it is conventional to denote the above state by $3^2S_{1/2}$ with the superscript 2 indicating a doublet, even though there is only a single possibility for J since $L = 0$.

Problems

1. If atoms could contain electrons with principal quantum numbers up to and including $n = 6$, how many elements would there be?

2. The ionization energies of the elements of atomic numbers 20 through 29 are very nearly equal. Why should this be so when considerable variations exist in the ionization energies of other consecutive sequences of elements?

3. Find the S, L, and J values that correspond to each of the following states: $^1S_0$, $^3P_2$, $^2D_{3/2}$, $^5F_5$, $^6H_{5/2}$.

4. The carbon atom has two $2s$ electrons and two $2p$ electrons outside a filled inner shell.

Its ground state is $^3P_0$. What are the term symbols of the other allowed states, if any? Why would you think the $^3P_0$ state is the ground state?

5. The lithium atom has one $2s$ electron outside a filled inner shell. Its ground state is $^2S_{1/2}$. What are the term symbols of the other allowed states, if any? Why would you think the $^2S_{1/2}$ state is the ground state?

6. The magnesium atom has two $3s$ electrons outside filled inner shells. Find the term symbol of its ground state.

7. The aluminum atom has two $3s$ electrons and one $3p$ electron outside filled inner shells. Find the term symbol of its ground state.

8. Many years ago it was pointed out that the atomic numbers of the rare gases are given by the following scheme:

$$Z(\text{He}) = 2(1^2) = 2$$
$$Z(\text{Ne}) = 2(1^2 + 2^2) = 10$$
$$Z(\text{Ar}) = 2(1^2 + 2^2 + 2^2) = 18$$
$$Z(\text{Kr}) = 2(1^2 + 2^2 + 2^2 + 3^2) = 36$$
$$Z(\text{Xe}) = 2(1^2 + 2^2 + 2^2 + 3^2 + 3^2) = 54$$
$$Z(\text{Rn}) = 2(1^2 + 2^2 + 2^2 + 3^2 + 3^2 + 4^2) = 86$$

Explain the origin of this scheme in terms of atomic theory.

9. A beam of electrons enters a uniform magnetic field of flux density 1.2 webers/m². Find the energy difference between electrons whose spins are parallel and antiparallel to the field.

10. How does the agreement between observations of the normal Zeeman effect and the theory of this effect tend to confirm the existence of electrons as independent entities within atoms?

11. A sample of a certain element is placed in a magnetic field of flux density 0.3 weber/m². How far apart are the Zeeman components of a spectral line of wavelength 4,500 A?

12. Why does the normal Zeeman effect occur only in atoms with an even number of electrons?

13. The spin-orbit effect splits the $3P \rightarrow 3S$ transition in sodium (which gives rise to the yellow light of sodium-vapor highway lamps) into two lines, 5,890 A corresponding to $3P_{3/2} \rightarrow 3S_{1/2}$ and 5,896 A corresponding to $3P_{1/2} \rightarrow 3S_{1/2}$. Use these wavelengths to calculate the effective magnetic induction experienced by the outer electron in the sodium atom as a result of its orbital motion.

14. The magnetic moment μ_J of an atom in which LS coupling holds has the magnitude

$$\mu_J = \sqrt{J(J + 1)}g_J\mu_B$$

where $\mu_B = e\hbar/2m$ is the Bohr magneton and

$$g_J = 1 + \frac{J(J + 1) - L(L + 1) + S(S + 1)}{2J(J + 1)}$$

is the *Landé g factor*. (a) Derive this result with the help of the law of cosines starting from the fact that, averaged over time, only the components of μ_L and μ_S parallel to $\mathbf{J}$ contribute to μ_J. (b) Consider an atom that obeys LS coupling that is in a weak magnetic field $\mathbf{B}$ in which the coupling is preserved. How many substates are there for a given value of J? What is the energy difference between different substates?

15. The ground state of chlorine is $^2P_{3/2}$. Find its magnetic moment (see previous problem). Into how many substates will the ground state split in a weak magnetic field?

16. Show that, if the angle between the directions of $\mathbf{L}$ and $\mathbf{S}$ in Fig. 10.8 is θ.

$$\cos\theta = \frac{j(j + 1) - l(l + 1) - s(s + 1)}{2\sqrt{l(l + 1)s(s + 1)}}$$

Chapter 11
Atomic Spectra

The quantum theory of the atom was developed in close contact with the experimental information provided by the line spectra of the elements, and the spectacular ability of the theory to explain the details of these spectra testifies to its validity as a description of nature. In this chapter we shall first examine the production of spectral lines in terms of the quantum theory, and then we shall consider several typical examples of atomic spectra to see how they fit into the picture of the atom developed in Chaps. 9 and 10.

11.1 Origin of Spectral Lines

In formulating his theory of the hydrogen atom, Bohr was obliged to postulate that the frequency ν of the radiation emitted by an atom dropping from an energy level E_m to a lower level E_n is

$$\nu = \frac{E_m - E_n}{h}$$

It is not difficult to show that this relationship arises naturally from the quantum theory of the atom we have been discussing. Our starting point is the straightforward assertion that an atomic electron whose average position relative to the nucleus is constant in time does not radiate, while if its average position relative to the nucleus oscillates, electromagnetic waves are emitted whose frequency is the same as that of the oscillation. For simplicity we shall consider the electron's motion in the x direction only.

The time-dependent wave function Ψ_n of an electron in a state of quantum number n and energy E_n is the product of a time-independent wave function ψ_n and a time-varying function whose frequency is

$$\nu_n = \frac{E_n}{h}$$

Hence

11.1 $$\Psi_n = \psi_n e^{-(iE_n/\hbar)t}$$

and

11.2 $$\Psi_n^* = \psi_n^* e^{+(iE_n/\hbar)t}$$

The average position of such an electron is the expectation value (Sec. 7.6) of x, namely

11.3
$$\bar{x} = \int_{-\infty}^{\infty} \Psi_n^* x \Psi_n \, dx$$

Inserting the wave functions of Eqs. 11.1 and 11.2,

$$\bar{x} = \int_{-\infty}^{\infty} \psi_n^* x \psi_n e^{[(iE_n/\hbar) - (iE_n/\hbar)]t} \, dx$$

11.4
$$= \int_{-\infty}^{\infty} \psi_n^* x \psi_n \, dx$$

which is constant in time since ψ_n and ψ_n^* are, by definition, functions of position only. The electron does not oscillate, and no radiation occurs. Thus quantum mechanics predicts that an atom in a specific quantum state does not radiate; this agrees with observation, though not with classical physics.

We are now in a position to consider an electron changing from one energy state to another. Let us formulate a definite problem: an atom is in its ground state when, at $t = 0$, an excitation process of some kind (a beam of radiation, say, or collisions with other particles) begins to act upon it. Subsequently we find that the atom emits radiation corresponding to a transition from an excited state of energy E_m to the ground state, and we conclude that at some time during the intervening period the atom existed in the state m. The wave function Ψ of an electron capable of existing in states n or m may be written

11.5
$$\Psi = a\Psi_n + b\Psi_m$$

where a^*a is the probability that the electron is in state n and b^*b the probability that it is in state m. Of course, it must always be true that $a^*a + b^*b = 1$. At $t = 0$, $a = 1$ and $b = 0$ by hypothesis; when the electron is in the excited state, $a = 0$ and $b = 1$; and ultimately $a = 1$ and $b = 0$ once more. While the electron is in either state, there is no radiation, but when it is in the midst of the transition from m to n (that is, when both a and b have nonvanishing values), electromagnetic waves are produced. Substituting the composite wave function of Eq. 11.5 into Eq. 11.3, we obtain for the average electron position

$$\bar{x} = \int_{-\infty}^{\infty} (a^*\Psi_n^* + b^*\Psi_m^*) \, x \, (a\Psi_n + b\Psi_m) \, dx$$

11.6
$$= \int_{-\infty}^{\infty} x(a^2\Psi_n^*\Psi_n + b^*a\Psi_m\Psi_n + a^*b\Psi_n^*\Psi_m + b^2\Psi_m^*\Psi_m) \, dx$$

(Here, as before, we let $a^*a = a^2$ and $b^*b = b^2$.) The first and last integrals are constants, according to Eq. 11.4, and so the second and third integrals are the only ones capable of contributing to a time variation in $\bar{x}$. Because x is not an operator, its position in each integral has no significance.

With the help of Eqs. 11.1 to 11.3 we may expand Eq. 11.6 to obtain

11.7
$$\bar{x} = a^2 \int_{-\infty}^{\infty} x\psi_n^*\psi_n \, dx + b^*a \int_{-\infty}^{\infty} x\psi_m^* e^{+(iE_m/\hbar)t}\psi_n e^{-(iE_n/\hbar)t} \, dx$$

$$+ a^*b \int_{-\infty}^{\infty} x\psi_n^* e^{+(iE_n/\hbar)t}\psi_m e^{-(iE_m/\hbar)t} \, dx + b^2 \int_{-\infty}^{\infty} x\psi_m^*\psi_m \, dx$$

In the case of a finite bound system of two states, which is what we have here,

$$\psi_n^*\psi_m = \psi_m^*\psi_n \qquad \text{and} \qquad a^*b = b^*a$$

and so we can combine the time-varying terms of Eq. 11.7 into the single term

11.8
$$a^*b \int_{-\infty}^{\infty} x\psi_n^*\psi_m \left[e^{(i/\hbar)(E_m-E_n)t} + e^{-(i/\hbar)(E_m-E_n)t} \right] dx$$

Now

$$e^{i\theta} = \cos\theta + i\sin\theta$$
$$e^{-i\theta} = \cos\theta - i\sin\theta$$

so that

$$e^{i\theta} + e^{-i\theta} = 2\cos\theta$$

Hence Eq. 11.8 simplifies to

$$2a^*b \cos\left(\frac{E_m - E_n}{\hbar}\right) t \int_{-\infty}^{\infty} x\psi_n^*\psi_m \, dx$$

which contains the time-varying factor

$$\cos\left(\frac{E_m - E_n}{\hbar}\right) t = \cos 2\pi \left(\frac{E_m - E_n}{h}\right) t$$

$$= \cos 2\pi\nu t$$

The electron's position therefore oscillates sinusoidally at the frequency

11.9
$$\nu = \frac{E_m - E_n}{h}$$

and the full expression for $\bar{x}$, the average position of the electron, is

11.10
$$\bar{x} = a^2 \int_{-\infty}^{\infty} x\psi_n^*\psi_n \, dx + b^2 \int_{-\infty}^{\infty} x\psi_m^*\psi_m \, dx + 2a^*b \cos 2\pi\nu t \int_{-\infty}^{\infty} x\psi_n^*\psi_m \, dx$$

When the electron is in state n or state m, the probabilities b^2 or a^2 respectively are zero, and the electron is, on the average, stationary. When the electron is undergoing a transition between these states, its average position oscillates with the frequency v. This frequency is identical with that postulated by Bohr and verified by experiment, and, as we have seen, Eq. 11.9 can be derived using quantum mechanics without making any special assumptions.

It is interesting to note that the frequency of the radiation is the same frequency as the beats we might imagine to be produced if the electron simultaneously existed in both the n and m states, whose characteristic frequencies are respectively E_n/h and E_m/h.

11.2 Selection Rules

It was not necessary for us to know the values of the probabilities a and b as functions of time, nor the electron wave functions ψ_n and ψ_m, in order to determine v. We must know these quantities, however, if we wish to compute the actual rate at which radiation is emitted. The uncertainty principle prohibits us from determining a and b with any precision for a particular atom, but their average values can be ascertained for a group of many atoms. Even when both a and b are finite, though, in order for a transition between the two states to take place, the integral

$$\int_{-\infty}^{\infty} x \, \psi_n^* \, \psi_m \, dx$$

cannot be zero, since the intensity of the radiation is proportional to it. Transitions for which this integral is finite are called *allowed transitions*, while those for which it is zero are called *forbidden transitions*.

In the case of the hydrogen atom, three quantum numbers are needed to specify the initial and final states involved in a radiative transition, and the integral must be over all space. If the total, orbital, and magnetic quantum numbers of the initial state are n', l', m_l' respectively and those of the final state are n, l, m_l, and the coordinate u represents either the x, y, or z coordinate, the condition for an allowed transition is

11.11
$$\int_{-\infty}^{\infty} u\psi_{n,l,m_l} \, \psi_{n',l',m_l'}^* \, dV \neq 0$$

When u is taken as x, for example, the radiation referred to is that which would be produced by an ordinary dipole antenna lying along the x axis. Since the wave functions ψ_{n,l,m_l} for the hydrogen atom are known, Eq. 11.11 can be evaluated for $u = x$, $u = y$, and $u = z$ for all pairs of states differing in one or more quantum numbers. When this is done, it is found that the only transitions that can occur are those in which the orbital quantum number l changes by $+1$ or -1 and the magnetic quantum number m_l does not change or changes by $+1$ or -1; in other words, the condition for an allowed transition is that

11.12
$$\Delta l = \pm 1$$

Selection rules

11.13
$$\Delta m_l = 0, \pm 1$$

The change in total quantum number n is not restricted. Equations 11.12 and 11.13 are known as the *selection rules* for allowed transitions.

In order to get an intuitive idea of the physical basis for these selection rules, let us refer to Fig. 9.13. There we see that, for instance, a transition from a $2p$ state to a $1s$ state involves a change from one probability-density distribution to another such that the oscillating charge during the transitions behaves like an electric dipole antenna. On the other hand, a transition from a $2s$ state to a $1s$ state involves a change from a spherically symmetric probability-density distribution to another spherically symmetric distribution, which means that the oscillations that take place are like those of a charged sphere that alternately expands and contracts. Oscillations of this kind do not lead to the radiation of electromagnetic waves.

The selection rule requiring that l change by ± 1 if an atom is to radiate means that an emitted photon carries off angular momentum equal to the difference between the angular momenta of the atom's initial and final states. The classical analog of a photon with angular momentum is a circularly polarized electromagnetic wave, so that this notion is not unique with quantum theory.

The selection rule for the magnetic quantum number m_l is quite easy to derive for the hydrogen atom. We start by recalling that x, y, and z are given in spherical polar coordinates by the formulas

$$x = r \sin \theta \cos \phi$$
$$y = r \sin \theta \sin \phi$$
$$z = r \cos \theta$$

and that the element of volume dV in these coordinates is

$$dV = r^2 \sin \theta \, dr \, d\theta \, d\phi$$

Hence

$$\bar{x} = \int_{-\infty}^{\infty} x \, \psi_{n,l,m_l}^* \, \psi_{n,l,m_l'}' \, dV$$

11.14
$$= \int_0^{\infty} \int_0^{\pi} \int_0^{2\pi} r \sin \theta \cos \phi \, \psi_{n,l,m_l}^* \, \psi_{n,l,m_l'}' \, r^2 \sin \theta \, dr \, d\theta \, d\phi$$

For the hydrogen atom, as we know from Chap. 9, the wave function ψ_{n,l,m_l} can be expressed as

11.15
$$\psi_{n,l,m_l}(r,\theta,\phi) = R_{n,l}(r)\Theta_{l,m_l}(\theta)\Phi_{m_l}(\phi)$$

and so

11.16 $$\bar{x} = \int_0^{\infty} r^3 R_{n,l}^* \, R_{n'l}' dr \int_0^{\pi} \sin^2\theta \, \Theta_{l,m_l}^* \Theta_{l',m_l'} \, d\theta \int_0^{2\pi} \Phi_{m_l}^* \Phi_{m_l'} \cos \phi \, d\phi$$

From Eq. 9.44 we have for the normalized azimuthal wave function

$$\Phi_{m_l} = \frac{1}{\sqrt{2\pi}} e^{im_l\phi}$$

and so

11.17

$$\Phi_{m_l}^* = \frac{1}{\sqrt{2\pi}} e^{-im_l\phi}$$

11.18

$$\Phi_{m_l'} = \frac{1}{\sqrt{2\pi}} e^{im_l'\phi}$$

Also,

$$\cos\phi = \frac{e^{i\phi} + e^{-i\phi}}{2}$$

The third integral of Eq. 11.16 accordingly becomes

11.19 $$\int_0^{2\pi} \Phi_{m_l}^* \Phi_{m_l'} \cos\phi \, d\phi = \frac{1}{4\pi} \int_0^{2\pi} [e^{-i(m_l-m_l'+1)\phi} + e^{-i(m_l-m_l'-1)\phi}] \, d\phi$$

where $(m_l - m_l' + 1)$ and $(m_l - m_l' - 1)$ are equal either to 0 or to $k = \pm 1, \pm 2, \pm 3, \ldots$ because m_l and m_l' are respectively limited to $0, \pm 1, \pm 2, \ldots, \pm l$ and $0, \pm 1, \pm 2, \ldots, \pm l'$. We note that

$$e^{-ik\phi} = \cos k\phi - i \sin k\phi$$

and so

$$\int_0^{2\pi} e^{-ik\phi} \, d\phi = \int_0^{2\pi} \cos k\phi \, d\phi - i \int_0^{2\pi} \sin k\phi \, d\phi$$

11.20 $$= 0 \qquad k = \pm 1, \pm 2, \pm 3, \ldots$$

Hence Eqs. 11.19 and 11.16 equal 0 and no transition can occur unless

$$m_l - m_l' + 1 = 0 \qquad \text{or} \qquad m_l - m_l' - 1 = 0$$

which is the same as

11.21 $$\Delta m_l = m_l - m_l' = \pm 1$$

This selection rule is also obtained by evaluating Eq. 11.11 with $u = y$.

For $u = z$ we have

$$\bar{z} = \int_{-\infty}^{\infty} z \, \psi_{n,l,m_l}^* \, \psi_{n',l',m_l'} \, dV$$

11.22 $$= \int_0^{\infty} r^3 R_{n,l}^* R_{n',l'} \, dr \int_0^{\pi} \sin\theta \cos\theta \, \Theta_{l,m} \Theta_{l',m_l'} \, d\theta \int_0^{2\pi} \Phi_{m_l}^* \Phi_{m_l'} \, d\phi$$

The third integral is

$$\int_0^{2\pi} \Phi_{m_l}^* \Phi_{m_l'} \, d\phi = \frac{1}{2\pi} \int_0^{2\pi} e^{-i(m_l - m_l')\phi} \, d\phi$$

and is equal to 0 in view of Eq. 11.20 unless

$$m_l = m_l' \qquad \Delta m_l = 0$$

The complete selection rule for m_l in the hydrogen atom is therefore

11.23 $$\Delta m_l = 0, \pm 1$$

as stated in Eq. 11.13.

Although we have verified Eq. 11.13 only for the hydrogen atom, both it and Eq. 11.12 ($\Delta l = \pm 1$) hold for transitions that involve a single outer-shell electron in any atom.

11.3 One-electron Spectra

Let us now consider the chief features of the spectra of the various elements. Before we examine some representative examples, it should be mentioned that further complications exist which have not been considered here, for instance those that originate in relativistic effects and in the coupling between electrons and vacuum fluctuations in the electromagnetic field (see Sec. 16.7). These additional factors split certain energy states into closely spaced substates and therefore represent other sources of fine structure in spectral lines.

Figure 11.1 shows the various states of the hydrogen atom classified by their total quantum number n and orbital angular-momentum quantum number l. The selection rule for allowed transitions here is $\Delta l = \pm 1$, which is illustrated by the transitions shown. To indicate some of the information absent from a simple diagram of this kind, the detailed structures of the $n = 2$ and $n = 3$ levels are pictured; not only are all substates of the same n and different j separated in energy, but the same is true of states of the same n and j but with different l. The latter effect is most marked for states of small n and l, and was first established in the "Lamb shift" of the $2^2S_{1/2}$ state relative to the $2^2P_{1/2}$ state. The various separations conspire to split the H_α spectral line (Fig. 6.5) into seven closely spaced components.

The sodium atom has a single 3s electron outside closed inner shells, and so, if we assume that the 10 electrons in its inner core completely shield $+10e$ of nuclear charge, the outer electron is acted upon by an effective nuclear charge of $+e$ just as in the hydrogen atom. Hence we expect, as a first approximation, that the energy levels of sodium will be the same as those of hydrogen except that the lowest one will correspond to $n = 3$ instead of $n = 1$ because of the exclusion principle. Figure 11.2 is the energy-level diagram for sodium and, by comparison with the hydrogen levels also shown, there is indeed agreement

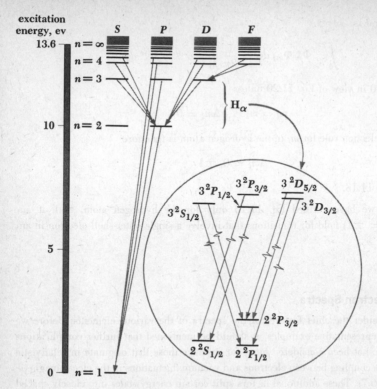

FIGURE 11.1 Energy-level diagram for hydrogen showing the origins of some of the more prominent spectral lines. The detailed structures of the $n = 2$ and $n = 3$ levels and the transitions that lead to the various components of the Hα line are pictured in the inset.

for the states of highest l, that is, for the states of highest angular momentum. To understand the reason for the discrepancies at lower values of l, we need only refer to Fig. 9.12 to see how the probability for finding the electron in a hydrogen atom varies with distance from the nucleus. The smaller the value of l for a given n, the closer the electron gets to the nucleus on occasion; this is perhaps even more obvious in Fig. 9.13. Although the sodium wave functions are not identical with those of hydrogen, their general behavior is similar, and accordingly we expect the outer electron in a sodium atom to penetrate the core of inner electrons most often when it is in an s state, less often when it is in a p state, still less often when it is in a d state, and so on. The less shielded an outer electron is from the full nuclear charge, the greater the average force acting on it, and the smaller (that is, the more negative) its total energy. For this reason the states of small l in sodium are displaced downward from their equivalents in hydrogen, as in Fig. 11.2, and there are pronounced differences in energy between states of the same n but different l.

11.4 Two-electron Spectra

A single electron is responsible for the energy levels of both hydrogen and sodium. However, there are two 1s electrons in the ground state of helium, and it is interesting to consider the effect of LS coupling on the properties and behavior of the helium atom. To do this we first note the selection rules for allowed transitions under LS coupling:

11.24 $$\Delta L = 0, \pm 1$$

11.25 $$\Delta J = 0, \pm 1$$ LS selection rules

11.26 $$\Delta S = 0$$

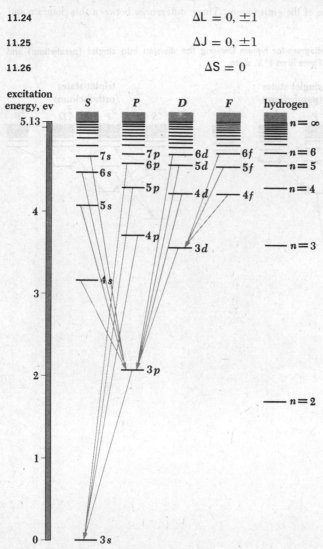

FIGURE 11.2 Energy-level diagram for sodium. The energy levels of hydrogen are included for comparison.

When only a single electron is involved, $\Delta L = 0$ is prohibited and $\Delta L = \Delta l = \pm 1$ is the only possibility. Furthermore, J must change when the initial state has $J = 0$, so that $J = 0 \nrightarrow J = 0$ is prohibited.

The helium energy-level diagram is shown in Fig. 11.3. The various levels represent configurations in which one electron is in its ground state and the other is in an excited state, but because the angular momenta of the two electrons are coupled, it is proper to consider the levels as characteristic of the entire atom. Three differences between this diagram and

FIGURE 11.3 Energy-level diagram for helium showing the division into singlet (parahelium) and triplet (orthohelium) states. There is no 1^3S_1 state.

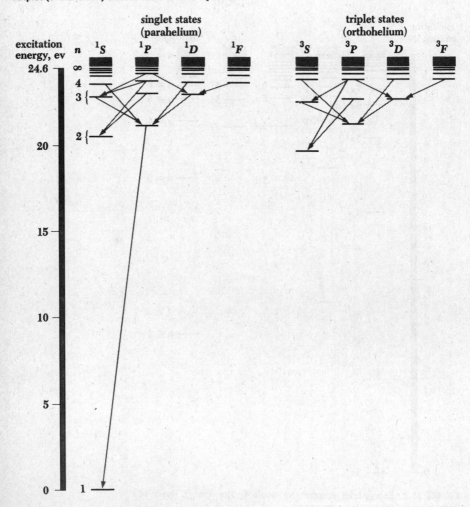

the corresponding ones for hydrogen and sodium are conspicuous. First, there is the division into singlet and triplet states, which are, respectively, states in which the spins of the two electrons are antiparallel (to give $S = 0$) and parallel (to give $S = 1$). Because of the selection rule $\Delta S = 0$, no allowed transitions can occur between singlet states and triplet states, and the helium spectrum arises from transitions in one set or the other. Helium atoms in singlet states (antiparallel spins) constitute *parahelium* and those in triplet states (parallel spins) constitute *orthohelium*. An orthohelium atom can lose excitation energy in a collision and become one of parahelium, while a parahelium atom can gain excitation energy in a collision and become one of orthohelium; ordinary liquid or gaseous helium is therefore a mixture of both. The lowest triplet states are called *metastable* because, in the absence of collisions, an atom in one of them can retain its excitation energy for a relatively long time (a second or more) before radiating.

The second obvious peculiarity in Fig. 11.3 is the absence of the 1^3S state. The lowest triplet state is 2^3S, although the lowest singlet state is 1^1S. The 1^3S state is missing as a consequence of the exclusion principle, since in this state the two electrons would have parallel spins and therefore identical sets of quantum numbers. Third, the energy difference between the ground state and the lowest excited state is relatively large, which reflects the tight binding of closed-shell electrons discussed in Chap. 10. The ionization energy of helium—the work that must be done to remove an electron from a helium atom—is 24.6 ev, the highest of any element.

The last energy-level diagram we shall consider is that of mercury, which has two electrons outside an inner core of 78 electrons in closed shells or subshells (Table 10.2). We expect a division into singlet and triplet states as in helium, but because the atom is so heavy we might also expect signs of a breakdown in the LS coupling of angular momenta. As Fig. 11.4 reveals, both of these expectations are realized, and several prominent lines in the mercury spectrum arise from transitions that violate the $\Delta S = 0$ selection rule. The transition $^3P_1 \rightarrow {}^1S_0$ is an example and is responsible for the strong 2,537-A line in the ultraviolet. To be sure, this does not mean that the transition probability is necessarily very high, since the three $^3P_1$ states are the lowest of the triplet set and therefore tend to be highly populated in excited mercury vapor. The $^3P_0 \rightarrow {}^1S_0$ and $^3P_2 \rightarrow {}^1S_0$ transitions, respectively, violate the rules that forbid transitions from $J = 0$ to $J = 0$ and that limit ΔJ to 0 or ± 1, as well as violating $\Delta S = 0$, and hence are considerably less likely to occur than the $^3P_1 \rightarrow {}^1S_0$ transition. The $^3P_0$ and $^3P_2$ states are therefore metastable and, in the absence of collisions, an atom can persist in either of them for a relatively long time. The strong spin-orbit interaction in mercury that leads to the partial failure of LS coupling is also responsible for the wide spacing of the elements of the ^3P triplets.

11.5 X-ray Spectra

In Chap. 3 we learned that the X-ray spectra of targets bombarded by fast electrons exhibit narrow spikes at wavelengths characteristic of the target material in addition to a continuous

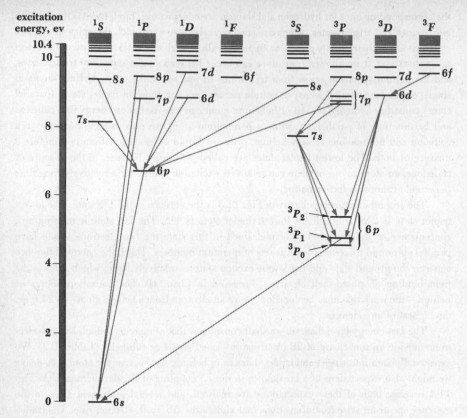

FIGURE 11.4 Energy-level diagram for mercury. In each excited level one outer electron is in the ground state, and the designation of the levels in the diagram corresponds to the state of the other electron.

distribution of wavelengths down to a minimum wavelength inversely proportional to the electron energy. The continuous X-ray spectrum is the result of the inverse photoelectric effect, with electron kinetic energy being transformed into photon energy $h\nu$. The discrete spectrum, on the other hand, has its origin in electronic transitions within atoms that have been disturbed by the incident electrons.

Transitions involving the outer electrons of an atom usually involve only a few electron volts of energy, and even removing an outer electron requires at most 24.6 ev (for helium). These transitions accordingly are associated with photons whose wavelengths lie in or near the visible part of the electromagnetic spectrum, as is evident from the diagram in the back endpapers of this book. The inner electrons of heavier elements are a quite different matter, because these electrons experience all or much of the full nuclear charge without shielding by intervening electron shells and in consequence are very tightly bound. In sodium, for example, only 5.13 ev is needed to remove the outermost 3s electron, while the corresponding figures for the inner ones are 31 ev for each 2p electron, 63 ev for each 2s electron, and 1.041 ev for each 1s electron. Transitions that involve the inner electrons in an atom are what give rise to discrete X-ray spectra because of the high photon energies involved.

Figure 11.5 shows the energy levels (not to scale) of a heavy atom classed by total quantum number n; energy differences between angular-momentum states within a shell are minor compared with the energy differences between shells. Let us consider what happens

FIGURE 11.5 The origin of X-ray spectra.

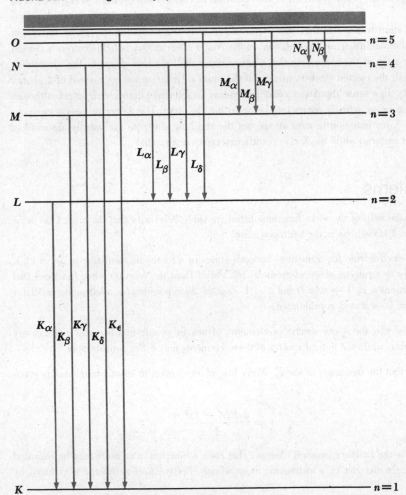

when an energetic electron strikes the atom and knocks out one of the K-shell electrons. (The K electron could also be elevated to one of the unfilled upper quantum states of the atom, but the difference between the energy needed to do this and that needed to remove the electron completely is insignificant, only 0.2 percent in sodium and still less in heavier atoms.) An atom with a missing K electron gives up most of its considerable excitation energy in the form of an X-ray photon when an electron from an outer shell drops into the "hole" in the K shell. As indicated in Fig. 11.5, the K *series* of lines in the X-ray spectrum of an element consists of wavelengths arising in transitions from the L, M, N, . . . levels to the K level. Similarly the longer-wavelength L series originates when an L electron is knocked out of the atom, the M series when an M electron is knocked out, and so on. The two spikes in the X-ray spectrum of molybdenum in Fig. 3.7 are the K_α and K_β lines of its K series.

An atom with a missing inner electron can also lose excitation energy by the *Auger effect* without emitting an X-ray photon. In the Auger effect an outer-shell electron is ejected from the atom at the same time that another outer-shell electron drops to the incomplete inner shell; the ejected electron carries off the atom's excitation energy instead of a photon doing this. In a sense the Auger effect represents an internal photoelectric effect, although the photon never actually comes into being within the atom. The Auger process is competitive with X-ray emission in most atoms, but the resulting electrons are usually absorbed in the target material while the X rays readily emerge to be detected.

Problems

1. With the help of the wave functions listed in Table 9.1 verify that $\Delta l = \pm 1$ for $n = 2 \rightarrow n = 1$ transitions in the hydrogen atom.

2. The selection rule for transitions between states in a harmonic oscillator is $\Delta n = \pm 1$. (*a*) Justify this rule on classical grounds. (*b*) Verify from the relevant wave functions that the transitions $n = 1 \rightarrow n = 0$ and $n = 1 \rightarrow n = 2$ are possible for a harmonic oscillator while $n = 1 \rightarrow n = 3$ is prohibited.

3. Explain why the X-ray spectra of elements of nearby atomic numbers are qualitatively very similar, while the optical spectra of these elements may differ considerably.

4. Show that the frequency of the K_α X-ray line of an element of atomic number Z is given by

$$\nu = \frac{3cR(Z-1)^2}{4}$$

where R is the Rydberg constant. Assume that each L electron in an atom may be regarded as the single electron in a hydrogenic atom whose effective nuclear charge is reduced by the presence of whatever K electrons are present. [The proportionality between ν and $(Z - 1)^2$ was used by Moseley in 1913 to establish the atomic numbers of the elements from their X-ray spectra. This proportionality is referred to as *Moseley's law*.]

5. What element has a K_α X-ray line of wavelength 1.785 A? Of wavelength 0.712 A?

6. Verify that the energy of a K_α photon from an atom of atomic number Z is given approximately by $10 \, (Z - 1)^2$ ev.

Chapter 12
The
Chemical
Bond

What is the nature of the forces that bond atoms together to form molecules? This question, of fundamental importance to the chemist, is hardly less important to the physicist, whose theory of the atom cannot be correct unless it provides a satisfactory answer. The ability of the quantum theory of the atom not only to explain chemical bonding but to do so partly in terms of an effect that has no classical analog is further testimony to the power of this approach.

12.1 Molecular Formation

A molecule is a stable arrangement of two or more atoms. By "stable" is meant that a molecule must be given energy from an outside source in order to break up into its constituent atoms. In other words, a molecule exists because the energy of the joint system is less than that of the system of separate noninteracting atoms. If the interactions among a certain group of atoms reduce their total energy, a molecule can be formed; if the interactions increase their total energy, the atoms repel one another.

Let us consider what happens when two atoms are brought closer and closer together. Three extreme situations may occur:

1. A *covalent bond* is formed. One or more pairs of electrons are shared by the two atoms. As these electrons circulate between the atoms, they spend more time between the atoms than elsewhere, which produces an attractive force. An example is H_2, the hydrogen molecule, whose two electrons belong jointly to the two protons (Fig. 12.1a).

2. An *ionic bond* is formed. One or more electrons from one atom may transfer to the other, and the resulting positive and negative ions attract each other. An example is NaCl, where the bond exists between Na^+ and Cl^- ions and not between Na and Cl atoms (Fig. 12.1b).

3. No bond is formed. When the electron structures of two atoms overlap, they constitute a single system, and according to the exclusion principle no two electrons in such a system can exist in the same quantum state. If some of the interacting electrons are thereby

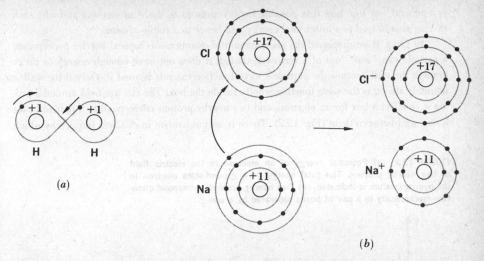

FIGURE 12.1 (a) Covalent bonding. The shared electrons spend more time on the average between their parent nuclei and therefore lead to an attractive force. (b) Ionic bonding. Sodium and chlorine combine chemically by the transfer of electrons from sodium atoms to chlorine atoms; the resulting ions attract electrostatically.

forced into higher energy states than they occupied in the separate atoms, the system may have much more energy than before and be unstable. To visualize this effect, we may regard the electrons as fleeing as far away from one another as possible to avoid forming a single system, which leads to a repulsive force between the nuclei. (Even when the exclusion principle can be obeyed with no increase in energy, there will be an electrostatic repulsive force between the various electrons; this is a much less significant factor than the exclusion principle in influencing bond formation, however.)

In H_2 the bond is purely covalent and in NaCl it is purely ionic, but in many other molecules an intermediate type of bond occurs in which the atoms share electrons to an unequal extent. An example is the HCl molecule, where the Cl atom attracts the shared electrons more strongly than the H atom. A strong argument can be made for thinking of the ionic bond as no more than an extreme case of the covalent bond, but it is customary to analyze them separately.

12.2 Electron Sharing

The simplest possible molecular system is H_2^+, the hydrogen molecular ion, in which a single electron bonds two protons. Before we consider the bond in H_2^+ in detail, let us look in a general way into how it is possible for two protons to share an electron and why such sharing should lead to a lower total energy and hence to a stable system.

In Chap. 8 we discussed the phenomenon of quantum-mechanical barrier penetration: a particle can "leak" out of a box even though it does not have enough energy to break through the wall because the particle's wave function extends beyond it. Only if the wall is infinitely strong is the wave function wholly inside the box. The electric field around a proton is in effect a box for an electron, and two nearby protons correspond to a pair of boxes with a wall between them (Fig. 12.2). There is no mechanism in classical physics by which

FIGURE 12.2 (a) Potential energy of an electron in the electric field of two nearby protons. The total energy of a ground-state electron in the hydrogen atom is indicated. (b) Two nearby protons correspond quantum-mechanically to a pair of boxes separated by a wall.

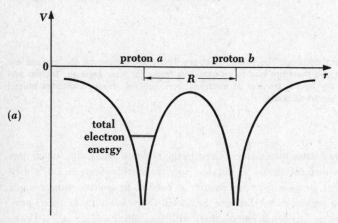

(a)

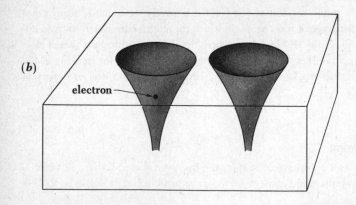

(b)

the electron in a hydrogen atom can transfer spontaneously to a neighboring proton more distant than its parent proton. In quantum physics, however, such a mechanism does exist. There is a certain probability that an electron trapped in one box will tunnel through the wall and get into the other box, and once there it has the same probability for tunneling back. This situation can be described by saying that the electron is shared by the protons.

To be sure, the likelihood that an electron will pass through the region of high potential energy—the "wall"—between two protons depends strongly upon how far apart the protons are. If the proton-proton distance is 1 A, the electron may be regarded as going from one proton to the other about every 10^{-15} sec, which means that we can legitimately consider the electron as being shared by both. If the proton-proton distance is 10 A, however, the electron shifts across an average of only about once per second, which is practically an infinite time on an atomic scale. Since the effective radius of the 1s wave function in hydrogen is 0.53 A, we conclude that electron sharing can take place only between atoms whose wave functions overlap appreciably.

Granting that two protons can share an electron, there is a simple argument that shows why the energy of such a system could be less than that of a separate hydrogen atom and proton. According to the uncertainty principle, the smaller the region to which we restrict a particle, the greater must be its momentum and hence kinetic energy. An electron shared by two protons is less confined than one belonging to a single proton, which means that it has less kinetic energy. The total energy of the electron in H_2^+ is therefore less than that of the electron in $H + H^+$, and provided the magnitude of the proton-proton repulsion in H_2^+ is not too great, H_2^+ ought to be stable.

The preceding arguments are quantum-mechanical ones, while we normally tend to consider the interactions between charged particles in terms of electrostatic forces. There is a very important theorem, independently proved by Feynman and by Hellmann and named after them, which states in essence that both types of approach always yield identical results. According to the Feynman-Hellmann theorem, if the electron probability distribution in a molecule is known, the calculation of the system energy can proceed classically and will lead to the same conclusions as a purely quantum-mechanical calculation. The Feynman-Hellmann theorem is not an obvious one, because treating a molecule in terms of electrostatic forces does not explicitly take into account electron kinetic energy, while a quantum treatment involves the total electron energy; nevertheless, once the electron wave function ψ has been determined, either way of proceeding may be used.

12.3   The H_2^+ Molecular Ion

What we would like to know is the wave function ψ of the electron in H_2^+, since from ψ we can calculate the energy of the system as a function of the separation R of the protons. If $E(R)$ has a minimum, we will know that a bond can exist, and we can also determine the bond energy and the equilibrium spacing of the protons.

Schrödinger's equation for an electron is

12.1
$$\nabla^2\psi + \frac{2m}{\hbar^2}(E - V)\psi = 0$$

If the distance between the electron and proton a is r_a and that between the electron and proton b is r_b (Fig. 12.3), the potential energy of the electron is

12.2
$$V = -\frac{e^2}{4\pi\varepsilon_0 r_a} - \frac{e^2}{4\pi\varepsilon_0 r_b}$$

and Schrödinger's equation becomes

12.3
$$\nabla^2\psi + \frac{2m}{\hbar^2}\left[E + \frac{e^2}{4\pi\varepsilon_0}\left(\frac{1}{r_a} + \frac{1}{r_b}\right)\right]\psi = 0$$

In order to solve Eq. 12.3 it is necessary to use a special coordinate system that involves what are called prolate spheroidal coordinates. These coordinates are ζ, η, and ϕ, where

$$\zeta = \frac{r_a + r_b}{R} \qquad \eta = \frac{r_a - r_b}{R}$$

and ϕ is the azimuthal angle about the molecular axis. Surfaces of constant ζ are prolate ellipsoids whose foci are at the positions of the protons, surfaces of constant η are two-sheeted hyperboloids with the same foci, and surfaces of constant ϕ are planes which contain the molecular axis (Fig. 12.4). The point to expressing Schrödinger's equation for the electron in terms of these coordinates is that it is then possible to separate the equation into three subsidiary ones, each involving only one coordinate. We recall that an analogous procedure, making use of spherical polar coordinates, permitted us to separate Schrödinger's equation for the hydrogen atom into individual equations in R, θ, and ϕ that could then be solved exactly.

The exact solution of Eq. 12.3 is important because its results can then be checked with experiment to test the applicability of quantum mechanics to molecular systems. Further, the validity of the approximate methods needed for more complex molecules can be verified by comparing their results for H_2^+ with those of the exact treatment. However, solving Eq. 12.3 is a lengthy and complicated affair whose details are of limited interest.

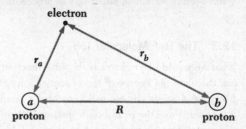

FIGURE 12.3 A system of two protons and an electron.

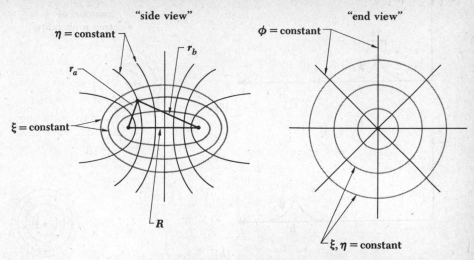

"side view"
$\eta = $ constant
r_a
r_b
$\xi = $ constant
R

"end view"
$\phi = $ constant
$\xi, \eta = $ constant

FIGURE 12.4 Prolate spheroidal coordinates. Surfaces of constant ζ are prolate ellipsoids whose foci are the positions of the protons, surfaces of constant η are two-sheeted hyperboloids with the same foci, and surfaces of constant ϕ are planes which contain the molecular axis.

Instead we shall use an approximate approach to arrive at the electron wave function and bond energy of H_2^+.

Let us try to predict the H_2^+ electron wave function ψ when R, the distance between the protons, is large compared with a_0, the radius of the smallest Bohr orbit in the hydrogen atom. In this event ψ near each proton must closely resemble the $1s$ wave function of the hydrogen atom, as pictured in Fig. 12.5 where the $1s$ wave function around proton a is called ψ_a and that around proton b is called ψ_b. To support the assertion that ψ resembles ψ_a near proton a and ψ_b near proton b, we note that Eq. 12.3 approaches Schrödinger's equation for the hydrogen atom when $r_a \gg r_b$ and when $r_b \gg r_a$.

We also know what ψ looks like when R is 0, that is, when the protons are imagined to be fused together. Here the situation is that of the He^+ ion, since the electron is now in the presence of a single nucleus whose charge is $+2e$. The $1s$ wave function of He^+ has the same form as that of H but with a greater amplitude at the origin, as in Fig. 12.5e. Evidently ψ is going to be something like the wave function sketched in Fig. 12.5d when R is comparable with a_0. There is an enhanced likelihood of finding the electron in the region between the protons, which we have spoken of in terms of sharing of the electron by the protons. Thus there is on the average an excess of negative charge between the protons, and this attracts them together. We have still to establish whether the magnitude of this attraction is enough to overcome the mutual repulsion of the protons.

The combination of ψ_a and ψ_b in Fig. 12.5 is symmetric, since exchanging a and b

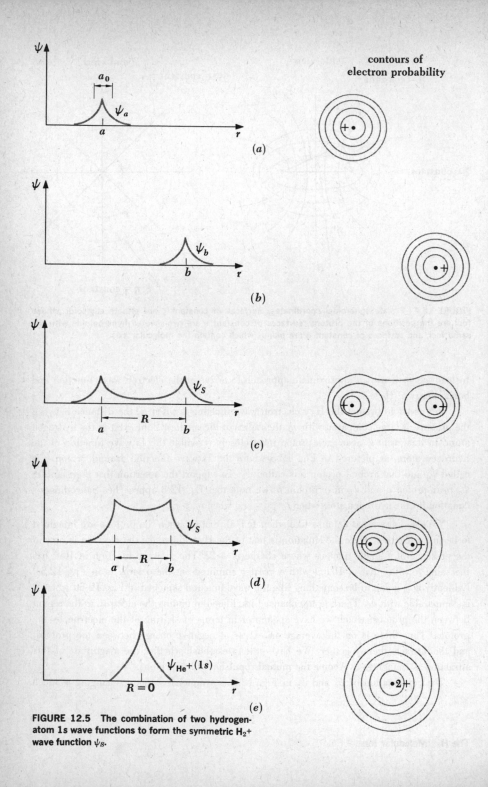

contours of
electron probability

(a)

(b)

(c)

(d)

(e)

FIGURE 12.5 The combination of two hydrogen-atom 1s wave functions to form the symmetric H_2^+ wave function ψ_S.

does not affect ψ (see Sec. 10.3). However, it is equally possible to have an *antisymmetric* combination of ψ_a and ψ_b, as in Fig. 12.6. Here there is a node between a and b where $\psi = 0$, which implies a diminished likelihood of finding the electron between the protons. Now there is on the average a deficiency of negative charge between the protons, and in consequence a repulsive force. With only repulsive forces acting, bonding cannot occur.

An interesting question concerns the behavior of the antisymmetric H_2^+ wave function ψ_A as $R \to 0$. Obviously ψ_A does not become the $1s$ wave function of He^+ when $R = 0$. However, ψ_A *does* approach the $2p$ wave function of He^+ (Fig. 12.6e), which has a node at the origin. Since the $2p$ state of He^+ is an excited state while the $1s$ state is the ground state, H_2^+ in the antisymmetric state ought to have more energy than when it is in the symmetric state, which agrees with our inference from the shapes of the wave functions ψ_A and ψ_S that in the former case there is a repulsive force and in the latter an attractive one.

A line of reasoning similar to the preceding one enables us to estimate how the total energy of the H_2^+ system varies with R. We first consider the symmetrical state. When R is large, the electron energy E_S must be the -13.6-ev energy of the hydrogen atom, while the electrostatic potential energy V_p of the protons,

12.4
$$V_p = \frac{e^2}{4\pi\varepsilon_0 R}$$

falls to 0 as $R \to \infty$. (V_p is a positive quantity, corresponding to a repulsive force.) When $R = 0$, the electron energy must equal that of the He^+ ion, which is Z^2 or 4 times that of the H atom. (See Eq. 6.35; the same formula is obtained from the quantum theory of one-electron atoms.) Hence $E_S = -54.4$ ev when $R = 0$. Also, when $R \to 0$, $V_p \to \infty$ as $1/R$. Both E_S and V_p are sketched in Fig. 12.7 as functions of R; the shape of the curve for E_S can only be approximated without a detailed calculation, but we do have its value for both $R = 0$ and $R = \infty$ and, of course, V_p obeys Eq. 12.4.

The total energy E_S^{total} of the system is the sum of the electron energy E_S and the potential energy V_p of the protons. Evidently E_S^{total} has a minimum, which corresponds to a stable molecular state. This result is confirmed by the experimental data on H_2^+ which indicate a bond energy of 2.65 ev and an equilibrium separation R of 1.06 A. By "bond energy" is meant the energy needed to break H_2^+ into $H + H^+$; the *total* energy of H_2^+ is the -13.6 ev of the hydrogen atom plus the -2.65-ev bond energy, or -16.3 ev in all.

In the case of the antisymmetric state, the analysis proceeds in the same way except that the electron energy E_A when $R = 0$ is that of the $2p$ state of He^+. This energy is, from Eq. 6.35 with $Z = 2$ and $n = 2$, just equal to the -13.6 ev of the ground-state hydrogen atom. Since $E_A \to -13.6$ ev also as $R \to \infty$, we might think that the electron energy is constant, but actually there is a small dip at intermediate distances. However, the dip is not nearly enough to yield a minimum in the total energy curve for the antisymmetric state, as indicated in Fig. 12.7, so in this state no bond is formed.

The H_2^+ Molecular Ion

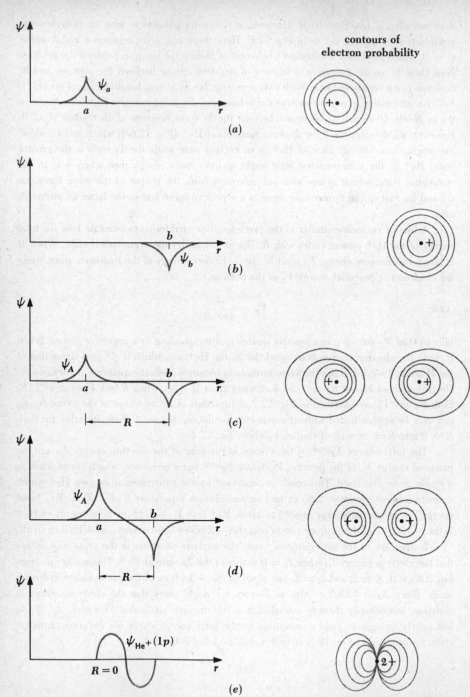

contours of
electron probability

FIGURE 12.6 The combination of two hydrogen-atom 1s wave functions to form the antisymmetric
H_2^+ wave function ψ_A.

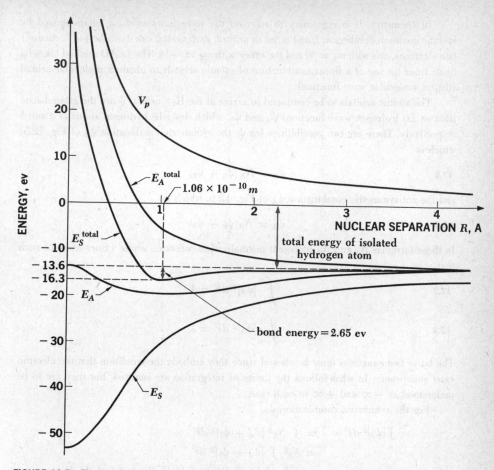

FIGURE 12.7 Electronic, proton repulsion, and total energy in H_2^+ as a function of nuclear separation R for the symmetric and antisymmetric states. The antisymmetric state has no minimum in its total energy.

12.4 The LCAO Method

In the previous section we saw how far it is possible to go in determining molecular wave functions by considering appropriate atomic wave functions. A widely used formal version of this approach, called the LCAO method, begins by assuming that molecular wave functions can be satisfactorily approximated by adding and subtracting unmodified atomic wave functions. Detailed calculations are then made on this basis, instead of merely the drawing of plausible graphs. As might be expected, the LCAO method is essentially exact when R is large, it is usually fairly accurate when the atomic wave functions begin to overlap appreciably, and it becomes progressively poorer as $R \rightarrow 0$.

The LCAO Method

In chemistry, it is customary to refer to the wave function ψ_{nlm_l} characterized by specific quantum numbers n, l, and m_l as an *orbital*; each orbital can describe (or "contain") two electrons, one with $m_s = \frac{1}{2}$ and the other with $m_s = -\frac{1}{2}$. The LCAO method takes its name from the use of a *l*inear *c*ombination of *a*tomic *o*rbitals to obtain a molecular orbital (that is, molecular wave function).

The atomic orbitals to be combined to arrive at the H_2^+ orbital ψ are the ground-state (that is, $1s$) hydrogen wave functions ψ_a and ψ_b, which describe hydrogen atoms at a and b respectively. There are two possibilities for ψ, the symmetric combination ψ_S of Fig. 12.5, which is

12.5
$$\psi_S = N_S (\psi_a + \psi_b)$$

and the antisymmetric combination ψ_A of Fig. 12.6, which is

12.6
$$\psi_A = N_A(\psi_a - \psi_b)$$

In these formulas N_S and N_A represent normalization constants whose values must be such that

12.7
$$\int_{-\infty}^{\infty} |\psi_S|^2 \, dV = 1$$

12.8
$$\int_{-\infty}^{\infty} |\psi_A|^2 \, dV = 1$$

The latter two equations must be obeyed since they embody the condition that the electron exist somewhere. In what follows the limits of integration are omitted, but they are to be understood as $-\infty$ and $+\infty$ in each case.

For the symmetric combination ψ_S,

$$\begin{aligned} \int |\psi_S|^2 \, dV = 1 &= \int N_S^2 \, |\psi_a + \psi_b|^2 \, dV \\ &= N_S^2 \int |\psi_a + \psi_b|^2 \, dV \\ &= N_S^2 \left(\int |\psi_a|^2 \, dV + \int |\psi_b|^2 \, dV + 2 \int \psi_a \psi_b \, dV \right) \end{aligned}$$

Since ψ_a and ψ_b are $1s$ hydrogen wave functions, $\psi_a = \psi_a^*$ and $\psi_b = \psi_b^*$, which entitles us to write the last term of the preceding equation as we did. Furthermore, both ψ_a and ψ_b are already normalized, so that

$$\int |\psi_a|^2 \, dV = \int |\psi_b|^2 \, dV = 1$$

and

$$\int |\psi_S|^2 \, dV = 1 = N_S^2 \, (2 + 2 \int \psi_a \psi_b \, dV)$$

If we let

12.9
$$S = \int \psi_a \psi_b \, dV$$

we have

12.10

$$N_S = \frac{1}{2(1 + S)}$$

A similar calculation for N_A yields

12.11

$$N_A = \frac{1}{2(1 - S)}$$

The quantity $S = \int \psi_a \psi_b \, dV$ is called the *overlap integral* because it is a measure of the extent to which the wave functions ψ_a and ψ_b mesh together. In order to carry out the integration, the spheroidal coordinates described in Sec. 12.3 must be used. The result is

12.12

$$S = \left(1 + R' + \frac{R'^2}{3}\right) e^{-R'}$$

where R' is the distance between the protons a and b in units of a_0, the radius of the first Bohr orbit in the hydrogen atom. When $R' \gg 1$, $S \approx 0$, which corresponds to the negligible overlap between ψ_a and ψ_b at large separations. When $R' = 0$, $S = 1$, which corresponds to the fusing together of the two protons to form an atomic rather than a molecular system. In the case of $H_2{}^+$, the equilibrium separation of the protons is $R' = 2$ (that is, $R = 2a_0$) and $S = 0.6$.

Now that the approximate molecular orbitals ψ_S and ψ_A have been obtained, we can proceed to determine the respective electron energies E_S and E_A. To simplify the arithmetic involved in the calculation, it is convenient to express lengths throughout in units of a_0 and energies in rydbergs (1 rydberg equals the ground-state energy of the hydrogen atom). From Chap. 6 we have

$$a_0 = \frac{4\pi \, \varepsilon_0 \, \hbar^2}{me^2} = 5.3 \times 10^{-11} \text{ m} = 0.53 \text{ A}$$

$$1 \text{ ry} = \frac{me^4}{32\pi^2 \, \varepsilon_0{}^2 \, \hbar^2} = 2.2 \times 10^{-18} \text{ joule} = 13.6 \text{ ev}$$

Schrödinger's equation for the electron in $H_2{}^+$ is, from Eq. 12.3,

$$\nabla^2\psi + \frac{2m}{\hbar^2}\left[E + \frac{e^2}{4\pi \, \varepsilon_0}\left(\frac{1}{r_a} + \frac{1}{r_b}\right)\right]\psi = 0$$

If we use primes to represent quantities expressed in units of a_0 and ry, we find that

$$\left(\frac{me^2}{4\pi \, \varepsilon_0 \, \hbar^2}\right)^2 \nabla^2\psi + \frac{2m}{\hbar^2}\left[\frac{me^4}{32\pi^2 \, \varepsilon_0{}^2 \, \hbar^2} E' + \frac{e^2}{4\pi \, \varepsilon_0}\left(\frac{me^2}{4\pi \, \varepsilon_0 \, \hbar^2}\right)\left(\frac{1}{r_a'} + \frac{1}{r_b'}\right)\right]\psi = 0$$

12.13

$$\nabla^2\psi + \left[E' + \left(\frac{2}{r_a'} + \frac{2}{r_b'}\right)\right]\psi = 0$$

The procedure for solving Eq. 12.13 for the electron energy E' is simplified by the fact that we already know the wave function ψ. We start by rewriting Eq. 12.13 in the form

$$-\nabla^2\psi - \left(\frac{2}{r_a'} + \frac{2}{r_b'}\right)\psi = E'\psi$$

Next we multiply through by ψ^* to give

$$\psi^*\left[-\nabla^2\psi - \left(\frac{2}{r_a'} + \frac{2}{r_b'}\right)\psi\right] = \psi^* E'\psi = E'|\psi|^2$$

and then integrate over all space:

$$\int \psi^*\left[-\nabla^2\psi - \left(\frac{2}{r_a'} + \frac{2}{r_b'}\right)\psi\right]dV = E'\int|\psi|^2\,dV$$

Since ψ is normalized, $\int|\psi|^2\,dV = 1$ and we have what we want,

12.14
$$E' = -\int \psi^*\left[\nabla^2\psi + \left(\frac{2}{r_a'} + \frac{2}{r_b'}\right)\psi\right]dV$$

For the total energy of the H_2^+ molecule we must add the mutual potential energy of the two protons to the electron energy E', which we shall do later.

Let us evaluate Eq. 12.14 for the symmetric wave function

$$\psi_S = N_S(\psi_a + \psi_b)$$

We have

12.15
$$E_S' = -N_S^2\int (\psi_a^* + \psi_b^*)\left(\nabla^2\psi_a + \nabla^2\psi_b\right.$$

$$\left. + \frac{2}{r_a'}\psi_a + \frac{2}{r_a'}\psi_b + \frac{2}{r_b'}\psi_a + \frac{2}{r_b'}\psi_b\right)dV$$

This expression can be simplified because ψ_a and ψ_b are each ground-state wave functions of the hydrogen atom corresponding to its ground-state energy E_1'. That is, E_1' is the eigenvalue of each of the eigenfunctions ψ_a and ψ_b, and so

$$-\nabla^2\psi_a - \frac{2}{r_a'}\psi_a = E_1'\psi_a$$

$$-\nabla^2\psi_b - \frac{2}{r_b'}\psi_b = E_1'\psi_b$$

By definition $E_1' = -1$ since $E_1 = -1$ ry, and we have

$$-\nabla^2\psi_a - \frac{2}{r_a'}\psi_a = -\psi_a$$

$$-\nabla^2\psi_b - \frac{2}{r_b'}\psi_b = -\psi_b$$

Equation 12.15 therefore becomes

$$E'_s = -N_s^2 \int (\psi_a^* + \psi_b^*) \left[\left(1 + \frac{2}{r_b'}\right)\psi_a + \left(1 + \frac{2}{r_a'}\right)\psi_b \right] dV$$

12.16
$$= -N_s^2 \int |\psi_a + \psi_b|^2 \, dV - 2N_s^2 \int \frac{1}{r_b'} |\psi_a|^2 \, dV$$

$$- 2N_s^2 \int \frac{1}{r_a'} |\psi_b|^2 \, dV - 2N_s^2 \int \frac{1}{r_b'} \psi_a \psi_b^* \, dV - 2N_s^2 \int \frac{1}{r_a'} \psi_a^* \psi_b \, dV$$

In view of Eqs. 12.5 and 12.7, the first term of the preceding formula is just -1, and so this term contributes an energy equal to that of an isolated ground-state hydrogen atom. In the present situation, the energy is to be interpreted as consisting of the potential energy in the field of proton a of the electron cloud around a, the potential energy in the field of proton b of the electron cloud around b, and the kinetic energy of the electron.

The next two terms are called *coulomb integrals*. The first of them represents the potential energy in the field of proton b of the electron cloud around a, and the second represents the potential energy in the field of proton a of the electron cloud around b. Of course, there is only one electron, and what we are really considering is the distribution of its probability density. It is nevertheless appropriate to think of the electron as a cloud whose charge density corresponds to the probability density $|\psi|^2$ in order to be able to interpret what is going on in familiar terms.

The last two terms in Eq. 12.16 are called *exchange integrals*, and they involve the mixed product of the wave functions ψ_a and ψ_b instead of the probability densities $|\psi_a|^2$ and $|\psi_b|^2$. Unlike the other three terms, the energies represented by the exchange integrals cannot be directly interpreted on the basis of classical notions, but represent a purely quantum-mechanical phenomenon.

The coulomb integrals of Eq. 12.16 are equal, since protons a and b can be interchanged without affecting anything, and each has the value

$$N_s^2 C = -2 N_s^2 \int \frac{1}{r_b'} |\psi_a|^2 \, dV$$

12.17
$$= -2N_s^2 \left[-\frac{1}{R'} + \left(\frac{1}{R'} + 1\right) e^{-2R'} \right]$$

The exchange integrals are also equal, and each has the value

$$N_s^2 A = -2N_s^2 \int \frac{1}{r_b'} \psi_a^* \psi_b \, dV$$

12.18
$$= -2N_s^2 (1 + R') e^{-R'}$$

The electron energy in ry may therefore be written

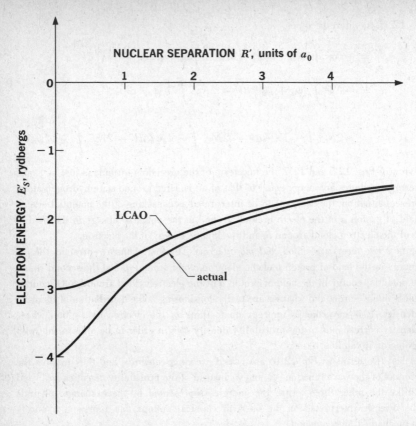

NUCLEAR SEPARATION R', units of a_0

ELECTRON ENERGY E'_S, rydbergs

LCAO

actual

FIGURE 12.8 A comparison between the actual variation of E'_S with R' and the variation predicted by the LCAO approximation.

$$E'_S = -1 + N_S{}^2(2C + 2A)$$

12.19
$$= -1 + \frac{C + A}{1 + S}$$

Each of the quantities C, A, and S is a function of R', the proton-proton distance in units of a_0.

With the help of Eqs. 12.12, 12.17, and 12.18 we can see how E'_S varies with R'. First we note that, as $R' \to \infty$, $E'_S \to -1$, the energy of an isolated hydrogen atom. However, when $R' = 0$, $E'_S = -3$, which is not the -4 energy of a He$^+$ ion as it should be. The reason for the discrepancy is the inadequacy of the LCAO method when R' is small. A comparison between $E'_S(R')$ as given by Eq. 12.19 and the actual $E'_S(R')$ is given in Fig. 12.8; evidently the LCAO approximation is still reasonably good when $R' = 2$, the equilibrium separation of the protons in H_2^+, but it becomes less and less accurate as R' decreases from that value.

A calculation similar to the preceding one for the antisymmetric case yields

12.20
$$E_A' = -1 + \frac{C - A}{1 - S}$$

To find the total energy of the H_2^+ molecule, we must add the mutual potential energy V_p' of the two protons to the electron energy E'. Since

$$V_p = \frac{e^2}{4\pi\varepsilon_0 R}$$

when R is in meters and V_p in joules, in the "atomic" units we are using here we have

12.21
$$V_p' = \frac{2}{R'}$$

Hence

12.22
$$E_{S,A}'^{\text{total}} = -1 + \frac{2}{R'} + \frac{C \pm A}{1 \pm S}$$

where the $+$ sign is to be used to find $E_S'^{\text{total}}$ and the $-$ sign to find $E_A'^{\text{total}}$. Equation 12.22 is plotted in Fig. 12.9. The curve for the symmetric state indicates a binding energy of

FIGURE 12.9 Total energy of H_2^+ for the symmetric and antisymmetric states according to Eq. 12.22. The correct curves are given in Fig. 12.7.

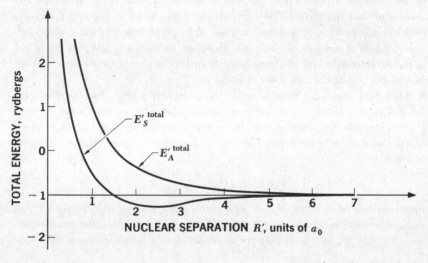

0.13 ry (1.8 ev) at an equilibrium separation of $R' = 2.5$, while the curve for the anti-symmetric state has no minimum. The correct figures for the bond in H_2^+ are 0.195 ry (2.65 ev) and $R' = 2$; the approximate nature of the LCAO method is responsible for the differences between these sets of figures.

Two important aspects of the LCAO method are illustrated by its application to H_2^+. The first is that it does not yield entirely accurate results in its simplest version, and more elaborate procedures must be employed where exact answers are required. The second point is that, despite its failings, the LCAO method *does* provide qualitative insight into many examples of molecular bonding, which is especially valuable in molecules so complex as to prohibit quantitative calculations.

12.5   The H₂ Molecule

The H_2 molecule contains two electrons instead of the single electron of H_2^+. According to the exclusion principle, both electrons can share the same orbital (that is, be described by the same wave function ψ_{nlm_l}) provided their spins are antiparallel. With two electrons to contribute to the bond, H_2 ought to be more stable than H_2^+—at first glance, twice as stable, with a bond energy of 5.3 ev compared with 2.65 ev for H_2^+. However, the H_2 orbitals are not quite the same as those of H_2^+ because of the electrostatic repulsion between the two electrons in H_2, a factor absent in the case of H_2^+. The latter repulsion weakens the bond in H_2, so that the actual bond energy is 4.72 ev instead of 5.3 ev. For the same reason, the bond length in H_2 is 0.74 A, which is somewhat larger than the use of unmodified H_2^+ wave functions would indicate. The general conclusion in the case of H_2^+ that the symmetric wave function ψ_S leads to a bound state and the antisymmetric wave function ψ_A to an unbound one remains valid for H_2.

In Sec. 10.3 the exclusion principle was formulated in terms of the symmetry and antisymmetry of wave functions, and it was concluded that systems of electrons are always described by antisymmetric wave functions (that is, by wave functions that reverse sign upon the exchange of any pair of electrons). However, we have just said that the bound state in H_2 corresponds to both electrons being described by a symmetrical wave function ψ_S, which seems to contradict the above conclusion.

A closer look shows that there is really no contradiction here. The *complete* wave function $\Psi(1,2)$ of a system of two electrons is the product of a spatial wave function $\psi(1,2)$ which describes the coordinates of the electrons and a spin function $s(1,2)$ which describes the orientations of their spins. The exclusion principle requires that the complete wave function

$$\Psi(1,2) = \psi(1,2)s(1,2)$$

be antisymmetric to an exchange of both coordinates and spins, not $\psi(1,2)$ by itself, and what we have been calling a molecular orbital is the same as $\psi(1,2)$. An antisymmetric complete wave function Ψ_A can result from the combination of a symmetric coordinate wave function ψ_S and an antisymmetric spin function s_A or from the combination of an anti-symmetric coordinate wave function ψ_A and a symmetric spin function s_S. That is, only

$$\Psi = \psi_S s_A$$

and

$$\Psi = \psi_A s_S$$

are acceptable. If the spins of the two electrons are parallel, their spin function is symmetric since it does not change sign when the electrons are exchanged. Hence the coordinate wave function ψ for two electrons whose spins are parallel must be antisymmetric; we may express this by writing

$$\psi_{\uparrow\uparrow} = \psi_A$$

On the other hand, if the spins of the two electrons are antiparallel, their spin function is antisymmetric since it reverses sign when the electrons are exchanged. Hence the coordinate wave function ψ for two electrons whose spins are antiparallel must be symmetric, and we may express this by writing

$$\psi_{\uparrow\downarrow} = \psi_S$$

Schrödinger's equation for the H_2 molecule has no exact solution. In fact, only for H_2^+ is an exact solution possible, and all other molecular systems must be treated approximately. The most detailed analysis of the H_2 molecule was carried out by James and Coolidge using a version of the molecular-orbital approach. Their results are shown in Fig. 12.10 for the

FIGURE 12.10   The variation of the energy of the system H + H showing their distances apart when the electron spins are parallel and antiparallel.

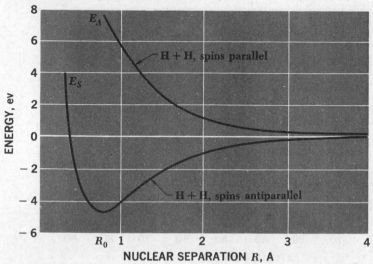

case when the electrons have their spins parallel and the case when their spins are antiparallel. The difference between the two curves is due to the exclusion principle, which prevents two electrons in the same quantum state in a system from having the same spin and therefore leads to a dominating repulsion when the spins are parallel.

12.6 The Ionic Bond

As stated earlier, molecular bonds range from the purely covalent at one extreme, as in H_2 where two electrons are equally shared by two protons, to the purely ionic at the other extreme, as in NaCl where an electron shifts from the Na atom to the Cl atom and the Na^+ and Cl^- ions then attract electrostatically with no electron sharing at all. We shall now examine the ionic bond in NaCl in detail, a much simpler task than in the case of the covalent bond in H_2.

The most stable electron configuration of an atom consists of closed subshells (Sec. 10.5). Atoms with incomplete outer subshells tend to gain or lose electrons in order to attain stable configurations, becoming negative or positive ions in the process. The *ionization energy* (sometimes called *ionization potential*) of an element is the energy needed to remove an electron from one of its atoms; it is therefore a measure of how tightly bound its outermost electron or electrons are. Table 12.1 contains the ionization energies of the elements, and Fig. 12.11 shows how these energies vary with atomic number.

It is not hard to see why the ionization energies of the elements vary as they do. For

FIGURE 12.11 The variation of ionization energy with atomic number.

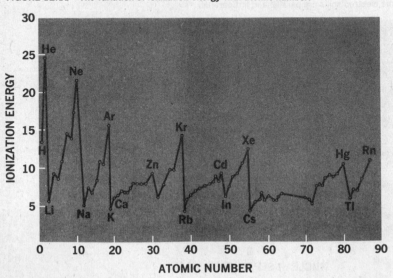

TABLE 12.1. IONIZATION ENERGIES OF THE ELEMENTS. In electron volts.

1																	2
H																	He
13.6																	24.6

3	4											5	6	7	8	9	10
Li	Be											B	C	N	O	F	Ne
5.4	9.3											8.3	11.3	14.5	13.6	17.4	21.6

11	12											13	14	15	16	17	18
Na	Mg											Al	Si	P	S	Cl	Ar
5.1	7.6											6.0	8.1	11.0	10.4	13.0	15.8

19	20	21	22	23	24	25	26	27	28	29	30	31	32	33	34	35	36
K	Ca	Sc	Ti	V	Cr	Mn	Fe	Co	Ni	Cu	Zn	Ga	Ge	As	Se	Br	Kr
4.3	6.1	6.6	6.8	6.7	6.8	7.4	7.9	7.9	7.6	7.7	9.4	6.0	7.9	9.8	9.8	11.8	14.0

37	38	39	40	41	42	43	44	45	46	47	48	49	50	51	52	53	54
Rb	Sr	Y	Zr	Nb	Mo	Tc	Ru	Rh	Pd	Ag	Cd	In	Sn	Sb	Te	I	Xe
4.2	5.7	6.5	7.0	6.8	7.1	7.3	7.4	7.5	8.3	7.6	9.0	5.8	7.3	8.6	9.0	10.5	12.1

55	56	*	72	73	74	75	76	77	78	79	80	81	82	83	84	85	86
Cs	Ba		Hf	Ta	W	Re	Os	Ir	Pt	Au	Hg	Tl	Pb	Bi	Po	At	Rn
3.9	5.2		5.5	7.9	8.0	7.9	8.7	9.2	9.0	9.2	10.4	6.1	7.4	7.3	8.4	—	10.7

87	88	†
Fr	Ra	
—	5.3	

*	57	58	59	60	61	62	63	64	65	66	67	68	69	70	71
	La	Ce	Pr	Nd	Pm	Sm	Eu	Gd	Tb	Dy	Ho	Er	Tm	Yb	Lu
	5.6	6.9	5.8	6.3	—	5.6	5.7	6.2	6.7	6.8	—	6.1	5.8	6.2	5.0

†	89	90	91	92	93	94	95	96	97	98	99	100	101	102	103
	Ac	Th	Pa	U	Np	Pu	Am	Cm	Bk	Cf	Es	Fm	Md	No	Lw
	—	7.0	—	6.1	—	5.1	6.0	—	—	—	—	—	—	—	—

instance, an atom of any of the alkali metals of group I has a single *s* electron outside a closed subshell. The electrons in the inner shells partially shield the outer electron from the nuclear charge $+Ze$, so that the effective charge holding the outer electron to the atom is just $+e$ rather than $+Ze$. Relatively little work must be done to detach an electron from such an atom, and the alkali metals form positive ions readily. The larger the atom, the farther the outer electron is from the nucleus and the weaker is the electrostatic force on it; this is why the ionization energy generally decreases as we go down any group. The increase in ionization energy from left to right across any period is accounted for by the increase in nuclear charge while the number of inner shielding electrons stays constant. There are two

The Ionic Bond

TABLE 12.2. ELECTRON AFFINITIES OF THE
HALOGENS.   In electron volts.

Fluorine	3.45
Chlorine	3.61
Bromine	3.36
Iodine	3.06

electrons in the common inner shell of period 2 elements, and the effective nuclear charge acting on the outer electrons of these atoms is therefore $+(Z-2)e$. The outer electron in a lithium atom is held to the atom by an effective charge of $+e$, while each outer electron in beryllium, boron, carbon, etc., atoms is held to its parent atom by effective charges of $+2e$, $+3e$. $+4e$, etc.

At the other extreme from alkali metal atoms, which tend to lose their outermost electrons, are halogen atoms, which tend to complete their outer p subshells by picking up an additional electron each. The *electron affinity* of an element is defined as the energy released when an electron is added to an atom of each element. The greater the electron affinity, the more tightly bound is the added electron. Table 12.2 shows the electron affinities of the halogens. In general, electron affinities decrease going down any group of the periodic table and increase going from left to right across any period. The experimental determination of electron affinities is quite difficult, and those for only a few elements are accurately known.

An ionic bond between two atoms can occur when one of them has a low ionization energy, and hence a tendency to become a positive ion, while the other one has a high electron affinity, and hence a tendency to become a negative ion. Sodium, with an ionization energy of 5.1 ev, is an example of the former and chlorine, with an electron affinity of 3.6 ev, is an example of the latter. When a Na^+ ion and a Cl^- ion are in the same vicinity and are free to move, the attractive electrostatic force between them brings them together. The condition that a stable molecule of NaCl result is simply that the total energy of the system of the two ions be less than the total energy of a system of two atoms of the same elements; otherwise the surplus electron on the Cl^- ion would transfer to the Na^+ ion, and the neutral Na and Cl atoms would no longer be bound together. Let us see how this criterion is met by NaCl.

Let us consider a Na atom and a Cl atom infinitely far apart. In order to remove the outer electron from the Na atom, leaving it a Na^+ ion, 5.1 ev of work must be done. That is,

12.23 $$Na + 5.1 \text{ ev} \rightarrow Na^+ + e^-$$

When this electron is brought to the Cl atom, the latter absorbs it to complete its outer electron subshell and thereby becomes a Cl^- ion. Since the electron affinity of chlorine is 3.6 ev, signifying that this amount of work must be done to liberate the odd electron from a Cl^- ion, the formation of Cl^- *evolves* 3.6 ev of energy. Hence

12.24 $$Cl + e^- \rightarrow Cl^- + 3.6 \text{ ev}$$

The net result of these two events is the sum of Eqs. 12.23 and 12.24:

12.25 $$Na + Cl + 1.5 \text{ ev} \rightarrow Na^+ + Cl^-$$

Our net expenditure of energy to form Na^+ and Cl^- ions from Na and Cl atoms has been only 1.5 ev.

What happens when the electrostatic attraction between the Na^+ and Cl^- ions brings them together, let us say (for the sake of argument) to 4 A of each other? The energy given off when this occurs is equal to the potential energy of the system of a $+e$ charge a distance of 4 A from a $-e$ charge. This energy is

$$V = -\frac{e^2}{4\pi\varepsilon_0 R}$$

$$= -\frac{9 \times 10^9 \text{ n-m}^2/\text{coulomb}^2 \times (1.6 \times 10^{-19} \text{ coulomb})^2}{4 \times 10^{-10} \text{ m}}$$

$$= -5.8 \times 10^{-19} \text{ joule}$$

$$= -3.6 \text{ ev}$$

Schematically,

12.26 $$Na^+ \quad + \quad Cl^- \rightarrow Na^+ + Cl^- + 3.6 \text{ ev}$$
$$\longleftarrow\infty\longrightarrow \qquad \leftarrow 4\text{ A}\rightarrow$$

If we shift an electron from a Na atom to an infinitely distant Cl atom and allow the resulting ions to come together, then, the entire process evolves an energy of 3.6 ev − 1.5 ev = 2.1 ev:

12.27 $$Na^+ \quad + \quad Cl \rightarrow Na^+ + Cl^- + 2.1 \text{ ev}$$
$$\longleftarrow\infty\longrightarrow \qquad \leftarrow 4\text{ A}\rightarrow$$

Evidently the NaCl molecule formed by the electrostatic attraction of Na^+ and Cl^- ions is stable, since work must be done on such a molecule in order to separate it into Na and Cl atoms.

The two ions do not, of course, approach each other until their electron structures mesh together. It is easy to see that, if such a meshing were to take place, the positively charged nuclei of the ions would no longer be completely shielded by their surrounding electrons and would simply repel each other electrostatically. There is another phenomenon that is even more effective in keeping the ions apart, as mentioned earlier in this chapter. According to the Pauli exclusion principle, no two electrons in the same atomic system can exist in the same quantum state. If the electron structures of Na^+ and Cl^- overlap, they constitute a single atomic system rather than separate, independent systems. If such a system is to obey the exclusion principle, some electrons will have to go to higher quantum states than they would otherwise occupy. These states have more energy than those corresponding to the normal electron configuration of the ions, and so the total energy of the

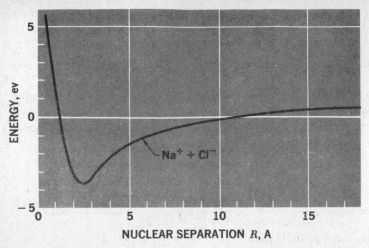

FIGURE 12.12 The variation of the energy of the system Na+ and Cl- with their distance apart.

NaCl molecule increases sharply when the Na+ and Cl- ions approach each other too closely.

An increasing potential energy when two bodies are brought together means that the force between them is repulsive, just as a decreasing potential energy under the same circumstances means an attractive force. Figure 12.12 contains a curve showing how the potential energy V of the system of a Na+ ion and a Cl- ion varies with separation distance r. At $r = \infty$, $V = +1.5$ ev. When the ions are closer together, the attractive electrostatic force between them takes effect, and the potential energy drops. Finally, when they are about 4 A apart, their electron structures begin to interact, a repulsive force comes into play, and the potential energy soon starts to rise. The minimum in the potential-energy curve occurs at $R = 2.4$ A, where $V = -4.2$ ev. At this separation distance the mutually attractive and repulsive forces on the ions exactly balance, and the system is in equilibrium. To *dissociate* a NaCl molecule into Na and Cl atoms requires an energy of 4.2 ev.

Ionic bonds usually do not result in the formation of molecules. Strictly speaking, a molecule is an electrically neutral aggregate of atoms that is held together strongly enough to be experimentally observable as a particle. Thus the individual units that constitute gaseous hydrogen each consist of two hydrogen atoms, and we are entitled to regard them as molecules. On the other hand, the crystals of rock salt (NaCl) are aggregates of sodium and chlorine ions which, although invariably arranged in a certain definite structure, do not pair off into discrete molecules consisting of one Na+ ion and one Cl- ion; rock salt crystals may in fact be of almost any size. While there are always equal numbers of Na+ and Cl- ions in rock salt, so that the formula NaCl correctly represents its composition, these ions form molecules rather than crystals only in the gaseous state. Despite the absence of individual NaCl molecules in solid NaCl, there is a very specific interaction between adjacent Na^+ and Cl^- ions in the latter, and it is the properties of this interaction that make NaCl as characteristic an example of chemical bonding as H_2.

Problems

1. Using the figures in this chapter, find the ionization energy of H_2 and compare it with the ionization energy of H. Explain the difference.

2. Derive Eq. 12.20.

3. Sodium atoms exhibit more pronounced chemical activity than sodium ions. Why?

4. Chlorine atoms exhibit more pronounced chemical activity than chlorine ions. Why?

5. Why are lithium and sodium so similar chemically?

6. Why are fluorine and chlorine so similar chemically?

7. (a) How much energy is required to form a K^+ and Cl^- ion pair from a pair of these atoms? (b) What must the separation be between a K^+ and a Cl^- ion if their total energy is to be zero?

8. (a) How much energy is required to form a K^+ and I^- ion pair from a pair of these atoms? (b) What must the separation be between a K^+ and an I^- ion if their total energy is to be zero?

9. (a) How much energy is required to form a Li^+ and Br^- ion pair from a pair of these atoms? (b) What must the separation be between a Li^+ and a Br^- ion if their total energy is to be zero?

Chapter 13
Molecular
Structure

In the previous chapter the origins of covalent and ionic bonds were discussed in terms of the especially simple examples of H_2 and NaCl respectively. Bonding in other molecules, diatomic as well as polyatomic, is usually a more complicated story, and we shall now consider how the ideas introduced in connection with H_2 and NaCl can be extended to provide an understanding of molecular systems in general.

13.1   Theories of Bonding

Two different theoretical approaches have been used to investigate covalent bonding. Although they differ in the procedures they follow to arrive at a description of the electronic structure of a molecule, in principle both approaches yield the same result when carried through far enough. In practice the calculations are so complex that they cannot be pursued anywhere near this far, and in a given case one or the other in simple form usually turns out to be more appropriate.

In the *valence-bond* description, a molecule is considered as a set of individual atoms held together by localized covalent bonds. A covalent bond is supposed to consist of two electrons of opposite spins, one provided by each of the atoms involved; the electrons spend more time on the average between the atoms than elsewhere, which leads to a net force that holds the atoms together. In considering a molecule by the valence-bond method, each bond is treated separately, and the orbitals of each pair of atoms participating in a bond are at first regarded as being the same as those of the isolated atoms. In order to obtain quantitative results, these orbitals must be modified to take into account the various interactions that occur.

In the *molecular-orbital* description, on the other hand, a molecule is regarded from the start as a single, unified entity whose outermost electrons are part of the entire system; the orbitals belong to the molecule, not to the individual atoms. The molecule exists as such because the electron charge distribution in the system has concentrations between the atomic nuclei that hold them together. This is precisely the same mechanism as in the valence-bond theory, of course, but the analysis starts out in a rather different way.

A major justification for the valence-bond picture is that it is in accord with the observation that in a great many molecules the bonds between pairs of atoms are more or less independent of one another. Thus the C—H bonds in hydrocarbon molecules all have almost identical properties, regardless of the particular molecule involved. However, the valence-bond method does not in practice yield as much detailed information on molecular structure as does the molecular-orbital method; the latter is, as a rule, the more realistic approach to understanding the various aspects of molecular bonding.

Both the valence-bond and molecular-orbital methods are simplified by the fact that any alteration in the electronic structure of an atom due to the proximity of another atom is confined to its outermost (or *valence*) electron shell. There are two reasons for this. First, the inner electrons are much more tightly bound and hence less responsive to external influences, partly because they are closer to their parent nucleus and partly because they are shielded from the nuclear charge by fewer intervening electrons. Second, the repulsive interatomic forces in a molecule become predominant while the inner shells of its atoms are still relatively far apart. Direct evidence in support of the idea that only the valence electrons are involved in chemical bonding is available from the X-ray spectra that arise from transitions to inner-shell electron states; it is found that these spectra are virtually independent of how the atoms are combined in molecules or solids.

13.2 The Valence-bond Approach

A virtue of the valence-bond description is that it permits a convenient notation for molecular structures. In this notation each shared pair of electrons is represented either by two dots or by a dash between the atoms bonded by them. Thus the molecules H_2, O_2, and N_2, which are respectively bonded by one, two, and three electron pairs, are represented

$$H:H \qquad O::O \qquad N:::N$$

or, equivalently,

$$H—H \qquad O=O \qquad N≡N$$

According to the valence-bond picture, only atoms with one or more unpaired electron spins can form molecules. Since Hund's rule (Sec. 10.6) states that electrons in the same subshell (same orbital quantum number l) have as many of their spins parallel as possible, all atoms except those with closed subshells have unpaired electrons and so are reactive, that is, able to form bonds. Let us examine the bonding into diatomic molecules of the elements in the first and second periods of the periodic table in terms of the valence-bond approach. Table 13.1 lists these elements and their atomic structures. We recall that there is only one s orbital, corresponding to $m_l = 0$ (since $l = 0$), while there are three p orbitals, corresponding to $m_l = +1, 0, -1$ (since $l = 1$). Each orbital can contain two electrons, but these electrons must have paired—antiparallel—spins. Hence the maximum number of unpaired spins in a p subshell is three, as is the case in nitrogen whose structure is $1s^2 2s^2 2p^3$. The N_2 molecule accordingly has three bonds, N≡N. Oxygen, with four $2p$ electrons, must have two of them paired in one of the three $2p$ orbitals, and therefore has only two unpaired electrons. The O_2 molecule accordingly has two bonds, O=O.

TABLE 13.1.   ATOMIC STRUCTURES OF FIRST- AND SECOND-PERIOD ELEMENTS AND THE STRUCTURES OF THE DIATOMIC MOLECULES THEY FORM AS SUGGESTED BY THE VALENCE-BOND APPROACH.

Element	Atomic number	Atomic structure	Occupancy of orbitals					Unpaired electrons	Molecular structure	Bond energy, ev
			$1s$	$2s$	$2p_x$	$2p_y$	$2p_z$			
Hydrogen, H	1	$1s$	↑					1 $(1s)$	H—H	4.72
Helium, He	2	$1s^2$	↑↓					0	No molecule	
Lithium, Li	3	$1s^22s$	↑↓	↑				1 $(2s)$	Li—Li	1.03
Beryllium, Be	4	$1s^22s^2$	↑↓	↑↓				0	No molecule	
Boron, B	5	$1s^22s^22p$	↑↓	↑↓	↑			1 $(2p)$	B—B	3.0
Carbon, C	6	$1s^22s^22p^2$	↑↓	↑↓	↑	↑		2 $(2p)$	C=C	6.5
Nitrogen, N	7	$1s^22s^22p^3$	↑↓	↑↓	↑	↑	↑	3 $(2p)$	N≡N	9.8
Oxygen, O	8	$1s^22s^22p^4$	↑↓	↑↓	↑↓	↑	↑	2 $(2p)$	O=O	5.1
Fluorine, F	9	$1s^22s^22p^5$	↑↓	↑↓	↑↓	↑↓	↑	1 $(2p)$	F—F	1.6
Neon, N	10	$1s^22s^22p^6$	↑↓	↑↓	↑↓	↑↓	↑↓	0	No molecule	

Although beryllium atoms, with the structures $1s^22s^2$, are unable to form valence bonds with each other, the unfilled $2p$ orbitals of only slightly greater energy than the $2s$ orbital enable them to join together to form a solid through the mechanism of the metallic bond, as discussed in Chap. 17. The unfilled $2p$ orbitals also permit Be atoms to form bonds with certain other atoms, for instance with O atoms in BeO and with Cl atoms in $BeCl_2$.

The bond energies listed in Table 13.1 substantiate the valence-bond approach. Molecules held together by single bonds that involve L-shell ($n = 2$) electrons have bond energies of from 1 to 3 ev, those held together by double bonds have bond energies of 5 or 6 ev, and the N_2 molecule, held together by a triple bond, has a bond energy of almost 10 ev. The more bonds there are in a diatomic molecule, the more firmly it is held together. Internuclear distances, too, are shorter the greater the number of bonds.

The valence-bond picture seems attractive, with observed bond energies correlating nicely with the predicted number of bonds per molecule. But a more detailed look reveals a serious discrepancy. In the valence-bond model of a diatomic molecule, all the electrons in the molecule are paired off with others of opposite spin. There is no net magnetic moment, and the molecule is diamagnetic. However, B_2 and O_2 are experimentally found to be *paramagnetic,* with the data indicating two unpaired electrons per molecule. We shall find that the molecular-orbital approach readily accounts for these unpaired electrons, as well as predicting the same pattern of single, double, and triple bonds as Table 13.1, although in a less direct manner. (The paramagnetism in O_2 is quite conspicuous: When liquid oxygen is poured over a magnet, it is attracted to the poles in the same way that iron filings are.)

13.3   Molecular Orbitals

In discussing chemical bonding on the basis of molecular orbitals, it is helpful to be able to visualize the distributions in space of the various atomic orbitals, which qualitatively resemble those of hydrogen. The pictures in Fig. 9.13 are limited to two dimensions and hence are not suitable for this purpose. It is more appropriate here to draw boundary surfaces of constant $|\psi|^2$ in each case that outline the regions within which the total probability of finding the electron has some definite value, say 90 or 95 percent. Further, the sign of the wave function ψ can be indicated in each lobe of such a drawing, even though what is being represented is $|\psi|^2$. Figure 13.1 contains boundary-surface diagrams for s, p, and d orbitals. These diagrams show $|\Theta\Phi|^2$ in each case; for the corresponding radial probability densities $|R|^2$, Figs. 9.10 and 9.12 can be consulted. The total probability density $|\psi|^2$ is, of course, equal to the product of $|\Theta\Phi|^2$ and $|R|^2$.

When two atoms come together, their orbitals overlap and the result will be either an increased electron probability density between them that signifies a *bonding molecular orbital* or a decreased concentration that signifies an *antibonding molecular orbital*. In the previous chapter we saw how the $1s$ orbitals of two hydrogen atoms could combine to form either the bonding orbital ψ_S or the antibonding orbital ψ_A. In the terminology of molecular physics, ψ_S is referred to as a $1s\,\sigma_g$ orbital and ψ_A as a $1s\,\sigma_u^*$ orbital. The "$1s$" identifies the atomic orbitals that are imagined to combine to form the molecular orbital on the basis of the LCAO scheme. The Greek letter σ signifies that the molecular state has no angular momentum about the bond axis (which is taken to be the z axis). This component of the angular momentum of a molecule is quantized, and is restricted to the values $\lambda\hbar$ where $\lambda = 0, 1, 2, \ldots$. Molecular states for which $\lambda = 0$ are denoted by σ, those for which $\lambda = 1$ by π, those for which $\lambda = 2$ by δ, and so on in alphabetical order. A symmetrical orbital is labeled g (from the German *gerade*, meaning "even") and an antisymmetrical orbital is labeled u (from the German *ungerade*, or "odd"). By "symmetric" is meant, as before, that the orbital does not change sign when inverted through the midpoint between the nuclei, while an antisymmetric orbital does change sign upon inversion. (The g and u distinction is significant only for homonuclear diatomic molecules, for instance H_2, N_2, O_2.) Finally, an antibonding orbital is labeled with an asterisk, as in $1s\,\sigma_u^*$ for the antibonding H_2 orbital ψ_A.

Figure 13.2 contains boundary-surface diagrams that show the formation of σ and π molecular orbitals from s and p atomic orbitals in homonuclear diatomic molecules. Evidently σ orbitals show rotational symmetry about the bond axis, while π orbitals change sign upon a $180°$ rotation about the bond axis. Since the lobes of p_z orbitals are on the bond axis, these atomic orbitals form σ molecular orbitals. The p_x and p_y orbitals both form π molecular orbitals.

We now need to know the relative energies of the various molecular orbitals in order to determine the order in which they are occupied in a molecule. The arrangement of molecular energy levels can be inferred by the same method used in Sec. 12.3 to estimate the energy of the H_2^+ molecular ion. What we do here is consider the two limits of nuclear separation R in a molecule, the *united atom* when $R = 0$ and the *separated atoms* when $R = \infty$. Then we plot the known sequence of energy levels in the united atom on one side of a *correlation diagram* and the known sequence of energy levels in the separated atoms on the other side. States with the same values of λ are then connected, and the sequence of molecular energy levels appears as a function of R. For homonuclear molecules the sym-

Molecular Orbitals

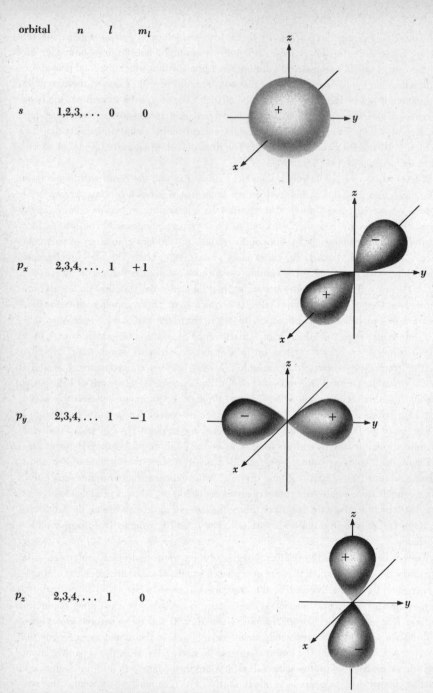

orbital	n	l	m_l
s	$1,2,3,\ldots$	0	0
p_x	$2,3,4,\ldots$	1	$+1$
p_y	$2,3,4,\ldots$	1	-1
p_z	$2,3,4,\ldots$	1	0

FIGURE 13.1 Boundary surface diagrams for s, p, and d atomic orbitals. The $+$ and $-$ signs

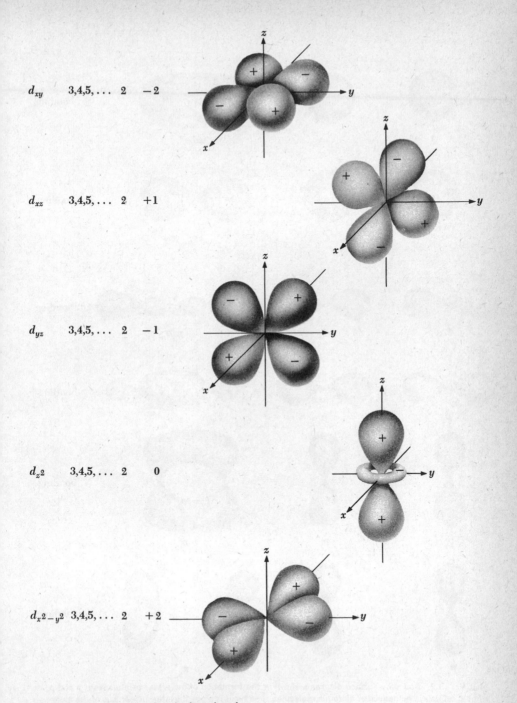

d_{xy} 3,4,5, ...   2   −2

d_{xz} 3,4,5, ...   2   +1

d_{yz} 3,4,5, ...   2   −1

d_{z^2} 3,4,5, ...   2   0

$d_{x^2-y^2}$ 3,4,5, ...   2   +2

refer to the sign of the wave function in each region.

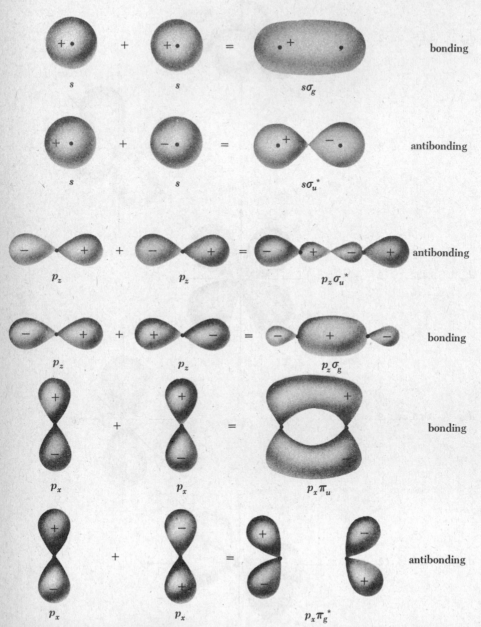

FIGURE 13.2 Boundary surface diagrams showing the formation of molecular orbitals from *s* and *p* atomic orbitals in homonuclear diatomic molecules. The *z* axis is along the internuclear axis of the molecule in each case, and the plane of the paper is the *xy* plane. The $p_y\pi_u$ and $p_y\pi_g$ orbitals are the same as the $p_x\pi_u$ and $p_x\pi_g$ orbitals except that they are rotated through 90°.

metric (g) or antisymmetric (u) character of the orbitals must be preserved, which is done by joining united atom and separated atom orbitals of the same character.

We can determine which united-atom orbital corresponds to which separated-atom orbital by comparing Figs. 13.1 and 13.2. Evidently two separated s atomic orbitals that form a $s\sigma_g$ molecular orbital yield an s united-atom orbital, while two separated s atomic orbitals that form a $s\sigma_u^*$ molecular orbital yield a p_z united-atom orbital. Figure 13.3 shows some of the relationships among H separated-atom orbitals, H_2 molecular orbitals, and He united-atom orbitals.

A correlation diagram for homonuclear diatomic molecules can be constructed on the basis of these considerations. Such a diagram is given in Fig. 13.4. The various united-atom levels are lower than the corresponding separated-atom levels because of the greater nuclear charges in the former, which leads to greater electron binding energies and hence lower total energies. The precise sequence of energy levels in a particular molecule evidently depends upon its internuclear distance R.

We are now in a position to investigate the details of bonding in the homonuclear molecules formed by the first 10 elements of the periodic table. The H_2 molecule we expect to be stable, since both of its electrons can occupy the $1s\sigma_g$ bonding orbital whose energy is less than that of the separated-atom orbital. The electronic configuration of the H_2 molecule is written $1s\sigma_g^2$, where the superscript 2 means that this orbital is occupied by two electrons. The exclusion principle limits the $1s\sigma_g$ bonding orbital to two electrons, so the other two electrons in He_2 would have to go into the $1s\sigma_u^*$ antibonding orbital, whose energy is greater than that of the $1s$ He atomic orbital. Hence the He_2 molecule would have more energy than a separated pair of He atoms, and He_2 cannot exist. If He_2 did occur, its configuration would be written $1s\sigma_g^2 1s\sigma_u^{*2}$.

In the molecular-orbital approach, as in the valence-bond approach, it is useful to consider that a full bond between two atoms involves two electrons. Also, electrons in bonding and antibonding orbitals are considered to have roughly comparable though opposite effects on molecular stability, so that each occupied antibonding orbital in a molecule cancels out one occupied bonding orbital. The number of bonds in a molecule is therefore given by the formula

Number of bonds = ½(electrons in bonding orbitals − electrons in antibonding orbitals)

In H_2, according to this formula, there is a single bond, and in He_2, no bond at all.

Table 13.2 shows the electron configurations of the molecules listed in Table 13.1 on the basis of the molecular-orbital approach. The number of bonds per molecule is the same as in the valence-bond picture, but now the presence of unpaired electrons in B_2 and O_2 follows directly from Hund's rule that, whenever possible, electrons in an atom or molecule remain unpaired. The $2p_x\pi_u$ and $2p_y\pi_u$ bonding orbitals have the same energy, and so of the two electrons available to them in B_2, one goes into each orbital. The two unpaired electrons per molecule lead to the observed paramagnetism of B_2. In O_2 these orbitals are both filled with two electrons each, but now there are two more electrons for the $2p_x\pi_g^*$ and

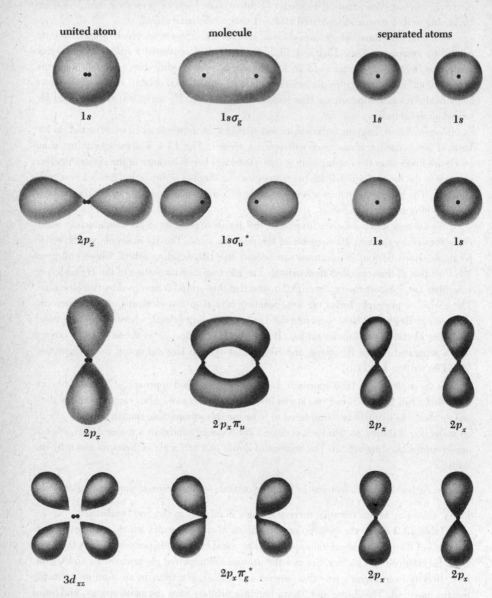

united atom

molecule

separated atoms

$1s$

$1s\sigma_g$

$1s$

$1s$

$2p_z$

$1s\sigma_u{}^*$

$1s$

$1s$

$2p_x$

$2p_x\pi_u$

$2p_x$

$2p_x$

$3d_{xz}$

$2p_x\pi_g{}^*$

$2p_x$

$2p_x$

FIGURE 13.3 Relationships among H separated-atom orbitals, H_2 molecular orbitals, and He united-atom orbitals.

$2p_y\pi_g^*$ antibonding orbitals. Each of the latter gets one electron, again leading to two un-paired electrons per molecule.

The properties of the ions O_2^+, O_2^-, and O_2^{--} further confirm the molecular-orbital analysis of this molecule. When an electron is removed from O_2 to give O_2^+, it comes from either the $2p_x\pi_g^*$ or $2p_y\pi_g^*$ antibonding orbital, so that now there is one less antibonding electron. The net number of bonds therefore *increases* to 2½, which is reflected in an inter-nuclear distance of $R = 1.12$ A in O_2^+ as compared with $R = 1.21$ A in O_2: O_2^+, with one *less* electron, is more tightly bound. Adding one or two electrons to O_2 has the opposite effect, since these electrons go into antibonding orbitals. Hence O_2^- has only 1½ bonds and

FIGURE 13.4 Correlation diagram for homonuclear diatomic molecules. The internuclear distances in several molecules are indicated. The system energy decreases for bonding orbitals and increases for antibonding orbitals.

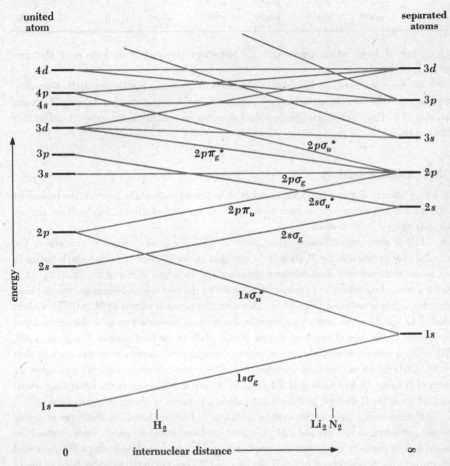

TABLE 13.2. ELECTRONIC STRUCTURES OF HOMONUCLEAR DIATOMIC MOLECULES AS SUGGESTED BY THE MOLECULAR-ORBITAL APPROACH.   Each singly occupied orbital represents an unpaired electron.

Molecule	$1s$	$2s$	$2p_x$	$2p_y$	$2p_z$	Net Bonds	Molecular Structure
H_2	σ^2					1	H—H
(He_2)	$\sigma^2\sigma^{*2}$					0	No molecule
Li_2	$\sigma^2\sigma^{*2}$	σ^2				1	Li—Li
(Be_2)	$\sigma^2\sigma^{*2}$	$\sigma^2\sigma^{*2}$				0	No molecule
B_2	$\sigma^2\sigma^{*2}$	$\sigma^2\sigma^{*2}$	π	π		1	B—B
C_2	$\sigma^2\sigma^{*2}$	$\sigma^2\sigma^{*2}$	π^2	π^2		2	C=C
N_2	$\sigma^2\sigma^{*2}$	$\sigma^2\sigma^{*2}$	π^2	π^2	σ^2	3	N≡N
O_2	$\sigma^2\sigma^{*2}$	$\sigma^2\sigma^{*2}$	$\pi^2\pi^*$	$\pi^2\pi^*$	σ^2	2	O=O
F_2	$\sigma^2\sigma^{*2}$	$\sigma^2\sigma^{*2}$	$\pi^2\pi^{*2}$	$\pi^2\pi^{*2}$	σ^2	1	F—F
(Ne_2)	$\sigma^2\sigma^{*2}$	$\sigma^2\sigma^{*2}$	$\pi^2\pi^{*2}$	$\pi^2\pi^{*2}$	$\sigma^2\sigma^{*2}$	0	No molecule

$O_2{}^{--}$ only 1 bond, which agrees with the observed values of $R = 1.26$ A in $O_2{}^-$ and $R = 1.49$ A in $O_2{}^{--}$.

On the other hand, all the valence electrons in N_2 are in bonding orbitals. Removing an electron from N_2 to form the ion $N_2{}^+$ means one less bonding electron, leaving 2½ bonds instead of 3. Thus $N_2{}^+$ is *less* tightly bound than N_2, in contrast to the tighter bond in $O_2{}^+$ than in O_2.

13.4   Electronegativity

A heteronuclear diatomic molecule consists of two unlike atoms. In general, the molecular orbitals are not symmetric in such molecules, so that the electrons they contain are not equally shared by both atoms.

LiH is the simplest heteronuclear molecule and is a good example of this effect. The normal configuration of the H atom is $1s$ and that of the Li atom is $1s^2 2s$, which means in each case that there is a single valence electron. The $1s$ orbital of H and the $2s$ orbital of Li form a σ bonding orbital in LiH that is occupied by the two valence electrons, thereby constituting a single net bond (Fig. 13.5). However, the electron affinity of H is 0.75 ev while that of Li is 0.54 ev; in both atoms the effective nuclear charge acting on a valence electron is $+e$ (in Li the core of two $1s$ electrons shields $+2e$ of the total nuclear charge of $+3e$), but in Li a valence electron is on the average several times farther from the nucleus than in H. The respective ionization energies also reflect this difference, with the ionization energy of H being 13.6 ev while it is 5.4 ev in Li. Hence the electrons in the σ bonding orbital of LiH favor the H nucleus, and there is a partial separation of charge in the molecule.

If there were a complete separation of charge in LiH, as there is in NaCl, the molecule would consist of an Li^+ ion and a H^- ion, and the bond would be purely ionic. Instead the bond is only partially ionic, with the two bonding electrons spending perhaps 80 percent of the time in the neighborhood of the H nucleus and 20 percent in the neighborhood of the Li nucleus. In contrast, the bonding electrons in a homonuclear molecule such as H_2 or O_2 spend 50 percent of the time near each nucleus. Molecules whose bonds are neither purely covalent nor purely ionic are sometimes called *polar covalent*, since they possess electric dipole moments.

The relative tendency of an atom to attract shared electrons when it is part of an atom is called its *electronegativity*. In the LiH molecule, for instance, H is more electronegative than Li. A scale of electronegativities has been devised, principally through the work of Linus Pauling, on the basis of observed bond energies. A partially ionic character to a bond means a higher bond energy—greater stability—than if the bond were purely covalent with equal electron sharing, since the attractive force due to the negative charge density between the atoms is supplemented by a direct mutual attraction between the more positive atom and the more negative one. Thus the difference between the actual bond energy in a molecule and the bond energy calculated by assuming equal electron sharing is a measure of the difference between the electronegativities of the atoms involved.

The preceding idea is put on a quantitative basis in the following way. Let us consider the homonuclear diatomic molecules A_2 and B_2 and the heteronuclear molecule AB. Both A_2 and B_2 are purely covalent, sharing their valence electrons equally. If the electron sharing in AB is also equal, the bond energy E_{AB} in AB ought to be the average of the bond energies E_{A_2} of A_2 and E_{B_2} of B_2. If the electron sharing is unequal, the bond is stronger and E_{AB} will exceed the average of E_{A_2} and E_{B_2} by some amount Δ. Originally the average of E_{A_2} and E_{B_2} was considered to be the arithmetic average $\frac{1}{2}(E_{A_2} + E_{B_2})$, but subsequently it was found that more consistent electronegativity values are obtained by taking the geometric mean $\sqrt{E_{A_2} \times E_{B_2}}$ instead. Hence

13.1
$$\Delta = E_{AB} - \sqrt{E_{A_2} \times E_{B_2}}$$

If the electronegativity of atom A in the molecule AB is X_A and that of atom B is X_B, the difference between them is defined as

13.2
$$X_A - X_B = \sqrt{\Delta}$$

where Δ is in ev. This procedure only establishes relative values of electronegativity; to construct a definite scale, an arbitrary value of X must be assigned to one atom. Pauling chose this atom to be fluorine and set $X_F = 4.0$, which is the highest value of X since fluorine is the most electronegative element.

Bond energies in chemistry are usually expressed in kcal/mole instead of in ev/molecule. Since 1 ev/molecule = 23 kcal/mole, Eq. 13.2 becomes

FIGURE 13.5 The bonding electrons in LiH occupy a σ molecular orbital formed from the 1s orbital of the H atom and the 2s orbital of the Li atom.

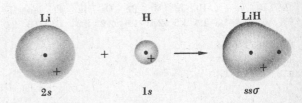

13.3

$$X_A - X_B = \sqrt{\frac{\Delta}{23}} = 0.208 \sqrt{\Delta}$$

when Δ is in kcal/mole.

As an example, let us calculate the electronegativity of H from the bond energies of H_2, F_2, and HF. The latter are 103 kcal/mole, 38 kcal/mole, and 134 kcal/mole respectively. Therefore

$$\Delta = E_{HF} - \sqrt{E_{H_2} \times E_{F_2}}$$
$$= 134 - \sqrt{103 \times 38} \text{ kcal/mole}$$
$$= 71 \text{ kcal/mole}$$

and

$$X_F - X_H = 0.208 \sqrt{71} = 1.8$$

Since $X_F = 4.0$ by definition, the electronegativity of H is 2.2. Table 13.3 is a list of atomic electronegativities obtained in this way.

TABLE 13.3 ELECTRONEGATIVITIES OF THE ELEMENTS

1																	2
H																	He
2.2																	—
3	**4**											**5**	**6**	**7**	**8**	**9**	**10**
Li	Be											B	C	N	O	F	Ne
1.0	1.6											2.0	2.6	3.0	3.4	4.0	—
11	**12**											**13**	**14**	**15**	**16**	**17**	**18**
Na	Mg											Al	Si	P	S	Cl	Ar
0.9	1.3											1.6	1.9	2.2	2.6	3.1	—
19	**20**	**21**	**22**	**23**	**24**	**25**	**26**	**27**	**28**	**29**	**30**	**31**	**32**	**33**	**34**	**35**	**36**
K	Ca	Sc	Ti	V	Cr	Mn	Fe	Co	Ni	Cu	Zn	Ga	Ge	As	Se	Br	Kr
0.8	1.0	1.4	1.5	1.6	1.7	1.6	1.8	1.9	1.9	1.9	1.7	1.8	2.0	2.2	2.6	3.0	—
37	**38**	**39**	**40**	**41**	**42**	**43**	**44**	**45**	**46**	**47**	**48**	**49**	**50**	**51**	**52**	**53**	**54**
Rb	Sr	Y	Zr	Nb	Mo	Tc	Ru	Rh	Pd	Ag	Cd	In	Sn	Sb	Te	I	Xe
0.8	1.0	1.2	1.3	1.6	2.2	1.9	2.2	2.3	2.2	1.9	1.7	1.8	2.0	2.0	2.1	2.7	—
55	**56**	**57–71**	**72**	**73**	**74**	**75**	**76**	**77**	**78**	**79**	**80**	**81**	**82**	**83**	**84**	**85**	**86**
Cs	Ba	—	Hf	Ta	W	Re	Os	Ir	Pt	Au	Hg	Tl	Pb	Bi	Po	At	Rn
0.8	0.9	1.1–1.3	1.3	1.5	2.4	1.9	2.2	2.2	2.3	2.5	2.0	2.0	2.3	2.0	2.0	2.2	—
87	**88**	**89–94**															
Fr	Ra	—															
0.7	0.9	1.1–1.4															

Table 13.3 enables us to predict how the electrons in a heteronuclear molecule are shared. Thus the electronegativity of H is 2.2 and that of Li is 1.0, in agreement with the fact that the shared electrons favor the H atom. On the other hand, in HF the electronegativity of F is 4.0, so here the shared electrons favor the F atom. In BH the respective electronegativities are 2.0 and 2.2, and the sharing is more nearly equal than in LiH or HF.

We might reasonably expect the electronegativity of an atom to be proportional to the sum of its ionization energy and electron affinity: A high ionization energy means little tendency to lose an electron and a high electron affinity means a considerable tendency to gain an electron. From Table 12.1 we find that the ionization energies of the halogens are 17.4, 13.0, 11.8, and 10.5 ev for F, Cl, Br, and I respectively, while from Table 12.2 we find that the corresponding electron affinities are 3.45, 3.61, 3.36, and 3.06 ev. Hence the relative electronegativities of the halogens should be F:Cl:Br:I $= 20.9:16.6:15.2:13.6$, and these ratios are indeed consistent with the F:Cl:Br:I $= 4.0:3.1:3.0:2.7$ of Table 13.3. A proportionality constant of about 0.19 brings the two scales into approximate coincidence. That is,

13.4 $$X \approx 0.19 \text{ (ionization energy + electron affinity)}$$

The coincidence is not exact because electronegativity is not too precisely defined a concept (though nevertheless a useful one), and the two ways of calculating X actually involve slightly different physical properties.

13.5 Polyatomic Molecules

In a heteronuclear molecule the atomic orbitals that are imagined to combine to form a molecular orbital may be of different character in each atom. An example is HF, where the $1s$ atomic orbital of H joins with the $2p_z$ orbital of F. There are two possibilities, as in Fig. 13.6, a bonding $sp\sigma$ molecular orbital and an antibonding $sp\sigma^*$ orbital. Since the $1s$

FIGURE 13.6 Bonding and antibonding molecular orbitals in HF.

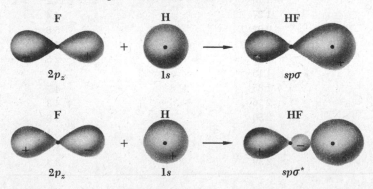

Polyatomic Molecules

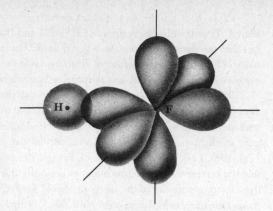

FIGURE 13.7 Valence atomic orbitals in HF. The atomic orbitals shown as overlapping form a σ bonding molecular orbital.

orbital of H and the $2p_z$ orbital of F each contain one electron (Table 13.1), the $sp\sigma$ orbital in HF is occupied by two electrons, and we may regard HF as being held together by a single bond. The electron structure of the HF molecule is shown in Fig. 13.7.

The configurations of the three p atomic orbitals together with their ability to join with s orbitals to form bonding molecular orbitals make it possible to understand the geometries of many polyatomic molecules. The water molecule H_2O is an example. Offhand we might expect a linear molecule, H—O—H, since oxygen is more electronegative than hydrogen and each H atom in H_2O accordingly exhibits a small positive charge. The resulting repulsion between the H atoms should keep them as far apart as possible, namely

TABLE 13.4. BOND LENGTHS AND BOND ENERGIES OF SOME HETERONUCLEAR DIATOMIC MOLECULES
1 ev/molecule = 23 kcal/mole

Molecule	Bond length, A	Bond energy, ev
BN	1.28	4.0
CO	1.13	11.2
HCl	1.27	4.4
HF	0.92	5.8
KCl	2.67	4.4
LiH	1.60	2.5
NO	1.15	7.0
NaF	1.85	4.7
NaCl	2.36	4.3
NaBr	2.50	3.8
NaI	2.71	3.1
PbO	1.92	4.1
PbS	2.39	3.3

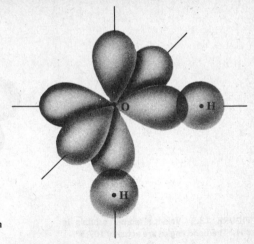

FIGURE 13.8 Valence atomic orbitals in
H_2O. The bond angle is actually 104.5°.

on opposite sides of the O atom. In reality, however, the water molecule has a structure
closer to O—H with an angle of 104.5° between the two O—H bonds.
|
H

The bent shape of the water molecule is not hard to account for. From Table 13.1
we find that the $2p_y$ and $2p_z$ orbitals in O are only singly occupied, so that each can join
the 1s orbital of an H atom to form a $sp\sigma$ bonding orbital (Fig. 13.8). The y and z axes
are 90° apart, and the larger 104.5° angle that is actually found may plausibly be attributed to the mutual repulsion of the H atoms. In support of the latter idea is the fact
that the bond angles in the otherwise similar molecules H_2S and H_2Se are 92° and 90°
respectively, which we may ascribe to the greater separation of the H atoms around the
larger atoms S (atomic number $Z = 16$) and Se ($Z = 34$).

A similar argument explains the pyramidal shape of the ammonia molecule NH_3.
From Table 13.1 we find that the $2p_x$, $2p_y$, and $2p_z$ atomic orbitals in N are singly occupied,
which means that each of them can form a $sp\sigma$ bonding orbital with the 1s orbital of an
H atom. The bonding molecular orbitals in NH_3 should therefore be centered along the
x, y, and z axes (Fig. 13.9) with N—H bonds 90° apart. As in H_2O, the actual bond angles
in NH_3 are somewhat greater than 90°, in this case 107.5°, owing to repulsions among the
H atoms. The similar hydrides of the larger atoms P ($Z = 15$) and As ($Z = 33$) exhibit
the smaller bond angles of 94° and 90° respectively, again in consequence of the reduced
mutual repulsions among the more distant H atoms.

13.6 Hybrid Orbitals

The straightforward way in which the shapes of the H_2O and NH_3 molecules are explained
is a conspicuous failure in the case of methane, CH_4. A carbon atom has two electrons in

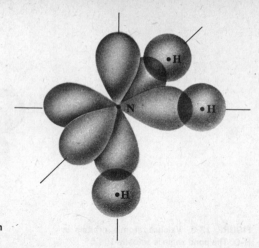

FIGURE 13.9 Valence atomic orbitals in NH_3. The bond angles are actually 107.5°.

its $2s$ orbital and one each in its $2p_x$ and $2p_y$ orbitals. Thus we would expect the hydride of carbon to be CH_2, with two $sp\sigma$ bonding orbitals and a bond angle of 90° or a little more. (In the valence-bond picture, the same conclusion is reached, since C atoms have two unpaired electrons each.) Yet CH_4 exists, and, furthermore, it is perfectly symmetrical in structure with tetrahedral molecules whose C—H bonds are exactly equivalent to one another.

We cannot consider carbon as an isolated exception, a sort of freak atom in which a fortuitous combination of circumstances leads to CH_4 instead of CH_2, because the same phenomenon occurs in other atoms as well. A boron atom, for example, has the configuration $1s^2 2s^2 2p$, and it forms BF_3 and BCl_3 instead of BF and BCl.

Clearly, what is happening in carbon and boron is that the $2s$ orbitals, despite being occupied by electron pairs and having less energy—more stability—than the $2p$ orbitals, somehow enter into the formation of molecular orbitals and thereby permit the $2s$ electrons to contribute to the bonds formed by C and B with other atoms. A C atom has four $n = 2$ electrons in all and forms CH_4, while a B atom has three $n = 2$ electrons in all and forms BF_3, respectively sharing four and three electrons with their partners.

The easiest way to explain the existence of CH_4 is to assume that one of the two $2s$ electrons in C is "promoted" to the vacant $2p_z$ orbital, so that there is now one electron each in the $2s$, $2p_x$, $2p_y$, and $2p_z$ orbitals and four bonds can be formed. To raise an electron from a $2s$ to a $2p$ state means increasing the energy of the C atom, but it is reasonable to suppose that the formation of four bonds (to yield CH_4) in place of two (to yield CH_2) lowers the energy of the resulting molecule more than enough to compensate for this. The foregoing picture suggests that three of the bonds in CH_4 are $sp\sigma$ bonds and one of them is a $ss\sigma$ bond involving the $1s$ orbital of H and now singly occupied $2s$ orbital of C; experimentally, however, all four bonds are found to be identical.

The correct explanation for CH_4 is based on a phenomenon called *hybridization* which can occur when the 2s and 2p states of an atom in a molecule are very close together in energy. In this case the atom can contribute a linear combination of *both* its 2s and 2p atomic orbitals to *each* molecular orbital if in this way the resulting bonds are more stable than otherwise. That such composite atomic orbitals can occur follows from the nature of Schrödinger's equation, which is a partial differential equation. The 2s and 2p wave functions of an atom are both solutions of the same equation if the corresponding energies are the same, and a linear combination of solutions of a partial differential equation is always itself a solution. In an isolated atom, an electron in a 2s orbital has less energy (is more tightly bound) than an electron in a 2p orbital, and there is accordingly no tendency for hybrid atomic orbitals to occur. On the other hand, when an atom in a certain molecule contributes superposed s and p orbitals to the molecular orbitals, the resulting bonds may be stronger than the bonds that the s and p orbitals by themselves would lead to, even though the p parts of the hybrids had higher energies in the separated atom. Hybrid orbitals thus occur when the bonding energy they give rise to is greater than that which pure orbitals would produce, which happens in practice when the s and p levels of an atom are close together.

In CH_4, then, carbon has four equivalent hybrid orbitals which participate in bonding. These four orbitals are hybrids of one 2s and three 2p orbitals, and we may consider each one as a combination of $\frac{1}{4}s$ and $\frac{3}{4}p$. This particular combination is therefore called a sp^3 hybrid. Its configuration can be visualized in terms of boundary-surface diagrams as shown in Fig. 13.10. Evidently a sp^3 hybrid orbital is strongly concentrated in a single direction, which accounts for its ability to produce an exceptionally strong bond—strong enough to compensate for the need to promote a 2s electron to a 2p state.

It must be kept in mind that hybrid orbitals do not exist in an isolated atom, even when it is in an excited state, but arise while that atom is interacting with others to form a molecule.

FIGURE 13.10   In sp^3 hybridization, an s orbital and three p orbitals in the same atom combine to form four sp^3 hybrid orbitals.

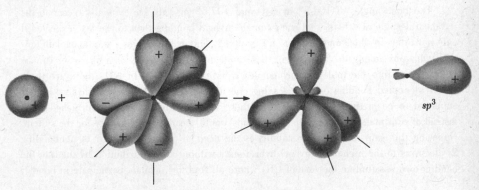

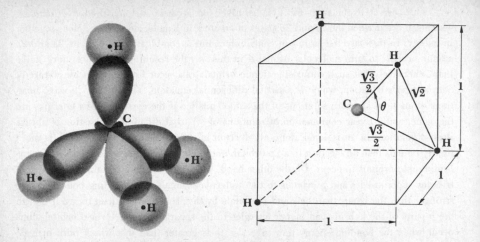

FIGURE 13.11   The tetrahedral methane (CH_4) molecule. The overlapping sp^3 hybrid orbitals of the C atom and $1s$ orbitals of the four H atoms form bonding molecular orbitals.

Figure 13.11 is a representation of the CH_4 molecule. Also shown is a model of this molecule that consists of a C atom in the center of a unit cube which has H atoms at alternate corners. A triangle with the C atom at one vertex and any two H atoms at the other vertexes has sides $\sqrt{3/2}$, $\sqrt{3/2}$, and $\sqrt{2}$ in length. If the angle between the C—H bonds is θ, from the law of cosines ($a^2 = b^2 + c^2 - 2bc \cos \theta$) we have

$$\cos \theta = -\frac{a^2 - b^2 - c^2}{2bc}$$

$$= -\frac{2 - \frac{3}{4} - \frac{3}{4}}{2 \times \frac{3}{4}} = -\frac{1}{3}$$

$$\theta = 109.5°$$

which is what is determined experimentally.

The bond angles of $104.5°$ in H_2O and $107.5°$ in NH_3 are evidently closer to the tetrahedral angle of $109.5°$ that occurs in sp^3 hybrid bonding than to the $90°$ expected if only p orbitals in the O and N atoms are involved. This fact provides a way to explain how the repulsions among the H atoms in these molecules that were spoken of earlier can be incorporated into the molecular-orbital description of bonding. In NH_3 there are three doubly occupied bonding orbitals, leaving one unshared pair of electrons that we earlier supposed to be in the $2s$ orbital of N. If there were sp^3 hybrid orbitals furnished by N instead of p orbitals, the more separated bonds would mean a lower energy for the system. Opposing this gain in molecular stability is the need to promote the pair of nonbonding $2s$ electrons to the higher-energy sp^3 hybrid state without any contribution by them to the bonding process (unlike the case of CH_4 where all four sp^3 orbitals participate in bonds).

Hence we may regard the $107.5°$ bond angle in NH_3 as the result of a compromise between the two extremes of four sp^3 hybrid orbitals in N with one of them nonbonding and three $2p$ bonding orbitals and one $2s$ nonbonding (but low-energy) orbital. Figure 13.12 is a representation of the NH_3 molecule on the basis of sp^3 hybridization, which may be compared with Fig. 13.9 which was drawn on the basis of p orbitals.

In H_2O, since there are two bonding orbitals and two nonbonding orbitals in O, the tendency to form hybrid sp^3 orbitals is less than in NH_3 where there are three bonding orbitals and only one nonbonding orbital. The smaller bond angle in H_2O is in agreement with this conclusion.

13.7 Carbon-Carbon Bonds

Two other types of hybrid orbital in addition to sp^3 can occur in carbon atoms. In sp^2 hybridization, one valence electron is in a pure p orbital and the other three are in hybrid orbitals that are $\frac{1}{3}s$ and $\frac{2}{3}p$ in character. In sp hybridization, two valence electrons are in pure p orbitals and the other two are in hybrid orbitals that are $\frac{1}{2}s$ and $\frac{1}{2}p$ in character.

Ethylene, C_2H_4, is an example of sp^2 hybridization in which the two carbon atoms are joined by two bonds. Figure 13.13 contains a boundary-surface diagram showing the three sp^2 hybrid orbitals, which are $120°$ apart in the plane of the paper, and the pure p_x orbital in each C atom. Two of the sp^2 orbitals in each C atom overlap s orbitals in H atoms to form σ bonding orbitals, and the third sp^2 orbital in each C atom forms a σ bonding orbital with the same orbital in the other C atom. The p_x orbitals of the C atoms form a π bond with each other, so that one of the bonds between the carbon atoms is a σ bond and the other is a π bond. The conventional structural formula of ethylene is accordingly

Ethylene

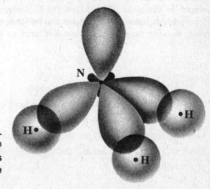

FIGURE 13.12 Valence atomic orbitals in the ammonia (NH_3) molecule based on the assumption of sp^3 hybridization in the N orbitals. One of the sp^3 orbitals is occupied by two N electrons and does not contribute to bonding.

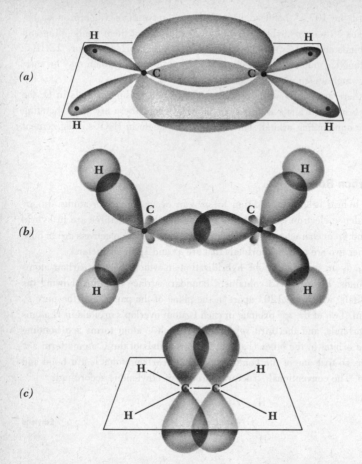

FIGURE 13.13   (a) The ethylene (C_2H_4) molecule. All the atoms lie in a plane perpendicular to the plane of the paper. (b) Top view, showing the sp^2 hybrid orbitals that form σ bonds between the C atoms and between each C atom and two H atoms. (c) Side view, showing the pure p_z orbitals that form a π bond between the C atoms.

Acetylene, C_2H_2, is an example of sp hybridization in which the two carbon atoms are joined by three bonds. One sp hybrid orbital in each C atom forms a σ bond with an H atom, and the second forms a σ bond with the other atom. The $2p_x$ and $2p_y$ orbitals in each C atom form π bonds, so that one of the three bonds between the carbon atoms is a σ_z bond and the others are π_x and π_y bonds (Fig. 13.14). The conventional structural formula of acetylene is

$$H—C\equiv C—H$$

In both ethylene and acetylene the electrons in the π orbitals are "exposed" on the outside of the molecules. These compounds are much more reactive chemically than molecules with only single σ bonds between carbon atoms, such as ethane,

$$
\begin{array}{c}
\text{H}\ \ \text{H} \\
|\ \ \ | \\
\text{H—C—C—H} \\
|\ \ \ | \\
\text{H}\ \ \text{H}
\end{array}
$$

Ethane

in which all the bonds are formed from sp^3 hybrid orbitals in the carbon atoms. Carbon compounds with double and triple bonds are said to be *unsaturated* because they can add other atoms to their molecules in such reactions as

$$
\begin{array}{ccc}
\text{H}\diagdown\ \ \diagup\text{H} & & \text{H}\ \ \text{H} \\
\text{C}=\text{C} & +\ \text{HCl} \longrightarrow & \text{H—C—C—H} \\
\text{H}\diagup\ \ \diagdown\text{H} & & \text{H}\ \ \text{Cl}
\end{array}
$$

In a *saturated* compound such as methane or ethane only single bonds are present.

13.8 The Benzene Ring

Benzene, C_6H_6, is an extremely interesting hydrocarbon because its six C atoms are arranged in a flat hexagonal ring. If we think in terms of valence bonds, there are three single

FIGURE 13.14 The acetylene (C_2H_2) molecule. There are three bonds between the C atoms, one σ bond between sp hybrid orbitals and two π bonds between pure p_x and p_y orbitals.

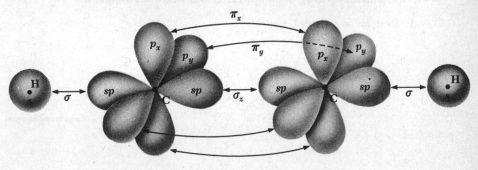

and three double bonds between the C atoms in benzene, so that its structure can be described in the following two ways:

(A)   (B)   **Benzene**

There are a number of major objections to this simple valence-bond picture of benzene, of which we shall mention three. The first is that double bonds are not particularly stable, so that benzene, like ethylene, should form addition compounds readily. What actually happens, on the contrary, is that benzene forms substitution compounds in which one or more of the H atoms are replaced by other atoms or atom groups, with the ring structure remaining intact. In other words, the benzene ring is stronger than the simple structural formula would indicate. Substantiating this observation, the heat of formation of C_6H_6 turns out to be about 40 kcal/mole—*1.7 ev/molecule*—*more* than would be expected if there were indeed six C—H, three C—C, and three C=C bonds in each molecule. (The heat of formation of a molecule is the energy given off when it is formed from isolated atoms.)

The second objection arises because C—C bonds are 1.54 A long on the average and C=C bonds are 1.35 A long; the shorter length of the C=C bond is consistent with the participation of four electrons in a double bond, instead of two as in C—C. But benzene molecules are regular hexagons in shape, with all bonds having the same length and all bond angles being 120°.

Finally, if C_6H_6 occurred with C—C and C=C bonds alternating, compounds such as dibromobenzene, $C_6H_4Br_2$, ought to occur in four forms, as shown below.

1. Br atoms on opposite C atoms:

Paradibromobenzene

2. Br atoms on alternate C atoms:

Br

H—C=C—H

H—C=C—Br

H

Metadibromobenzene

3. Br atoms on adjacent double-bonded C atoms:

H

H—C=C—Br

H—C=C—Br

H

Orthodibromobenzene

4. Br atoms on adjacent single-bonded C atoms:

H

Br—C=C—H

Br—C=C—H

H

Orthodibromobenzene

Only three forms of $C_6H_4Br_2$ are actually found, however—there is only one kind of orthodibromobenzene molecule, not two. Hence in nature there is no distinction between the double and single bonds we have used to represent the benzene molecule.

To resolve these serious conflicts, each benzene molecule is regarded in the valence-bond approach as resonating between the two possible states A and B. On the average there are 1½ bonds between adjacent C atoms, instead of either one or two. To describe the situation formally, we would say that the wave function for the entire system is proportional to $\psi_A + \psi_B$, where ψ_A and ψ_B are the wave functions of the valence-bond structures A and B. A detailed calculation shows that the combined wave function indeed leads to a lower energy for the system of six C atoms and six H atoms than either ψ_A or ψ_B separately. (This calculation is analogous to that carried out in the previous chapter for the H+ molecular ion, in which it turned out that the system of two protons and an electron has least energy when the electron is described by a wave function that is a superposition of the wave functions of two alternative but equivalent states.) The benzene molecule is thus in a sense a *resonance hybrid* of the two configurations A and B.

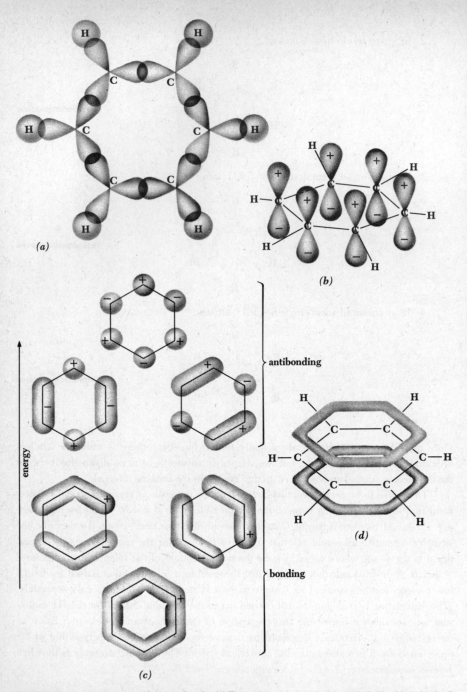

FIGURE 13.15 The benzene (C_6H_6) molecule. (*a*) The overlaps between the sp^2 hybrid orbitals in the C atoms with each other and with the *s* orbitals of the H atoms lead to σ bonds. (*b*) Each C atom has a pure p_z orbital occupied by one electron. (*c*) The six possible π molecular orbitals in benzene. (*d*) The three bonding orbitals constitute a continuous electron probability distribution around the molecule that contains six delocalized electrons.

A picture of the nature of the benzene molecule that is both closer to physical reality and more intuitively meaningful is provided by considering the molecular orbitals involved. Because the carbon-carbon bonds in the benzene ring are 120° apart, we conclude that the basic structure of the molecule is the result of bonding by sp^2 hybrid orbitals. Of the three sp^2 orbitals per C atom, one forms a σ bonding orbital with the $1s$ orbital of an H atom and the other two form σ bonding orbitals with the corresponding sp^2 orbitals of the C atoms on either side (Fig. 13.15). This leaves one $2p_x$ orbital per C atom, which has lobes above and below the plane of the ring. The total of six $2p_x$ orbitals in the molecule can be combined in six possible ways to produce three bonding and three antibonding π orbitals. In the actual benzene molecule the bonding π orbitals, having the lowest energies, are occupied by the available electrons. The result is a continuous electron probability distribution above and below the plane of the ring whose six electrons belong to the molecule as a whole and not to any particular pair of atoms; these electrons are *delocalized*.

With six delocalized electrons in π bonding orbitals and twelve electrons in the σ bonds between the carbon atoms, each carbon-carbon bond consists of 1½ net bonds, one σ bond and half of a π bond. Confirming this result, each C$=$C bond in benzene turns out to be 1.40 A long, intermediate between the 1.54-A average length of a C—C bond and the 1.35-A average length of a C=C bond. In the valence-bond view, the extra stability of benzene arises because the system resonates between two different states of equal energy; in the molecular-orbital view, the extra stability is due to the delocalization of the six $2p_x$ electrons.

Problems

1. Distinguish between electron affinity and electronegativity. Are they related in any way?

2. The electric dipole moment of the H_2O molecule is 6.13×10^{-30} coulomb-m. Find the dipole moment of each O—H bond.

3. The outer electronic configuration of the CO molecule may be written as $\sigma_s^2\sigma_s^{*2}\pi_{x,y}^4\sigma_z^2$ and that of the CO$^+$ molecular ion as $\sigma_s^2\sigma_s^{*2}\pi_{x,y}^4\sigma_z$. How many bonds are present in each case and what is their nature? Is CO or CO$^+$ more strongly bound?

4. The outer electronic configuration of the NO molecule may be written as $\sigma_s^2\sigma_s^{*2}\pi_{x,y}^4$ $\pi_{x,y}^*\sigma_z^2$ and that of the NO$^+$ molecular ion as $\sigma_s^2\sigma_s^{*2}\pi_{x,y}^4\sigma_z^2$. How many bonds are present in each case and what is their nature? Is NO or NO$^+$ more strongly bound?

5. Write down the outer electronic configurations of F_2, F_2^+, and F_2^- and determine the number of net bonds in each. Which has the highest bond energy and which the lowest?

6. Although the molecule He$_2$ is unstable and does not occur, the molecular ion He$_2^+$ is stable and has a bond energy about equal to that of H_2^+. Explain this observation in terms of the molecular orbital approach.

Chapter 14
Molecular Spectra

Molecular energy states arise from the rotation of a molecule as a whole and from the vibrations of its constituent atoms relative to one another as well as from changes in its electronic configuration. Rotational states are separated by quite small energy intervals (10^{-3} ev is typical), and the spectra that arise from transitions between these states are in the microwave region with wavelengths of 0.1 mm—1 cm.. Vibrational states are separated by somewhat larger energy intervals (0.1 ev is typical), and vibrational spectra are in the infrared region with wavelengths of 10,000 A—0.1 mm. Molecular electronic states have higher energies, with typical separations between the energy levels of valence electrons of several electron volts and spectra in the visible and ultraviolet regions. A detailed picture of a particular molecule can often be obtained from its spectra, including bond lengths, force constants, and bond angles.

14.1 Rotational Energy Levels: Diatomic Molecules

We shall first consider the rotational energy levels of a diatomic molecule. We may picture such a molecule as consisting of atoms of masses m_1 and m_2 a distance R apart, as in Fig. 14.1. The moment of inertia of this molecule about an axis passing through its center of mass and perpendicular to a line joining the atoms is

14.1
$$I = m_1 r_1{}^2 + m_2 r_2{}^2$$

where r_1 and r_2 are the distances of atoms 1 and 2 respectively from the center of mass. Since

14.2
$$m_1 r_1 = m_2 r_2$$

by definition, the moment of inertia may be written

$$I = \left(\frac{m_1 m_2}{m_1 + m_2}\right)(r_1 + r_2)^2$$

14.3
$$= m'R^2$$

where

14.4
$$m' = \frac{m_1 m_2}{m_1 + m_2}$$

Reduced mass

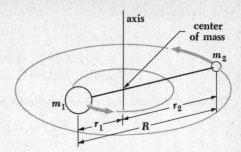

FIGURE 14.1 A diatomic molecule can rotate about its center of mass.

is the reduced mass of the molecule as discussed in Sec. 6.7. Equation 14.3 states that the rotation of a diatomic molecule is equivalent to the rotation of a single particle of mass m' about an axis located a distance of R away.

The angular momentum $\mathbf{L}$ of the molecule has the magnitude

14.5
$$L = I\omega$$

where ω is its angular velocity. Angular momentum is always quantized in nature, as we know. If we denote the *rotational quantum number* by J, we have here

14.6
$$L = \sqrt{J(J + 1)}\ \hbar \qquad J = 0, 1, 2, 3, \ldots$$

The energy of a rotating molecule is $\frac{1}{2}I\omega^2$, and so its energy levels are specified by

$$E_J = \tfrac{1}{2}I\omega^2$$
$$= \frac{L^2}{2I}$$

14.7
$$= \frac{J(J + 1)\ \hbar^2}{2I} \qquad \text{Rotational energy levels}$$

Let us see what sorts of energies and angular velocities are involved in molecular rotation. The carbon monoxide (CO) molecule has a bond length R of 1.13 A and the masses of the C^{12} and O^{16} atoms are respectively 1.99×10^{-26} kg and 2.66×10^{-26} kg. The reduced mass m' of the CO molecule is therefore

$$m' = \frac{m_1 m_2}{m_1 + m_2}$$
$$= \frac{1.99 \times 2.66}{1.99 + 2.66} \times 10^{-26} \text{ kg}$$
$$= 1.14 \times 10^{-26} \text{ kg}$$

Rotational Energy Levels: Diatomic Molecules

and its moment of inertia I is

$$I = m'R^2$$
$$= 1.14 \times 10^{-26} \text{ kg} \times (1.13 \times 10^{-10} \text{ m})^2$$
$$= 1.46 \times 10^{-46} \text{ kg-m}^2$$

The lowest rotational energy level corresponds to $J = 1$, and for this level in CO

$$E_{J=1} = \frac{J(J + 1)\ \hbar^2}{2I} = \frac{\hbar^2}{I}$$

$$= \frac{(1.054 \times 10^{-34} \text{ joule-sec})^2}{1.46 \times 10^{-46} \text{ kg-m}^2}$$

$$= 7.61 \times 10^{-23} \text{ joule}$$

$$= 5.07 \times 10^{-4} \text{ ev}$$

This is not a great deal of energy, and at room temperature, when $kT \approx 2.6 \times 10^{-2}$ ev, nearly all of the molecules in a sample of CO are in excited rotational states (see Sec. 15.6). The angular velocity of the CO molecule when $J = 1$ is

$$\omega = \sqrt{\frac{2E}{I}}$$

$$= \sqrt{\frac{2 \times 7.61 \times 10^{-23} \text{ joule}}{1.46 \times 10^{-46} \text{ kg-m}^2}}$$

$$= 3.23 \times 10^{11} \text{ rad/sec}$$

The preceding analysis was based on the assumption that the molecule is rigid and does not "stretch" at high angular velocities. Since molecular bonds are elastic, a certain amount of centrifugal distortion does in fact occur; the bond length in HF is 0.929 A in the $J = 1$ rotational state, for instance, while it is 0.941 A in the $J = 5$ state and 0.969 in the $J = 10$ state. To determine the effect of bond elongation on rotational energy levels, we shall consider a diatomic molecule held together by a bond that obeys Hooke's law. If the force constant is k, the restoring force that comes into being when the bond is stretched from its normal length R_0 to a length R is therefore $k(R - R_0)$. This restoring force provides the centripetal force acting on the atoms as they rotate. The centripetal force on a particle of mass m' rotating in a circle of radius R with the angular velocity ω is $m'\omega^2 R$, and so

$$k(R - R_0) \doteq m'\omega^2 R$$

14.8
$$R = \frac{kR_0}{k - m'\omega^2}$$

The total energy of the rotating molecule is its kinetic energy of $\frac{1}{2}I\omega^2$ plus the elastic potential energy of the stretched bond, which is $\frac{1}{2}k(R - R_0)^2$. Hence

$$E = \tfrac{1}{2}I\omega^2 + \tfrac{1}{2}k(R - R_0)^2$$

$$= \frac{1}{2}I\omega^2 + \frac{1}{2}\frac{m'^2\omega^4 R^2}{k}$$

$$= \frac{1}{2}I\omega^2 + \frac{1}{2}\frac{(I\omega^2)^2}{kR^2}$$

$$= \frac{L^2}{2I} + \frac{L^4}{2I^2 R^2 k}$$

where L is the magnitude of the angular momentum of the molecule. Since L can have only the values $\sqrt{J(J+1)}\ \hbar$, the rotational energy levels are specified by

14.9
$$E_J = \frac{J(J+1)\ \hbar^2}{2I} + \frac{J^2(J+1)^2\ \hbar^4}{2I^2 R^2 k}$$

In this equation both the moment of inertia I and the bond length R refer to the rotating molecule, so these quantities are different for each value of J. What we would like instead is an expression for E_J in terms of the reduced mass m' and the equilibrium bond length R_0. With the help of Eq. 14.8 and the fact that $I\omega = m'R^2\omega = \sqrt{J(J+1)}\ \hbar$ we find that

14.10
$$E_J \approx \frac{J(J+1)\ \hbar^2}{2m'R_0^2} - \frac{J^2(J+1)^2\ \hbar^4}{2m'^2 R_0^6 k}$$

Comparison with Eq. 14.7 reveals that the rotational energy levels of a nonrigid molecule are lower than the corresponding levels of a rigid molecule, with the discrepancy increasing with increasing J (Fig. 14.2). The larger the value of the force constant k, the more nearly rigid the bond and the less the molecule deforms during rotation, which is confirmed by the dependence on $1/k$ of the second term of Eq. 14.10.

Thus far we have been considering only rotation about an axis perpendicular to the axis of symmetry of a diatomic molecule, as in Fig. 14.1—end-over-end rotations. What about rotations about the axis of symmetry itself? The reason the latter can be neglected is that the mass of an atom is almost entirely concentrated in its nucleus, whose radius is only $\sim 10^{-4}$ of the radius of the atom itself. The principal contribution to the moment of inertia of a diatomic molecule about its symmetry axis therefore comes from its electrons, which are concentrated in a region whose radius about the axis is roughly half the bond length R but whose total mass is only about $1/4,000$ of the total molecular mass. Since the allowed rotational energy levels are proportional to $1/I$, rotation about the symmetry axis must involve energies $\sim 10^4$ times the E_J values for end-over-end rotations. Hence energies of at least several ev would be involved in any rotation about the symmetry axis of a diatomic molecule. Since bond energies are of this order of magnitude too, the molecule would be likely to dissociate in any environment in which such a rotation could be excited. In CO, for example, the moment of inertia of the nuclei about the symmetry axis is roughly 1.6×10^{-55} kg-m^2, while that of the electron cloud is roughly 1.4×10^{-50} kg-m^2—10^5 times greater. The moment of inertia of the molecule about a perpendicular axis is, as we found

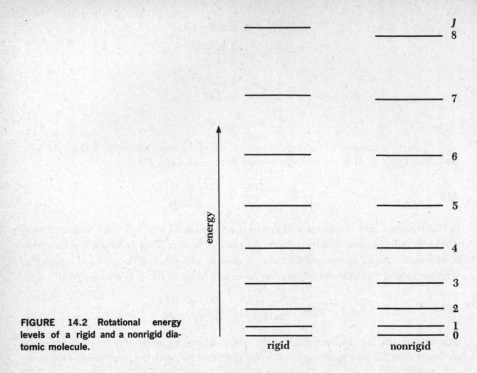

FIGURE 14.2 Rotational energy levels of a rigid and a nonrigid diatomic molecule.

rigid   nonrigid

earlier, 1.46×10^{-46} kg·m², which means that the energy of the lowest rotational state of CO about its symmetry axis is ~5 ev. To have $kT = 5$ ev, a temperature of 58,000°K is required! The bond energy in CO is 11 ev (Table 13.4), an atypically high figure that results from the presence of three bonds (one σ and two π) between the atoms; in the dynamically similar molecule NO it is only 7 ev.

14.2   Rotational Energy Levels: Polyatomic Molecules

Most of what has been said in the previous section also applies to polyatomic molecules such as carbon oxysulfide (OCS) and chloroacetylene (HC≡CCl) whose atoms lie along a straight line. The major difference is that the moments of inertia of linear polyatomic molecules may be considerably greater than those of diatomic molecules, leading to more closely spaced energy levels. The moment of inertia of OCS, for instance, is 9.48 times that of CO, and as a result the lowest rotational energy of OCS is 0.53×10^{-4} ev, compared with 5.07×10^{-4} ev for CO.

Most polyatomic molecules are not linear, however, and their rotations can be quite complicated. As in classical physics, it is appropriate to identify three mutually perpendicular *principal axes* in a polyatomic molecule and to resolve all rotational motions it undergoes into rotations about these axes. Labeling these axes A, B, and C the total rotational energy of a polyatomic molecule (assumed rigid) is

14.11
$$E = \frac{L_A{}^2}{2I_A} + \frac{L_B{}^2}{2I_B} + \frac{L_C{}^2}{2I_C}$$

The total angular momentum **L** is given by

$$\mathbf{L} = \mathbf{L_A} + \mathbf{L_B} + \mathbf{L_C}$$

Since the three angular momentum components are perpendicular to one another, the magnitude of **L** is $L_A{}^2 + L_B{}^2 + L_C{}^2$. Introducing the quantum condition on L yields

14.12
$$L = \sqrt{L_A{}^2 + L_B{}^2 + L_C{}^2} = \sqrt{J(J+1)}\,\hbar \quad J = 0, 1, 2, \ldots$$

In the most general case, the moments of inertia I_A, I_B, and I_C are all different, and the molecule's dynamical behavior resembles that of an asymmetric top. Before we consider such molecules, we shall examine the simpler symmetric-top molecules in which two of the moments of inertia are equal. If the axis of symmetry is A, then $I_B = I_C$ in a symmetric-top molecule, and its rotational energy is

14.13
$$E = \frac{L_A{}^2}{2I_A} + \frac{L_B{}^2}{2I_B} + \frac{L_C{}^2}{2I_B}$$

The axis of symmetry of a symmetric-top molecule constitutes a unique, identifiable direction, and the component L_A of the total angular momentum in this direction is accordingly limited to the values

14.14
$$L_A = K\hbar \quad K = 0, \pm 1, \pm 2, \ldots, \pm J$$

The A axis here plays the same role as the z axis in an atom, so that K corresponds to the atomic magnetic quantum number. The $+$ and $-$ values of K refer to the two possible directions of rotation. The maximum value of $|L_A|$ is less than L in conformity with the uncertainty principle, because if $|L_A| = L$ the direction of **L** would be specified with infinite precision.

Since we know that

14.15
$$L_A{}^2 = K^2\hbar^2$$

and

14.16
$$L_A{}^2 + L_B{}^2 + L_C{}^2 = J(J+1)\hbar^2$$

it follows that

14.17
$$L_B{}^2 + L_C{}^2 = J(J+1)\hbar^2 - K^2\hbar^2$$

The rotational energy of a symmetric-top molecule therefore can be written

$$E_{J,K} = \frac{K^2\hbar^2}{2I_A} + \frac{J(J+1)\hbar^2}{2I_B} - \frac{K^2\hbar^2}{2I_B}$$

14.18
$$= \frac{J(J+1)\hbar^2}{2I_B} + K^2\left(\frac{\hbar^2}{2I_A} - \frac{\hbar^2}{2I_B}\right) \qquad \text{Symmetric top}$$

Symmetric tops are of three kinds: prolate ($I_B > I_A$), spherical ($I_B = I_A$), and oblate ($I_B < I_A$). A football is a prolate spheroid, and a pumpkin is an oblate spheroid. Figure 14.3 shows a molecular example of each of these tops. From Eq. 14.18 we see that, for a prolate molecule, the coefficient of K^2 is positive, and the higher the value of K for a given J, the greater the rotational energy. The opposite is true for an oblate molecule, where the coefficient of K^2 is negative and the higher the value of K for a given J, the lower the rotational energy. A spherical-top molecule such as methane behaves like a linear molecule as far as rotation is concerned, with E depending on J only.

Centrifugal distortion occurs in rotating polyatomic molecules, and, as in the case of diatomic molecules, the result is a slight lowering of the energy levels. The correction to Eq. 14.18 involves a term proportional to $J^2(J+1)^2$, as for diatomic molecules (Eq. 14.10), but now there are also terms proportional to $J(J+1)K^2$ and to K^4. Figure 14.4 illustrates the effect of rotation about the A axis on the symmetric-top molecule CH_3F; evidently I_B is altered in addition to I_A, which we shall find to have important spectroscopic significance.

In an asymmetric-top molecule, there is no unique direction, and therefore none of the components L_A, L_B, L_C is quantized. When Schrödinger's equation for such a molecule is solved, there turn out to be $2J+1$ different eigenfunctions and eigenvalues for each J; there is no general formula for asymmetric-top molecules such as Eq. 14.7 for linear

FIGURE 14.3 Examples of symmetric-top ($I_B = I_C$) molecules. (a) Prolate, $I_A < I_B$; (b) oblate, $I_A > I_B$; (c) spherical, $I_A = I_B$.

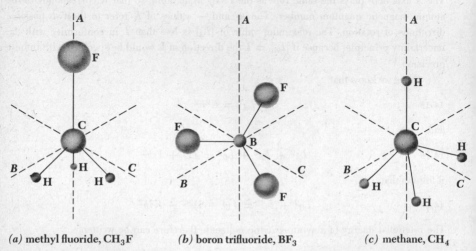

(a) methyl fluoride, CH_3F *(b)* boron trifluoride, BF_3 *(c)* methane, CH_4

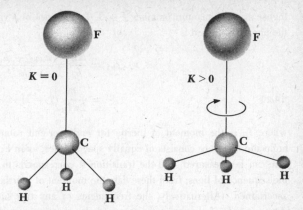

FIGURE 14.4 Centrifugal distortion in a rotating molecule.

molecules or Eq. 14.18 for symmetric-top molecules, and each molecule must be analyzed individually. In practice, the energy levels of an asymmetric-top molecule are intermediate between those of comparable prolate and oblate molecules, and a reasonable picture of the asymmetric-top levels can be obtained by appropriate interpolation.

14.3 Rotational Spectra

Rotational spectra arise from transitions between rotational energy states. Only molecules that have electric dipole moments can absorb or emit electromagnetic photons in such transitions, which means that nonpolar diatomic molecules such as H_2, symmetric linear molecules such as CO_2 ($O\!=\!C\!=\!O$), and spherical-top polyatomic molecules such as CH_4 do not exhibit rotational spectra. (Transitions between rotational states in molecules like H_2, CO_2, and CH_4 can take place during collisions, however.) Furthermore, even in molecules that possess permanent dipole moments, not all transitions between rotational states involve radiation. As in the case of atomic spectra (Sec. 11.2), certain *selection rules* summarize the conditions for a radiative transition between rotational states to be possible. They are

14.19
$$\Delta J = \pm 1$$
$$\Delta K = 0$$

Selection rules for rotational transitions

The origin of the selection rule $\Delta J = \pm 1$ is the same as that of the selection rule $\Delta l = \pm 1$ for transitions in the hydrogen atom, since the angular parts $\Theta\Phi$ of the hydrogen wave functions are simply angular-momentum wave functions. The selection rule $\Delta K = 0$ comes from the fact that the quantum number K concerns rotations about the axis of symmetry of a symmetric-top molecule, and by definition there is no dipole moment about this axis in such a molecule. Hence electromagnetic radiation cannot interact with the rotations of a symmetric-top molecule about its A axis, which is expressed by saying that $\Delta K = 0$.

In practice, rotational spectra are always obtained in absorption, so that each transition that is found involves a change from some initial state of quantum number J to the next higher state of quantum number $J + 1$. In the case of a rigid molecule, the frequency of the photon absorbed is

$$\nu_{J \to J+1} = \frac{\Delta E}{h} = \frac{E_{J+1} - E_J}{h}$$

14.20
$$= \frac{\hbar}{2\pi I_B} (J + 1) \qquad \text{Rotational spectra}$$

where I_B is the moment of inertia for end-over-end rotations. The spectrum of a rigid molecule therefore consists of equally spaced lines, as in Fig. 14.5. The frequency of each line can be measured, and the transition it corresponds to can often be ascertained from the sequence of lines; from these data the moment of inertia of the molecule can be readily ascertained. (Alternatively, the frequencies of any two successive lines may be used to determine I_B if the spectrometer used does not record the lowest-frequency lines in a particular spectral sequence.) In CO, for instance, the $J = 0 \to J = 1$ absorption line occurs at a frequency of 1.153×10^{11} cycles/sec. Hence

$$I_{CO} = \frac{\hbar}{2\pi\nu} (J + 1)$$

$$= \frac{1.054 \times 10^{-34} \text{ joule-sec}}{2\pi \times 1.153 \times 10^{11} \text{ sec}^{-1}}$$

$$= 1.46 \times 10^{-46} \text{ kg-m}^2$$

Since the reduced mass of the CO molecule is 1.14×10^{-26} kg, the bond length R_{CO} is $\sqrt{I/m'} = 1.13$ A. This is the way in which the bond length for CO quoted earlier in this chapter was determined.

The spectrum of a nonrigid linear molecule is the same as that of a rigid one except that the spacing of the lines decreases as J increases, because the nonrigid energy levels are all displaced to lower energies (Fig. 14.2). From the second term of Eq. 14.10 we find that the shifts in the spectral lines are proportional to $(J + 1)^3$ and hence become increasingly significant at large values of J. It is worth recalling here that, since the spacing of rotational energy levels is small compared with kT, states of high J are quite likely to occur and to lead to absorption lines. At room temperature, more CO molecules have $J = 7$ than any other value, and $J = 20$ is not uncommon. In molecules with larger moments of inertia, still higher J values occur: thus the most frequent J value in N_2O at room temperature is $J = 15$, and more molecules with $J = 40$ are present than with $J = 0$. In atomic spectra, on the other hand, $h\nu \geq kT$, and only lines due to transitions from the ground state of an atom are found in absorption spectra obtained at room temperature.

The effect of centrifugal distortion on the spectrum of a symmetric-top molecule is quite interesting. Now there is a dependence on K as well as on J, even though the selection rules require that $\Delta K = 0$ and hence prohibit radiative changes in the energy of rotation about the A axis. The reason is evident from Fig. 14.4; the shape of the molecule is affected by rotation about the A axis, and I_B as well as I_A accordingly varies with K. As mentioned

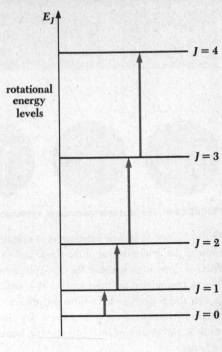

rotational
energy
levels

E_J

$J = 4$

$J = 3$

$J = 2$

$J = 1$
$J = 0$

FIGURE 14.5 Energy levels and spectrum of molecular rotation.

rotational spectrum

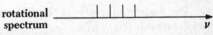

ν

in the previous section, the formula for the energy levels of a nonrigid symmetric-top molecule includes terms proportional to $J(J + 1)K^2$ and to K^4. Since $\Delta K = 0$, the latter term has no spectroscopic consequence, while the former term means that the frequency of a particular absorption line depends upon $(J + 1)K^2$ as well as upon $J + 1$ and $(J + 1)^3$ as in the case of a nonrigid linear molecule. Because $K = 0, \pm1, \pm2, \ldots, \pm J$, K^2 can have $J + 1$ different values, and each spectral line due to a transition from a level of a given J is therefore split into $J + 1$ components. The effect of centrifugal distortion on the rotational spectrum of a symmetric-top molecule is shown in Fig. 14.6. The importance of the splitting is threefold: (1) It provides a criterion for establishing whether a certain molecule is linear or symmetric-top in character; (2) from the multiplicity of each line, the initial and final J values of the transition can be identified; and (3) the precise variation of the observed frequencies with K as well as with J provides additional information on the structural parameters of the molecule.

14.4 Isotopic Substitution

In a polyatomic molecule there is more than one bond, and the moment of inertia obtained from its rotational spectrum is insufficient by itself to establish the geometry of the molecule.

Isotopic Substitution

325

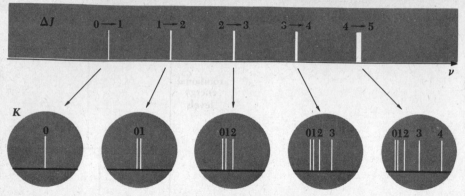

FIGURE 14.6 The rotational spectrum of a symmetric-top molecule.

The technique of *isotopic substitution* is valuable in such cases. An illustration of this technique is the determination of the C—O and C—S bond lengths in carbon oxysulfide, OCS. What is done is to measure the absorption lines in both $O^{16}C^{12}S^{32}$ and $O^{16}C^{12}S^{34}$, and then to assume that the bond lengths R_{CO} and R_{CS} are the same in the two molecules, even though the S isotopes have different masses. (The assumption is reasonable because the electron structure of an atom governs its chemical behavior, not its nuclear mass.) The result is two equations involving the two quantities R_{CO} and R_{CS}, which enables them to be found.

The moment of inertia of the OCS molecule for end-over-end rotations about its center of mass is (Fig. 14.7)

14.21
$$I = m_O r_O^2 + m_C r_C^2 + m_S r_S^2$$

where r_O, r_C, and r_S are the distances of the O, C, and S atoms from the center of mass. To express these distances in terms of the bond lengths R_{CO} and R_{CS} we note that, from the figure,

14.22
$$r_O = R_{CO} + r_C$$

14.23
$$r_S = R_{CS} - r_C$$

Because the center of mass is defined by the condition

$$m_O r_O + m_C r_C = m_S r_S$$

we have, with the help of Eqs. 14.22 and 14.23,

$$(m_O + m_C + m_S)r_C = m_S R_{CS} - m_O R_{CO}$$

14.24
$$r_C = \frac{m_S R_{CS} - m_O R_{CO}}{m_O + m_C + m_S} = \frac{m_S R_{CS} - m_O R_{CO}}{M}$$

where $M = m_O + m_C + m_S$ is the total mass of the molecule. By substituting Eqs. 14.22

to 14.24 into Eq. 14.21 we obtain

14.25
$$I = m_O R_{CO}{}^2 + m_S R_{CS}{}^2 - \frac{(m_O R_{CO} - m_S R_{CS})^2}{M}$$

If we consider Eq. 14.25 to apply to the OCS^{32} molecule, we can write for the moment of inertia I' of the OCS^{34} molecule

14.26
$$I' = m_O R_{CO}{}^2 + m'_S R_{CS}{}^2 - \frac{(m_O R_{CO} - m'_S R_{CS})^2}{M'}$$

where m'_S is the mass of the S^{34} atom and M' is the mass of the OCS^{34} molecule. Experimentally the spacing of the rotational lines in OCS^{32} is found to be 1.216×10^{10} cycles/sec and that in OCS^{34} is found to be 1.187×10^{10} cycles/sec, from which is obtained

$$I = 1.380 \times 10^{-47} \text{ kg-m}^2$$

$$I' = 1.414 \times 10^{-47} \text{ kg-m}^2$$

The only unknown quantities in Eqs. 14.25 and 14.26 are R_{CO} and R_{CS}, and solving them simultaneously yields

$$R_{CO} = 1.161 \text{ A}$$

$$R_{CS} = 1.559 \text{ A}$$

More complicated polyatomic molecules than OCS can be studied with the help of isotopic substitution, and quite precise information on bond lengths and angles has been obtained in this way. In the methyl fluoride (CH_3F) molecule shown in Fig. 14.3, for instance, the bond lengths have been found to be $R_{CH} = 1.109$ A and $R_{CF} = 1.385$ A and the H—C—H angle has been found to be $110°$.

14.5   Vibrational Energy Levels: Diatomic Molecules

When sufficiently excited, a molecule can vibrate as well as rotate. As before, we shall begin by considering a diatomic molecule. Figure 14.8 shows how the potential energy

FIGURE 14.7   The carbon oxysulfide (OCS) molecule.

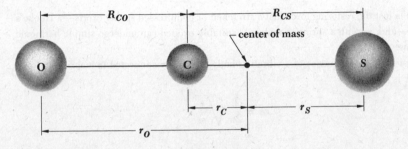

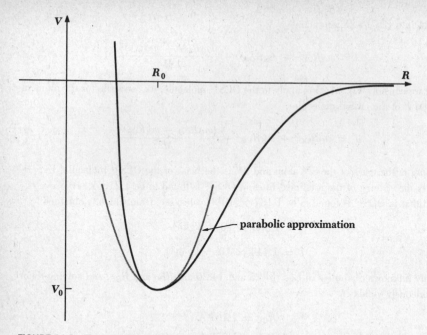

FIGURE 14.8 The potential energy of a diatomic molecule as a function of internuclear distance.

of a molecule varies with the internuclear distance R. In the neighborhood of the minimum of this curve, which corresponds to the normal configuration of the molecule, the shape of the curve is very nearly a parabola. In this region, then,

14.27
$$V = V_0 + \tfrac{1}{2}k(R - R_0)^2$$

where R_0 is the equilibrium separation of the atoms. The interatomic force that gives rise to this potential energy may be found by differentiating V:

$$F = -\frac{dV}{dR}$$

14.28
$$= -k(R - R_0)$$

The force is just the restoring force that a stretched or compressed spring exerts—a Hooke's law force—and, as with a spring, a molecule suitably excited can undergo simple harmonic oscillations.

Classically, the frequency of a vibrating body of mass m connected to a spring of force constant k is

14.29
$$\nu_0 = \frac{1}{2\pi}\sqrt{\frac{k}{m}}$$

What we have in the case of a diatomic molecule is the somewhat different situation of two bodies of masses m_1 and m_2 joined by a spring, as in Fig. 14.9. In the absence of external forces the linear momentum of the system remains constant, and the oscillations of the bodies therefore cannot affect the motion of their center of mass. For this reason m_1 and m_2 vibrate back and forth relative to their center of mass in opposite directions, and both reach the extremes of their respective motions at the same times. The length of the spring at any time is $x_2 - x_1$, which equals the equilibrium length l of the spring plus its displacement x:

$$l + x = x_2 - x_1$$

14.30
$$x = x_2 - x_1 - l$$

The displacement x is positive when the spring is extended and negative when it is compressed.

The restoring forces F_1 and F_2 which the spring exerts on each body are equal in magnitude and opposite in direction, and so

$$F_1 = kx$$
$$F_2 = -kx$$

where k is the force constant of the spring. From the second law of motion, $\mathbf{F} = m\mathbf{a}$, we have

14.31
$$m_1 a_1 = m_1 \frac{d^2 x_1}{dt^2} = kx$$

14.32
$$m_2 a_2 = m_2 \frac{d^2 x_2}{dt^2} = -kx$$

These equations can be combined by multiplying the first by m_2 and the second by m_1 and then subtracting the first from the second. The result is

$$m_1 m_2 \frac{d^2 x_2}{dt^2} - m_1 m_2 \frac{d^2 x_1}{dt^2} = -m_1 kx - m_2 kx$$

14.33
$$\frac{m_1 m_2}{m_1 + m_2} \left(\frac{d^2 x_2}{dt^2} - \frac{d^2 x_1}{dt^2} \right) = -kx$$

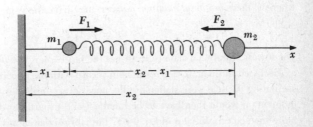

FIGURE 14.9 A two-body oscillator.

Vibrational Energy Levels: Diatomic Molecules

Next we differentiate Eq. 14.30 twice with respect to time, which yields

14.34
$$\frac{d^2x}{dt^2} = \frac{d^2x_2}{dt^2} - \frac{d^2x_1}{dt^2}$$

since l, the equilibrium spring length, is a constant. The equation of motion of the system can therefore be expressed entirely in terms of x, the spring displacement, by substituting Eq. 14.34 into Eq. 14.33 to give

$$\frac{m_1 m_2}{m_1 + m_2} \frac{d^2x}{dt^2} = -kx$$

or

14.35
$$m' \frac{d^2x}{dt^2} = -kx$$

where

14.36
$$m' = \frac{m_1 m_2}{m_1 + m_2}$$

is again the reduced mass of the system.

The equation of motion of an ordinary harmonic oscillator is

$$m \frac{d^2x}{dt^2} = -kx$$

whose solution yields the frequency of oscillation stated in Eq. 14.29. Hence we conclude that the frequency of oscillation of a two-body oscillator is given by the same formula but with the reduced mass m' substituted for m:

14.37
$$\nu_0 = \frac{1}{2\pi} \sqrt{\frac{k}{m'}}$$
Two-body oscillator

When the harmonic-oscillator problem is solved quantum mechanically, as was done in Chap. 8, the energy of the oscillator is found to be restricted to the values

14.38
$$E_v = (v + \tfrac{1}{2})\, h\nu_0$$

where v, the *vibrational quantum number*, may have the values

$$v = 0, 1, 2, 3, \ldots$$

The lowest vibrational state ($v = 0$) has the finite energy $\tfrac{1}{2}h\nu_0$, not the classical value of 0; as discussed in Chap. 8, this result is in accord with the uncertainty principle, because if the oscillating particle were stationary, the uncertainty in its position would be $\Delta x = 0$ and its momentum would then have to be infinite—and a particle with $E = 0$ cannot have an infinite momentum. In view of Eq. 14.37 the vibrational energy levels of a diatomic molecule are specified by

$$E_v = (v + \tfrac{1}{2})\, \hbar \sqrt{\dfrac{k}{m'}} \qquad \textbf{Vibrational energy levels}$$

Let us calculate the frequency of vibration of the CO molecule and the spacing between its vibrational energy levels. The force constant k of the bond in CO is 187 n/m (which is 10 lb/in.—not an exceptional figure for an ordinary spring) and, as we found in Sec. 14.1, the reduced mass of the CO molecule is $m' = 1.14 \times 10^{-26}$ kg. The frequency of vibration is therefore

$$\nu_0 = \frac{1}{2\pi} \sqrt{\frac{k}{m'}}$$

$$= \frac{1}{2\pi} \sqrt{\frac{187 \text{ n/m}}{1.14 \times 10^{-26} \text{ kg}}}$$

$$= 2.04 \times 10^{13} \text{ cycles/sec}$$

The separation ΔE between the vibrational energy levels in CO is

$$\Delta E = E_{v+1} - E_v = h\nu_0$$

$$= 6.63 \times 10^{-34} \text{ joule-sec} \times 2.04 \times 10^{13} \text{ sec}^{-1}$$

$$= 8.44 \times 10^{-2} \text{ ev}$$

which is considerably more than the spacing between its rotational energy levels. Because $\Delta E > kT$ for vibrational states in a sample at room temperature, most of the molecules in such a sample exist in the $v = 0$ state with only their zero-point energies. This situation is very different from that characteristic of rotational states, where the much smaller energies mean that the majority of the molecules in a room-temperature sample are excited to higher states.

The higher vibrational states of a diatomic molecule do not obey Eq. 14.39 because the parabolic approximation to its potential-energy curve becomes less and less valid with increasing energy. A better analytic approximation to actual molecular potential-energy curves is the *Morse potential*,

$$V(R) = D[1 - e^{-a(R-R_0)}]^2$$

in which D is the dissociation energy of the molecule and a is a measure of the width of the potential well. Figure 14.10 is a comparison between the actual $V(R)$ for a diatomic molecule and the Morse curve that best fits the data; evidently the Morse potential is an excellent approximation in the region of interest here, namely where $V(R) < 0$. The signal merit of the Morse potential, however, lies not so much in the accuracy with which it mimics $V(R)$— other empirical formulas can be devised that do as well—as in the fact that Schrödinger's equation for the vibrational energy levels of a diatomic molecule can be solved exactly when the Morse potential is used. The result is

Vibrational Energy Levels: Diatomic Molecules

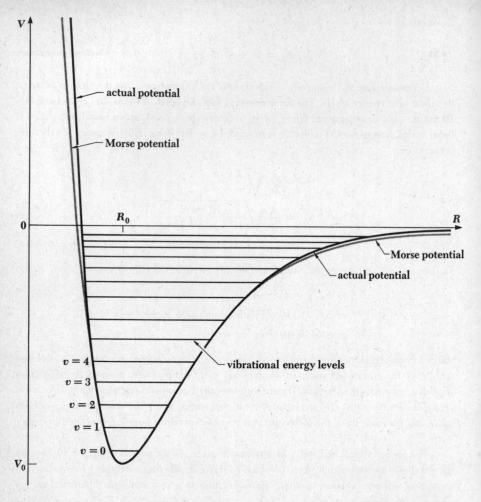

FIGURE 14.10 The Morse potential is an excellent approximation to the actual potential of a diatomic molecule. Because the vibrations of a diatomic molecule are anharmonic, the vibrational energy levels are not evenly spaced.

14.41
$$E_v = (v + \frac{1}{2})\, \hbar \sqrt{\frac{k}{m'}} - (v + \frac{1}{2})^2\, \frac{\hbar^2 k}{4Dm'}$$

Evidently the spacing of the energy levels decreases with increasing quantum number v. This result is what we would expect from the quantum theory of a particle in a box (Chap. 8): the larger the box, which means here the more the potential-energy curve departs from a parabola, the closer together the energy levels are.

14.6 Vibrational Energy Levels: Polyatomic Molecules

Three independent coordinates are needed to describe the position of a free atom, and such an atom is accordingly said to have three *degrees of freedom*. An assembly of N atoms has $3N$ degrees of freedom, three for each atom. When the N atoms are linked together in a molecule, there are still $3N$ degrees of freedom for the system, but it is more appropriate in this situation to consider them in terms of the molecule as a whole instead of in terms of the individual atoms. Thus we can regard the molecule itself as having three degrees of translational freedom, corresponding to the three coordinates needed to describe the position of its center of mass, and if it is a nonlinear molecule, it also has three degrees of rotational freedom, corresponding to the three angular coordinates needed to describe its orientation in space. The remaining $3N - 6$ degrees of freedom refer to changes in the relative positions of the atoms within the molecule; they are *internal* degrees of freedom. Because rotations about the symmetry axis of a linear molecule can be ignored, there are only two degrees of rotational freedom for such a molecule and $3N - 5$ internal degrees of freedom.

Each internal degree of freedom in a molecule corresponds to a particular mode of vibration. Some of these modes involve the stretching and compressing of bonds (as in the back-and-forth vibrations of a diatomic molecule) and the rest involve bending or distorting the molecule. A diatomic molecule is linear and has $N = 2$, and so, since $3N - 5 = 1$, there is only a single degree of freedom and only one mode of vibration. The H_2O molecule is nonlinear and therefore has $3N - 6 = 3$ modes of vibration, which are shown in Fig. 14.11. These modes consist of a symmetric bending, a symmetric stretching, and an asymmetric stretching, in order of increasing frequency. Each mode has its own set of energy levels, as indicated in the figure.

FIGURE 14.11 The normal modes of vibration of the H_2O molecule and the energy levels of each mode.

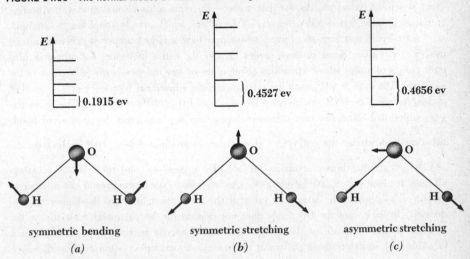

symmetric bending
(a)

symmetric stretching
(b)

asymmetric stretching
(c)

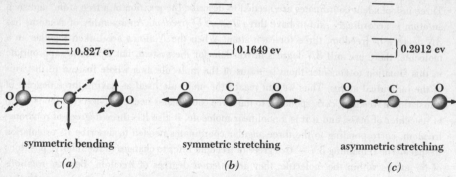

symmetric bending
(a)

symmetric stretching
(b)

asymmetric stretching
(c)

FIGURE 14.12   The normal modes of vibration of the CO_2 molecule and the energy levels of each mode. The symmetric bending mode is degenerate.

The CO_2 molecule is linear, and it has $3N - 5$ modes of vibration. Two of these are a symmetric and an asymmetric stretching, as in the case of H_2O, and two are bending modes (Fig. 14.12). The latter are degenerate, since they consist of identical motions in two perpendicular planes which involve the same energies.

The modes of vibration shown in Figs. 14.11 and 14.12 are the *normal modes* for the H_2O and CO_2 molecules. More than one of these modes may be excited at a particular time, and the atoms then execute more complicated motions that consist of superposed normal vibrations. The normal modes of vibration of a system have the property that the particles participating in the vibrations all move in phase with one another at the same frequency; if the forces and torques obey Hooke's law, the vibrations will be simple harmonic in character. The virtue of describing the internal motions of a molecule—or any other system of linked objects, for that matter—in terms of its normal modes of vibration is that each mode occurs independently of the others, which greatly simplifies the analysis.

A molecule that consists of many atoms may have a large number of different normal modes of vibration. Some of these modes involve the entire molecule, but others involve only groups of atoms whose vibrations occur more or less independently of the rest of the molecule. Thus the —OH group has a characteristic vibrational frequency of 1.1×10^{14} cycles/sec and the —NH_2 group has a frequency of 1.0×10^{14} cycles/sec. The characteristic vibrational frequency of a carbon-carbon group depends upon the number of bonds between the C atoms: the $\geq$C—C$\leq$ group vibrates at about 3.3×10^{13} cycles/sec, the $\geq$C=C$\leq$ group vibrates at about 5.0×10^{13} cycles/sec, and the —C≡C— group vibrates at about 6.7×10^{13} cycles/sec. (As we would expect, the greater the number of carbon-carbon bonds, the larger the value of the force constant k and the higher the frequency.) In each case the frequency does not depend on the particular molecule or the location in the molecule of the group. This independence makes vibrational spectra a valuable tool in determining molecular structures. An example is thioacetic acid, whose structure might conceivably be either CH_3CO—SH or CH_3CS—OH. The infrared absorption spectrum of thioacetic acid contains lines at frequencies equal to the vibrational frequencies of the $>C=O$ and —SH groups, but no lines corresponding to the $>C=S$ or —OH groups, so the former alternative is the correct one.

14.7 Vibration-Rotation Spectra

The selection rule for transitions between vibrational states is $\Delta v = \pm 1$ in the harmonic oscillator approximation. This rule is easy to understand. An oscillating dipole whose frequency is ν_0 can only absorb or emit electromagnetic radiation of the same frequency, and all quanta of frequency ν_0 have the energy $h\nu_0$. The oscillating dipole accordingly can only absorb $\Delta E = h\nu_0$ at a time, in which case its energy increases from $(v + \frac{1}{2})h\nu_0$ to $(v + \frac{1}{2} + 1)h\nu_0$, and it can only emit $\Delta E = h\nu_0$ at a time, in which case its energy decreases from $(v + \frac{1}{2})h\nu_0$ to $(v + \frac{1}{2} - 1)h\nu_0$. Hence the selection rule $\Delta v = \pm 1$.

The selection rule for the harmonic oscillator is obtained formally by evaluating the integral

$$\int_{-\infty}^{\infty} \psi_n^* x \psi_m \, dx$$

where ψ_n^* and ψ_m are the harmonic-oscillator wave functions for states n and m. As discussed in Sec. 11.2, the probability that the oscillator will undergo a transition from state n to state m is proportional to this integral, whose value turns out to equal 0 except for $m = n \pm 1$ (see Prob. 2 of Chap. 11).

The selection rules for an anharmonic oscillator are $\Delta v = \pm 1, \pm 2, \pm 3, \ldots$ instead of just $\Delta v = \pm 1$. However, the greater the change in v, the smaller the likelihood of the transition, so that in practice transitions for which $\Delta v = \pm 1$ are by far the most common. Furthermore, because $h\nu_0 > kT$, few molecules in a room-temperature sample are in excited vibrational states, and so only transitions from the $v = 0$ to $v = 1$ states are conspicuous in infrared absorption spectra. Transitions from $v = 0$ to $v = 2$ and $v = 3$ and from $v = 1$ to $v = 2$ do occur, to be sure, but they are relatively infrequent and lead to far weaker absorption lines than the $v = 0$ to $v = 1$ transition.

Pure vibrational spectra are observed only in liquids, where interactions between adjacent molecules inhibit rotation. Because the excitation energies involved in molecular rotation are considerably smaller than those involved in vibration, the freely moving molecules in a gas or vapor nearly always are rotating, regardless of their vibrational state. The spectra of such molecules do not show isolated lines corresponding to each vibrational transition, but instead a large number of closely spaced lines due to transitions between the various rotational states of one vibrational level and the rotational states of the other. In spectra obtained using a spectrometer with inadequate resolution, the lines appear as a broad streak called a vibration-rotation band.

To a first approximation, the vibrations and rotations of a molecule take place independently of each other, and we can also ignore the effects of centrifugal distortion and anharmonicity. Under these circumstances the energy levels of a diatomic molecule are specified by

14.42
$$E_{v,J} = (v + \tfrac{1}{2})\, \hbar \sqrt{\frac{k}{m'}} + J(J + 1)\,\frac{\hbar^2}{2I}$$

Figure 14.13 shows the $J = 0, 1, 2, 3,$ and 4 levels of a diatomic molecule for the $v = 0$ and $v = 1$ vibrational states, together with the spectral lines in absorption that are consistent with the selection rules $\Delta v = +1$ and $\Delta J = \pm 1$. The $v = 0 \rightarrow v = 1$ transitions fall into two categories, the *P branch* in which $\Delta J = -1$ (that is, $J \rightarrow J - 1$) and the *R branch* in which $\Delta J = +1$ $(J \rightarrow J + 1)$. From Eq. 14.42 the frequencies of the spectral lines in each branch are given by

$$\nu_P = E_{1,J-1} - E_{0,J}$$

$$= \frac{1}{2\pi}\sqrt{\frac{k}{m'}} + [(J - 1)J - J(J + 1)]\,\frac{\hbar}{4\pi I}$$

14.43
$$= \nu_0 - J\,\frac{\hbar}{2\pi I} \qquad J = 1, 2, 3, \dots \qquad\qquad \text{\small P branch}$$

and

$$\nu_R = E_{1,J+1} - E_{0,J}$$

$$= \frac{1}{2\pi}\sqrt{\frac{k}{m'}} + [(J + 1)(J + 2) - J(J + 1)]\,\frac{\hbar}{4\pi I}$$

14.44
$$= \nu_0 + (J + 1)\,\frac{\hbar}{2\pi I} \qquad J = 0, 1, 2, \dots \qquad\qquad \text{\small R branch}$$

There is no line at $\nu = \nu_0$ because transitions for which $\Delta J = 0$ are forbidden in diatomic molecules. The spacing between the lines in both the P and the R branch is $\Delta \nu = \hbar/2\pi I$; hence the moment of inertia of a molecule can be ascertained from its infrared vibration-rotation spectrum as well as from its microwave pure-rotation spectrum.

The spacings of the component lines in the vibration-rotation spectrum of a diatomic molecule are actually not quite identical owing to a coupling between the vibrations and rotations. (The effects of centrifugal distortion and anharmonicity also contribute to the discrepancies, but to a smaller degree than the coupling.) A molecule vibrates many ($\sim 10^3$) times during each rotation, so the bond length R varies constantly. In a simple harmonic oscillator, the average value of R is always R_0 and does not depend upon the vibrational energy, but the quantity $\hbar/2\pi I$ in Eq. 14.42 is proportional to $\overline{1/R^2}$, which *does* depend upon the vibrational energy. To illustrate the latter statement, we can consider an imaginary bond which oscillates between 0.9 and 1.1 A in one vibrational state and between 0.8 and 1.2 A in another state of more energy. In the first state

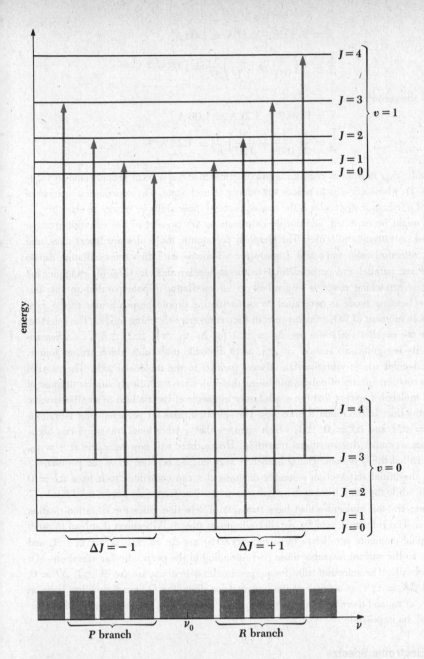

FIGURE 14.13   The rotational structure of the $v = 0 \rightarrow v = 1$ transitions in a diatomic molecule. There is no line at $v = v_0$ (the Q branch) because of the selection rule $\Delta J = \pm 1$.

Vibration-Rotation Spectra

$$\bar{r} = \frac{1}{2} \, (0.9 + 1.1) \, A = 1.00 \, A$$

$$\frac{\overline{1}}{r^2} = \frac{1}{2} \left(\frac{1}{0.9^2} + \frac{1}{1.1^2} \right) \frac{1}{A^2} = 1.03 A^{-2}$$

while in the second state

$$\bar{r} = \frac{1}{2} \, (0.8 + 1.2) \, A = 1.00 \, A$$

$$\frac{\overline{1}}{r^2} = \frac{1}{2} \left(\frac{1}{0.8^2} + \frac{1}{1.2^2} \right) \frac{1}{A^2} = 1.12 \, A^{-2}$$

As a result, $E_{v,J}$ in a more realistic approximation includes a term proportional to $(v + \frac{1}{2})J(J + 1)$, whose effect is to reduce the energy of each state. The nonconstant spacing of spectra of polyatomic molecules is the consequence of these shifts in energy levels.

As might be expected, additional complications are present in the vibration-rotation spectra of polyatomic molecules. The simplest polyatomic molecules are linear ones, and different selection rules for v and J apply for vibrations such that the oscillating dipole moments are parallel and perpendicular to the molecular axis. In CO_2, for example, the asymmetric stretching mode is equivalent to an oscillating dipole parallel to the axis while the bending mode is equivalent to an oscillating dipole perpendicular to the axis. (The dipole moment of CO_2 remains zero in the symmetric stretching mode.) The selection rules for the parallel vibrations are $\Delta v = \pm 1$ (or $\Delta v = \pm 1, \pm 2, \pm 3, \ldots$ when anharmonicity is significant) and $\Delta J = \pm 1$, as in diatomic molecules which are, of course, linear molecules whose vibrations are always parallel to the molecular axis. The parallel vibration-rotation spectra of polyatomic linear molecules are accordingly similar to those of diatomic molecules, except that the much larger moments of inertia lead to smaller spacing between the lines in the P and R branches. The selection rules for perpendicular vibrations are $\Delta v = \pm 1$ and $\Delta J = 0, \pm 1$, which signifies that a vibrational transition can occur without an accompanying rotational transition. Hence there will now be a line at $\nu = \nu_0$, which is called the Q branch. The Q branch is very intense because all of the populated J levels in the initial states of an assembly of molecules can contribute to it by a $\Delta J = 0$ transition, while the other lines each originate in transitions from only one of the J levels.

Symmetric-top molecules also have two sets of selection rules for vibration-rotation transitions. The rules that hold for parallel vibrations (that is, vibrations that lead to oscillating dipole moments parallel to the symmetry axis) are $\Delta v = \pm 1$, $\Delta J = 0, \pm 1$, and $\Delta K = 0$, so the spectra resemble those corresponding to the perpendicular vibrations of a linear molecule. The selection rules for perpendicular vibrations are $\Delta v = \pm 1$, $\Delta J = 0, \pm 1$, and $\Delta K = \pm 1$; as a result of the last of these rules, the Q branch now has lines on both sides of ν_0, and there are many sets of P and R branch lines so close together that resolution may be impossible.

14.8 Electronic Spectra

The energies of rotation and vibration in a molecule are due to the motion of its atomic nuclei, since the nuclei contain essentially all of the molecule's mass. The molecular electrons also can be excited to higher energy levels than those corresponding to the ground state of the molecule, though the spacing of these levels is much greater than the spacing of rotational or vibrational levels. Electronic transitions involve radiation in the visible or

FIGURE 14.14 Vibrational energy levels and associated probability densities $|\psi|^2$ for the electronic ground state and first excited state of a diatomic molecule.

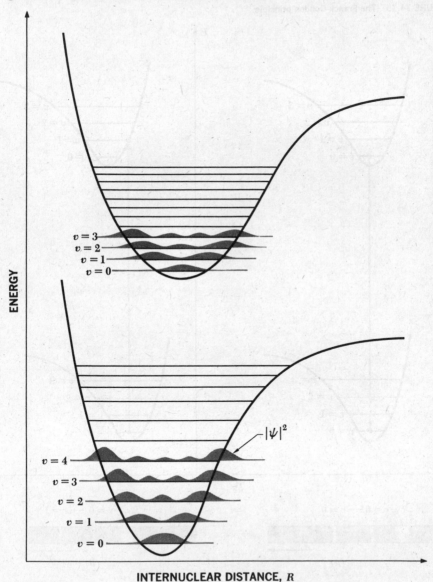

ultraviolet parts of the spectrum, with each transition appearing as a series of closely spaced lines, called a band, due to the presence of different rotational and vibrational states in each electronic state (see Fig. 6.3). All molecules exhibit electronic spectra, since a dipole moment change always accompanies a change in the electronic configuration of a molecule.

FIGURE 14.15   The Franck-Condon principle.

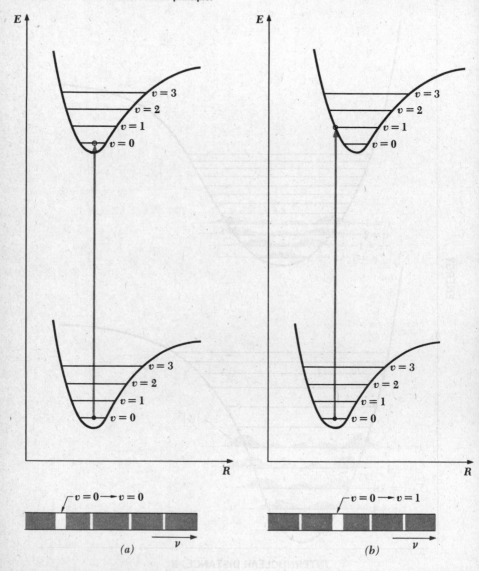

Therefore homonuclear molecules, such as H_2 and N_2, which have neither rotational nor vibrational spectra because they lack permanent dipole moments, nevertheless have electronic spectra which possess rotational and vibrational fine structures that permit their moments of inertia and bond force constants to be ascertained.

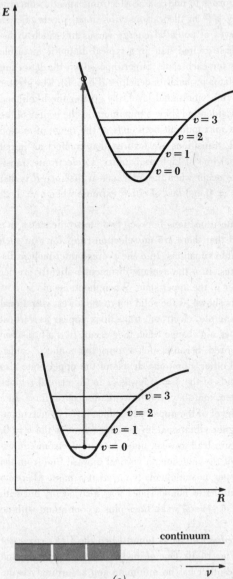

(c)

Electronic transitions occur in a small fraction of the time required for a vibrating molecule to execute a cycle. Such a transition is so rapid, in fact, that we can consider the instantaneous internuclear distances in a molecule as being unchanged while it takes place. This observation is known as the *Franck-Condon principle*.

Another general principle we need in order to understand electronic spectra concerns the distributions of the probability density $|\psi|^2$ in the various vibrational levels of each electronic state. Figure 14.14 contains curves of potential energy versus internuclear distance for the electronic ground state and first excited state in a typical diatomic molecule together with the vibrational energy levels for each state. Superimposed on the latter are graphs of the corresponding harmonic-oscillator probability densities (Chap. 8). The greater the value of $|\psi|^2$ at a particular value of R in a vibrational level, the greater the likelihood that the nuclei will be that distance apart. Evidently $|\psi|^2$ is a maximum in the center of the range of motion in each $v = 0$ level and a maximum at both ends of the range of motion in the higher vibrational levels. In general, transitions are favored that connect an initial configuration of high probability with a final one of high probability, so electronic transitions in a diatomic molecule are most apt to occur when the internuclear distance R is that of the center of the range of motion when $v = 0$ and that of either extreme when $v = 1, 2, 3, \ldots$.

With these ideas in mind we can examine transitions between two electronic states in a diatomic molecule; it should also be noted that there are no selection rules for v in such transitions. Figure 14.15 shows three possible situations. In a the average internuclear distances are the same in both electronic states, in b the average distance is slightly greater in the upper state, and in c it is much greater in the upper state. When the molecule is in its ground state, the most probable value of R is shown by the solid dot in the lowest vibrational level. According to the Franck-Condon principle, electronic transitions appear as vertical lines on diagrams of this kind because R does not change when they occur. In a a transition to the $v = 0$ level of the upper state is favored, because such a transition connects configurations of high probability. Transitions to other vibrational levels of the upper state also occur (while small, $|\psi|^2$ is not 0 near the ends of the $v = 0$ levels or in the central regions of the $v = 1, 2, 3, \ldots$ levels), but less often, and they lead to weak absorption lines. In b the favored transition is now to the $v = 1$ level of the upper state, since $|\psi|^2$ is a maximum at the ends of the range of motion in the higher vibrational levels. Transitions to the $v = 0$, 2, 3, and 4 levels take place as well, but again lead to weak lines. In c an electronic transition is most likely to produce dissociation of the molecule; a vertical upward line from the center of the lowest level intersects the upper potential-energy curve at a point where no bound states are possible. However, transitions to bound states may also occur, but with small probability, so the spectrum consists of several weak lines plus a continuum with no line structure.

An interesting phenomenon called *predissociation* is found when the $V(R)$ curves of two different excited states intersect, as in Fig. 14.16. One of the states has a minimum and the usual set of vibrational levels, while the other has no minimum and accordingly is unstable. A transition from the ground state to level d leads to the possibility that the molecule

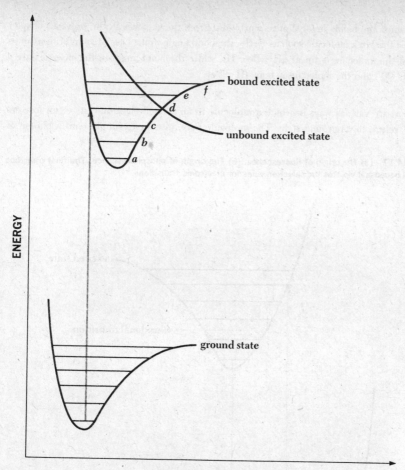

FIGURE 14.16 Predissociation can occur after a transition to the vibrational level *d* of the bound excited state.

will shift to the unbound state at the high-R end of its vibrations. However, a transition to one of the *higher* levels $e, f, \ldots$ will usually *not* lead to dissociation, because the value of $|\psi|^2$ is small where the $V(R)$ curve of the unbound state intersects these vibrational levels; hence the term predissociation.

Electronic excitation in a polyatomic molecule often leads to a change in its shape, which can be determined from the rotational fine structure in its band spectrum. The origin of such changes lies in the different characters of the wave functions of electrons in different states, which lead to correspondingly different types of bond. For example, a possible electronic transition in a molecule whose bonds involve sp hybrid orbitals is to a higher-energy

Electronic Spectra

343

state in which the bonds involve pure p orbitals. From the sketches in the previous chapter we can see that, in a molecule such as BeH_2, the bond angle in the case of sp hybridization is 180° and the molecule is linear (H—Be—H), while the bond angle in the case of pure p orbitals is 90° and the molecule is bent (H—Be).
$$\begin{matrix} \\ | \\ H \end{matrix}$$

There are various ways in which a molecule in an excited electronic state can lose energy and return to its ground state. The molecule may, of course, simply emit a photon of

FIGURE 14.17 (*a*) **The origin of fluorescence.** (*b*) **The origin of phosphorescence. The final transition is delayed because it violates the selection rules for electronic transitions.**

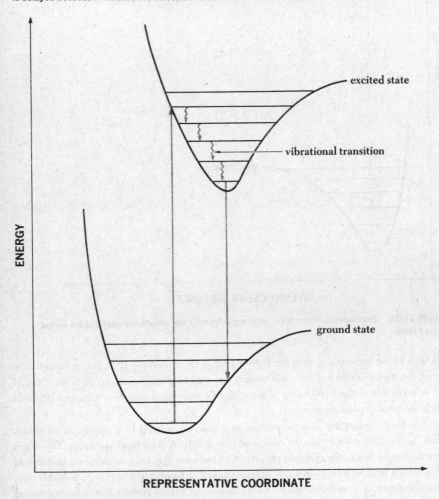

(*a*)

the same frequency as that of the photon it absorbed, thereby returning to the ground state in a single step. Another possibility is *fluorescence;* the molecule may give up some of its vibrational energy in collisions with other molecules, so that the downward radiative transition originates from a lower vibrational level in the upper electronic state (Fig. 14.17*a*). Fluorescent radiation is therefore of lower frequency than that of the absorbed radiation.

In molecular spectra, as in atomic spectra, radiative transitions between electronic states of different total spin are prohibited (see Sec. 11.4). Figure 14.17*b* shows a situation in which the molecule in its singlet ($S = 0$) ground state absorbs a photon and is elevated to a singlet excited state. In collisions the molecule can undergo radiationless transitions to a lower vibrational level that may happen to have about the same energy as one of the levels

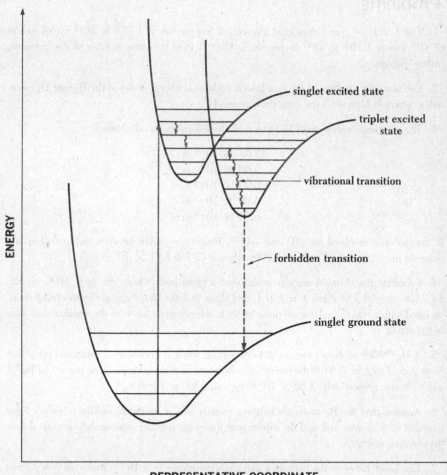

ENERGY

singlet excited state

triplet excited state

vibrational transition

forbidden transition

singlet ground state

REPRESENTATIVE COORDINATE

(b)

in the triplet ($S = 1$) excited state, and there is then a certain probability for a shift to the triplet state to occur. Further collisions in the triplet state bring the molecule's energy below that of the crossover point, so that it is now trapped in the triplet state and ultimately reaches the $v = 0$ level. A radiative transition from a triplet to a singlet state is "forbidden" by the selection rules, which really means not that it is impossible to occur but that it has only a minute likelihood of doing so. Such transitions accordingly have very long half lives, and the resulting *phosphorescent radiation* may be emitted minutes or even hours after the initial absorption.

Problems

1. The $J = 0 \rightarrow J = 1$ rotational absorption line occurs at 1.153×10^{11} cycles/sec in $C^{12}O^{16}$ and at 1.102×10^{11} cycles/sec in $C^?O^{16}$. Find the mass number of the unknown carbon isotope.

2. Calculate the energies of the four lowest rotational energy states of the H_2 and D_2 molecules, where D represents the deuterium atom $_1H^2$.

3. The rotational spectrum of HCl contains the following wavelengths:

$$12.03 \times 10^{-5} \text{ m}$$
$$9.60 \times 10^{-5} \text{ m}$$
$$8.04 \times 10^{-5} \text{ m}$$
$$6.89 \times 10^{-5} \text{ m}$$
$$6.04 \times 10^{-5} \text{ m}$$

If the isotopes involved are $_1H^1$ and $_{17}Cl^{35}$, find the distance between the hydrogen and chlorine nuclei in an HCl molecule. (The mass of Cl^{35} is 5.81×10^{-26} kg.)

4. Calculate the classical angular velocity of a rigid body whose energy is given by Eq. 14.7 for states of $J = J$ and $J = J + 1$, and show that the frequency of the spectral line associated with a transition between these states is intermediate between the angular velocities of the states.

5. A $Hg^{200}Cl^{35}$ molecule emits a 4.4-cm photon when it undergoes a rotational transition from $J = 1$ to $J = 0$. Find the interatomic distance in this molecule. (The masses of Hg^{200} and Cl^{35} are, respectively, 3.32×10^{-25} kg and 5.81×10^{-26} kg.)

6. Assume that the H_2 molecule behaves exactly like a harmonic oscillator with a force constant of 573 n/m and find the vibrational quantum number corresponding to its 4.5-ev dissociation energy.

7. The bond between the hydrogen and chlorine atoms in a HCl^{35} molecule has a force constant of 516 n/m. It is likely that a HCl molecule will be vibrating in its first excited vibrational state at room temperature?

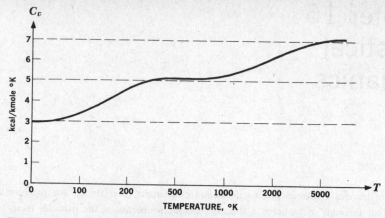

FIGURE 14.18 Molar specific heat of hydrogen at constant volume.

8. The hydrogen isotope deuterium has an atomic mass approximately twice that of ordinary hydrogen. Does H_2 or HD have the greater zero-point energy? How does this affect the binding energies of the two molecules?

9. The force constant of the HF^{19} molecule is 966 n/m. Find the frequency of vibration of the molecule.

10. The observed molar specific heat of hydrogen gas at constant volume is plotted in Fig. 14.18 versus absolute temperature. (The temperature scale is logarithmic.) Since each degree of freedom (that is, each mode of energy possession) in a gas molecule contributes ~1 kcal/kmole °K to the specific heat of the gas, this curve is interpreted as indicating that only translational motion, with three degrees of freedom, is possible for hydrogen molecules at very low temperatures. At higher temperatures the specific heat rises to ~5 kcal/kmole °K, indicating that two more degrees of freedom are available, and at still higher temperatures the specific heat is ~7 kcal/kmole °K, indicating two further degrees of freedom. The additional pairs of degrees of freedom represent, respectively, rotation, which can take place about two independent axes perpendicular to the axis of symmetry of the H_2 molecule, and vibration, in which the two degrees of freedom correspond to the kinetic and potential modes of energy possession by the molecule. (a) Verify this interpretation of Fig. 14.18 by calculating the temperatures at which kT is equal to the minimum rotational energy and to the minimum vibrational energy a H_2 molecule can have. Assume that the force constant of the bond in H_2 is 573 n/m and that the H atoms are 7.42×10^{-11} m apart. (At these temperatures, approximately half the molecules are rotating or vibrating, respectively, though in each case some are in higher states than $J = 1$ or $v = 1$.) (b) To justify considering only two degrees of rotational freedom in the H_2 molecule, calculate the temperature at which kT is equal to the minimum rotational energy a H_2 molecule can have for rotation about its axis of symmetry. (c) How many rotations does a H_2 molecule with $J = 1$ and $v = 1$ make per vibration?

Problems

Chapter 15
Statistical
Mechanics

The branch of physics known as *statistical mechanics* attempts to relate the macroscopic properties of an assembly of particles to the microscopic properties of the particles themselves. Statistical mechanics, as its name implies, is not concerned with the actual motions or interactions of individual particles, but investigates instead their *most probable* behavior. While statistical mechanics cannot help us determine the life history of a particular particle, it *is* able to inform us of the likelihood that a particle (exactly which one we cannot know in advance) has a certain position and momentum at a certain instant. Because so many phenomena in the physical world involve assemblies of particles, the value of a statistical rather than deterministic approach is clear. Owing to the generality of its arguments, statistical mechanics can be applied with equal facility to classical problems (such as that of molecules in a gas) and quantum-mechanical problems (such as those of free electrons in a metal or photons in a box), and it is one of the most powerful tools of the theoretical physicist.

15.1 Phase Space

The state of a system of particles is completely specified classically at a particular instant if the position and momentum of each of its constituent particles are known. Since position and momentum are vectors with three components apiece, we must know six quantities,

$$x, y, z, p_x, p_y, p_z$$

for each particle.

The position of a particle is a point having the coordinates x, y, z in ordinary three-dimensional space. It is convenient to generalize this conception by imagining a six-dimensional space in which a point has the six coordinates x, y, z, p_x, p_y, p_z. This combined position and momentum space is called *phase space*. The notion of phase space is introduced to enable us to develop statistical mechanics in a geometrical framework, thereby permitting a simpler and more straightforward method of analysis than an equivalent one wholly abstract in character. A point in phase space corresponds to a particular position *and* momentum, while a point in ordinary space corresponds to a particular position only. Thus every particle is completely specified by a point in phase space, and the state of a system of particles corresponds to a certain distribution of points in phase space.

The uncertainty principle compels us to elaborate what we mean by a "point" in phase space. Let us divide phase space into tiny six-dimensional cells whose sides are dx, dy, dz, dp_x, dp_y, dp_z. As we reduce the size of the cells, we approach more and more closely to the limit of a point in phase space. However, the volume of each of these cells is

$$\tau = dx \, dy \, dz \, dp_x \, dp_y \, dp_z$$

and, according to the uncertainty principle,

$$dx \, dp_x \geqslant \hbar$$
$$dy \, dp_y \geqslant \hbar$$
$$dz \, dp_z \geqslant \hbar$$

Hence we see that

$$\tau \geqslant \hbar^3$$

A "point" in phase space is actually a cell whose minimum volume is of the order of $\hbar^3$. We must think of a particle in phase space as being located somewhere in such a cell centered at some location x, y, z, p_x, p_y, p_z instead of being precisely at the point itself.

A more detailed analysis shows that each cell in phase space actually has the volume h^3, which does not contradict the uncertainty-principle argument since $h^3 > \hbar^3$. In general, each cell in a phase space consisting of k coordinates and k momenta occupies a volume of h^k. We can readily verify this statement for a particle in a one-dimensional box with infinite walls, where $k = 1$. The momentum eigenfunctions of a particle in a box L wide are $\pm n\pi\hbar/L$ according to Eq. 8.11. Figure 15.1 shows the possible paths in x, p_x phase space that the particle is able to follow. Each path is separated from its neighbors on either side by a rectangle of area $L \times \pi\hbar/L = \pi\hbar$, and we can regard each momentum eigenstate as occupying an area of $\pi\hbar$ as indicated. However, there are *two* momentum eigenstates for every energy eigenstate, one for motion in the $+x$ direction and one for motion in the $-x$ direction. The total area in phase space occupied by a particle in a quantum state of any quantum number n is therefore $2\pi\hbar$, which is h. The fact that the "cells" here each have two compartments is of no concern here, since what is pictured in Fig. 15.1 is phase space and not actual space.

While the notion of a point of infinitesimal size in phase space can have no physical significance, since it violates the uncertainty principle, the notion of a point of infinitesimal size in either position space or momentum space alone is perfectly acceptable: we *can* in principle determine the position of a particle with as much precision as we like merely by accepting an unlimited uncertainty in our knowledge of its momentum, and vice versa.

It is the task of statistical mechanics to determine the state of a system by investigating how the particles constituting the system distribute themselves in phase space. If we can find

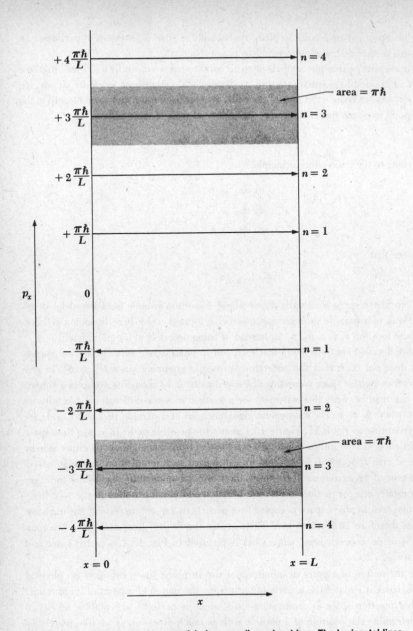

FIGURE 15.1 Phase space for a particle in a one-dimensional box. The horizontal lines represent the possible paths of the particle in phase space, with two paths (one for each direction of motion) for every energy eigenstate of given n. Every energy eigenstate therefore corresponds to two regions whose total area is $2\pi\hbar = h$.

the probabilities of occurrence w of all possible distributions that are permitted by the nature of the system, we can immediately select the most probable one and assert that the system tends to behave according to this distribution of particle positions and momenta. That is, we assert that the state of a system when it is in thermal equilibrium corresponds to the most probable distribution of particles in phase space.

15.2 The Probability of a Distribution

We shall now consider an elementary example in order to introduce the mathematical methods that will be necessary. Let us suppose that we have a large box divided into k cells whose areas are $a_1, a_2, a_3, \ldots, a_k$, as in Fig. 15.2. We proceed to throw N balls into the box in a completely random manner, so that no part of the box is favored. We note how many balls fall in each cell, and then repeat the experiment. After a great many determinations of this kind we will find that a certain distribution of balls among the various cells occurs more often than any other; we call this the most probable distribution. While the most probable distribution may or may not actually be found after any particular throw of the balls, it is the one *most likely* to be found. Clearly the most probable distribution is that in which the number of balls in each cell is proportional to the size of the cell: a large cell is more apt to be hit than a small cell. Let us obtain this result by actually calculating the probabilities of the various possible distributions. Our procedure will resemble the use of a steamroller to crack a peanut, but for the moment our interest is in the operation of the steamroller.

The probability W that the balls be distributed in a certain way among the cells depends upon two factors: (1) the *a priori probability* G of the distribution, which is based upon the properties of each cell, and (2) the *thermodynamic probability* M of the distribution, which is the number of different sequences in which the balls may be distributed among the cells

FIGURE 15.2 A box of area A divided into k cells.

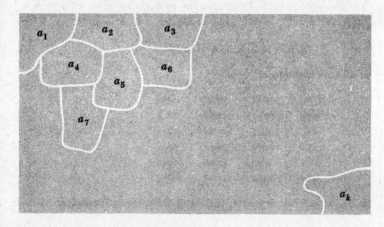

without changing the number in each cell. We thus explicitly assume that the balls are identical but distinguishable.

Here the a priori probability g_i that a ball fall into the ith cell is the ratio between the area a_i of the cell and the total area A of the entire box. That is,

15.1
$$g_i = \frac{a_i}{A}$$

where

15.2
$$A = a_1 + a_2 + \cdots + a_k$$

The sum of the a priori probabilities for all of the cells must be 1, since the ball, by hypothesis, falls into the box somewhere:

15.3
$$\Sigma g_i = g_1 + g_2 + \cdots + g_k = 1$$

The probability that *two* balls fall in the ith cell is g_i^2, the product of the probabilities g_i and g_i that each one separately fall in. (If there is a 10 percent chance that one ball will fall in a given cell, there is only a 1 percent chance that two will fall in.) The a priori probability for n_i balls to fall in the ith cell is $(g_i)^{n_i}$. The a priori probability G of any particular *distribution* of the N balls among the k cells is therefore the product of k probabilities of the form $(g_i)^{n_i}$, namely,

15.4
$$G = (g_1)^{n_1}(g_2)^{n_2} \cdots (g_k)^{n_k}$$

subject to the condition that the total number of balls equal N:

15.5
$$\Sigma n_i = n_1 + n_2 + \cdots + n_k = N$$

If the cells are equal in size, they all have the same a priori probability g and $G = g^N$.

The total number of permutations possible for N balls is $N!$; in other words, N balls can be arranged in $N!$ different sequences. As an example, we might have four balls, a, b, c, and d. The value of $4!$ is

$$4! = 4 \times 3 \times 2 \times 1 = 24$$

and there are indeed 24 ways of arranging them:

abcd	bacd	cabd	dabc
abdc	badc	cadb	dacb
acbd	bcad	cbad	dbac
acdb	bcda	cbda	dbca
adbc	bdac	cdab	dcab
adcb	bdca	cdba	dcba

When more than one ball is in a cell, however, permuting them among themselves has no significance in this situation. For instance, if balls a, b, and c happen to be in cell j, it does not matter here whether we enumerate them as abc, acb, bca, bac, cab, or cba; these six distributions are equivalent, since all we care about is the fact that $n_j = 3$. Thus the n_i balls in the ith cell contribute $n_i!$ irrelevant permutations. If there are n_1 balls in cell 1, n_2 balls in cell 2, and so on through n_k balls in cell k, there are $n_1!n_2!n_3! \ldots n_k!$ irrelevant permutations. Therefore the thermodynamic probability M of the distribution is the total number of possible permutations $N!$ divided by the total number of irrelevant permutations, or

15.6
$$M = \frac{N!}{n_1!n_2! \ldots n_k!}$$

The total probability W of the distribution is the product GM of the a priori probability (Eq. 15.4) and the thermodynamic probability (Eq. 15.6):

15.7
$$W = \frac{N!}{n_1!n_2! \ldots n_k!} (g_1)^{n_1}(g_2)^{n_2} \cdots (g_k)^{n_k}$$

To verify that Eq. 15.7 is correct, we add together the probabilities for all possible distributions and obtain

$$\Sigma W = \Sigma \frac{N!}{n_1!n_2! \ldots n_k!} (g_1)^{n_1}(g_2)^{n_2} \cdots (g_k)^{n_k}$$
$$= (g_1 + g_2 + \cdots + g_k)^N$$

with the help of the multinomial theorem of algebra. (This theorem is a generalization of the binomial theorem.) Since, from Eq. 15.3, the sum of the a priori probabilities g_i is 1 and $1^N = 1$, the sum of the probabilities for all possible distributions is 1. The meaning of

$$\Sigma W = 1$$

is that all N balls are certain to fall in the box somewhere, which agrees with our initial hypothesis and confirms the correctness of Eq. 15.7.

15.3 The Most Probable Distribution

What we now must do is determine just which distribution of the balls is most probable, that is, which distribution yields the largest value of W. Our first step is to obtain a suitable analytic approximation for the factorial of a large number. We note that, since

$$n! = n(n - 1)(n - 2) \cdots (4)(3)(2)$$

the natural logarithm of $n!$ is

$$\ln n! = \ln 2 + \ln 3 + \ln 4 + \cdots + \ln (n - 1) + \ln n$$

Figure 15.3 is a plot of $\ln n$ versus n. The area under the stepped curve is $\ln n!$ When n is

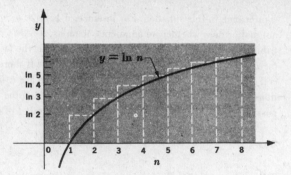

FIGURE 15.3 The area under the stepped curve is ln n! When n is very large, the smooth curve is a good approximation of the stepped curve, and ln n! can be found by integrating ln n from $n = 1$ to $n = n$.

very large, the stepped curve and the smooth curve of ln n become indistinguishable, and we can find ln n! by merely integrating ln n from $n = 1$ to $n = n$:

$$\ln n! = \int_1^n \ln n \, dn$$

$$= n \ln n - n + 1$$

Because we are assuming that $n \gg 1$, we may neglect the 1 in the above result, and so we obtain

15.8 $$\ln n! = n \ln n - n \qquad n \gg 1 \qquad \text{Stirling's formula}$$

Equation 15.8 is known as *Stirling's formula*.

The natural logarithm of Eq. 15.7 is

$$\ln W = \ln N! - \Sigma \ln n_i! + \Sigma n_i \ln g_i$$

Stirling's formula enables us to write this expression as

$$\ln W = N \ln N - N - \Sigma n_i \ln n_i + \Sigma n_i + \Sigma n_i \ln g_i$$

Since $\Sigma n_i = N$,

15.9 $$\ln W = N \ln N - \Sigma n_i \ln n_i + \Sigma n_i \ln g_i$$

While we have an equation for ln W rather than for W itself, this is no handicap since

$$(\ln W)_{\max} = \ln W_{\max}$$

The condition for a distribution to be the most probable one is that small changes δn_i in any of the n_i's not affect the value of W. (If the n_i's were continuous variables instead of being restricted to integral values, we could express this condition in the usual way as $\partial W/\partial n_i = 0$.) If the change in ln W corresponding to a change in n_i of δn_i is $\delta \ln W$, from Eq. 15.9 we see that

15.10 $$\delta \ln W_{\text{max}} = -\Sigma\, n_i \delta \ln n_i - \Sigma \ln n_i \delta n_i + \Sigma \ln g_i \delta n_i = 0$$

since $N \ln N$ is constant. Now

$$\delta \ln n_i = \frac{1}{n_i}\, \delta n_i$$

and so

$$\Sigma\, n_i \delta \ln n_i = \Sigma\, \delta n_i$$

Because the total number of balls is constant, the sum $\Sigma\, \delta n_i$ of all the changes in the number of balls in each cell must be 0, which means that

$$\Sigma\, n_i \delta \ln n_i = 0$$

Hence Eq. 15.10 becomes

15.11 $$-\Sigma \ln n_i \delta n_i + \Sigma \ln g_i \delta n_i = 0$$

While Eq. 15.11 must be fulfilled by the most probable distribution of the balls, it does not by itself completely specify this distribution. We must take account of the fact that the variations $\delta n_1,\ \delta n_2,\ \ldots$ in the number of balls in each of the k cells are not independent, since the total number of balls is fixed, but instead obey the relationship

15.12 $$\Sigma\, \delta n_i = \delta n_1 + \delta n_2 + \cdots + \delta n_k = 0$$

To do so, we make use of Lagrange's method of undetermined multipliers. We let α be a quantity that does not depend upon any of the n_i's and multiply Eq. 15.12 by α to give

15.13 $$\Sigma\, \alpha \delta n_i = 0$$

Now we add this equation to Eq. 15.11 and obtain

15.14 $$-\Sigma \ln n_i \delta n_i + \Sigma \ln g_i \delta n_i + \Sigma\, \alpha \delta n_i = \Sigma\, (-\ln n_i + \ln g_i + \alpha)\delta n_i = 0$$

In each of the k equations added together to give Eq. 15.14, the variation δn_i is effectively an independent variable. In order for Eq. 15.14 to be true, then, the quantity in parentheses must always be 0 regardless of the values given to the δn_i's. Thus we have

$$-\ln n_i + \ln g_i + \alpha = 0$$

15.15 $$n_i = g_i e^\alpha$$

Adding together the n_i's and noting that α does not depend upon i,

$$\Sigma\, n_i = e^\alpha \Sigma\, g_i$$

We recall that g_i is the a priori probability that a ball fall into the ith cell and, from Eq. 15.3,

The Most Probable Distribution 355

$$\Sigma \, g_i = 1$$

Hence

$$\Sigma \, n_i = e^{\alpha}$$

and, since the total number of balls $\Sigma \, n_i$ is N, we see that e^{α} is just

$$e^{\alpha} = N$$

Equation 15.15 therefore becomes

$$n_i = N g_i$$

The most probable number of balls in any cell is proportional both to the total number of balls N and to the a priori probability g_i, which is equal to the relative size of the cell. More precisely, we note from Eq. 15.1 that

$$g_i = \frac{a_i}{A}$$

and so

$$n_i = \frac{N}{A} a_i$$

The most probable number of balls in any cell is equal to the average density of balls N/A multiplied by the area of the cell. This is the result we set out to derive.

15.4 Maxwell-Boltzmann Statistics

We shall now apply the methods of statistical mechanics to determine how a fixed total amount of energy is distributed among the various members of an assembly of identical particles. That is, our problem is to find out how many particles (on the average) have the energy u_1, how many have the energy u_2, and so on. We shall consider assemblies of three kinds of particles:

A. Identical particles of any spin that are sufficiently widely separated to be distinguished. The molecules of a gas are particles of this kind.

B. Identical particles of 0 or integral spin that cannot be distinguished. These are Bose particles and do not obey the exclusion principle.

C. Identical particles of spin ½ that cannot be distinguished. These are Fermi particles and obey the exclusion principle.

In this section our efforts will be directed to obtaining the distribution law for particles of type A, which we shall refer to in general as *molecules* because gas molecules constitute the most important class of such particles.

Let us consider an assembly of N molecules whose energies are limited to the k values

$u_1, u_2, \ldots u_k$, arranged in order of increasing energy. These energies may represent either discrete energy states or average energies within a sequence of energy intervals. If there are n_i molecules of energy u_i and the total energy of the assembly is U, the most probable distribution of molecules among these k energies is subject to two conditions, namely, conservation of molecules,

15.16
$$\Sigma\, n_i = n_1 + n_2 + \cdots + n_k = N \qquad \text{Conservation of particles}$$

and conservation of energy,

15.17
$$\Sigma\, n_i u_i = n_1 u_1 + n_2 u_2 + \cdots + n_k u_k = U \qquad \text{Conservation of energy}$$

If the a priori probability for a molecule to have the energy u_i is g_i, the probability W for any distribution is given by Eq. 15.7. As before, the distribution of maximum probability must obey Eq. 15.11, but now there are *two* conditions the δn_i must fulfill. These conditions follow from Eq. 15.16 and 15.17; they are

15.18
$$\Sigma\, \delta n_i = \delta n_1 + \delta n_2 + \cdots + \delta n_k = 0$$

15.19
$$\Sigma\, u_i \delta n_i = u_i \delta n_i + u_2 \delta n_2 + \cdots + u_k \delta n_k = 0$$

To incorporate these conditions into Eq. 15.11, we multiply Eq. 15.18 by $-\alpha$ and Eq. 15.19 by $-\beta$, where α and β are quantities independent of the n_i's, and add these expressions to Eq. 15.11. We obtain

$$\Sigma\, (-\ln n_i + \ln g_i - \alpha - \beta u_i)\delta n_i = 0$$

Again, the δn_i's are now effectively independent, and the quantity in parentheses must be 0 for each value of i. Hence

$$-\ln n_i + \ln g_i - \alpha - \beta u_i = 0$$

15.20
$$n_i = g_i e^{-\alpha} e^{-\beta u_i} \qquad \text{Maxwell-Boltzmann distribution law}$$

This result is known as the *Maxwell-Boltzmann distribution law*.

We shall now evaluate $e^{-\alpha}$ and β. To do so, it is convenient to consider a continuous distribution of molecular energies, rather than the discrete set $u_1, u_2, \ldots, u_k$, so that Eq. 15.20 becomes

15.21
$$n(u)\, du = g(u)e^{-\alpha} e^{-\beta u}\, du$$

(This is a perfectly valid approximation for the molecules in a gas, where energy quantization is inconspicuous and the total number of molecules may be very large.) In Eq. 15.21 $n(u)\, du$ is interpreted as the number of molecules whose energies lie between u and $u + du$. In terms of molecular momentum, since

$$u = \frac{p^2}{2m}$$

we have

$$n(p)\, dp = g(p)e^{-\alpha}e^{-\beta p^2/2m}\, dp$$

The a priori probability $g(p)$ that a molecule have a momentum between p and $p + dp$ is equal to the number of cells in phase space within which such a molecule may exist. If each cell has the infinitesimal volume h^3,

$$g(p)\, dp = \frac{\int\int\int\int\int dx\, dy\, dz\, dp_x\, dp_y\, dp_z}{h^3}$$

where the numerator is the phase-space volume occupied by particles with the specified momenta. Here

$$\int\int\int dx\, dy\, dz = V$$

where V is the volume occupied by the gas in ordinary position space, and

$$\int\int dp_x\, dp_y\, dp_z = 4\pi p^2\, dp$$

where $4\pi p^2\, dp$ is the volume of a spherical shell of radius p and thickness dp in momentum space (see Fig. 9.11). Hence

15.22
$$g(p)\, dp = \frac{4\pi V p^2\, dp}{h^3}$$

and

15.23
$$n(p)\, dp = \frac{4\pi V p^2 e^{-\alpha}e^{-\beta p^2/2m}}{h^3}\, dp$$

We are now able to find $e^{-\alpha}$. Since

$$\int_0^\infty n(p)\, dp = N$$

we find by integrating Eq. 15.23 that

$$N = \frac{4\pi e^{-\alpha}V}{h^3}\int_0^\infty p^2 e^{-\beta p^2/2m}\, dp$$

$$= \frac{e^{-\alpha}V}{h^3}\left(\frac{2\pi m}{\beta}\right)^{3/2}$$

where we have made use of the definite integral

$$\int_0^\infty x^2 e^{-ax^2}\, dx = \frac{1}{4}\sqrt{\frac{\pi}{a^3}}$$

Hence

$$e^{-\alpha} = \frac{Nh^3}{V}\left(\frac{\beta}{2\pi m}\right)^{3/2}$$

and

15.24
$$n(p)\ dp = 4\pi N \left(\frac{\beta}{2\pi m}\right)^{3/2} p^2 e^{-\beta p^2/2m}\ dp$$

To find β, we compute the total energy U of the assembly of molecules. Since

$$p^2 = 2mu \qquad \text{and} \qquad dp = \frac{m\ du}{\sqrt{2mu}}$$

we can write Eq. 15.24 in the form

15.25
$$n(u)\ du = \frac{2N\beta^{3/2}}{\sqrt{\pi}}\ \sqrt{u}\ e^{-\beta u}\ du$$

The total energy is

$$\begin{aligned}
U &= \int_0^\infty un(u)\ du \\
&= \frac{2N\beta^{3/2}}{\sqrt{\pi}} \int_0^\infty u^{3/2} e^{-\beta u}\ du \\
&= \frac{3}{2}\frac{N}{\beta}
\end{aligned}$$

15.26

where we have made use of the definite integral

$$\int_0^\infty x^{3/2} e^{-ax}\ dx = \frac{3}{4a^2}\sqrt{\frac{\pi}{a}}$$

According to the kinetic theory of gases, the total energy U of N molecules of an ideal gas (which is what we have been considering) at the absolute temperature T is

15.27
$$U = \frac{3}{2}\ NkT$$

where k is Boltzmann's constant

$$k = 1.380 \times 10^{-23}\ \text{joule/molecule-degree}$$

Equations 15.26 and 15.27 agree if

15.28
$$\beta = \frac{1}{kT}$$

Maxwell-Boltzmann Statistics

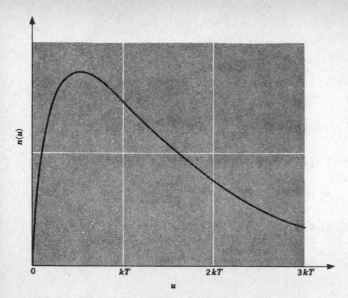

FIGURE 15.4   Maxwell-Boltzmann energy distribution.

Now that the parameters α and β have been evaluated, we can write the Boltzmann distribution law in its final form,

15.29
$$n(u)\ du = \frac{2\pi N}{(\pi kT)^{3/2}} \sqrt{u}\ e^{-u/kT}\ du \qquad \textbf{Boltzmann distribution of energies}$$

This equation gives the number of molecules with energies between u and $u + du$ in a sample of an ideal gas that contains a total of N molecules and whose absolute temperature is T. The Boltzmann energy distribution is plotted in Fig. 15.4 in terms of kT. The curve is not symmetrical because the lower limit to u is $u = 0$ while there is, in principle, no upper limit (although the likelihood of energies many times greater than kT is small).

According to Eq. 15.26, the total energy U of an assembly of N molecules is

$$U = \frac{3}{2}\frac{N}{\beta}$$

The *average energy* $\bar{u}$ per molecule is U/N, so that

$$\bar{u} = \frac{3}{2}\frac{1}{\beta}$$

15.30
$$= \frac{3}{2}kT \qquad \textbf{Average molecular energy}$$

At $300\,°\text{K}$, which is approximately room temperature,

$$\bar{u} = 6.21 \times 10^{-21} \text{ joule/molecule}$$

$$\approx \text{⅟}_{25} \text{ ev/molecule}$$

This average energy is the same for all molecules at $300°\text{K}$, regardless of their mass.

15.5 Molecular Speeds

The Boltzmann distributions of molecular momenta and speeds can be obtained from Eq. 15.29 by noting that

$$u = \frac{p^2}{2m} = \frac{1}{2}mv^2$$

$$du = \frac{p}{m}\,dp = mv\,dv$$

We find that

15.31
$$n(p)\,dp = \frac{\sqrt{2}\pi N}{(\pi mkT)^{3/2}}p^2 e^{-p^2/2mkT}\,dp \qquad \text{Boltzmann distribution of momenta}$$

is the number of molecules having momenta between p and $p + dp$, and

15.32
$$n(v)\,dv = \frac{\sqrt{2}\pi Nm^{3/2}}{(\pi kT)^{3/2}}v^2 e^{-mv^2/2kT}\,dv \qquad \text{Boltzmann distribution of speeds}$$

is the number of molecules having speeds between v and $v + dv$. The last formula, which was first obtained by Maxwell in 1859, is plotted in Fig. 15.5.

The speed of a molecule with the average energy of $\tfrac{3}{2}kT$ is

15.33
$$v_{\text{rms}} = \sqrt{\overline{v^2}} = \sqrt{\frac{3kT}{m}} \qquad \text{Rms speed}$$

since $\tfrac{1}{2}mv^2 = \tfrac{3}{2}kT$. This speed is denoted v_{rms} because it is the square root of the average of the squared molecular speeds—the *root-mean-square* speed—and is not the same as the simple arithmetical average speed $\bar{v}$. An example will illustrate the difference between the two kinds of average. Let us consider an assembly of two molecules, one with a speed of 1 m/sec and the other with a speed of 3 m/sec. The average speed of the two molecules is

$$\bar{v} = \frac{v_1 + v_2}{2} = \frac{1 + 3}{2}\text{ m/sec} = 2\text{ m/sec}$$

whereas the rms speed is

$$v_{\text{rms}} = \sqrt{\frac{v_1^2 + v_2^2}{2}} = \sqrt{\frac{1^2 + 3^2}{2}}\text{ m/sec} = 2.24\text{ m/sec}$$

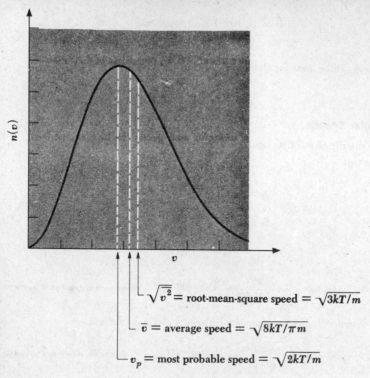

$$\sqrt{\overline{v^2}} = \text{root-mean-square speed} = \sqrt{3kT/m}$$

$$\overline{v} = \text{average speed} = \sqrt{8kT/\pi m}$$

$$v_p = \text{most probable speed} = \sqrt{2kT/m}$$

FIGURE 15.5   Maxwell-Boltzmann velocity distribution.

The relationship between $\overline{v}$ and v_{rms} depends upon the distribution law that governs the molecular speeds being considered. For a Boltzmann distribution,

$$v_{\text{rms}} = \sqrt{\frac{3\pi}{8}}\,\overline{v} \approx 1.09\overline{v}$$

so that the rms speed is about 9 percent greater than the arithmetical average speed.

Because the Boltzmann distribution of speeds is not symmetrical, the most probable speed v_p is smaller than either $\overline{v}$ or v_{rms}. To find v_p, we set equal to zero the derivative of $n(v)$ with respect to v and solve the resulting equation for v. We obtain

15.34
$$v_p = \sqrt{\frac{2kT}{m}}$$
Most probable speed

Molecular speeds in a gas vary considerably on either side of v_p. Figure 15.6 shows the distribution of molecular speeds in oxygen at $73°\,K$ $(-200°C)$, in oxygen at $273°\,K$ $(0°C)$, and in hydrogen at $273°\,K$. The most probable molecular speed increases with temperature

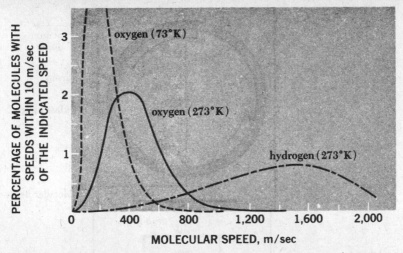

FIGURE 15.6   The distributions of molecular speeds in oxygen at 73°K, in oxygen at 273°K, and in hydrogen at 273°K.

and decreases with molecular mass. Accordingly molecular speeds in oxygen at 73°K are on the whole less than at 273°K, and at 273°K molecular speeds in hydrogen are on the whole greater than in oxygen at the same temperature. (The average molecular *energy* is the same in both oxygen and hydrogen at 273°K, of course.)

A beam of molecules that emerged from a slit in the wall of an oven provided the basis of a direct test of the Maxwell-Boltzmann velocity distribution. This experiment, performed in the early 1930s by I. F. Zartman and C. C. Ko, made use of the apparatus shown in Fig. 15.7. The oven contains bismuth vapor at about 800°C, some of which escapes through a slit and is collimated by another slit a short distance away. Above the second slit is a drum that rotates about a horizontal axis at 6,000 rpm. At those instants when the slit in the drum faces the bismuth beam, a burst of molecules enters the drum. These molecules reach the opposite face of the drum, where a glass plate is attached, at various times, depending upon their speeds. Because the drum is turning, the faster and slower molecules strike different parts of the plate. From the resulting distribution of deposited bismuth on the plate it is possible to infer the distribution of speeds in the beam, and this distribution agrees with the prediction of Maxwell-Boltzmann statistics.

15.6   Rotational Spectra

A continuous distribution of energies occurs only in the translational motions of molecules. As we saw in Chap. 14, molecular rotations and vibrations are quantized, with only certain specific energies E_i being possible. The Boltzmann distribution law for modes of energy possession of the latter sort may be written

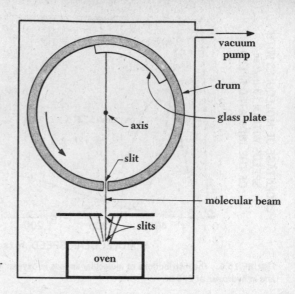

vacuum pump

drum

glass plate

axis

slit

molecular beam

slits

oven

FIGURE 15.7 The Zartman-Ko experiment.

15.35

$$n_i = n_0 \, g_i \, e^{-E_i/kT}$$

which is simply Eq. 15.20 with n_0 replacing $e^{-\alpha}$ and E_i/kT replacing βu_i. The factor $e^{-E_i/kT}$, often called the *Boltzmann factor*, expresses the relative probability that a quantum state of energy E_i be occupied at the temperature T. The factor g_i, the multiplicity (or *statistical weight*) of the level, is the number of quantum states that have the same energy E_i. All quantum states are assumed to have the same a priori probability—that is, the cells that represent them in phase space have the same volumes—so the a priori probability g_i that the energy level E_i be occupied is equal to the number of states that are included in the level.

Let us apply Eq. 15.35 to the rotational energy levels of a molecule. (The relative populations of atomic energy levels can be treated in the same way.) As we know, more than one rotational state may correspond to a particular rotational quantum number J. The degeneracy arises because the component L_z in any specified direction of the angular momentum **L** may have any value in multiples of $\hbar$ from $J\hbar$ through 0 to $-J\hbar$, for a total of $2J + 1$ possible values. That is, there are $2J + 1$ possible orientations of **L** relative to the specified (z) direction, with each of these orientations constituting a separate quantum state. Hence an energy level whose rotational quantum number is J has a statistical weight of

$$g_J = 2J + 1$$

For a rigid diatomic molecule,

$$E_J = J(J + 1)\frac{\hbar^2}{2I}$$

and so the Boltzmann factor corresponding to the quantum number J is

$$e^{-J(J+1)\hbar^2/2IkT}$$

The Boltzmann distribution formula for the probabilities of occupancy of the rotational energy levels of a rigid diatomic molecule is therefore

15.36
$$n_J = (2J + 1)\, n_0\, e^{-J(J+1)\hbar^2/2IkT}$$

Here the quantity n_0 is the number of molecules in the $J = 0$ rotational state.

In Sec. 14.1 we found that the moment of inertia of the CO molecule is 1.46×10^{-46} kg-m^2. For a sample of carbon monoxide gas at room temperature (293°K, which is 20°C)

$$\frac{\hbar^2}{2IkT} = \frac{(1.054 \times 10^{-34}\ \text{joule-sec})^2}{2 \times 1.46 \times 10^{-46}\ \text{kg-m}^2 \times 1.38 \times 10^{-23}\ \text{joule/°K} \times 293°\text{K}}$$

$$= 0.00941$$

and so

$$n_J = (2J + 1)\, n_0\, e^{-0.00941\, J(J+1)}$$

Figure 15.8 contains graphs of the statistical weight $2J + 1$, the Boltzmann factor $e^{-0.00941J(J+1)}$, and the relative population n_J/n_0 for CO at 20°C, all as functions of J. The $J = 7$ rotational energy level is evidently the most highly populated, and about as many molecules in a sample of CO at room temperature are in the $J = 19$ level as are in the $J = 0$ level.

To find analytically the energy level with the highest population, we need only differentiate n_J with respect to J and then set dn_J/dJ equal to 0. The resulting expression for $J_{\text{max pop}}$ is

15.37
$$J_{\text{max pop}} = \frac{\sqrt{IkT}}{\hbar} - \frac{1}{2}$$

and, of course, we are to take the integer closest to the calculated figure. In the case of CO, Eq. 15.37 yields $J = 7.29$, and $J = 7$ is the nearest quantum number.

The intensities of the rotational lines in a molecular spectrum are proportional to the relative populations of the various rotational energy levels. Thus we expect the highest intensities for lines corresponding to initial levels in the vicinity of $J_{\text{max pop}}$, with the intensities dropping off on either side. Figure 15.9a shows the vibration-rotation band of CO for the $v = 0 \rightarrow v = 1$ vibrational transition under high resolution; lines are identified according to the J value of the initial rotational level. The P and R branches both have their maxima at $J = 7$. At low resolution, the same band appears as in Fig. 15.9b, with no rotational fine structure visible. However, the frequency difference $\Delta\nu$ between the maxima in the P and R branches can be measured, and given $\Delta\nu$ the moment of inertia of the molecule can be found with fair accuracy with the help of Eqs. 14.43, 14.44, and 15.37.

Rotational Spectra

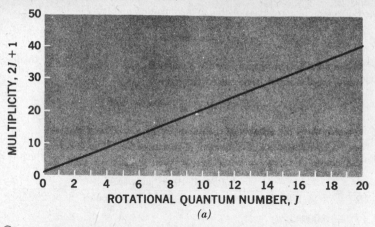

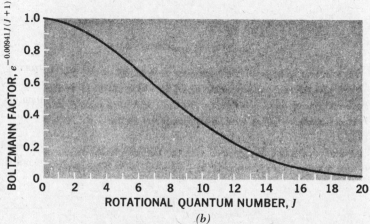

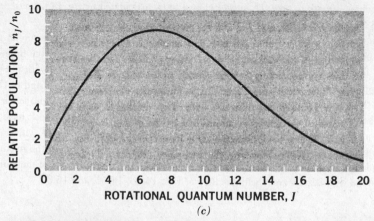

FIGURE 15.8 The multiplicities (a), Boltzmann factors (b), and relative populations (c) of the rotational energy levels of the CO molecule at 20°C.

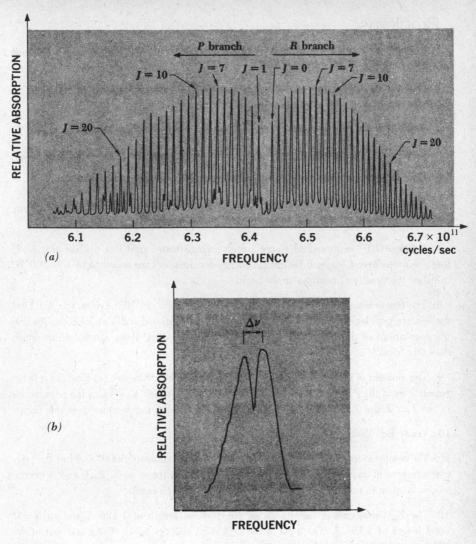

FIGURE 15.9 (a) The $v = 0 \rightarrow v = 1$ vibration-rotation absorption band in CO under high resolution. The lines are identified by the value of J in the initial rotational state. (b) The same band under low resolution.

Problems

1. Find the thermodynamic probability of the most probable distribution of 10^6 identical particles among 5×10^5 identical cells.

2. Find the thermodynamic probability of the least probable distribution of 10^6 identical particles among 5×10^5 identical cells.

3. Verify that the most probable speed of a molecule of an ideal gas is equal to $\sqrt{2kT/m}$.

4. Verify that the average speed of a molecule of an ideal gas is equal to $\sqrt{8kT/\pi m}$.

5. Find the average value of $1/v$ in a gas obeying Maxwell-Boltzmann statistics.

6. What proportion of the molecules of an ideal gas have components of velocity in any particular direction greater than twice the most probable speed?

7. A flux of 10^{12} neutrons/m² emerges each second from a port in a nuclear reactor. If these neutrons have a Maxwell-Boltzmann energy distribution corresponding to $T = 300°K$, calculate the density of neutrons in the beam.

8. The frequency of vibration of the H_2 molecule is 1.32×10^{14} cycles/sec. (a) Find the relative populations of the $v = 0, 1, 2, 3,$ and 4 vibrational states at $5000°K$. (b) Can the populations of the $v = 2$ and $v = 3$ states ever be equal? If so, at what temperature does this occur?

9. The moment of inertia of the H_2 molecule is 4.64×10^{-48} kg-m². (a) Find the relative populations of the $J = 0, 1, 2, 3,$ and 4 rotational states at $300°K$. (b) Can the populations of the $J = 2$ and $J = 3$ states ever be equal? If so, at what temperature does this occur?

10. Verify Eq. 15.37.

11. The temperature of the sun's chromosphere is approximately $5000°K$. Find the relative numbers of hydrogen atoms in the chromosphere in the $n = 1, 2, 3,$ and 4 energy levels. Be sure to take into account the multiplicity of each level.

12. The N_2O molecule is linear with an N—N bond length of 1.126 A and an N—O bond length of 1.191 A. The mass of the O^{18} atom is 2.66×10^{-26} kg and that of the N^{14} atom is 2.32×10^{-26} kg. (a) What is the quantum number of the most populated rotational energy level at $300°K$? (b) Plot n_J/n_0 versus J at $300°K$.

Chapter 16
Quantum Statistics

Although the Maxwell-Boltzmann distribution law was derived in the previous chapter in a manner consistent with quantum mechanics, exactly the same results are obtained in a purely classical treatment. As we saw, there is no essential difference between a situation in which a continuous spread of energies is possible, as in the translational motions of molecules in a gas, and one in which the energies are limited to a discrete set of specific levels, as in the rotational states of a molecule. In either case the particles involved were assumed to be distinguishable from one another, which is true for molecules in a gas but *not* true for, say, photons in an enclosure or electrons in a metal. Quantum considerations become important for indistinguishable particles, and the distribution laws they obey differ significantly from the Maxwell-Boltzmann law. In this chapter the Bose-Einstein and Fermi-Dirac statistical distribution laws will be derived; these laws apply respectively to identical, indistinguishable particles to which the exclusion principle does not apply (for instance, photons) and to such particles to which it does apply (for instance, electrons).

16.1 Bose-Einstein Statistics

The basic distinction between Maxwell-Boltzmann statistics and Bose-Einstein statistics is that the former governs identical particles which can be distinguished from one another in some way, while the latter governs identical particles which cannot be distinguished apart, though they can be counted. In Bose-Einstein statistics all quantum states are assumed to have equal a priori probabilities, so that g_i represents the number of states that have the same energy u_i. (This assumption was also made in Sec. 15.6 in connection with the rotational states of a molecule.) Each quantum state corresponds to a cell in phase space, and our first step is to determine the number of ways in which n_i indistinguishable particles can be distributed in g_i cells.

To carry out the required enumeration, we consider a series of $n_i + g_i - 1$ objects placed in a line (Fig. 16.1). We note that $g_i - 1$ of the objects can be regarded as partitions separating a total of g_i intervals, with the entire series therefore representing n_i particles arranged in g_i cells. In the picture $g_i = 12$ and $n_i = 20$; 11 partitions separate the 20 particles into 12 cells. The first cell contains two particles, the second none, the third one particle, the fourth three particles, and so on. There are $(n_i + g_i - 1)!$ possible permuta-

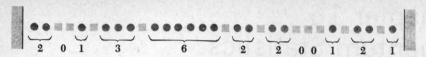

$$\underbrace{\quad}_{2} \quad 0 \quad 1 \quad \underbrace{\quad}_{3} \quad \underbrace{\quad}_{6} \quad \underbrace{\quad}_{2} \quad \underbrace{\quad}_{2} \quad 0 \quad 0 \quad 1 \quad \underbrace{\quad}_{2} \quad \underbrace{\quad}_{1}$$

● particle
▨ partition

number of indistinguishable particles $= n_i = 20$
number of partitions $= g_i - 1 = 11$
number of cells $= g_i = 12$

FIGURE 16.1 A series of n_i indistinguishable particles separated by g_i-1 partitions into g_i cells.

tions among $n_i + g_i - 1$ objects, but of these the $n_i!$ permutations of the n_i particles among themselves and the $(g_i - 1)!$ permutations of the $g_i - 1$ partitions among themselves do not affect the distribution and are irrelevant. Hence there are

$$\frac{(n_i + g_i - 1)!}{n_i!(g_i - 1)!}$$

possible distinguishably different arrangements of the n_i indistinguishable particles among the g_i cells.

The probability W of the entire distribution of N particles is the product

16.1
$$W = \Pi \frac{(n_i + g_i - 1)!}{n_i!(g_i - 1)!}$$

of the numbers of distinct arrangements of particles among the states having each energy. We now assume that

$$(n_i + g_i) \gg 1$$

so that $(n_i + g_i - 1)$ can be replaced by $(n_i + g_i)$, and take the natural logarithm of both sides of Eq. 16.1 to give

$$\ln W = \Sigma \left[\ln (n_i + g_i)! - \ln n_i! - \ln (g_i - 1)! \right]$$

Stirling's formula

$$\ln n! = n \ln n - n$$

permits us to rewrite $\ln W$ as

16.2 $$\ln W = \Sigma \left[(n_i + g_i) \ln (n_i + g_i) - n_i \ln n_i - \ln (g_i - 1)! - g_i \right]$$

As before, the condition that this distribution be the most probable one is that small changes

δn_i in any of the individual n_i's not affect the value of W. If a change in $\ln W$ of $\delta \ln W$ occurs when n_i changes by δn_i, the above condition may be written

$$\delta \ln W_{max} = 0$$

Hence, if the W of Eq. 16.2 represents a maximum,

16.3 $\qquad \delta \ln W_{max} = \Sigma \left[\ln (n_i + g_i) - \ln n_i \right] \delta n_i = 0$

where we have made use of the fact that

$$\delta \ln n = \frac{1}{n} \delta n$$

As in Sec. 15.4, we incorporate the conservation of particles, expressed in the form

16.4 $\qquad \Sigma \, \delta n_i = 0$

and the conservation of energy, expressed in the form

16.5 $\qquad \Sigma \, u_i \, \delta n_i = 0$

by multiplying the former equation by $-\alpha$ and the latter by $-\beta$ and adding to Eq. 16.3. The result is

$$\Sigma \left[\ln (n_i + g_i) - \ln n_i - \alpha - \beta u_i \right] \delta n_i = 0$$

Since the δn_i's are independent, the quantity in brackets must vanish for each value of i. Hence

$$\ln \frac{n_i + g_i}{n_i} - \alpha - \beta u_i = 0$$

$$1 + \frac{g_i}{n_i} = e^{\alpha} e^{\beta u_i}$$

and

16.6 $\qquad n_i = \dfrac{g_i}{e^{\alpha} e^{\beta u_i} - 1}$

Substituting for β from Eq. 15.28,

16.7 $\qquad \beta = \dfrac{1}{kT}$

we arrive at the *Bose-Einstein distribution law:*

16.8 $\qquad n_i = \dfrac{g_i}{e^{\alpha} e^{u_i/kT} - 1}$ $\qquad$ **Bose-Einstein distribution law**

Bose-Einstein Statistics

16.2 Black-body Radiation

Every substance emits electromagnetic radiation, the character of which depends upon the nature and temperature of the substance. We have already discussed the discrete spectra of excited gases which arise from electronic transitions within isolated atoms. At the other extreme, dense bodies such as solids radiate continuous spectra in which all frequencies are present; the atoms in a solid are so close together that their mutual interactions result in a multitude of adjacent quantum states indistinguishable from a continuous band of permitted energies.

The ability of a body to radiate is closely related to its ability to absorb radiation. This is to be expected, since a body at a constant temperature is in thermal equilibrium with its surroundings and must absorb energy from them at the same rate as it emits energy. It is convenient to consider as an ideal body one that absorbs *all* radiation incident upon it, regardless of frequency. Such a body is called a *black body*.

It is easy to show experimentally that a black body is a better emitter of radiation than anything else. The experiment, illustrated in Fig. 16.2, involves two identical pairs of dissimilar surfaces. No temperature difference is observed between surfaces I' and II'. At a given temperature the surfaces I and I' radiate at the rate of e_1 watts/m^2, while II and II' radiate at the different rate e_2. The surfaces I and I' absorb some fraction a_1 of the radiation falling on them, while II and II' absorb some other fraction a_2. Hence I' absorbs energy from II at a rate proportional to a_1e_2, and II' absorbs energy from I at a rate proportional to a_2e_1. Because I' and II' remain at the same temperature, it must be true that

$$a_1e_2 = a_2e_1$$

and

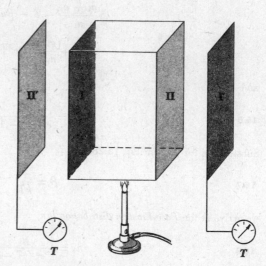

FIGURE 16.2 Surfaces *I* and *I'* are identical to each other and are different from the identical pair of surfaces *II* and *II'*.

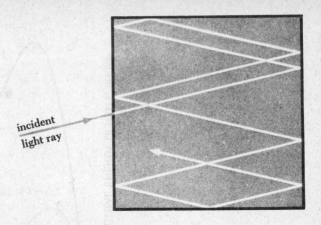

incident
light ray

FIGURE 16.3 A hole in the wall of a hollow object is an excellent approximation of a black body.

16.9
$$\frac{e_1}{a_1} = \frac{e_2}{a_2}$$

The ability of a body to emit radiation is proportional to its ability to absorb radiation. Let us suppose that I and I′ are black bodies, so that $a_1 = 1$, while II and II′ are not, so that $a_2 < 1$. Hence

$$e_1 = \frac{e_2}{a_2}$$

and, since $a_2 < 1$, $e_1 > e_2$. A black body at a given temperature radiates energy at a faster rate than any other body.

The point of introducing the idealized black body in a discussion of thermal radiation is that we can now disregard the precise nature of whatever is radiating, since all black bodies behave identically. In the laboratory a black body can be approximated by a hollow object with a very small hole leading to its interior (Fig. 16.3). Any radiation striking the hole enters the cavity, where it is trapped by reflection back and forth until it is absorbed. The cavity walls are constantly emitting and absorbing radiation, and it is in the properties of this radiation (*black-body radiation*) that we are interested. Experimentally we can sample black-body radiation simply by inspecting what emerges from the hole. The results agree with our everyday experience; a black body radiates more when it is hot than when it is cold, and the spectrum of a hot black body has its peak at a higher frequency than the peak in the spectrum of a cooler one. We recall the familiar behavior of an iron bar as it is heated to progressively higher temperatures: at first it glows dull red, then bright orange-red, and eventually becomes "white hot." The spectrum of black-body radiation is shown in Fig. 16.4 for two temperatures.

Certain aspects of black-body radiation can be understood by considering a black body as a thermodynamic system whose working substance is electromagnetic radiation. On this basis it is easy to show, for example, that the energy density of black-body radiation

Black-body Radiation

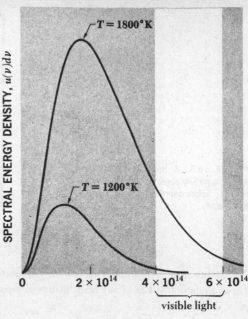

FIGURE 16.4 Black-body spectra. The spectral distribution of energy in the radiation depends only upon the temperature of the body.

within a cavity depends only upon the temperature of its walls. Let us suppose that this statement is *not* true, and that it is possible to have two cavities, A and B, whose walls are at the same temperature while cavity A has a higher energy density than cavity B (Fig. 16.5a). If the cavities are joined together, as in Fig. 16.5b, radiant energy flows from A to B until the energy densities in both cavities are the same. Now the cavities are separated and resealed. When this is done, heat begins to flow from the walls of cavity A to the radiation inside it in order to make up for the radiation lost to cavity B, so that when equilibrium is restored cavity A will be at a lower temperature than its initial one. Meanwhile some of the radiation that had flowed into cavity B is absorbed by its walls, and when equilibrium is restored cavity B will be at a higher temperature than its initial one (Fig. 16.5c). Thus we now have two heat reservoirs at different temperatures, which means that we can connect them to a heat engine and obtain work from the system until the two cavities are again at the same temperature, which is a lower one than originally. Next we disconnect the cavities from the heat engine, join them together once more as in Fig. 16.5a, and again repeat the entire sequence. We can continue the process until no energy is left in the cavities. However, the second law of thermodynamics states that it is impossible to construct an engine that, operating in a cycle, produces no effect other than the extraction of heat from a reservoir and the performance of an equivalent amount of work—and cavities A and B, which are initially at the same temperature, constitute a single reservoir from the point of view of thermodynamics. The conclusion is that the energy density of the radiation within a cavity depends *solely* upon the temperature of the cavity, since we have proved the only alternative to be false. A black body is equivalent to a hole in the wall of a cavity, which means that the rate at which a black body radiates energy also depends solely upon the body's temperature.

16.3   The Rayleigh-Jeans Formula

What we would like to have is a general formula for the spectrum of black-body radiation; that is, we would like to account for empirical curves such as those shown in Fig. 16.4.

As a first step, we can verify that the shape of the spectrum—the distribution of energy among the various wavelengths of the radiation—is a function only of the temperature of the black body. The argument is the same as that used in the preceding section to show that the total radiation rate is a function only of the temperature, except that we insert a filter between the cavities when they are joined together that allows only radiation in a narrow wavelength interval to go from cavity A to cavity B. We conclude that the rate at which energy is radiated in this wavelength interval is the same for all black bodies at any given temperature. Hence a single formula ought to describe all black-body spectra.

The problem of the black-body spectrum was examined at the end of the nineteenth century by Rayleigh and Jeans using classical physics, since the notion of electromagnetic quanta was as yet unknown. They considered the radiation inside a cavity of temperature T whose walls are perfect reflectors as a series of standing electromagnetic waves, in essence a three-dimensional generalization of standing waves in a stretched string. The condition

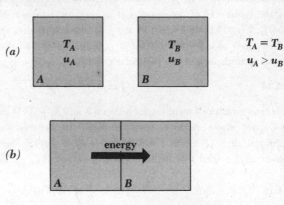

FIGURE 16.5 The energy density u of black-body radiation depends only upon the temperature T of the body. If this were not so, it would be possible for radiant energy to flow from one body to another when both have the same initial temperature, which violates the second law of thermodynamics. The hypothetical situation shown here cannot occur.

for standing waves in such a cavity is that the path length from wall to wall, whatever the direction, must be an integral number of half-wavelengths, so that a node occurs at each reflecting surface. If the cavity is a cube L long on each edge, this condition means that, for standing waves in the x, y, and z directions respectively, the possible wavelengths are such that

16.10
$$j_x = \frac{2L}{\lambda} = 1, 2, 3, \ldots$$

16.11
$$j_y = \frac{2L}{\lambda} = 1, 2, 3, \ldots$$

16.12
$$j_z = \frac{2L}{\lambda} = 1, 2, 3, \ldots$$

For a standing wave in any arbitrary direction, it must be true that

16.13
$$j_x{}^2 + j_y{}^2 + j_z{}^2 = \left(\frac{2L}{\lambda}\right)^2 \qquad \begin{array}{l} j_x = 0, 1, 2, \ldots \\ j_y = 0, 1, 2, \ldots \\ j_z = 0, 1, 2, \ldots \end{array}$$

in order that the wave terminate in a node at its ends. (Of course, if $j_x = j_y = j_z = 0$, there is no wave, though it is possible for any one or two of the j's to equal 0.)

To count the number of standing waves $n(\lambda)\, d\lambda$ per unit volume within the cavity whose wavelengths lie between λ and $\lambda + d\lambda$, what we have to do is count the number of permissible sets of j_x, j_y, j_z values that yield wavelengths in this interval and then divide by the volume L^3 of the cavity. Let us imagine a j-space whose coordinate axes are j_x, j_y, and j_z; Fig. 16.6 shows part of the j_x-j_y plane of such a space. Each point in the j-space corresponds to a permissible set of j_x, j_y, j_z values and thus to a standing wave. If j is a vector from the origin to a particular point j_x, j_y, j_z, its magnitude is

16.14
$$j = \sqrt{j_x{}^2 + j_y{}^2 + j_z{}^2}$$

The total number of wavelengths between λ and $\lambda + d\lambda$ is the same as the number of points in j-space whose distances from the origin lie between j and $j + dj$. The volume of a spherical shell of radius j and thickness dj is $4\pi j^2\, dj$, but we are only interested in the octant of this shell that includes positive values of j_x, j_y, and j_z. Hence

16.15
$$N(j)\, dj = \frac{1}{8} \times 4\pi j^2\, dj = \frac{\pi}{2} j^2\, dj$$

From Eqs. 16.13 and 16.14 we have

16.16
$$j = \frac{2L}{\lambda}$$

16.17
$$dj = -\frac{2L}{\lambda^2}\, d\lambda$$

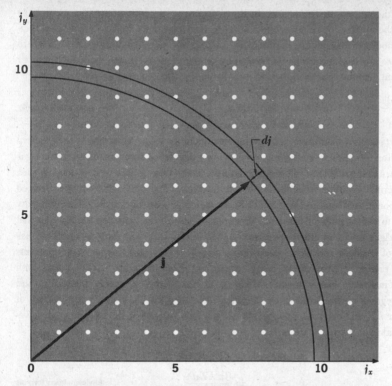

FIGURE 16.6   Each point in j-space corresponds to a possible standing wave.

Since an increase in λ corresponds to a decrease in j, the total number of permissible wavelengths in the cavity is

$$N(\lambda)\, d\lambda = -\, N(j)\, dj$$

$$= \frac{\pi}{2}\left(\frac{2L}{\lambda}\right)^2 \frac{2L}{\lambda^2}\, d\lambda$$

16.18
$$= \frac{4\pi L^3\, d\lambda}{\lambda^4}$$

The cavity volume is L^3, which means that the number of standing waves per unit volume is

$$n(\lambda)\, d\lambda = \frac{1}{L^3}\, N(\lambda)\, d\lambda$$

16.19
$$= \frac{4\pi\, d\lambda}{\lambda^4}$$

Equation 16.19, which was obtained by Rayleigh, is independent of the shape of the

The Rayleigh-Jeans Formula

cavity, even though we used a cubical cavity to facilitate the derivation. The smaller the wavelength, the greater the number of standing waves that occur, in agreement with intuition. Jeans modified this picture by pointing out that for each standing wave, counted in this way, there are two perpendicular directions of polarization. Hence the actual number of independent standing waves in the cavity is twice the above figure, or

16.20
$$n(\lambda) \, d\lambda = \frac{8\pi \, d\lambda}{\lambda^4}$$
Standing waves in a cavity

The next step is to find the average energy per standing wave. According to the theorem of equipartition of energy, the average energy per degree of freedom of an entity (gas molecule, harmonic oscillator, rigid rotor, and so forth) that is part of a system of such entities in thermal equilibrium at the absolute temperature T is $\frac{1}{2}kT$. (This result is derived for the harmonic oscillator in Sec. 19.2.) Each standing wave in a radiation-filled cavity corresponds to two degrees of freedom. This conclusion follows from the observation that each wave originates in an atomic oscillator in the cavity wall, and such an oscillator has two degrees of freedom, one that represents its kinetic energy and another that represents its potential energy. (Alternatively, we note that an electromagnetic wave contains both electric and magnetic energy, again adding up to two modes of energy possession and hence two degrees of freedom.) Each standing wave therefore has an average energy of $2 \times \frac{1}{2}kT$ or kT, and the total energy $u(\lambda) \, d\lambda$ per unit volume in the cavity in the wavelength interval from λ to $\lambda + d\lambda$ is

$$u(\lambda) \, d\lambda = kT \, n(\lambda) \, d\lambda$$

16.21
$$= \frac{8\pi \, kT \, d\lambda}{\lambda^4}$$
Rayleigh-Jeans formula

Equation 16.21, the *Rayleigh-Jeans formula*, is what we have been seeking, an expression for the shape of the black-body spectrum.

To express the Rayleigh-Jeans formula in terms of frequency ν rather than wavelength, we observe that, since $\lambda = c/\nu$,

$$d\lambda = -\frac{c \, d\nu}{\nu^2}$$

and, since an increase in frequency corresponds to a decrease in wavelength,

$$u(\lambda) \, d\lambda = -u(\nu) \, d\nu$$

Therefore

16.22
$$u(\nu) \, d\nu = \frac{8\pi \, \nu^2 kT \, d\nu}{c^3}$$
Rayleigh-Jeans formula

A glance at either statement of the Rayleigh-Jeans formula—which is a rigorous consequence of classical physics—reveals that it is hopelessly invalid. As ν increases, corresponding to the ultraviolet region of the spectrum, Eq. 16.22 suggests that the rate at which energy is radiated increases as ν^2, and in the limit of infinitely high frequency, $u(\nu)\, d\nu$ approaches infinity. In reality, as we can see from Fig. 16.4, $u(\nu)\, d\nu \to 0$ as $\nu \to 0$. The discrepancy between theory and experiment was at once recognized as crucial, and it became known as the "ultraviolet catastrophe."

16.4   The Planck Radiation Formula

Where did Rayleigh and Jeans go wrong? The answer is that the equipartition theorem is only valid for a continuous distribution of possible energies, while the energy content of an electromagnetic wave of frequency ν is (as we now know) actually quantized in units of $h\nu$. Let us determine the average energy $\overline{E}$ of an oscillator limited to the energies $E_n = nh\nu$, where $n = 0, 1, 2, \ldots, \infty$. The *relative* probability that the oscillator have the energy E_n at the temperature T is given by the Boltzmann factor $e^{-E_n/kT}$. To find $\overline{E}$, what we must do is sum $E_n e^{-E_n/kT}$ over all n from $n = 0$ to $n = \infty$ and then divide the result by the sum of $e^{-E_n/kT}$ over all n to normalize the result. In this way we obtain

$$\overline{E} = \frac{\sum\limits_{n=0}^{\infty} E_n\, e^{-E_n/kT}}{\sum\limits_{n=0}^{\infty} e^{-E_n/kT}}$$

$$= \frac{\sum\limits_{n=0}^{\infty} nh\nu\, e^{-nh\nu/kT}}{\sum\limits_{n=0}^{\infty} e^{-nh\nu/kT}}$$

$$= \frac{h\nu\, (e^{-h\nu/kT} + 2e^{-2h\nu/kT} + \cdots)}{(1 + e^{-h\nu/kT} + e^{-2h\nu/kT} + \cdots)}$$

$$= h\nu\, \frac{d}{d(-h\nu/kT)} \ln (1 + e^{-h\nu/kT} + e^{-2h\nu/kT} + \cdots)$$

The quantity in parenthesis in the preceding equation is a geometric series in which the ratio between successive terms is $e^{-h\nu/kT}$. The sum of the series is $1/(1 - e^{-h\nu/kT})$, and so

$$\overline{E} = h\nu\, \frac{d}{d(-h\nu/kT)} \ln \frac{1}{1 - e^{-h\nu/kT}}$$

16.23
$$= \frac{h\nu}{e^{h\nu/kT} - 1}$$ **Average energy of quantized oscillator**

The resulting formula for the spectral energy density in a cavity, which was obtained in 1900 by Max Planck, is

$$16.24 \qquad\qquad u(\nu)\, d\nu = \frac{8\pi h}{c^3} \frac{\nu^3\, d\nu}{e^{h\nu/kT} - 1} \qquad\qquad \textbf{Planck radiation formula}$$

Planck's formula is in excellent accord with the data, and its derivation, buttressed by the later (1905) quantum theory of the photoelectric effect due to Einstein, represented the initial step in the development of what has come to be called modern physics.

We shall now obtain Eq. 16.24 by means of the Bose-Einstein distribution law as an exercise in the methods of quantum statistics. As before, our model of a black body will consist of a cavity in some opaque material. This cavity has some volume V, and it contains a large number of indistinguishable photons of various frequencies. Photons do not obey the exclusion principle, and so they are Bose particles that follow the Bose-Einstein distribution law. The a priori probability $g(p)$ that a photon have a momentum between p and $p + dp$ is equal to twice the number of cells in phase space within which such a photon may exist. The reason for the possible double occupancy of each cell is that photons of the same frequency can have two different directions of polarization (circularly clockwise and circularly counterclockwise). Hence, using the argument that led to Eq. 15.22,

$$g(p)\, dp = \frac{8\pi V p^2\, dp}{h^3}$$

Since the momentum of a photon is $p = h\nu/c$,

$$p^2\, dp = \frac{h^3 \nu^2\, d\nu}{c^3}$$

and

$$16.25 \qquad\qquad g(\nu)\, d\nu = \frac{8\pi V}{c^3} \nu^2\, d\nu$$

We must now evaluate the lagrangian multiplier α in Eq. 16.8. To do this, we note that the number of photons in the cavity need *not* be conserved. Unlike gas molecules or electrons, photons may be created and destroyed, and so, while the total radiant energy within the cavity must remain constant, the number of photons that incorporate this energy can change. For instance, two photons of energy $h\nu$ can be emitted simultaneously with the absorption of a single photon of energy $2h\nu$. Hence

$$\Sigma\, \delta n_i \neq 0$$

which we can express by setting α equal to zero since it multiplies Eq. 16.4.

Substituting Eq. 16.25 for g_i and $h\nu$ for u_i, and letting $\alpha = 0$ in the Bose-Einstein distribution law (Eq. 16.8), we find that the number of photons with frequencies between ν and $\nu + d\nu$ in the radiation within a cavity of volume V whose walls are at the absolute temperature T is

$$16.26 \qquad\qquad n(\nu)\, d\nu = \frac{8\pi V}{c^3} \frac{\nu^2\, d\nu}{e^{h\nu/kT} - 1}$$

The corresponding spectral energy density $u(\nu)\,d\nu$, which is the energy per unit volume in radiation between ν and $\nu + d\nu$ in frequency, is given by

$$u(\nu)\,d\nu = \frac{h\nu n(\nu)\,d\nu}{V}$$

(16.24)
$$= \frac{8\pi h}{c^3}\,\frac{\nu^3\,d\nu}{e^{h\nu/kT} - 1}$$

This is the Planck radiation formula, which agrees with experiment.

Two interesting results can be obtained from the Planck radiation formula. To find the wavelength whose energy density is greatest, we express Eq. 16.24 in terms of wavelength and set

$$\frac{du(\lambda)}{d\lambda} = 0$$

and then solve for $\lambda = \lambda_{max}$. We obtain

$$\frac{hc}{kT\lambda_{max}} = 4.965$$

which is more conveniently expressed as

$$\lambda_{max}T = \frac{hc}{4.965k}$$

16.27
$$= 2.898 \times 10^{-3}\,\text{m}^\circ\text{K}$$

Equation 16.27 is known as *Wien's displacement law.* It quantitatively expresses the empirical fact that the peak in the black-body spectrum shifts to progressively shorter wavelengths (higher frequencies) as the temperature is increased.

Another result we can obtain from Eq. 16.24 is the total energy density u within the cavity. This is the integral of the energy density over all frequencies,

$$u = \int_0^\infty u(\nu)\,d\nu$$

$$= \frac{8\pi^5 k^4}{15c^3 h^3}\,T^4$$

$$= aT^4$$

where a is a universal constant. The total energy density is proportional to the fourth power of the absolute temperature of the cavity walls. We therefore expect that the energy e radiated by a black body per second per unit area is also proportional to T^4, a conclusion embodied in the *Stefan-Boltzmann law:*

16.28
$$e = \sigma T^4$$

The Planck Radiation Formula

The value of Stefan's constant σ is

$$\sigma = 5.67 \times 10^{-8} \ \text{watt/m}^2 \ {}^\circ\text{K}^4$$

Both Wien's displacement law and the Stefan-Boltzmann law are evident in qualitative fashion in Fig. 16.4; the maxima in the various curves shift to higher frequencies and the total areas underneath them increase rapidly with rising temperature.

16.5 Fermi-Dirac Statistics

Fermi-Dirac statistics apply to indistinguishable particles which are governed by the exclusion principle. Our derivation of the Fermi-Dirac distribution law will therefore parallel that of the Bose-Einstein distribution law except that now each cell (that is, quantum state) can be occupied by at most one particle.

If there are g_i cells having the same energy u_i and n_i particles, n_i cells are filled and $(g_i - n_i)$ are vacant. The g_i cells can be rearranged in $g_i!$ different ways, but the $n_i!$ permutations of the filled cells among themselves are irrelevant here since the particles are indistinguishable; the $(g_i - n_i)!$ permutations of the vacant cells among themselves are also irrelevant since the cells are not occupied. The number of distinguishable arrangements of the particles among the cells is therefore

$$\frac{g_i!}{n_i!(g_i - n_i)!}$$

The probability W of the entire distribution of particles is the product

16.29
$$W = \Pi \ \frac{g_i!}{n_i!(g_i - n_i)!}$$

Taking the natural logarithm of both sides,

$$\ln W = \Sigma \ [\ln g_i! - \ln n_i! - \ln (g_i - n_i)!]$$

which Stirling's formula

$$\ln n! = n \ln n - n$$

permits us to rewrite as

16.30
$$\ln W = \Sigma \ [g_i \ln g_i - n_i \ln n_i - (g_i - n_i) \ln (g_i - n_i)]$$

For this distribution to represent maximum probability, small changes δn_i in any of the individual n_i's must not alter W. Hence

16.31
$$\delta \ln W_{\text{max}} = \Sigma \ [-\ln n_i + \ln (g_i - n_i)] \ \delta n_i = 0$$

As before, we take into account the conservation of particles and of energy by adding

$$-\alpha \Sigma \, \delta n_i = 0$$

and

$$-\beta \Sigma \, u_i \delta n_i = 0$$

to Eq. 16.31, with the result that

16.32 $$\Sigma \, [-\ln n_i + \ln (g_i - n_i) - \alpha - \beta u_i] \, \delta n_i = 0$$

Since the δn_i's are independent, the quantity in brackets must vanish for each value of i, and so

$$\ln \frac{g_i - n_i}{n_i} - \alpha - \beta u_i = 0$$

$$\frac{g_i}{n_i} - 1 = e^\alpha e^{\beta u_i}$$

16.33 $$n_i = \frac{g_i}{e^\alpha e^{\beta u_i} + 1}$$

Substituting

$$\beta = \frac{1}{kT}$$

yields the *Fermi-Dirac distribution law,*

16.34 $$n_i = \frac{g_i}{e^\alpha e^{u_i/kT} + 1}$$ **Fermi-Dirac distribution law**

The most important application of the Fermi-Dirac distribution law is in the free-electron theory of metals, which we shall examine in Chap. 19.

16.6 Comparison of Results

The three statistical distribution laws are as follows:

16.35 $$n_i = \frac{g_i}{e^\alpha e^{u_i/kT}}$$ **Maxwell-Boltzmann**

16.36 $$n_i = \frac{g_i}{e^\alpha e^{u_i/kT} - 1}$$ **Bose-Einstein**

16.37 $$n_i = \frac{g_i}{e^\alpha e^{u_i/kT} + 1}$$ **Fermi-Dirac**

In these formulas n_i is the number of particles whose energy is u_i and g_i is the number of states that have the same energy u_i. The quantity

16.38
$$f(u_i) = \frac{n_i}{g_i}$$
Occupation index

called the *occupation index* of a state of energy u_i, is therefore the average number of particles in each of the states of that energy. The occupation index does not depend upon how the energy levels of a system of particles are distributed, and for this reason it provides a convenient way of comparing the essential natures of the three distribution laws.

The Maxwell-Boltzmann occupation index is plotted in Fig. 16.7 for three different values of T and α. This index is a pure exponential, dropping by the factor $1/e$ for each increase in u_i of kT. While $f(u_i)$ depends upon the parameter α, the *ratio* between the occupation indices $f(u_i)$ and $f(u_j)$ of the two energy levels u_i and u_j does not;

16.39
$$\frac{f(u_i)}{f(u_j)} = e^{(u_j - u_i)/kT}$$
Boltzmann factor

This formula is useful because under certain circumstances the Bose-Einstein and Fermi-Dirac distributions resemble the Maxwell-Boltzmann distribution, and it then permits us to determine the relative degrees of occupancy of two quantum states.

The Bose-Einstein occupation index is plotted in Fig. 16.8 for temperatures of 1000°K, 5000°K, and 10,000°K, in each case for $\alpha = 0$ (corresponding to a "gas" of photons). When $u_i \gg kT$, the Bose-Einstein distribution approaches the Maxwell-Boltzmann distribution, while when $u_i \ll kT$, the -1 term in the denominator of Eq. 16.36 causes the occupation index of the former distribution to be much greater.

The Fermi-Dirac occupation index is plotted in Fig. 16.9 for four values of T and α. The occupation index never goes above 1, signifying one particle per state, which is a consequence of the obedience of Fermi particles to the exclusion principle. At low temperatures virtually all of the lower energy states are filled, with the occupation index dropping rapidly

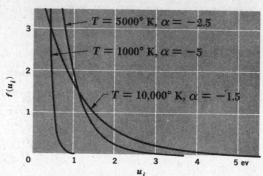

FIGURE 16.7 The occupation indexes for three Maxwell-Boltzmann distributions.

Maxwell-Boltzmann statistics

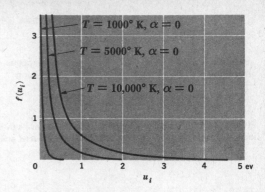

FIGURE 16.8 The occupation indexes
for three Bose-Einstein distributions.

Bose-Einstein statistics

near a certain critical energy known as the *Fermi energy*. At high temperatures the occupation index is sufficiently small at all energies for the effects of the exclusion principle to be unimportant, and the Fermi-Dirac distribution becomes similar to the Maxwell-Boltzmann one.

16.7 Transitions between States

In this chapter and the previous one we have been discussing particles of various kinds that can exist in different energy states and have considered the relative populations of these states in an assembly of particles of each kind. We shall now draw upon some of these ideas to look into the transitions between different energy states in the special case of atoms.

Let us consider two energy levels in a particular atom, a lower one i and an upper one j (Fig. 16.10). If the atom is initially in state i, it can be raised to state j by absorbing a photon of light whose frequency is

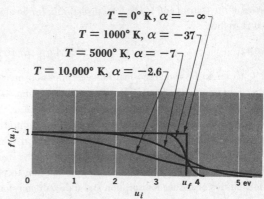

FIGURE 16.9 The occupation indexes for three Fermi-Dirac distributions.

Fermi-Dirac statistics

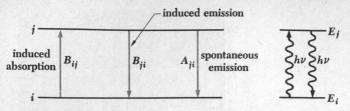

FIGURE 16.10 Transitions between two energy levels in an atom can occur by induced absorption, spontaneous emission, and induced emission.

16.40
$$\nu = \frac{E_j - E_i}{h}$$

(We shall neglect recoil effects in this analysis.) The likelihood that the atom will actually undergo the transition is proportional to the rate at which photons of frequency ν fall on it and therefore to the spectral energy density $u(\nu)$. Of course, the transition probability also depends upon the properties of states i and j, but we can include this dependence for any specific pair of states in some constant B_{ij}. Hence if we shine light of frequency ν and energy density $u(\nu)$ on the atom when it is in the lower state i, the probability for it to go to the higher state j is

16.41
$$P_{i \to j} = B_{ij} u(\nu) \qquad \text{Absorption}$$

If the atom is initially in the upper state j, it has a certain probability A_{ji} to spontaneously drop to state i by emitting a photon of frequency ν. Let us also suppose that shining light of frequency ν on the atom when it is in the upper state somehow helps induce its transition to the lower state. A spectral energy density of $u(\nu)$ therefore means a probability for *induced emission* of $B_{ji} u(\nu)$, where B_{ji}, like B_{ij} and A_{ji}, depends upon the detailed properties of the states i and j. The total probability for an atom in state j to fall to the lower state i is therefore

16.42
$$P_{j \to i} = A_{ji} + B_{ji} u(\nu) \qquad \text{Emission}$$

(It might be remarked that induced emission involves no novel concepts. Let us consider a harmonic oscillator, for instance a pendulum, which has a sinusoidal force applied to it whose period is the same as its natural period of vibration. If the applied force is exactly in phase with the pendulum swings, the amplitude of the latter increases; this corresponds to induced absorption of energy. However, if the applied force is 180° out of phase with the pendulum swings, the amplitude of the latter *decreases*; this corresponds to induced emission of energy. If the applied force is random in phase relative to the pendulum, the probabilities of induced absorption and induced emission are equal—that is, if a number of trials are made, on the average half the trials will result in absorption and half in emission of energy. This corresponds to the fact that $B_{ij} = B_{ji}$, which is demonstrated more formally in what follows. In any case, we have not assumed that induced emission *does* occur, but only that it *may* occur. If we have been mistaken, we should ultimately find merely that $B_{ji} = 0$.)

Next we consider an assembly of N_i atoms in state i and N_j atoms in state j, all in thermal equilibrium at the temperature T with light of frequency ν and energy density $u(\nu)$. The number of atoms in state i that absorb a photon and go to state j per second is

$$N_i P_{i \to j} = N_i B_{ij} u(\nu)$$

while the corresponding number in state j that drop to state i either by spontaneously emitting a photon or by being induced to do so is

$$N_j P_{j \to i} = N_j [A_{ji} + B_{ji} u(\nu)]$$

The opposite processes that participate in an equilibrium must occur equally often if that equilibrium is to continue, and so

$$N_i P_{i \to j} = N_j P_{j \to i}$$
$$N_i B_{ij} u(\nu) = N_j [A_{ji} + B_{ji} u(\nu)]$$

Dividing both sides of the latter equation by $N_j B_{ji}$ and solving it for $u(\nu)$, we obtain

$$\left(\frac{N_i}{N_j} \right) \left(\frac{B_{ij}}{B_{ji}} \right) u(\nu) = \frac{A_{ji}}{B_{ji}} + u(\nu)$$

16.43
$$u(\nu) = \frac{A_{ji}/B_{ji}}{\left(\dfrac{N_i}{N_j} \right) \left(\dfrac{B_{ij}}{B_{ji}} \right) - 1}$$

Finally we draw upon Eq. 16.39 for the ratio between the populations of states i and j to obtain

$$\frac{N_i}{N_j} = e^{(E_j - E_i)/kT} = e^{h\nu/kT}$$

with the result that

16.44
$$u(\nu) = \frac{A_{ji}/B_{ji}}{\left(\dfrac{B_{ij}}{B_{ji}} \right) e^{h\nu/kT} - 1}$$

Equation 16.44 is a formula for the energy density of photons of frequency ν in equilibrium at the temperature T with atoms whose possible energies are E_i and E_j. We can see at once that, if this formula is to be consistent with the Planck radiation formula of Eq. 16.24, it must be true that

16.45
$$B_{ij} = B_{ji}$$

16.46
$$\frac{A_{ji}}{B_{ji}} = \frac{8\pi h\nu^3}{c^3}$$

The preceding analysis, which was first carried out by Einstein in 1917, thus not only confirms that induced emission can occur, but also shows that its coefficient for a transition between two states is the same as the coefficient for induced absorption. In addition, there is a definite ratio between the spontaneous emission and induced emission coefficients that varies with ν^3, so that the relative likelihood of spontaneous emission increases rapidly with the energy difference between the two states. All we need know is one of the coefficients A_{ji}, B_{ij}, or B_{ji} to find the others.

From the discussion of Sec. 16.6 we note that, since $\alpha = 0$ for a photon gas, the average number N_ν of photons of frequency ν at the temperature T is

16.47
$$N_\nu = f(u_\nu) = \frac{1}{e^{h\nu/kT} - 1}$$ **Average number of photons**

Thus Planck's formula can be expressed in terms of N_ν as

16.48
$$u(\nu) = \frac{8\pi h\nu^3}{c^3} N_\nu$$

It is instructive to rewrite Eq. 16.41 and 16.42 for the transition probabilities $P_{i\to j}$ and $P_{j\to i}$ in terms of N_ν and one of the transition coefficients, say B_{ij}. The result is

16.49
$$P_{i\to j} = \left(\frac{8\pi h\nu^3}{c^3} B_{ij}\right) N_\nu$$

$$P_{j\to i} = \left(\frac{8\pi h\nu^3}{c^3} B_{ij}\right) + \left(\frac{8\pi h\nu^3}{c^3} B_{ij}\right) N_\nu$$

16.50
$$= \left(\frac{8\pi h\nu^3}{c^3} B_{ij}\right) (N_\nu + 1)$$

If there are N_ν photons present, the probability that an atom in the lower state i will absorb a photon is proportional to N_ν, and the probability that an atom in the upper state j will emit a photon is proportional to $(N_\nu + 1)$. Evidently the process of spontaneous emission, which contributes the additive factor 1 in the latter case, is intimately related to the processes of induced emission and induced absorption. Induced emission and absorption can be understood in a straightforward way by considering the interaction between an atom and an electromagnetic wave of frequency ν, but spontaneous emission occurs in the absence of any such wave, apparently by a comparable interaction to judge by the above formulas. This paradox is removed by a quantum-theoretical treatment of the electromagnetic field, which shows that actual fields constantly fluctuate about what would be expected on classical grounds. The fluctuations occur even when electromagnetic waves are absent and when, classically, $\mathbf{E} = \mathbf{B} = 0$, and it is these fluctuations (often called "vacuum fluctuations" and analogous in a sense to the zero-point vibrations of a harmonic oscillator) that induce the "spontaneous" emission of photons by atoms in excited states.

16.8 Masers and Lasers

Since $h\nu$ is normally much greater than kT for atomic and molecular radiations, at thermal equilibrium the population of upper energy states in an atomic system is considerably smaller than that of the lowest state. Suppose we shine light of frequency ν upon a system in which the energy difference between the ground state E_0 and an excited state E_1 is $E_1 - E_0 = h\nu$. With the upper state largely unoccupied, there will be little stimulated emission, and the chief events that occur will be absorption of incident photons by atoms in the ground state and the subsequent spontaneous reradiation of photons of the same frequency. (A certain proportion of excited atoms will give up their energies in collisions.)

Certain atomic systems can sustain inverted energy populations, with an upper state occupied to a greater extent than the ground state. (This situation corresponds to a negative absolute temperature T in Eq. 16.39, which is an interesting notion.) Figure 16.11 shows a three-level system in which the intermediate state 1 is metastable (Sec. 11.4), which means that the transition from it to the ground state is forbidden by selection rules. The system can be "pumped" to the upper state 2 by radiation of frequency $\nu' = (E_2 - E_0)/h$. Atoms in state 2 have lifetimes of about 10^{-8} sec against spontaneous emission via an allowed transition, so they fall to the metastable state 1 (or to the ground state) almost at once. Metastable states may have lifetimes of well over 1 sec against spontaneous emission, and it is therefore possible to continue pumping until there is a higher population in state 1 than there is in state 0. If now we direct radiation of frequency $\nu = (E_1 - E_0)/h$ on the system, the induced emission of photons of this frequency will exceed their absorption since more atoms are in the higher state, and the net result will be an output of radiation of frequency ν that exceeds the input. This is the principle of the *maser* (*m*icrowave *a*mplification by *s*timulated *e*mission of *r*adiation) and the laser (*l*ight *a*mplification by *s*timulated *e*mission of *r*adiation).

The radiated waves from spontaneous emission are, as might be expected, incoherent, with random phase relationships in space and time since there is no coordination among the atoms involved. The radiated waves from induced emission, however, are in phase with the inducing waves, which makes it possible for a maser or laser to produce a completely coherent beam. A typical laser is a gas-filled tube or a transparent solid that has mirrors at both ends, one of them partially transmitting to allow some of the light produced to emerge. The pumping light of frequency ν' is directed at the active medium from the sides of the tube, while the back-and-forth traversals of the trapped light stimulate emissions of frequency ν that maintain the emerging beam collimated. A wide variety of masers and lasers have been devised; usually, the required inverted energy distribution is obtained less directly than by the straightforward mechanism described above.

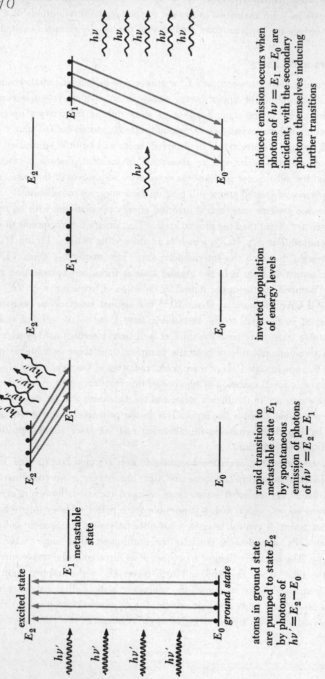

atoms in ground state
are pumped to state E_2
by photons of
$h\nu' = E_2 - E_0$

rapid transition to
metastable state E_1
by spontaneous
emission of photons
of $h\nu'' = E_2 - E_1$

inverted population
of energy levels

induced emission occurs when
photons of $h\nu = E_1 - E_0$ are
incident, with the secondary
photons themselves inducing
further transitions

FIGURE 16.11 Principle of masers and lasers.

Problems

1. Express the Planck radiation law in terms of wavelength rather than frequency and show that, in the limit of $\lambda \to \infty$, it reduces to the Rayleigh-Jeans formula.

2. If the tungsten filament of a light bulb is equivalent to a black body at $2900°K$, find the percentage of the emitted radiant energy in the form of visible light with frequencies between 4×10^{14} and 7×10^{14} cycles/sec.

3. Sunlight arrives at the earth at the rate of about $1,400$ watts/m^2 when the sun is directly overhead. The sun's radius is 6.96×10^8 m and the mean radius of the earth's orbit is 1.49×10^{11} m. From these data find the surface temperature of the sun on the assumption that it radiates like a black body. (The actual surface temperature of the sun is slightly less than this value.)

4. At what temperature are the probabilities for spontaneous and induced emission equal?

5. At the same temperature, will a gas of classical molecules, a gas of *bosons* (particles that obey Bose-Einstein statistics), or a gas of *fermions* (particles that obey Fermi-Dirac statistics) exert the greatest pressure? The least pressure? Why?

6. Derive the Stefan-Boltzmann law in the following way. Consider a Carnot engine that consists of a cylinder and piston whose inside surfaces are perfect reflectors and which uses electromagnetic radiation as its working substance. The operating cycle of this engine has four steps: an isothermal expansion at the temperature T during which the pressure remains constant at p; an adiabatic expansion during which the temperature drops by dT and the pressure drops by dp; an isothermal compression at the temperature $T - dT$ and pressure $p - dp$; and an adiabatic compression to the original temperature, pressure, and volume. The pressure exerted by radiation of energy density u in a container with reflecting walls is $u/3$, and the efficiency of all Carnot engines is $dW/Q = 1 - (T - dT)/T$, where Q is the heat input during the isothermal expansion and dW is the work done by the engine during the entire cycle. Calculate the efficiency of this particular engine in terms of u and T with the help of a p-V diagram and show that $u = aT^4$, where a is a constant.

7. In a continuous helium-neon laser, He and Ne atoms are pumped to metastable states respectively 20.61 and 20.66 ev above their ground states by electron impact. Some of the excited He atoms transfer energy to Ne atoms in collisions, with the 0.05 ev additional energy provided by the kinetic energy of the atoms. An excited Ne atom emits a 6328-A photon in the forbidden transition that leads to laser action. Then a 6680-A photon is emitted in an allowed transition to another metastable state, and the remaining excitation energy is lost in collisions with the tube walls. Find the excitation energies of the two intermediate states in Ne. Why are He atoms needed?

Chapter 17
Bonding
in Solids

A solid consists of atoms, ions, or molecules packed closely together, and their proximity is responsible for the characteristic properties of this state of matter. The ionic and covalent bonds that are involved in the formation of molecules have important counterparts in the solid state. In addition, *van der Waals* and *metallic* bonds provide the cohesive forces in, respectively, molecular crystals and metals. All these bonds are electrostatic in origin, so that the chief distinctions among them lie in the distribution of electrons around the various particles whose regular arrangement forms a crystal.

17.1 Amorphous Solids

The majority of solids are crystalline, with the atoms, ions, or molecules of which they are composed falling into regular, repeated three-dimensional patterns. The presence of *long-range order* is thus the defining property of a crystal. Other solids lack long-range order in the arrangements of their constituent particles and may properly be regarded as supercooled liquids whose stiffness is due to an exceptionally high viscosity. Glass, pitch, and many plastics are examples of such amorphous ("without form") solids.

Amorphous solids do exhibit *short-range order* in their structures, however. The distinction between the two kinds of order is nicely exhibited in boron trioxide (B_2O_3), which can occur in both crystalline and amorphous forms. In each case every boron atom is surrounded by three oxygen atoms, which represents a short-range order. In a B_2O_3 crystal the oxygen atoms are present in hexagonal arrays, as in Fig. 17.1, which is a long-range ordering, while amorphous B_2O_3, a vitreous or "glassy" substance, lacks this additional regularity.

The analogy between an amorphous solid and a liquid is worth pursuing as a means of better understanding both states of matter. Liquids are usually regarded as resembling gases more closely than solids; after all, liquids and gases are both fluids, and at temperatures above the critical point the two become indistinguishable. However, from a microscopic point of view, liquids and solids also have much in common. The density of a given liquid is usually close to that of the corresponding solid, for instance, which suggests that the degree of packing is similar, an inference supported by the compressibilities of these

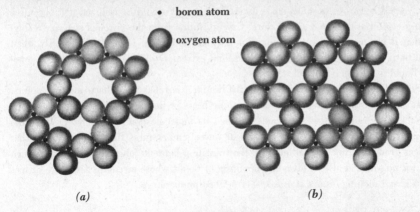

boron atom

oxygen atom

(a) *(b)*

FIGURE 17.1   *(a)* Amorphous B_2O_3 exhibits only short-range order. *(b)* Crystalline B_2O_3 exhibits long-range order as well.

states. Furthermore, X-ray diffraction indicates that many liquids have definite short-range structures at any instant, quite similar to those of amorphous solids except that the groupings of liquid molecules are continually shifting.

A conspicuous example of short-range order in a liquid occurs in water just above the melting point, where the result is a lower density than at higher temperatures because H_2O molecules are less tightly packed when linked in crystals than when free to move (Sec. 17.5). Even more striking are "liquid crystals" which exhibit double refraction— an attribute ordinarily confined to anisotropic crystals such as those of quartz and calcite— over a limited temperature range after melting. For instance, the fixed, regular arrangement of the molecules in crystalline *p*-azoxyanisole disappears at the melting point of 116°C, but liquid *p*-azoxyanisole retains its ability to produce double refraction until 135°C. The conclusion is that, while able to move about, the elongated *p*-azoxyanisole molecules nevertheless remain parallel to one another between 116°C and 135°C to yield an anisotropic liquid.

Now let us see what happens when a liquid is cooled to its freezing point. The transformation from liquid to solid begins with the appearance of minute crystals called *nuclei*, which proceed to grow in size until the transformation is complete. The nuclei may appear spontaneously, or foreign particles such as dust grains or impurity crystals may act as nuclei. If the liquid is pure and undisturbed, it may be possible to cool it below the freezing point without crystallization. Ordinarily such a supercooled liquid is unstable and will crystallize at the slightest disturbance, but in some cases the viscosity is so great at the lowered temperature that the molecules are unable to move about to arrange themselves into a crystalline pattern. The result is a solid material whose internal configuration is that of a liquid; no change of state has occurred, but instead one of the transient molecular arrangements of the liquid has been preserved by the reduction in molecular mobility. A

Amorphous Solids

vitreous solid is thus metastable, since its internal energy would be reduced if it became a crystalline solid at the same temperature. Crystallization from the vitreous state is so sluggish that it ordinarily does not occur, but it is not unknown: Glass may devitrify when heated until it has not quite begun to soften, and extremely old glass specimens are sometimes found to have crystallized.

Since amorphous solids are essentially liquids, they have no sharp melting points. We can interpret this behavior on a microscopic basis by noting that, since an amorphous solid lacks long-range order, the bonds between its molecules vary in strength. When the solid is heated, the weakest bonds rupture at lower temperatures than the others, so that it softens gradually. In a crystalline solid the transition between long-range and short-range order (or no order at all) involves the breaking of bonds whose strengths are more or less identical, and melting occurs at a precisely defined temperature.

17.2 Ionic Crystals

Ionic bonds in crystals are very similar to ionic bonds in molecules. Such bonds come into being when atoms which have low ionization energies, and hence lose electrons readily, interact with other atoms which have high electron affinities. The former atoms give up electrons to the latter, and they thereupon become positive and negative ions respectively. In an ionic crystal these ions come together in an equilibrium configuration in which the attractive forces between positive and negative ions predominate over the repulsive forces between similar ions. As in the case of molecules, crystals of all types are prevented from collapsing under the influence of the cohesive forces present by the action of the exclusion principle, which requires the occupancy of higher energy states when electron shells of different atoms overlap and mesh together.

Two common types of structure found in ionic crystals are shown in Figs. 17.2 and 17.3. The arrangement of Na^+ and Cl^- ions in a sodium chloride crystal is illustrated in Fig. 17.2. The ions of either kind may be thought of as being located at the corners and at the centers of the faces of an assembly of cubes, with the Na^+ and Cl^- assemblies interleaved. Each ion thus has six nearest neighbors of the other kind. Such a crystal structure is called *face-centered cubic*. In NaCl crystals the distance between like ions is 5.62 A. A different structure is found in cesium chloride crystals, where each ion is located at the center of a cube at whose corners are ions of the other kind (Fig. 17.3). This structure is called *body-centered cubic*, and each ion has eight nearest neighbors of the other kind. In CsCl crystals the distance between like ions is 4.11 A.

Most ionic solids are hard, owing to the strength of the bonds between their constituent ions, and have high melting points. They are usually brittle as well, since the slipping of atoms past one another that accounts for the ductility of metals is prevented by the ordering of positive and negative ions imposed by the nature of the bonds. Polar liquids such as water are able to dissolve many ionic crystals, but covalent liquids such as gasoline generally cannot.

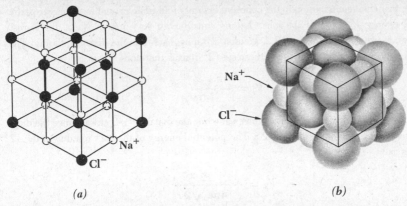

FIGURE 17.2   (a) The face-centered cubic structure of a NaCl crystal. The coordination number (number of nearest neighbors about each ion) is 6. (b) Scale model of NaCl crystal.

The *cohesive energy* of an ionic crystal is the energy that would be liberated by the formation of the crystal from individual neutral atoms. Cohesive energy is usually expressed in ev/atom, in ev/molecule, or in kcal/mole, where "molecule" and "mole" here refer to sets of atoms specified by the formula of the compound involved (for instance NaCl in the case of a sodium chloride crystal) even though molecules as such do not exist in the crystal. The bond energy of a molecule held together by an ionic bond is not the same as the cohesive energy of the corresponding crystal, because in the crystal each ion interacts with all the other ions present and not just with one, two, or three of the opposite sign.

FIGURE 17.3   (a) The body-centered cubic structure of a CsCl crystal. The coordination number is 8. (b) Scale model of CsCl crystal.

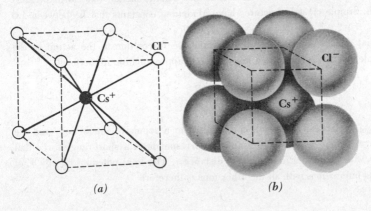

The principal contribution to the cohesive energy of an ionic crystal is the electrostatic potential energy $V_{coulomb}$ of the ions. Let us consider an Na^+ ion in NaCl. Its nearest neighbors, as we can see in Fig. 17.2, are six Cl^- ions, each one the distance r away. The potential energy of the Na^+ ion due to these six Cl^- ions is therefore

$$V_1 = -\frac{6e^2}{4\pi\epsilon_0 r}$$

The next nearest neighbors are 12 Na^+ ions, each one the distance $\sqrt{2}\, r$ away since the diagonal of a square r long on a side is $\sqrt{2}\, r$. The potential energy of the Na^+ ion due to the 12 Na^+ ions is

$$V_2 = +\frac{12e^2}{4\pi\epsilon_0 \sqrt{2}\, r}$$

When the summation is continued over all the + and − ions in a crystal of infinite size, the result is

$$V_{coulomb} = -\frac{e^2}{4\pi\epsilon_0 r}\left(6 - \frac{12}{\sqrt{2}} + \cdots\right)$$

$$= -1.748\,\frac{e^2}{4\pi\epsilon_0 r}$$

17.1 $$= -\alpha\,\frac{e^2}{4\pi\epsilon_0 r}$$ **Coulomb energy of ionic crystal**

(This result holds for the potential energy of a Cl^- ion as well, of course.) The quantity α is called the *Madelung constant* of the crystal, and it has the same value for all crystals of the same kind. Similar calculations for other crystal varieties yield different Madelung constants: crystals whose structures are like that of cesium chloride (Fig. 17.3), for instance, have $\alpha = 1.763$, and those with structures like that of zinc blende (one form of ZnS) have $\alpha = 1.638$. Simple crystal structures have Madelung constants that lie between 1.6 and 1.8.

The potential energy contribution of the repulsive forces due to the action of the exclusion principle can be expressed to a fair degree of approximation in the form

17.2 $$V_{repulsive} = \frac{B}{r^n}$$

The sign of $V_{repulsive}$ is positive, which corresponds to a repulsive interaction, and the dependence on r^{-n} (where n is a large number) corresponds to a short-range force that increases rapidly with decreasing internuclear distance r. The total potential energy V of each ion due to its interactions with all the other ions is therefore

$$V = V_{\text{coulomb}} + V_{\text{repulsive}}$$

17.3
$$= -\frac{\alpha e^2}{4\pi\varepsilon_0 r} + \frac{B}{r^n}$$

At the equilibrium separation r_0 of the ions, V is a minimum by definition, and so $(dV/dr) = 0$ when $r = r_0$. Hence

$$\left(\frac{dV}{dr}\right)_{r=r_0} = \frac{\alpha e^2}{4\pi\varepsilon_0 r_0^2} - \frac{nB}{r_0^{n+1}} = 0$$

$$\frac{\alpha e^2}{4\pi\varepsilon_0 r_0^2} = \frac{nB}{r_0^{n+1}}$$

17.4
$$B = \frac{\alpha e^2}{4\pi\varepsilon_0 n} r_0^{n-1}$$

and the total potential energy is

17.5
$$V = -\frac{\alpha e^2}{4\pi\varepsilon_0 r_0}\left(1 - \frac{1}{n}\right)$$

It is possible to evaluate the exponent n from the observed compressibilities of ionic crystals. The average result is $n \approx 9$, which means that the repulsive force varies quite sharply with r: the ions are "hard" rather than "soft" and strongly resist being packed too tightly. At the equilibrium ion spacing, the mutual repulsion due to the exclusion principle (as distinct from the electrostatic repulsion between like ions) decreases the potential energy by about 11 percent. A really precise knowledge of n is evidently not essential; if $n = 10$ instead of $n = 9$, V would change by only 1 percent.

In an NaCl crystal, the equilibrium distance r_0 between ions is 2.81 A. Since $\alpha = 1.748$ and $n = 9$, the potential energy of an ion of either sign is

$$V = -\frac{\alpha e^2}{4\pi\varepsilon_0 r_0}\left(1 - \frac{1}{n}\right)$$

$$= -\frac{9 \times 10^9 \text{ n-m}^2/\text{coulomb}^2 \times 1.748 \times (1.60 \times 10^{-19} \text{ coulomb})^2}{2.81 \times 10^{-10} \text{ m}}\left(1 - \frac{1}{9}\right)$$

$$= -1.27 \times 10^{-18} \text{ joule}$$

$$= -7.97 \text{ ev}$$

Because we may not count each ion more than once, only half of this potential energy, or -3.99 ev, represents the contribution *per ion* to the cohesive energy of the crystal.

We must also take into account the energy needed to transfer an electron from a Na atom to a Cl atom to yield a Na^+—Cl^- ion pair. This electron transfer energy is the difference between the $+5.14$-ev ionization energy of Na and the -3.61-ev electron affinity of

Cl, or +1.53 ev. Each atom therefore contributes +0.77 ev to the cohesive energy from this source. The total cohesive energy per atom is thus

$$E_{\text{cohesive}} = (-3.99 + 0.77) \text{ ev/atom} = -3.22 \text{ ev/atom}$$

As we learned in Sec. 12.6, the bond energy of an actual NaCl molecule is 4.2 ev, or 2.1 ev/atom. Because the cohesive energy in crystalline NaCl is so much greater than the bond energy in molecular NaCl, discrete NaCl molecules are found only in sodium chloride vapor.

An empirical figure for the cohesive energy of an ionic crystal can be obtained from measurements of its heat of vaporization, dissociation energy, and electron exchange energy. The result for NaCl is 3.28 ev, in close agreement with the calculated value. (Sometimes the *lattice energy*, rather than the cohesive energy, is tabulated; the lattice energy is that evolved when a crystal is formed from individual ions, rather than from individual atoms, and is equal to the V of Eq. 17.5. In the case of NaCl, the experimental lattice energy is found to be 183 kcal/mole, which is 7.96 ev/ion pair. The latter figure agrees with the 7.97 ev/ion pair calculated earlier.)

There are two other contributions to the cohesive energy of an ionic crystal that have not been considered in the foregoing: (1) the van der Waals attraction between adjacent ions, to be discussed in Sec. 17.4, which increases the cohesive energy; and (2) the zero-point oscillations of the ions about their equilibrium positions (Sec. 19.4), which, since they represent a mode of energy possession present in the solid and not in the individual atoms, decrease the cohesive energy. Both of these effects are small in ionic crystals, and usually can be neglected.

17.3 Covalent Crystals

The cohesive forces in covalent crystals arise from the presence of electrons between adjacent atoms. Each atom participating in a covalent bond contributes an electron to the bond, and these electrons are shared by both atoms rather than being the virtually exclusive property of one of them as in an ionic bond. Diamond is an example of a crystal whose atoms are linked by covalent bonds. Figure 17.4 shows the structure of a diamond crystal; the tetrahedral arrangement is a consequence of the ability of each carbon atom to form covalent bonds with four other atoms (see Fig. 13.11). The length of each bond in diamond is 1.54 A.

Purely covalent crystals are relatively few in number. In addition to diamond, some examples are silicon, germanium, and silicon carbide; in SiC each atom is surrounded by four atoms of the other kind in the same tetrahedral structure as that of diamond. All covalent crystals are hard (diamond is the hardest substance known, and SiC is the industrial abrasive carborundum), have high melting points, and are insoluble in all ordinary liquids, behavior which reflects the strength of the covalent bonds. Cohesive energies of 3 to 5 ev/atom are typical of covalent crystals, which is the same order of magnitude as the cohesive energies in ionic crystals.

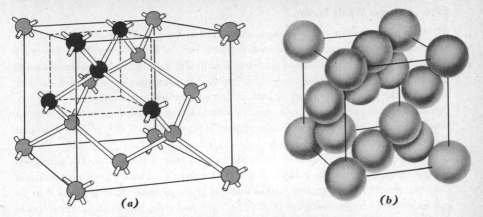

FIGURE 17.4 *(a)* **The tetrahedral structure of diamond. The coordination number is 4.** *(b)* **Scale model of diamond crystal.**

There are several ways to ascertain whether the bonds in a given nonmetallic, non-molecular crystal are predominantly ionic or covalent. In general, a compound of an element from group I or II of the periodic table (Table 10.1) with one from group VI or VII exhibits ionic bonding in the solid state. Another guide is the coordination number of the crystal, which is the number of nearest neighbors about each constituent particle. A high coordination number suggests an ionic crystal, since it is hard to see how an atom can form purely covalent bonds with six other atoms (as in a face-centered cubic structure like that of NaCl) or with eight other atoms (as in a body-centered cubic structure like that of CsCl). A co-ordination number of 4, however, as in the diamond structure, is compatible with an ex-clusively covalent character. To be sure, as with molecules, it is not always possible to classify a particular crystal as being wholly ionic or covalent: AgCl, whose structure is the same as that of NaCl, and CuCl, whose structure resembles that of diamond, both have bonds of intermediate character, as do a great many other solids.

More or less direct methods are also available for investigating the nature of the bonds in a crystal. One involves the mapping of electron density inside a crystal by means of very precise X-ray techniques; when light atoms are involved, for instance in the case of LiF, the presence of discrete positive and negative ions can be verified. Another technique is the elegant one of comparing the dielectric constant K_{lf} of a crystal as measured at low fre-quencies with its dielectric constant K_{hf} as measured at high frequencies, $> 10^{13}$ sec^{-1}. The latter is conveniently obtained from the optical index of refraction n, since $n = \sqrt{K_{hf}}$. The response of an ionic crystal to an applied electric field, and hence its dielectric constant, is different at low frequencies, when the ions themselves can move with the field oscillations, from what it is at high frequencies, when only the electron distribution is affected. In KCl, for example, $K_{lf} = 4.7$ and $K_{hf} = 2.1$. In a covalent crystal, on the other hand, the di-electric constant is independent of frequency.

Covalent Crystals

17.4   Van der Waals Forces

All atoms and molecules—even inert-gas atoms such as those of helium and argon—exhibit weak, short-range attractions for one another due to *van der Waals* forces. These forces are responsible for the condensation of gases into liquids and the freezing of liquids into solids despite the absence of ionic, covalent, or metallic bonding mechanisms. Such familiar aspects of the behavior of matter in bulk as friction, surface tension, viscosity, adhesion, cohesion, and so on also arise from van der Waals forces. As we shall find, the van der Waals attraction between two molecules r apart is proportional to r^{-7}, so it is significant only for molecules very close together.

We begin by noting that many molecules (called *polar molecules*) possess permanent electric dipole moments. An example is the H_2O molecule, in which the concentration of electrons around the oxygen atom makes that end of the molecule more negative than the end where the hydrogen atoms are (Fig. 17.5). Such molecules tend to align themselves so that ends of opposite sign are adjacent, as in Fig. 17.6, and in this orientation the molecules strongly attract each other.

A polar molecule is also able to attract molecules which do not normally have a permanent dipole moment. The process is illustrated in Fig. 17.7: the electric field of the polar molecule causes a separation of charge in the other molecule, with the induced moment the same in direction as that of the polar molecule. The result is an attractive force. (The effect is the same as that involved in the attraction of an unmagnetized piece of iron by a magnet.) It is not difficult to determine the characteristics of this attractive force. The electric field E a distance r from a dipole of moment p is given by

17.6
$$E = \frac{1}{4\pi\varepsilon_0}\left[\frac{p}{r^3} - \frac{3(p \cdot r)}{r^5}r\right]$$

($p \cdot r = pr\cos\theta$, where θ is the angle between p and r.) The field E induces in the other, normally nonpolar molecule an electric dipole moment p' proportional to E in magnitude and ideally in the same direction. Hence

17.7
$$p' = \alpha E$$

where α is a constant called the *polarizability* of the molecule. The energy of the induced dipole in the electric field E is

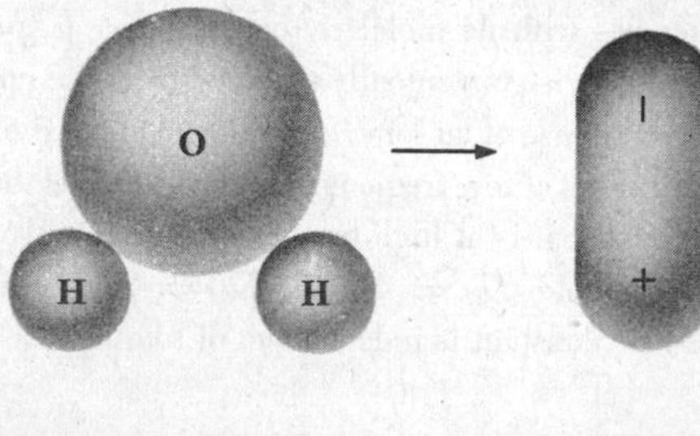

FIGURE 17.5   Water molecules are polar because the shared electrons lie between the oxygen and hydrogen atoms.

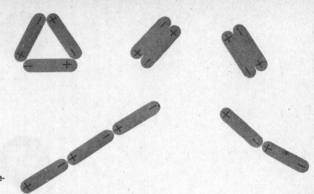

FIGURE 17.6 Polar molecules attract each other.

$$V = -\mathbf{p'} \cdot \mathbf{E}$$

$$= -\frac{\alpha}{(4\pi\varepsilon_0)^2} \left(\frac{p^2}{r^6} - \frac{3p^2}{r^6} \cos^2 \theta - \frac{3p^2}{r^6} \cos^2 \theta + \frac{9p^2}{r^6} \cos^2 \theta \right)$$

17.8
$$= -\frac{\alpha}{(4\pi\varepsilon_0)^2} (1 + 3 \cos^2 \theta) \frac{p^2}{r^6}$$

The mutual energy of the two molecules that arises from their interaction is thus negative, signifying that the force between them is attractive, and is proportional to $1/r^6$. The force itself is equal to $-dV/dr$ and so is proportional to $1/r^7$ which means that it drops rapidly with increasing separation. Doubling the distance between two molecules reduces the attractive force between them to only 0.8 percent of its original value.

Only a limited number of molecules are polar, while van der Waals forces act between *all* molecules. There is no contradiction here, however, because the interaction energy V is proportional to p^2, the square of the dipole moment. Even though the electron distribution in a nonpolar molecule is symmetrical *on the average*, at any instant it is asymmetrical and so possesses an electric dipole moment. This instantaneous moment fluctuates in magnitude

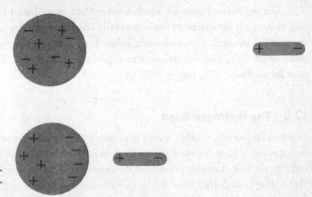

FIGURE 17.7 Polar molecules attract polarizable molecules.

Van der Waals Forces

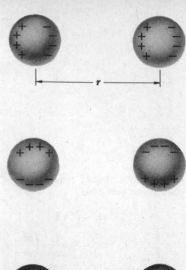

FIGURE 17.8 On the average, nonpolar molecules have symmetric charge distributions, but at any instant the distributions are asymmetric. The fluctuations in the charge distributions of nearby molecules are coordinated as shown, which leads to an attractive force between them whose magnitude is proportional to $1/r^7$.

and direction, but while the average moment $\bar{p}$ is zero, the average of the square of the moment $\overline{p^2}$ is *not* zero but has some finite value. The forces arising from such instantaneous molecular moments according to the mechanism of the previous paragraph are present not only between all molecules but also between all atoms, including those of the rare gases which do not otherwise interact (Fig. 17.8). The values of $\overline{p^2}$ and α are comparable for most molecules; this is part of the reason why the densities and heats of vaporization of liquids, properties that depend upon the strength of intermolecular forces, have a rather narrow range.

Van de Waals forces are much weaker than those found in ionic and covalent bonds, and as a result molecular crystals generally have low melting and boiling points and little mechanical strength. Cohesive energies are low, only 0.08 ev/atom in solid argon (melting point $-189°$C), 0.01 ev/molecule in solid hydrogen (mp $-259°$C), and 0.1 ev/molecule in solid methane, CH_4 (mp $-183°$C).

17.5 The Hydrogen Bond

Certain compounds, notably water, have much higher melting and boiling points than would be expected if they followed the general trend of these figures for other compounds of the same types. For example, in order of decreasing molecular weight, the nonmetallic hydrides H_2Te, H_2Se, and H_2S have the respective boiling points $-2°$C, $-42°$C, and $-60°$C, while H_2O, with a still smaller molecular weight, boils at $100°$C. Similar though less marked anomalies are exhibited by ammonia (NH_3) and hydrogen fluoride (HF). The inter-molecular forces in H_2O, NH_3, and HF are due to *hydrogen bonds*, which are stronger than ordinary van der Waals bonds but weaker than ionic or covalent bonds.

Hydrogen bonds occur in the hydrides of atoms so electronegative that we may think of the molecules as having virtually bare protons present on the outside. More precisely, the electron distribution in such a molecule is so distorted by the affinity of the "parent" atom for electrons that each hydrogen atom in essence has donated most of its negative charge to the parent atom, leaving behind a poorly shielded proton. In the case of HF, the result is a molecule with a strong, localized positive charge at one end which can link up with the less concentrated negative charge at the opposite end of another HF molecule. The key factor here is the small effective size of the poorly shielded proton, since electrostatic forces vary as r^{-2}. So strong are the hydrogen bonds in hydrogen fluoride that even in the gas phase such supermolecules as H_2F_2, H_3F_3, H_4F_4, H_5F_5, and H_6F_6—or, better, HF·HF, HF·HF·HF, and so on—occur frequently.

Water molecules are exceptionally prone to form hydrogen bonds because the four pairs of electrons around the O atom occupy sp^3 hybrid orbitals that project outward as though toward the vertexes of a tetrahedron (Fig. 17.9). Hydrogen atoms are at two of these

FIGURE 17.9 In an H_2O molecule, the four pairs of valence electrons around the oxygen atom (six contributed by the O atom and one each by the H atoms) occupy four sp^3 hybrid orbitals that form a tetrahedral pattern. Each H_2O molecule can form hydrogen bonds with four other H_2O molecules.

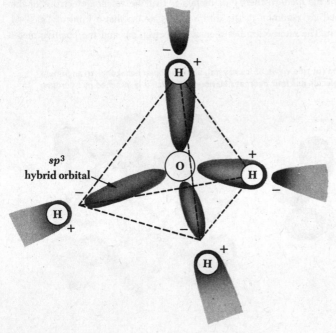

sp^3 hybrid orbital

The Hydrogen Bond

vertexes, which accordingly exhibit localized positive charges, while the other two vertexes exhibit somewhat more diffuse negative charges. Each H_2O molecule can therefore form hydrogen bonds with *four* other H_2O molecules; in two of these bonds the central molecule provides the bridging protons, and in the other two the attached molecules provide them. In the liquid state, the hydrogen bonds between adjacent H_2O molecules are continually being broken and reformed owing to thermal agitation, but even so at any instant the molecules are combined in definite clusters. In the solid state, these clusters are large and stable and constitute ice crystals.

The characteristic hexagonal pattern (Fig. 17.10) of an ice crystal arises from the tetrahedral arrangement of the four hydrogen bonds each H_2O molecule can participate in. With only four nearest neighbors around each molecule, ice crystals have extremely open structures, which is the reason for the exceptionally low density of ice. Because the molecular clusters are smaller and less stable in the liquid state, water molecules on the average are packed more closely together than are ice molecules, and water has the higher density: hence ice floats. The density of water increases from 0°C to a maximum at 4°C as large clusters of H_2O molecules are broken up into smaller ones that occupy less space in the aggregate; only past 4°C does the normal thermal expansion of a liquid manifest itself in a decreasing density with increasing temperature.

17.6 The Metallic Bond

The underlying theme of the modern theory of metals is that the valence electrons of the atoms comprising a metal are common to the entire aggregate, so that a kind of "gas" of free electrons pervades it. The interaction between this electron gas and the positive metal

FIGURE 17.10   (a) Top and (b) side views of ice crystal, showing open hexagonal arrangement of H_2O molecules. Each molecule has four nearest neighbors to which it is attached by hydrogen bonds.

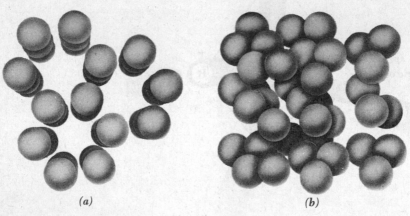

(a)   *(b)*

ions leads to a strong cohesive force. (There is a parallel here with the delocalized π electrons in the benzene molecule.) The presence of such free electrons accounts nicely for the high electrical and thermal conductivities, opacity, surface luster, and other unique properties of metals. To be sure, no electrons in any solid, even a metal, are able to move about its interior with total freedom. All of them are influenced to some extent by the other particles present, and when the theory of metals is refined to include these complications, there emerges a comprehensive picture that is in excellent accord with experiment. In this section we shall confine ourselves to some general observations on the metallic bond, leaving other aspects of the physics of metals to Chaps. 19 and 20.

Some insight into the ability of metal atoms to bond together to form crystals of unlimited size can be gained by viewing the metallic bond as an unsaturated covalent bond. Let us compare the bonding processes in hydrogen and in lithium, both members of group I of the periodic table. A H_2 molecule contains two $1s$ electrons with opposite spins, the maximum number of K electrons that can be present. The H_2 molecule is therefore saturated, since the exclusion principle requires that any additional electrons be in states of higher energy and the stable attachment of further H atoms cannot occur unless their electrons are in $1s$ states. Superficially lithium might seem obliged to behave in a similar way, having the electron configuration $1s^2 2s$. There are, however, six *unfilled* $2p$ states in every Li atom whose energies are only very slightly greater than those of the $2s$ states. When a Li atom comes near a Li_2 molecule, it readily becomes attached with a covalent bond without violating the exclusion principle, and the resulting Li_3 molecule is stable since all its valence electrons remain in L shells. There is no limit to the number of Li atoms that can join together in this way, since lithium forms body-centered cubic crystals (Fig. 17.3) in which each atom has eight nearest neighbors. With only one electron per atom available to enter into bonds, each bond involves one-fourth of an electron on the average instead of two electrons as in ordinary covalent bonds. Hence the bonds are far from being saturated; this is true of the bonds in other metals as well.

One consequence of the unsaturated nature of the metallic bond is the weakness of metals as compared with ionic and covalent crystals having saturated bonds. Another is the ease with which metals can be deformed. Having neither definite localized bonds between adjacent atoms, as in a covalent crystal, nor a configuration of alternating positive and negative ions, as in an ionic crystal, the atoms of a metal can be rearranged in position without rupturing the crystal. For the same reason the properties of a mixture of different metal atoms do not depend critically on the proportion of each kind of atom, provided their sizes are similar. Thus the characteristics of an alloy often vary smoothly with changes in its composition, in contrast to the specific atomic proportions found in ionic solids and in covalent solids such as SiC.

The most striking consequence of the unsaturated bonds in a metal is the ability of the valence electrons to wander freely from atom to atom. To understand this phenomenon intuitively, we can think of each valence electron as constantly moving from bond to bond. In solid Li, each electron participates in eight bonds, so it only spends a short time between any pair of Li⁺ ions. The electron cannot remember (so to speak) which of the two ions it really belongs to, and it is just as likely to move on to a bond that does not involve its parent ion at all. The valence electrons in a metal therefore behave in a manner quite similar to that of molecules in a gas.

As in the case of any other solid, metal atoms cohere because their collective energy is lower when they are bound together than when they exist as separate atoms. To understand why this reduction in energy occurs in a metallic crystal, we note that, because of the proximity of the ions, each valence electron is on the average closer to one nucleus or another than it would be if it belonged to an isolated atom. Hence the potential energy of the electron is less in the crystal than in the atom, and it is this decrease in potential energy that is responsible for the metallic bond.

There is another factor to be considered, however. Whereas the electron potential energy is reduced in a metallic crystal, the electron kinetic energy is increased. The free electrons in a metal constitute a single system of electrons, and the exclusion principle prohibits more than two of them (one with each spin) from occupying each energy level. It would seem at first glance that, again using lithium as an example, only eight valence electrons in an entire Li crystal could occupy $n = 2$ quantum states, with the rest being forced into higher and higher states of such great energy as to disrupt the entire structure. What actually happens is less dramatic. The valence energy levels of the various metal atoms are all slightly altered by their interactions, and an *energy band* comes into being that consists of as many closely spaced energy levels as the total number of valence energy levels in all the atoms in the crystal. The free electrons accordingly range in kinetic energy from 0 to some maximum u_F, called the *Fermi energy*; the Fermi energy in lithium, for example, is 4.72 ev, and the average kinetic energy of the free electrons in metallic lithium is 2.8 ev. (The energy distribution of the electrons in a metal is considered in detail in Chap. 19.) Since electron kinetic energy is a positive quantity, its increase in the metal over what it was in separate atoms leads to a repulsion.

Metallic bonding occurs when the attraction between the positive metal ions and the electron gas exceeds the mutual repulsion of the electrons in that gas; that is, when the reduction in electron potential energy exceeds in magnitude the concomitant increase in electron kinetic energy. The greater the number of valence electrons per atom, the higher the average kinetic energy will be in a metallic crystal, but without a commensurate drop in the potential energy. For this reason the metallic elements are nearly all found in the first three groups of the periodic table. Some elements are right on the border line and may form both metallic and covalent crystals. Tin is a notable example. Above 13.2°C the metal "white tin" exists whose atoms form a body-centered tetragonal structure with a coordination number of 6. Below 13.2°C the covalent solid "gray tin" exists whose structure is the same as that of diamond, with a coordination number of 4. Gray tin and white tin are quite different substances; they have the respective densities of 5.8 and 7.3 g/cm³, for instance, and gray tin is a semiconductor whereas white tin has the typically high electric conductivity of a metal.

TABLE 17.1. CRYSTAL TYPES

Type		Bond	Example	Properties
Ionic	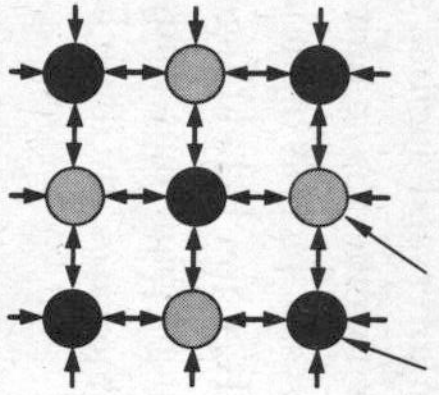 negative ion, positive ion	Electrostatic attraction	Sodium chloride NaCl $E_{cohesive} = 3.28$ ev/atom	Hard; high melting points; may be soluble in polar liquids such as water
Covalent	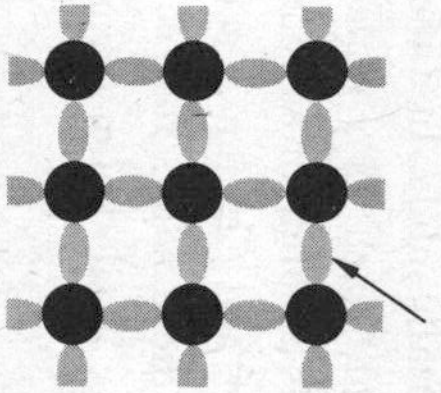 shared electrons	Shared electrons	Diamond C $E_{cohesive} = 7.4$ ev/atom	Very hard; high melting points; insoluble in nearly all solvents
Metallic	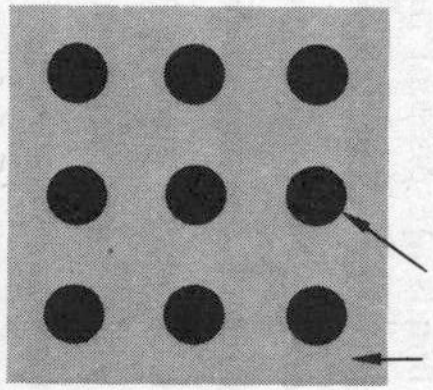 metal ion, electron gas	Electron gas	Sodium Na $E_{cohesive} = 1.1$ ev/atom	Ductile; metallic luster; high electrical and thermal conductivity
Molecular	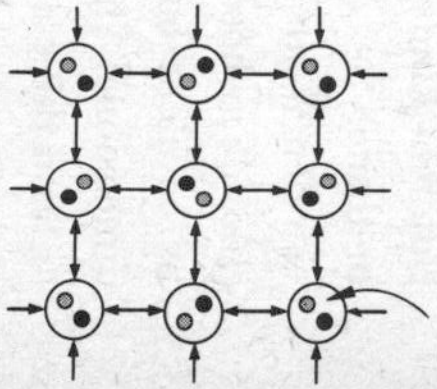 instantaneous charge separation in molecule	Van der Waals forces	Methane CH_4 $E_{cohesive} = 0.1$ ev/atom	Soft; low melting and boiling points; soluble in covalent liquids

The Metallic Bond

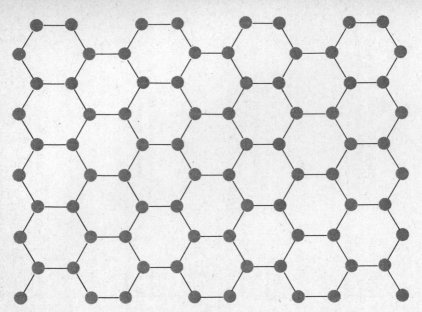

FIGURE 17.11 Graphite consists of layers of carbon atoms in hexagonal arrays, with each atom bonded to three others. The layers are bonded together by weak van der Waals forces.

In general, as has been mentioned, metallic bonds are not as strong as covalent or ionic bonds, though very much stronger than van der Waals bonds. Examples of cohesive energies in metals are 1.4 ev/atom in zinc, 1.6 ev/atom in lithium, 2.0 ev/atom in lead, and 3.5 ev/atom in copper.

17.7 One- and Two-dimensional Crystals

Different kinds of bonds are present in certain solids. In a molecular crystal such as solid methane, for instance, covalent bonds hold each CH_4 molecule together as a unit and van der Waals forces tie the molecules into a solid. In an ionic crystal such as ammonium chloride, whose structure consists of NH_4^+ and Cl^- ions, each NH_4^+ ion is held together by covalent bonds. More interesting are those solids that consist of giant one- and two-dimensional "molecules" joined by van der Waals bonds or, in some cases, by ionic bonds to form three-dimensional structures.

Graphite is composed of layers of carbon atoms with each layer being a flat, hexagonal array of atoms (Fig. 17.11). Each carbon atom is joined to three others, presumably by sp^2 hybrid bonds (Sec. 13.7) since the bonds are 120° apart. The remaining valence

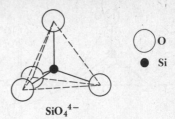

FIGURE 17.12 The Si—O bonds in silicate structures are partly covalent and partly ionic in nature.

SiO_4^{4-}

electrons (one per atom) occupy bonding π orbitals that have concentrations above and below the layer of carbon atoms. The π electrons are delocalized, as in the benzene molecule, and are able to move around the entire layer to which they belong with relative freedom. The bonding in each graphite layer is thus partly covalent and partly metallic in nature. A graphite sample has a substantial electrical conductivity parallel to its layers but a much smaller one across the layers, which agrees with the foregoing picture.

The van der Waals forces that bond the graphite layers together are quite weak, in contrast to the strong bonding forces within the layers. The layers consequently can slide past one another readily and are easily flaked apart, which is why graphite is so useful as a lubricant and in pencils. The internuclear separation in a graphite layer is 1.42 A and the spacing between layers is 3.40 A, while in diamond the internuclear spacing is 1.54 A throughout the crystal (Fig. 17.4). The more open structure of graphite is reflected in its density of 2.25 g/cm^3, as compared with 3.51 g/cm^3 for diamond.

In addition to the two crystalline forms of diamond and graphite, carbon occurs in an amorphous solid form called carbon black. Charcoal and coke are also essentially amorphous carbon, although traces of graphite structure sometimes appear in the carbon grains of which they consist.

The many different silicate structures include such familiar examples of one- and two-dimensional crystals weakly joined together as asbestos and mica. The fundamental structural unit in the silicates is the SiO_4 tetrahedron (Fig. 17.12) in which the quite strong bonds are partly covalent and partly ionic. Each oxygen atom effectively carries a negative charge in such tetrahedrons and can form a bond either with a metal atom or with another silicon atom. In the mineral forsterite (Mg_2SiO_4) the SiO_4 tetrahedra are linked by Mg^{++} ions to form an ionic crystal with alternate SiO_4^{4-} and Mg^{++} ions. In silica (SiO_2) each O atom belongs to two adjacent tetrahedra, so the entire assembly is held together by Si—O bonds and is quite hard as a result. Silica occurs in two forms, crystalline quartz, which exhibits a long-range order in its structure, and vitreous (or fused) quartz, a glass which has the same short-range order as crystalline quartz but without the fixed, repetitive pattern of the latter.

In between the two extremes of SiO_4 tetrahedra linked solely by metal ions and linked solely through the sharing of oxygen atoms are silicate structures with both types of bond. Figure 17.13 shows a chain of SiO_4 tetrahedra that each share two O atoms with their

One- and Two-dimensional Crystals

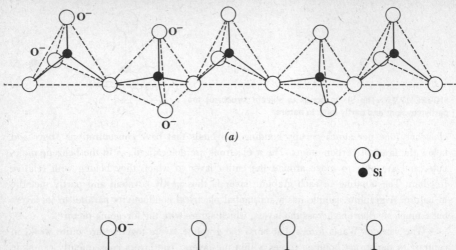

$$\bigcirc \; O$$
$$\bullet \; Si$$

(a)

(b)

FIGURE 17.13 (a) Single chain of SiO_4 tetrahedra. Each structural unit has the formula $[SiO_3]^=$. The O^- ions at the top and sides of the chain can form ionic bonds with metal ions. (b) A two-dimensional representation of the chain. The solid lines here represent Si—O bonds.

neighbors, leaving two O^- ions free to form ionic bonds with metal ions. In Fig. 17.14 such chains are shown bonded by Mg^{++} ions in the mineral enstatite, $MgSiO_3$. The ionic bonds between chains are weaker than the covalent-ionic bonds within each chain; this effect accounts for the fibrous nature of the various minerals jointly called asbestos.

Clay, talc, and mica are examples of two-dimensional silicate structures. Kaolinite, the simplest clay, is the chief constituent of kaolin (china clay), widely used in the manufacture of ceramics, paper, paint, and plastics. Kaolinite has a two-sheet layer structure that essentially consists of a sheet of SiO_4 tetrahedra (Fig. 17.15) and a sheet of $Al(OH)_3$ in which each Al^{+++} ion has six OH^- ions as nearest neighbors (Fig. 17.16). In the SiO_4 sheet three of the O atoms in each tetrahedron are shared by adjacent tetrahedra, leaving one O^- ion per tetrahedron exposed on one side of the sheet. Each exposed O^- ion replaces one OH^- ion in the $Al(OH)_3$ sheet, thereby tying the two sheets together into a firmly bonded flat crystal (Fig. 17.17). Adjacent layers are held together by van der Waals forces to form clay, whose plasticity is the result of the feeble bonds between layers.

The van der Waals forces in clay are actually stronger than otherwise because each

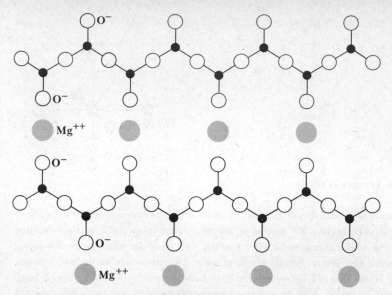

FIGURE 17.14   In MgSiO₃, adjacent [SiO₃]= chains are bonded by Mg++ ions.

FIGURE 17.15   Sheet of SiO₄ tetrahedra. Each structural unit has the formula [Si₂O₅]=. The O−
ions on the top of the sheet can form ionic bonds with metal ions.

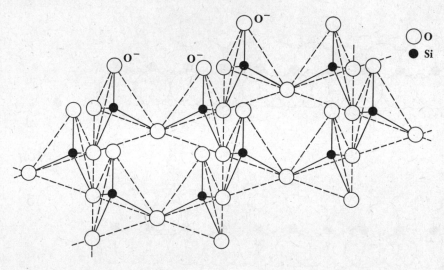

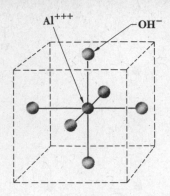

FIGURE 17.16 Structure of Al(OH)₃.

layer in kaolinite is polarized, with the oxygen atoms on the silicon side exhibiting a slight negative charge whereas the OH groups on the aluminum side exhibit a slight positive charge. For this reason clay absorbs water readily, with the polar water molecules fitting between adjacent clay layers. Several layers of water molecules may be trapped between adjacent clay layers (Fig. 17.18), which explains both the ability of clay to absorb large amounts of water into its structure and the extreme plasticity of wet clay. Talc, another layered silicate mineral, has a more symmetric structure that consists of an $Mg(OH)_2$ sheet with an SiO_4 sheet on both sides. The nonpolar talc layers accordingly show less tendency to stick together than clay layers do, and absorb water much less readily.

FIGURE 17.17 Side view of the structure of the clay mineral kaolinite. Successive layers are bonded together by weak van der Waals forces.

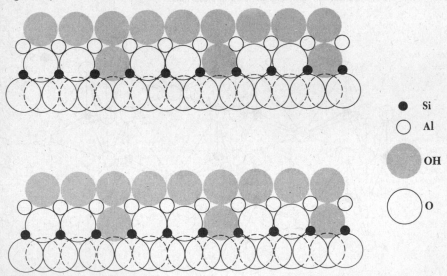

● Si

○ Al

◯ OH

◯ O

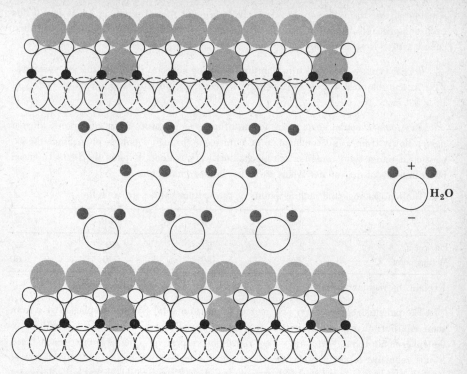

FIGURE 17.18 Water molecules are readily absorbed into the structure of kaolinite because the O atoms exhibit slight negative charges and the OH groups exhibit slight positive charges.

Problems

1. The binding energy of the hydrogen molecule is relatively high. What bearing does this have on the fact that hydrogen is not a metal?

2. Show that the first five terms in the series for the Madelung constant of NaCl are

$$\alpha = 6 - \frac{12}{\sqrt{2}} + \frac{8}{\sqrt{3}} - \frac{6}{2} + \frac{24}{\sqrt{5}} - \cdots$$

3. (a) The ionization energy of potassium is 4.34 ev and the electron affinity of chlorine is 3.61 ev. The Madelung constant for the KCl structure is 1.748 and the distance between ions of opposite sign is 3.14 A. On the basis of these data only, compute the cohesive energy of KCl. (b) The observed cohesive energy of KCl is 6.42 ev per ion pair. On the

assumption that the difference between this figure and that obtained in (a) is due to the exclusion-principle repulsion, find the exponent n in the formula Br^{-n} for the potential energy arising from this source.

4. Repeat Prob. 3 for LiCl, in which the Madelung constant is 1.748, the ion spacing is 2.57 A, and the observed cohesive energy is 6.8 ev per ion pair. The ionization energy of Li is 5.4 ev.

5. The *Joule-Thomson effect* refers to the drop in temperature a gas undergoes when it passes slowly from a full container to an empty one through a porous plug. Since the expansion is into a rigid container, no mechanical work is done. Explain the Joule-Thomson effect in terms of the van der Waals attraction between molecules.

6. The ion spacings and melting points of the sodium halides are as follows:

	NaF	NaCl	NaBr	NaI
Ion spacing, A	2.3	2.8	2.9	3.2
Melting point, °C	988	801	740	660

Explain the regular variation in these quantities with halogen atomic number.

7. The potential energy $V(x)$ of a pair of atoms in a solid that are displaced by x from their equilibrium separation at $0°K$ may be written $V(x) = ax^2 - bx^3 - cx^4$, where the anharmonic terms $-bx^3$ and $-cx^4$ represent, respectively, the asymmetry introduced by the repulsive forces between the atoms and the leveling off of the attractive forces at large displacements (see Fig. 14.8). At a temperature T the likelihood that a displacement x will occur relative to the likelihood of no displacement is $e^{-V/kT}$, so that the average displacement x at this temperature is

$$\bar{x} = \frac{\int_{-\infty}^{\infty} xe^{-V/kT}\, dx}{\int_{-\infty}^{\infty} e^{-V/kT}\, dx}$$

Show that, for small displacements, $\bar{x} \approx 3bkT/4a^2$. (This is the reason that the change in length of a solid when its temperature changes is proportional to ΔT.)

8. The van der Waals attraction between two He atoms leads to a binding energy of about 6×10^{-5} ev at an equilibrium separation of about 3 A. Use the uncertainty principle to show that, at ordinary pressures (<25 atm), solid He cannot exist.

Chapter 18
Crystal Structure

In the previous chapter the various types of bonding mechanisms responsible for the existence of solids were discussed. Now we turn to the actual crystals themselves, their structures and the imperfections in these structures that have such important bearing on the physical properties of solids. It must be kept in mind that, while we shall be concerned with individual crystals, most actual solids are aggregates of a great many minute crystals, and the behavior of a polycrystalline bulk solid may not be identical with that of one of its component units.

18.1 Bravais Lattices

The atoms in a crystal, by definition, are arranged in a periodic array. It is therefore possible to isolate a representative *unit cell* in each variety of crystal—a group of ions, atoms, or molecules, as the case may be—and construct the crystal by repeatedly translating the unit cell in three dimensions (Fig. 18.1).

Unit cells fall into 14 categories, corresponding to the 14 possible different space lattices. A space lattice is a three-dimensional array of points such that the neighboring points of each one surround it in an identical way. In the simplest crystals there is a single atom at each lattice point, but often an assembly of two or more atoms occupies each lattice

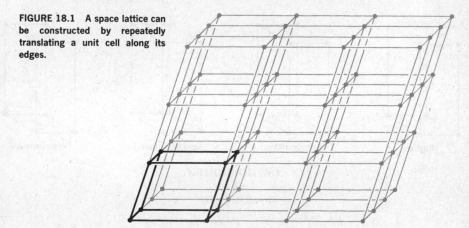

FIGURE 18.1 A space lattice can be constructed by repeatedly translating a unit cell along its edges.

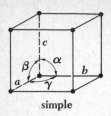
simple

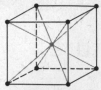
body-centered

face-centered

CUBIC
$$a = b = c$$
$$\alpha = \beta = \gamma = 90°$$

simple

TETRAGONAL
$$a = b \neq c$$
$$\alpha = \beta = \gamma = 90°$$

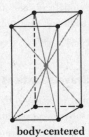
body-centered

simple

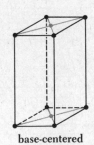
base-centered

body-centered

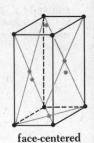
face-centered

ORTHORHOMBIC
$$a \neq b \neq c$$
$$\alpha = \beta = \gamma = 90°$$

point, where each assembly is the same in composition, arrangement, and orientation. There is consequently a distinction between the structure of a crystal, which is the actual ordering in space of its constituent atoms, ions, or molecules, and the corresponding crystal lattice, which is a geometrical abstraction useful for classifying the crystal. Figure 18.2 shows the 14 *Bravais lattices,* named after the man who showed that these are the only possible types of lattice.

From Fig. 18.2 it is clear that seven sets of axes are sufficient to construct the 14 Bravais lattices, which leads to the classification of all crystals into seven *crystal systems.* (Sometimes the trigonal and hexagonal systems are considered jointly as a single system.) These systems have the following properties:

1. *Cubic.* The crystal axes are perpendicular to one another, and the repetitive interval is the same along all three axes. Cubic lattices may be simple, body-centered, or face-centered.

2. *Tetragonal.* The crystal axes are perpendicular to one another. The repetitive intervals along two axes are the same, but the interval along the third axis is different. Tetragonal lattices may be simple or body-centered.

FIGURE 18.2 The fourteen Bravais space lattices and the seven crystal systems based upon them.

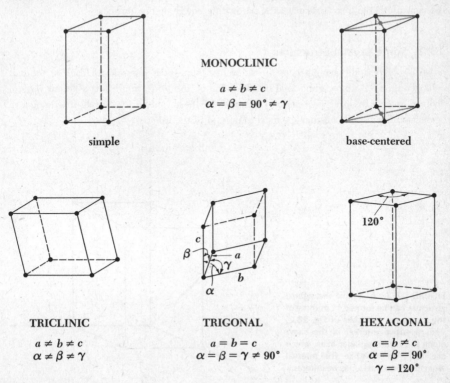

MONOCLINIC

$a \neq b \neq c$
$\alpha = \beta = 90° \neq \gamma$

simple

base-centered

TRICLINIC

$a \neq b \neq c$
$\alpha \neq \beta \neq \gamma$

TRIGONAL

$a = b = c$
$\alpha = \beta = \gamma \neq 90°$

HEXAGONAL

$a = b \neq c$
$\alpha = \beta = 90°$
$\gamma = 120°$

3. *Orthorhombic.* The crystal axes are perpendicular to one another, but the repetitive intervals are different along all three axes. Orthorhombic lattices may be simple, base-centered, body-centered, or face-centered.

4. *Monoclinic.* Two of the crystal axes are not perpendicular to each other, but the third is perpendicular to both of them. The repetitive intervals are different along all three axes. Monoclinic lattices may be simple or base-centered.

5. *Triclinic.* None of the crystal axes is perpendicular to any of the others, and the repetitive intervals are different along all three axes.

6. *Trigonal* (sometimes called *rhombohedral*). The angles between each pair of crystal axes are the same but are not 90°. The repetitive interval is the same along all three axes.

7. *Hexagonal.* Two of the crystal axes are 60° apart while the third is perpendicular to both of them. The repetitive intervals are the same along the axes that are 60° apart, but the interval along the third axis is different.

The hexagonal system is so called because a simple hexagonal structure can be formed from such unit cells, as shown in Fig. 18.3. It is usually more convenient to consider a hexagonal structure in terms of a unit hexagonal cell like that of Fig. 18.3 instead of the rhombic cell of Fig. 18.2. A unit hexagonal cell has four axes, of which three, 120° apart, lie in one plane with the fourth axis perpendicular to this plane. The repetitive interval is the same along the three coplanar axes, but is different along the fourth.

18.2 Some Crystal Structures

A variety of periodic arrangements of different symmetry is possible for each of the unit cells of Fig. 18.2, for a grand total of 230. That is, there are in all 230 different ways to arrange objects in space in a periodic manner. Of these, only a few are at all common in crystals, and we shall consider the most important of them here.

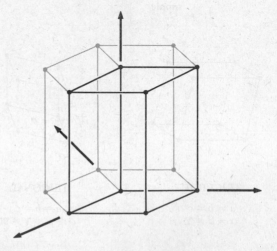

FIGURE 18.3 A simple hexagonal structure can be formed from three of the hexagonal unit cells of Fig. 18.2. The repetitive interval is the same along the three coplanar axes, which are 120° apart, while the interval along the fourth axis may be different.

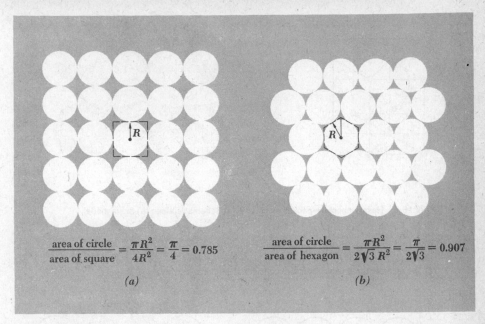

$$\frac{\text{area of circle}}{\text{area of square}} = \frac{\pi R^2}{4R^2} = \frac{\pi}{4} = 0.785$$

(a)

$$\frac{\text{area of circle}}{\text{area of hexagon}} = \frac{\pi R^2}{2\sqrt{3}\,R^2} = \frac{\pi}{2\sqrt{3}} = 0.907$$

(b)

FIGURE 18.4 Layers of identical spheres. (a) In a square array, the spheres occupy 78.5 percent of the available area. (b) In a hexagonal array, the spheres occupy 90.7 percent of the available area.

A crystal whose constituent atoms are so arranged as to occupy the least possible volume is said to have a *close-packed* structure. Close-packed structures occur when the bonding forces are spherically symmetric, as in the inert gases, or very nearly so, as in many metals. The coordination number (number of nearest neighbors) in a close-packed crystal is 12. To verify this figure, we note that a layer of identical spheres assumes a hexagonal configuration when the spheres are squeezed together as tightly as possible (Fig. 18.4), with each sphere in contact with six others. There are two different ways in which such hexagonal layers can be stacked to form a close-packed three-dimensional array. In both cases each sphere in a particular layer fits into the hollow formed by three spheres in the layer below it; hence each sphere is in contact with three spheres in the layer below it, and with three in the layer above it as well. Adding these six spheres to the six it touches in its own layer, each sphere in a close-packed array evidently has a coordination number of 12. In such an array 74 percent of the available volume is occupied by the spheres; the corresponding figure for a body-centered cubic array of identical spheres, where the coordination number is 8, is 68 percent, and still smaller figures hold for other arrays down to 34 percent for the diamond structure.

In a hexagonal close-packed structure, abbreviated hcp, the spheres in alternate layers are directly above one another, as in Fig. 18.5. Beryllium, magnesium, cobalt, and zinc are among the elements with hcp structures. In a face-centered cubic (fcc) structure, sometimes referred to as cubic close-packed, the stacking sequence repeats itself every three

Some Crystal Structures   **419**

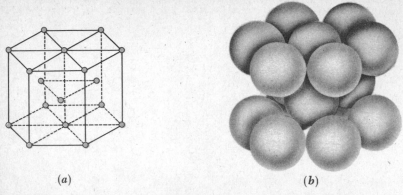

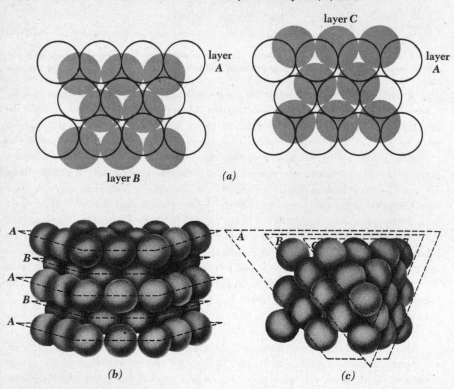

FIGURE 18.5   Hexagonal close-packed structure. (a) Schematic view. (b) Scale model.

FIGURE 18.6   (a) The atoms in close-packed layers *B* and *C* are located over different sets of hollows in close-packed layer *A*. (b) The hexagonal close-packed structure consists of alternate layers *A* and *B*. (c) The face-centered cubic structure consists of sequences of layers *A*, *B*, and *C*.

layers, instead of every two layers, so that each sphere in the third layer occupies hollows located above those hollows in the first layer which are not filled by the spheres of the second layer (Fig. 18.6). The resulting unit cell is shown in Fig. 18.7; it contain atoms that lie in four successive hexagonal layers, as indicated in part (a) of the figure. Copper, lead, gold, and argon are examples of solids with close-packed fcc structures.

In the previous chapter the structures of NaCl, CsCl, and diamond crystals were pictured. As we can see from Fig. 17.2, NaCl has a face-centered cubic structure in which each lattice point is occupied by a Na^+—Cl^- ion pair in the same orientation, although it is equally well described in terms of interleaved fcc Na^+ and Cl^- sublattices. The co-ordination number is 6. Other crystals with this structure include LiH, KBr, PbS, NH_4I, and AgBr.

A cesium chloride (CsCl), shown in Fig. 17.3, can also be thought of in two ways, either as a body-centered cubic (bcc) structure with alternating Cs^+ and Cl^- ions at the lattice points, or as interleaved simple cubic sublattices of Cs^+ and Cl^- ions. The coordina-tion number is 8. This structure is also exhibited by NH_4Cl, BeCu, and CuZn (a type of brass) among others.

The diamond structure, shown in Fig. 17.4, actually consists of interleaved fcc sub-lattices, one of which is shifted relative to the other by one-fourth the length of a unit

FIGURE 18.7 Face-centered cubic structure. (a) View showing close-packed layers. (b) Schematic view. (c) Scale model.

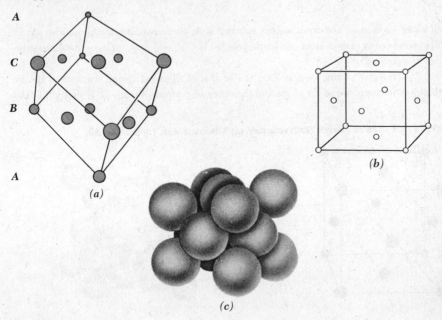

(a)

(b)

(c)

TABLE 18.1. CRYSTAL STRUCTURES OF REPRESENTATIVE ELEMENTS

Element	Structure	Lattice constants, A	
		a	c
Aluminum	Fcc	4.04	
Argon (20°K)	Fcc	5.43	
Beryllium	Hcp	2.27	3.59
Bismuth	Rhombohedral	4.74	($\alpha = 58°$)
Calcium	Fcc	5.56	
Carbon	Diamond	3.56	
Cobalt	Hcp	2.51	4.07
Copper	Fcc	3.61	
Helium (2°K)	Hcp	3.57	5.83
Iron	Bcc	2.86	
Lead	Fcc	4.94	
Lithium	Bcc	3.50	
Neon (20°K)	Fcc	4.52	
Silicon	Diamond	5.43	
Silver	Fcc	4.08	
Titanium	Hcp	3.45	5.51
Tungsten	Bcc	3.16	
Zinc	Hcp	2.66	4.94

cell along each axis. The coordination number is 4, corresponding to the number of co-valent bonds each carbon atom can participate in. Ge, Si, and gray tin have the same structures as diamond.

The structure of zinc blende, ZnS, is like that of diamond except that one of the fcc sublattices is composed of Zn atoms and the other of S atoms. Figure 18.8 shows the zinc-

FIGURE 18.8   The zinc-blende (ZnS) structure. (a) Schematic view. (b) Scale model.

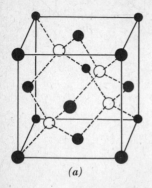

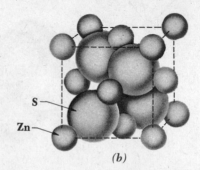

S

Zn

(a)   (b)

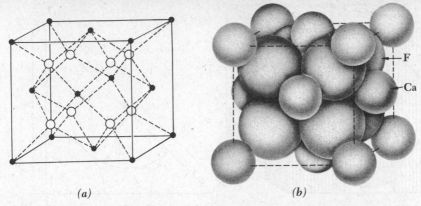

FIGURE 18.9   The fluorite (CaF$_2$) structure. (a) Schematic view. (b) Scale model.

blende structure. As in diamond, the coordination number is 4. CuCl, AgI, CdS, and SiC are among the compounds that have zinc-blende structures.

An example of a crystal in which the coordination number is different for each of the types of atom present is fluorite, CaF$_2$. As can be seen from Fig. 18.9, each Ca atom has eight F atoms around it, while each F atom is surrounded by only four Ca atoms. The space lattice here is fcc, with one Ca atom and two F atoms around each lattice point. UO$_2$, K$_2$O, and AuAl$_2$ are among the crystals with fluorite structures.

18.3   Atomic Radii

If it is assumed that the atoms in an elemental solid are spheres in contact, it is easy to calculate their radii from a knowledge of the crystal structure of the solid and the lattice parameters. (The latter information can be obtained by X-ray diffraction analysis.) For example, Fig. 18.10 shows a face-centered cubic unit cell whose lattice constant—the length of each edge of the cube—is a. If the radius of each atom is r, the length of the diagonal on any face is $4r$, and

$$(4r)^2 = a^2 + a^2$$

18.1
$$r = \frac{\sqrt{2}\,a}{4}$$
Fcc structure

For other cubic structures the atomic radii in terms of a are as follows:

18.2
$$r = \frac{a}{2}$$
Simple cubic structure

18.3
$$r = \frac{\sqrt{3}\,a}{4}$$
Bcc structure

18.4
$$r = \frac{\sqrt{3}\,a}{8}$$
Diamond structure

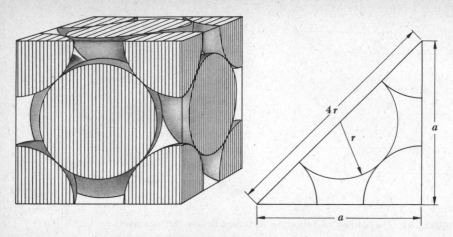

FIGURE 18.10   Geometry of a face-centered cubic unit cell.

FIGURE 18.11   Atomic radii of the elements. Several have two radii, corresponding to different crystal structures.

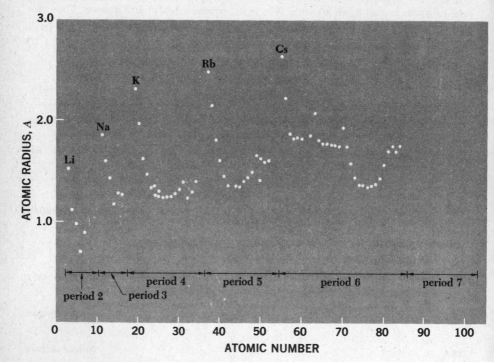

In the case of copper, which has an fcc structure with the lattice constant $a = 3.61$ A, the atomic radius is

$$r = \frac{\sqrt{2} \times 3.61 \text{ A}}{4} = 1.28 \text{ A}$$

Of course, what is really being calculated is half the internuclear distance of closest approach and not the "atomic radius," since it is hardly correct to consider an atom as a hard sphere, but the resulting figures are nevertheless interesting and useful. Figure 18.11 shows the atomic radii of the elements obtained in this way. The alkali metals are the largest in each period of the periodic table, since their atomic structures consist of a single electron outside closed inner shells that shield the electron from all but $+e$ of nuclear charge. There is then a regular decrease in size within each period as the nuclear charge increases, which pulls the outer electrons in closer to the nucleus. At the end of each period there is a small increase in size due to the mutual repulsion of the outer electrons.

TABLE 18.2. CRYSTAL STRUCTURES OF REPRESENTATIVE COMPOUNDS

Compound	Structure	Lattice constant, A a
AgBr	NaCl	5.77
AgI	ZnS	6.47
$AuAl_2$	CdF_2	6.00
BeCu	CsCl	2.70
CaF_2	CaF_2	5.45
CdS	ZnS	5.82
CsCl	CsCl	4.11
CuCl	ZnS	5.41
CuZn	CsCl	2.94
KBr	NaCl	6.59
K_2O	CaF_2	6.44
LiH	NaCl	4.08
NH_4Cl	CsCl	3.87
NH_4I	NaCl	7.24
NaCl	NaCl	5.63
PbS	NaCl	5.92
SiC	ZnS	4.35
UO_2	CaF_2	5.47

Ionic radii can be determined in a similar way from data on ionic crystals (Table 18.3). These radii differ from those of the corresponding neutral atoms, often quite considerably.

TABLE 18.3. SOME IONIC RADII

Positive ions	Radius, A	Negative ions	Radius, A
Li+	0.60	H−	2.08
Na+	0.95	F−	1.36
K+	1.33	Cl−	1.81
Rb+	1.48	Br−	1.95
Cs+	1.69	I−	2.16
Be++	0.31	O−	1.40
Mg++	0.65	S−	1.84
Ca++	0.99	N³−	1.71
Sr++	1.13	P³−	2.12
Ba++	1.35	C⁴−	2.60
Al³+	0.50	Si⁴−	2.71

The removal of one or more electrons to form a positive ion leads to a reduction in size because the remaining electrons are more tightly bound, and if all the electrons in the outermost shell are removed, the size shrinks to one smaller than that of the inert gas atom of the preceding period. Thus the radius of the Na atom is 1.86 A, while that of the Na+ ion is only 0.95 A, less than the 1.1 A radius of the Ne atom whose electron structure is the same as that of Na+. A negative ion, on the other hand, has an excess of electrons, which in consequence are more weakly bound to yield an ionic radius larger than that of the neutral atom. Thus the radius of the S atom is 1.27 A, while that of the S− ion is 1.84 A. These differences are especially conspicuous in carbon, where the radii of C, C⁴+, and C⁴− are respectively 0.72 A, 0.15 A, and 2.60 A.

In general, the structure of an ionic crystal is determined by the requirement that each ion be in contact (so to speak) with as many ions of the opposite sign as possible, which leads to the maximum bonding energy. The relative sizes of the ions involved therefore govern the type of structure that occurs. Figure 18.12 illustrates the significance of

FIGURE 18.12   Threefold coordination for various radius ratios.

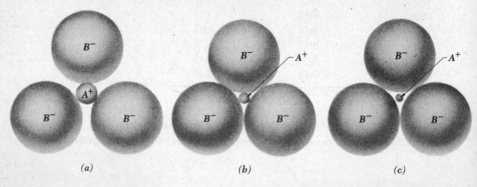

(a)   *(b)*   *(c)*

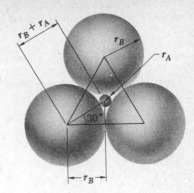

FIGURE 18.13 The minimum radius ratio for threefold coordination.

the radius ratio in a binary (AB) compound. In (a) the three negative B^- ions are in contact with the positive A^+ ion but not with each other; there is not enough room for a fourth B^- ion to be squeezed in and still touch the positive ion, so this is a stable configuration. In (b) the A^+ ion is smaller, and the B^- ions are now in contact with each other; the configuration is just at the limit of stability. In (c) the A^+ ion is so small that none of the B^- ions touch it; here the repulsive forces among the B^- ions exceed the attractive force of the A^+ ion, and the configuration is unstable.

To calculate the limiting ionic radius ratio for threefold coordination, the construction in Fig. 18.13 is helpful. Evidently

$$\cos 30° = \frac{r_B}{r_B + r_A}$$

$$\frac{r_A}{r_B} = \frac{1 - \cos 30°}{\cos 30°} = \frac{1 - 0.866}{0.866} = 0.155$$

The ratio between the radii of the ions must exceed 0.155 for threefold coordination to result in a stable structure. By a similar calculation the minimum radius ratio for fourfold coordination turns out to be 0.225, which means that binary ionic compounds with radius ratios between 0.155 and 0.225 crystallize into structures with coordination numbers of 3. Table 18.4 shows the possible range of ratios for each coordination number. Though there are a number of exceptions, the correlation between the predictions of Table 18.4 and

TABLE 18.4. RADIUS RATIOS AND COORDINATION NUMBERS IN IONIC CRYSTALS

Radius ratio	Coordination number
0.155–0.225	3
0.225–0.414	4
0.414–0.732	6
0.732–1.000	8
1.000	12

Atomic Radii

observed crystal structure is generally good. For example, the radius ratio in NaCl is

$$\frac{r_{Na^+}}{r_{Cl^-}} = \frac{0.95 \text{ A}}{1.81 \text{ A}} = 0.525$$

From this figure we infer a coordination number of 6, in agreement with the actual NaCl structure (Fig. 17.2).

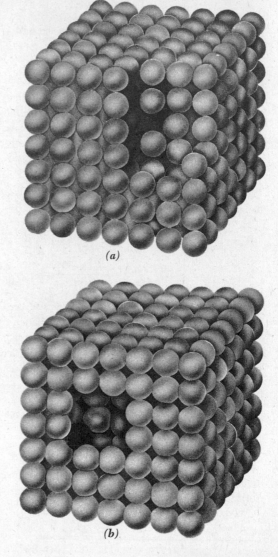

(a)

(b)

FIGURE 18.14 Three types of point defect in a crystal. (a) Vacancy; (b) interstitial; (c) impurity. The presence of a point defect disturbs the regularity of the crystal structure in its vicinity.

18.4 Point Defects

We have thus far been discussing ideal crystals in which each atom occupies a definite location in a regular array. Actual crystals are far from meeting this specification. For one thing, their atoms are in continual thermal vibration, a topic considered in the next chapter. However, such vibrations represent imperfections only in a limited sense, since the equilibrium positions of the atoms can still be described in terms of a perfect lattice. More drastic are actual defects in the structure of a crystal—missing atoms, atoms in interstitial positions, irregularities in the periodicity of the lattice along certain lines, and so forth. These defects often have considerable bearing on the physical properties of solids. For example, the mechanical strength of a solid is largely determined by the nature and concentration of defects in its structure, as is the electrical behavior of a semiconductor. On the other hand, such properties of solids as paramagnetism and the electrical conductivity of metals are practically independent of crystal structure.

The simplest category of crystal imperfection is the localized *point defect*. Figure 18.14 shows the three basic kinds of point defect: the vacancy, the interstitial, and the impurity. Both vacancies and interstitials occur in all crystals with a probability that increases rapidly with temperature. A consideration of the energies involved shows why this is so. To produce a vacancy, the bonds holding an atom in place must be broken, which in a typical crystal might require 0.5 ev per bond. If the coordination number is 8, a total of 4 ev is involved. However, the missing atom does not vanish, and we may imagine it to end up at the surface where it is held in place by a smaller number of bonds than in the interior, say four in this case. Hence the net energy needed is reduced to 2 ev. In addition, the atoms that border the vacancy tend to be partially displaced into it, as pictured in Fig. 18.14a, which lowers their energies somewhat and thereby further reduces the required energy $w_{\bar{v}}$ to the neighborhood of 1 ev.

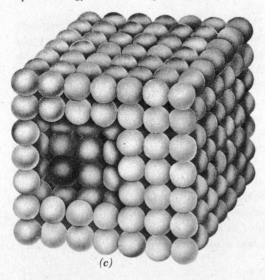

(c)

The relative probability that an atom in a crystal have an energy w_v of thermal origin in excess of its normal energy is $e^{-w_v/kT}$ at the temperature T. If there are N atoms in the crystal, the number n_v of vacancies to a first approximation is given by

18.5
$$n_v = N e^{-w_v/kT}$$

In copper, where $w_v \approx 0.8$ ev, at room temperature the predicted ratio n_v/N is 3×10^{-15}, while at $1350°K$, just under the melting point, $n_v/N = 6 \times 10^{-4}$. The observed occurrence of vacancies in metals turns out to be somewhat more frequent than Eq. 18.5 suggests, for various reasons, but the dependence on $e^{-w_v/kT}$ holds in general.

Energy is needed to insert an atom in an interstitial location in a crystal because the neighboring atoms must be squeezed more tightly together despite their mutual repulsions in order to make room. The required energy w_i depends upon the crystal structure, being greater in a close-packed crystal than in one with a more open structure, and ranges up to as much as 5 ev. In aluminum w_i is roughly 3 ev although w_v is only 0.76 ev. To be sure, the expected number of interstitials exceeds that suggested by a formula like Eq. 18.5 because there are more interstitial locations than there are atoms in a crystal, and hence more chance for an atom to enter one. If the number of interstitial locations per atom is A, the number n_i of interstitials at the temperature T in a crystal that consists of N atoms is

18.6
$$n_i = AN e^{-w_i/kT}$$

But A is not a large number, and the dominance of the exponential factor together with the fact that $w_v < w_i$ means that vacancies are more common in solids than interstitials.

Impurity atoms smaller in size than the host atoms in a crystal are able to fit more readily into interstitial locations than the latter and may be present in large numbers under certain circumstances. Hydrogen, carbon, oxygen, and nitrogen atoms are readily absorbed as interstitials in many metals, for example.

Point defects in an ionic crystal must be such as to leave the crystal electrically neutral. A *Schottky defect* is the simultaneous occurrence of a positive-ion vacancy and a negative-ion vacancy, as in Fig. 18.15a. The required energy is the sum of the energies for producing each vacancy separately. A *Frenkel defect* is the simultaneous occurrence of a positive-ion interstitial and a positive-ion vacancy, as in Fig.18.15b. Here the required energy is the sum of the appropriate w_v and w_i values. Other possibilities that preserve electrical neutrality are a negative-ion interstitial-vacancy pair, which is rarely found as the negative ions in a crystal are usually larger than the positive ones and hence are less apt to become interstitials, and a positive-ion–negative-ion interstitial pair, which never seems to occur.

Both Frenkel and Schottky defects are present in all ionic solids, but one or the other predominates in any given crystal because their energies of formation are usually different. Thus Schottky defects prevail in alkali halides and Frenkel defects in silver halides.

Point defects in excess of the number in thermal equilibrium may be created in a number of ways. A simple method is to heat a solid to a high temperature so that many

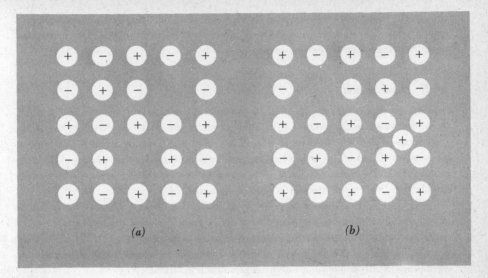

FIGURE 18.15 Point defects in an ionic crystal: (a) Schottky defect; (b) Frenkel defect. Both preserve electrical neutrality.

defects are present, and then to cool it quickly enough to retain some of the additional defects in the structure at the final temperature. Of much importance is the production of defects by particle radiation. In a nuclear reactor, for instance, high-energy neutrons produce defects by simply knocking atoms out of their normal locations. Intense bombardment severely affects the properties of nearly all solids; radiation damage leads to a reduction in ductility (embrittlement) in most metals, often accompanied by an increase in the yield strength.

The existence of point defects in a crystal makes it possible for the diffusion of atoms within the crystal to occur—a surprising notion at first glance, though less surprising when one thinks of the bonding of dissimilar metals in such processes as soldering, brazing, and galvanizing. When a vacancy is present, diffusion occurs by the jumping of an adjacent atom into the vacancy to leave a new vacancy behind, into which another atom may jump later, and so on. In effect, the vacancy moves from place to place through the crystal. When an interstitial is present, diffusion occurs as it migrates from one interstitial location to a neighboring one. In both cases each change in position requires a certain amount of extra energy, typically about 1 ev, in order for the moving atom to squeeze past the atoms that project in its path. As we might expect, diffusion in a solid is strongly temperature dependent, increasing from a usually negligible rate at room temperature to one not far from that characteristic of the corresponding liquid near the melting point.

Diffusion in an ionic crystal enables it to conduct electric current, since the moving defects are either charged interstitials or vacancies that behave as though they carry charges

Point Defects

opposite to those of the missing ions. In fact, the study of the electrical conductivity of ionic crystals has provided much detailed information on point defects, information that would have been difficult to obtain in any other way.

18.5 Dislocations

Line defects in a crystal are called *dislocations*. A dislocation is a line of atoms with different coordinations from those of the other atoms in the crystal. Dislocations are of two basic kinds. Figure 18.16 shows an *edge dislocation*, which can be visualized in terms of the removal of part of one layer of atoms and the subsequent accommodation of the array to the defect. The dislocation itself is indicated by the symbol ⊥, and in its immediate neighborhood the crystal structure is severely distorted.

The other kind of dislocation is the *screw dislocation*. We can visualize the formation of a screw dislocation by imagining that a cut is made a certain distance into a perfect crystal and one side of the cut then displaced relative to the other (Fig. 18.17). The atomic layers spiral around the dislocation, which accounts for its name. Actual dislocations in crystals are usually combinations of the edge and screw varieties.

Dislocations are important because they provide an explanation for the plastic behavior of solids. When an applied stress exceeds its elastic limit, a material no longer

FIGURE 18.16 An edge dislocation in a simple cubic crystal.

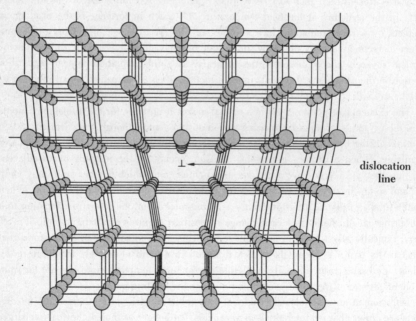

dislocation line

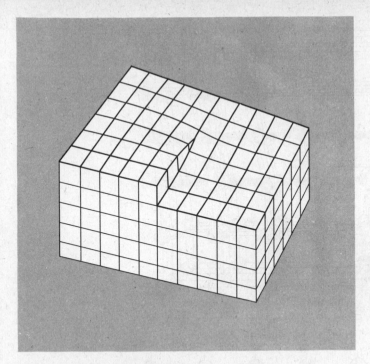

FIGURE 18.17 Geometry of a screw dislocation.

returns to its original shape but is permanently deformed. Most metals can undergo substantial plastic deformation before fracture, a property described as ductility; in other solids the plastic range is usually smaller. The elastic response of a solid is readily interpreted in terms of the bonding forces within it, which act like Hooke's law restoring forces for small displacements from the equilibrium configuration. But this direct approach fails to account for plastic behavior, since calculations of the force required to slide one layer of atoms in a crystal past another yield figures $\sim 10^3$ times higher than those actually observed.

The presence of dislocations makes it possible to understand why solids are only ~ 0.1 percent as strong as they ought to be on the basis of perfect crystal structures. Figure 18.18 shows how a crystal that contains an edge dislocation can be permanently deformed by a relatively modest shearing stress. The line of atoms below and to the right of the dislocation shifts its bonds to the line of atoms directly above it when the shear is applied, which causes the dislocation to move one atom spacing to the right. The process is repeated until the dislocation arrives at the edge of the crystal, and the deformation is then permanent. The entire process is called *slip*, and the plane along which the dislocation moves is called the *slip plane*. In slip, the atomic bonds holding one layer to the next are broken

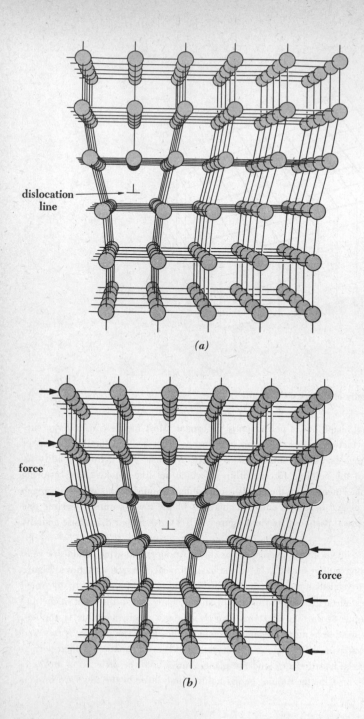

(a)

dislocation
line

⊥

force

⊥

force

force

force

(b)

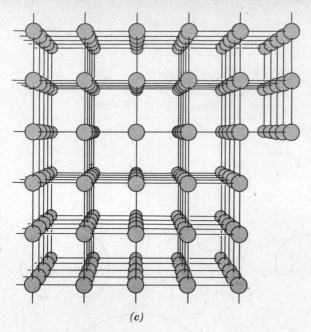

(c)

FIGURE 18.18 Slip results from the motion of a dislocation through a crystal under stress. (a) Initial configuration of crystal. (b) The dislocation moves to the right as the atoms in the layer under it successively shift their bonds with those of the upper layer one line at a time. (c) The crystal has acquired a permanent deformation.

only one line at a time, whereas in a perfect crystal all the bonds between two layers would have to be broken more or less simultaneously for plastic flow to occur, a much more formidable proposition.

The energy needed to produce a dislocation ranges from perhaps 3 to 10 ev per atom spacing. Since a dislocation may be tens of thousands of atom spacings in length, the energy requirement makes it essentially impossible for one to spontaneously appear (as point defects do) in a crystal, regardless of the temperature. The available evidence indicates that dislocations come into being naturally in crystals during their formation; in fact, it is exceedingly difficult to produce crystals free of dislocations. (Such perfect crystals turn out to be phenomenally strong, as we would expect. Perfect iron crystals have tensile strengths of about 10^6 lb/in.^{2}, for example.) Furthermore, dislocations multiply in number during plastic flow in a variety of more or less complicated ways. One of these is the *Frank-Read mechanism*, pictured in Fig. 18.19, which shows how an initial dislocation can act as a source for further dislocations in a crystal under stress.

Because dislocations multiply during plastic flow, the continued deformation of a

Dislocations

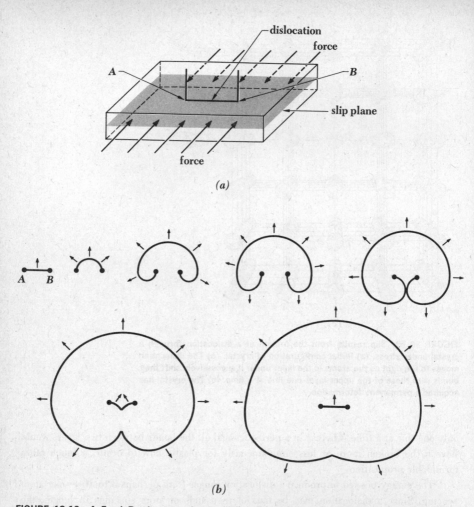

FIGURE 18.19 A Frank-Read source that multiplies dislocations in a crystal under stress. (a) The dislocation segment *AB* in the slip plane is initially a straight line. The ends of the segment are "anchored" at *A* and *B* since the dislocation changes direction at these points and slip only occurs in the indicated plane. (b) The applied shear causes the dislocation segment to move outward, and it eventually expands until the spiral arms come together at the rear to form a complete dislocation loop. The remaining cusp then becomes a straight segment again under the influence of the shear, and the process repeats itself.

solid increases its dislocation content. Eventually the dislocations become so numerous and so tangled together that they impede one another's motion, which *decreases* the plasticity of the material. This phenomenon is known as *work hardening* and becomes prominent when there are perhaps 10^8 dislocations per cm^2 in a slip plane. If a work-hardened crystal is heated, its disordered atomic array tends to return to regularity; the rise in temperature, which increases the vibrational energy of the atoms, provides the activation energy needed to initiate the release of the energy stored in the dislocations. (The stored energy is about 20 joules/cm^3 for a severely deformed metal.) The process of heating a work-hardened crystal to restore its ductility is called *annealing*. Steel bars and sheets formed by cold-rolling are much harder than those formed by hot-rolling.

Besides work hardening, another means of impeding slip in a metal is the deliberate introduction of foreign atoms to interfere with the motion of dislocations. Thus the addition of small amounts of such other elements as carbon, chromium, manganese, and tungsten to iron converts it into the vastly stronger steel by reducing slip.

Problems

1. Verify that $\pi/6$ of the available volume is occupied by hard spheres in contact in a simple cubic arrangement.

2. Verify that $\sqrt{3}\,\pi/8$ of the available volume is occupied by hard spheres in contact in a body-centered cubic arrangement.

3. Lithium, iron, and tungsten have bcc structures with the respective lattice constants 3.50 A, 2.86 A, and 3.16 A. Find their atomic radii. Why does lithium, with an atomic number of only 3 as compared with 26 for iron and 74 for tungsten, have the largest atomic radius?

4. The densities of beryllium, aluminum, and iron are respectively 1.85, 2.70, and 7.87 g/cm^3. From these figures together with the data in Table 18.1 compute the atomic masses of these elements.

5. Derive Eqs. 18.2 to 18.4.

6. Prove that the minimum ratio of ion radii that will permit a coordination number of 6 is 0.414.

7. An aluminum atom requires 1.48 ev to move from one interstitial location to another in solid aluminum. In an fcc structure such as that of aluminum, an interstitial atom has 12 equivalent neighboring sites to which it can jump. The frequency of vibration of an aluminum atom is about 6×10^{12} sec^{-1}. From these figures, calculate how often, on the average, an interstitial aluminum atom jumps to a new location (a) at room temperature (20°C) and (b) at the melting point of aluminum (660°C).

8. The maximum energy stored in a metal as a result of severe plastic deformation is about 20 joules/cm^3. (a) Assume that the linear energy density of a dislocation is 3 ev/A and find the maximum number of centimeters of dislocation line per cm^3 in a metal.

(*b*) If these lines are all parallel to one side of an aluminum cube 100 atoms on each edge, how many are there in the cube? (The density of aluminum is 2.70 g/cm^3.)

9. A copper rod is cold-worked until it has 10 joules/cm^3 of stored energy. The rod is heated to 200°C, at which temperature it recrystallizes and releases the stored energy. What is its final temperature? (The density of copper is 8.89 g/cm^3 and its specific heat is 0.093 kcal/kg °C.)

10. A sample of copper contains 0.001 percent by weight of nickel as an impurity. Copper has an fcc structure with a lattice constant of 3.61 A. If the nickel atoms are uniformly distributed and each replaces a copper atom in the lattice, find the distance separating nickel atoms from one another.

11. An iron sample is found to have an fcc structure with a lattice constant of 3.61 A. It contains 1.7 percent by weight of carbon, and its density is 7.88 g/cm^3. Do the carbon atoms occupy interstitial locations or do they replace iron atoms in the lattice?

Chapter 19
Specific Heats
of Solids

The specific heat of a substance is so elementary and familiar a property that it may be surprising to find a chapter devoted to this subject in a text on modern physics. In reality, as we shall see, it is impossible to understand the observed values of the specific heats of solids except by invoking quantum statistical mechanics. Apart from their intrinsic interest, specific heats are examined here both as a further exercise in statistical methods and, more important, because of the light such an examination sheds on the behavior of solids in general and of electrons in metals in particular.

19.1 Thermal Vibrations: Frequencies

The atoms in a solid are in constant thermal vibration. We can get an idea of the frequencies and amplitudes of these vibrations rather easily on the basis of the simple model illustrated in Fig. 19.1. In this model (1) all the atoms of the solid are fixed in place except the central

FIGURE 19.1 Simplified model of a solid. (a) Equilibrium configuration. (b) The central atom vibrates along one axis only; the other atoms are fixed in place.

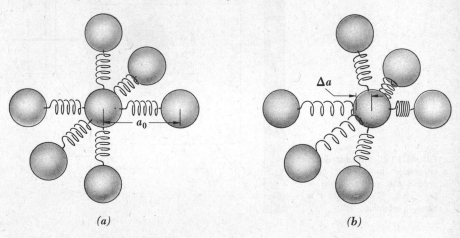

(a) (b)

one in the figure; (2) the latter atom is limited to motion in one dimension only; and (3) its displacements Δa from the equilibrium position are sufficiently small for Hooke's law to apply.

The frequency of vibration ν of the atom is given by the usual formula for a harmonic oscillator

19.1
$$\nu = \frac{1}{2\pi}\sqrt{\frac{K}{m}}$$

where m is the atom's mass and K is the force constant that governs its displacement in response to an applied force. To estimate K, we note that the stress F/A on a cube of the solid one atom spacing a_0 long on each edge is related to the resulting strain $\Delta a/a_0$ by

19.2
$$\frac{\Delta a}{a_0} = \frac{1}{Y}\frac{F}{A}$$

where Y is Young's modulus for the particular material (Fig. 19.2). Since the applied force F and the consequent displacement Δa, by hypothesis, follow Hooke's law,

$$F = K\Delta a$$

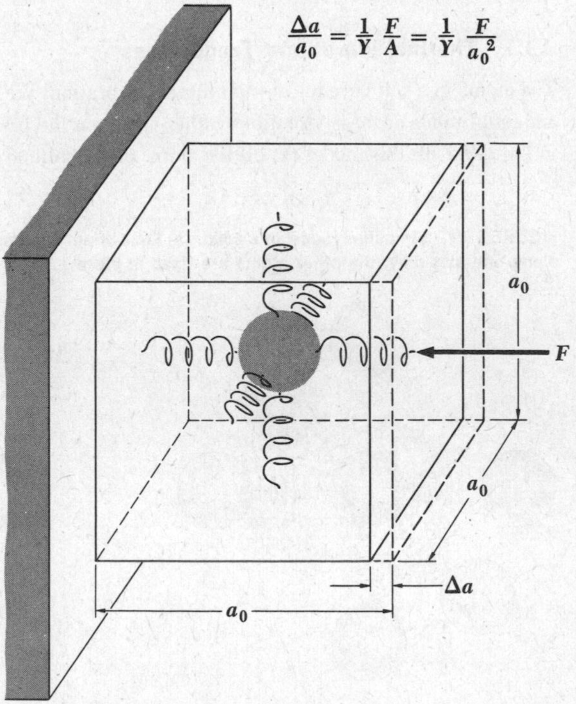

$$\frac{\Delta a}{a_0} = \frac{1}{Y}\frac{F}{A} = \frac{1}{Y}\frac{F}{a_0^2}$$

FIGURE 19.2 Stress-strain relationship for a unit cube around the central atom of Fig. 19.1. The quantity Y is Young's modulus.

and the area A of each face of the cube is a_0^2, Eq. 19.2 becomes

$$\frac{\Delta a}{a_0} = \frac{1}{Y} \frac{K \Delta a}{a_0^2}$$

and we have

19.3 $K = a_0 Y$

This formula gives the force constant K in terms of the atom spacing a_0 and Young's modulus Y, both of which can be readily measured.

In the case of aluminum, where $a_0 = 2.86$ A and $Y = 7.0 \times 10^{10}$ n/m²,

$$K = a_0 Y$$
$$= 2.86 \times 10^{-10} \text{ m} \times 7.0 \times 10^{10} \text{ n/m}^2$$
$$= 20 \text{ n/m}$$

which is 0.12 lb/in.; the interatomic bonds in aluminum in the elastic region are about as stiff as rubber bands. The corresponding frequency of vibration is

$$\nu = \frac{1}{2\pi} \sqrt{\frac{K}{m}}$$
$$= \frac{1}{2\pi} \sqrt{\frac{20 \text{ n/m}}{4.5 \times 10^{-26} \text{ kg}}}$$
$$= 3.4 \times 10^{12} \text{ sec}^{-1}$$

since the mass of an aluminum atom is 4.5×10^{-26} kg. The actual vibration frequency in aluminum is closer to 6.4×10^{12} sec^{-1}. The discrepancy is due to the crude model we have been using, in which, for one thing, each atom was assumed to vibrate independently in the midst of an array of stationary atoms. Actually, *all* the atoms are vibrating, and it is unrealistic to neglect the fluctuations in interatomic spacing that result from the motions of the atoms surrounding any particular one. In addition, a more sophisticated calculation shows that the force constant K is underestimated by Eq. 19.3. In spite of these flaws in the argument, however, the predicted frequency is of the right order of magnitude, which means that our elementary picture of atomic vibrations in a solid is not an unreasonable approximation.

19.2 Thermal Vibrations: Amplitudes

In order to estimate the amplitudes of the vibrations in a solid, we must first determine the average energy of a one-dimensional harmonic oscillator in thermal equilibrium with its surroundings. The *relative* probability that such an oscillator have the energy E at the temperature T is given by the Boltzmann factor $e^{-E/kT}$, and so its average energy is found by integrating $Ee^{-E/kT}$ over all possible energies and then dividing by the integral of $e^{-E/kT}$ in order to normalize the result.

The total energy of the oscillator at any time is the sum of its instantaneous kinetic and potential energies,

$$19.4 \qquad E = \tfrac{1}{2}mv^2 + \tfrac{1}{2}Kx^2$$

where v is the particle's speed and x is its displacement from the equilibrium position. The average energy is therefore

$$\overline{E} = \frac{\displaystyle\int_{-\infty}^{\infty}\int_{-\infty}^{\infty} E\, e^{-E/kT}\, dv\, dx}{e^{-E/kT}\, dv\, dx}$$

$$= \frac{\displaystyle\int_{-\infty}^{\infty}\int_{-\infty}^{\infty} (\tfrac{1}{2}mv^2 + \tfrac{1}{2}Kx^2)\, e^{-(mv^2 + Kx^2)/2kT}\, dv\, dx}{\displaystyle\int_{-\infty}^{\infty}\int_{-\infty}^{\infty} e^{-(mv^2 + Kx^2)/2kT}\, dv\, dx}$$

$$= \frac{\tfrac{1}{2}m\displaystyle\int_{-\infty}^{\infty} v^2\, e^{-mv^2/2kT}\, dv \int_{-\infty}^{\infty} e^{-Kx^2/2kT}\, dx}{\displaystyle\int_{-\infty}^{\infty} e^{-mv^2/2kT}\, dv \int_{-\infty}^{\infty} e^{-Kx^2/2kT}\, dx}$$

$$+ \frac{\tfrac{1}{2}K\displaystyle\int_{-\infty}^{\infty} x^2\, e^{-Kx^2/2kT}\, dx \int_{-\infty}^{\infty} e^{-mv^2/2kT}\, dv}{\displaystyle\int_{-\infty}^{\infty} e^{-mv^2/2kT}\, dv \int_{-\infty}^{\infty} e^{-Kx^2/2kT}\, dx}$$

$$= \frac{\tfrac{1}{2}m\displaystyle\int_{-\infty}^{\infty} v^2\, e^{-mv^2/2kT}\, dv}{\displaystyle\int_{-\infty}^{\infty} e^{-mv^2/2kT}\, dv}$$

$$19.5 \qquad + \frac{\tfrac{1}{2}K\displaystyle\int_{-\infty}^{\infty} x^2\, e^{-Kx^2/2kT}\, dx}{\displaystyle\int_{-\infty}^{\infty} e^{-Kx^2/2kT}\, dx}$$

We now let

$$19.6 \qquad y^2 = \frac{\tfrac{1}{2}mv^2}{kT}$$

$$19.7 \qquad z^2 = \frac{\tfrac{1}{2}Kx^2}{kT}$$

so that

$$v^2 = \frac{2kT}{m}y^2 \qquad dv = \sqrt{\frac{2kT}{m}}\, dy$$

and

$$x^2 = \frac{2kT}{K}z^2 \qquad dx = \sqrt{\frac{2kT}{K}}\, dz$$

With these substitutions Eq. 19.5 becomes

19.8
$$\overline{E} = \frac{kT \displaystyle\int_{-\infty}^{\infty} y^2\, e^{-v^2}\, dy}{\displaystyle\int_{-\infty}^{\infty} e^{-v^2}\, dy} + \frac{kT \displaystyle\int_{-\infty}^{\infty} z^2\, e^{-z^2}\, dz}{\displaystyle\int_{-\infty}^{\infty} e^{-z^2}\, dz}$$

Since

$$\int_{-\infty}^{\infty} \xi^2 e^{-\xi^2}\, d\xi = \frac{\sqrt{\pi}}{2}$$

and

$$\int_{-\infty}^{\infty} e^{-\xi^2}\, d\xi = \sqrt{\pi}$$

we have for the average energy of the oscillator

19.9 $$\overline{E} = \tfrac{1}{2}kT + \tfrac{1}{2}kT = kT \qquad \text{Average energy of harmonic oscillator}$$

The kinetic and potential energies each average $\tfrac{1}{2}kT$ for a total of $\overline{E} = kT$. We can interpret this result as signifying that a one-dimensional harmonic oscillator has two degrees of freedom, one corresponding to each of its two modes of energy possession, with $\tfrac{1}{2}kT$ of energy associated on the average with each degree of freedom. A free particle restricted to one-dimensional motion has only $\tfrac{1}{2}kT$ of thermal energy on the average, since it has only one degree of freedom.

The amplitude A of a harmonic oscillator is its maximum displacement on either side of the equilibrium position. When $x = A$, all the energy is potential, and

$$\overline{E} = \tfrac{1}{2}KA^2 = kT$$

In view of the other approximations we have made, we may disregard the distinction between $\overline{A^2}$ and $\overline{A}^2$, and hence obtain

19.10 $$\overline{A} = \sqrt{\frac{2kT}{K}} \qquad \text{Average amplitude of harmonic oscillator}$$

Thermal Vibrations: Amplitudes

for the average amplitude of a harmonic oscillator in thermal equilibrium at the temperature T. Evidently $\overline{A}$ depends only on K and T, and not on the mass m of the particle.

The Hooke's law constant K for an aluminum atom in solid aluminum was found earlier to be ~ 20 n/m. At $300°$K, which is approximately room temperature, the average amplitude of such an atom is

$$\overline{A} = \sqrt{\frac{2kT}{K}}$$

$$= \sqrt{\frac{2 \times 1.38 \times 10^{-23} \text{ joule/}°\text{K} \times 300°\text{K}}{20 \text{ n/m}}}$$

$$= 2.0 \times 10^{-11} \text{ m} = 0.20 \text{ A}$$

which is about 7 percent of the 2.86 A equilibrium bond length at that temperature. The actual value of $\overline{A}$ in aluminum is somewhat less than 0.20 A because, as mentioned earlier, K is greater than our estimated value of 20 n/m.

19.3 Specific Heats of Solids

When the temperature of a solid is raised, its internal energy increases. If the internal energy of a solid resides in the vibrations of its constituent atoms, as we would expect, we should be able to determine its specific heat directly from the results of the previous section. We shall consider the molar specific heat at constant volume, c_V, which is the energy that must be added to 1 kmole of a substance, whose volume is held constant, to raise its temperature by $1°$C. The specific heat at constant pressure, c_p, is 3 to 5 percent higher than c_V in solids because it includes the work associated with a volume change as well as the change in internal energy.

The vibrations of each atom in a solid may be resolved into components along three perpendicular axes, so that we may represent the atom by three harmonic oscillators. The average thermal energy of such an atom is $3kT$, since $\overline{E} = kT$ for its vibrations in each direction. A kmole of the solid contains Avogadro's number N_0 of atoms, and its total internal energy U at the temperature T is accordingly

19.11
$$U = 3N_0kT = 3RT$$

where

$$R = N_0 k$$

$$= 8.31 \times 10^3 \text{ joules/kmole }°\text{K}$$

$$= 1.99 \text{ kcal/kmole }°\text{K}$$

is the universal gas constant. The specific heat at constant volume is given in terms of U by

$$c_V = \left(\frac{\partial U}{\partial T}\right)_V$$

and so here

19.12 $c_V = 3R = 5.97 \text{ kcal/kmole °K}$ **Dulong-Petit law**

Over a century ago Dulong and Petit found empirically that, indeed, $c_V \approx 3R$ for most solids at room temperature and above, and the above result is known as the *Dulong-Petit law* in their honor.

However, the Dulong-Petit law fails for such light elements as boron, beryllium, and carbon (as diamond), for which c_V = 3.34, 3.85, and 1.46 kcal/kmole °K respectively at room temperature. Even worse, the specific heats of *all* solids drop sharply at low temperatures and approach 0 as T approaches 0°K. Figure 19.3 shows how c_V varies with T for various elements. Clearly something serious is wrong with the analysis leading up to Eq. 19.12, and it must be something quite fundamental because the curves of Fig. 19.3 share the same general character.

19.4   The Einstein Theory

In 1907 Einstein discerned that the basic error in the derivation of Eq. 19.12 lies in the figure of kT for the average energy per oscillator in a solid. This error is the same as that responsible for the incorrect Rayleigh-Jeans formula (Sec. 16.3): the spectrum of possible energies is not continuous, but is quantized in multiples of $h\nu$. Hence the average energy per oscillator is that given by Eq. 16.23,

FIGURE 19.3   The variation of the molar specific heat at constant volume c_V with temperature for several elements.

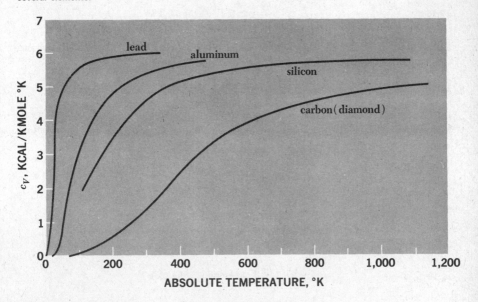

19.13
$$\overline{E} = \frac{h\nu}{e^{h\nu/kT} - 1}$$
Average energy per oscillator

and not $\overline{E} = kT$. The total internal energy of a kmole of a solid therefore becomes

19.14
$$U = 3N_0\overline{E} = \frac{3N_0h\nu}{e^{h\nu/kT} - 1}$$

and its molar specific heat is

$$c_V = \left(\frac{\partial U}{\partial T}\right)_V$$

19.15
$$= 3R\left(\frac{h\nu}{kT}\right)^2 \frac{e^{h\nu/kT}}{(e^{h\nu/kT} - 1)^2}$$
Einstein specific heat formula

We can see at once that this approach is at least on the right track. At high temperatures $h\nu \ll kT$, and

$$e^{h\nu/kT} \approx 1 + \frac{h\nu}{kT}$$

since

$$e^x = 1 + x + \frac{x^2}{2!} + \frac{x^3}{3!} + \cdots$$

Hence at high temperatures Eq. 19.13 becomes

19.16
$$\overline{E} \approx \frac{h\nu}{h\nu/kT} \approx kT \qquad h\nu \ll kT$$

and we have $c_V \approx 3R$, the Dulong-Petit value, which agrees with the data under such conditions. At low temperatures $h\nu \gg kT$ and $e^{h\nu/kT} \gg 1$; therefore

19.17
$$\overline{E} \approx h\nu e^{-h\nu/kT} \qquad h\nu \gg kT$$

As T decreases, $\overline{E}$ falls off rapidly; most of the oscillators are in their ground states, since the first excited state involves an energy increment of $h\nu$ and $h\nu \gg kT$. At low temperatures, then,

$$U = 3N_0\overline{E} \approx 3RT\left(\frac{h\nu}{kT}\right)e^{-h\nu/kT}$$

and

$$c_V = \left(\frac{\partial U}{\partial T}\right)_V$$

19.18
$$\approx 3R\left(\frac{h\nu}{kT}\right)^2 e^{-h\nu/kT} \qquad h\nu \gg kT$$

446 Specific Heats of Solids

Thus Einstein's formula for c_V approaches 0 at low temperatures, in contrast to the Dulong-Petit formula.

It is worth recalling that the possible energies of a harmonic oscillator are restricted to

$$E_n = \left(n + \frac{1}{2}\right) h\nu \qquad n = 0, 1, 2, \ldots$$

according to Eq. 8.34. Hence the ground-state energy of each oscillator in a solid is $E_0 = \frac{1}{2}h\nu$, the zero-point value, and not $E_0 = 0$. The presence of a zero-point energy does not affect the preceding analysis, however, since it merely adds a constant term of $U_0 = \frac{1}{2}N_0 h\nu$ that is independent of T to the molar energy of the solid, and this term vanishes when the partial derivative $(\partial U / \partial T)_V$ is taken.

Equation 19.15 is plotted in Fig. 19.4 for aluminum with $\nu = 6.4 \times 10^{12} \ \text{sec}^{-1}$. There is evidently good agreement with the data except at the lowest temperatures, where c_V is more nearly proportional to T^3 than to Eq. 19.18. We must seek further for a complete explanation of the specific heats of solids.

19.5 The Debye Theory

In Einstein's model of a solid, each atom is regarded as vibrating independently of its neighbors. In an effort to take into account the coupling between adjacent atoms, Peter

FIGURE 19.4 A comparison of specific-heat data for aluminum (dots) and the predicted curve of the Einstein theory, Eq. 19.15.

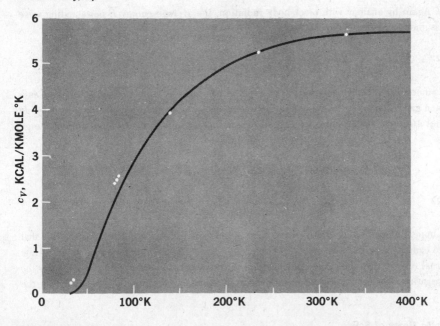

Debye went to the opposite extreme in 1912 and considered a solid as a continuous elastic body. Instead of residing in the vibrations of individual atoms, the internal energy of a solid is assumed to reside in elastic standing waves. These waves, like electromagnetic waves in a cavity, have quantized energy contents. A quantum of vibrational energy in a solid is called a *phonon*, and it travels with the speed of sound since sound waves are elastic in nature. The concept of phonons is quite general and has applications other than in connection with specific heats.

The number of possible elastic standing waves of any kind with wavelengths between λ and $\lambda + d\lambda$ in a solid per unit volume is given by Rayleigh's formula, Eq. 16.19, as

19.19
$$n(\lambda) \, d\lambda = \frac{4\pi \, d\lambda}{\lambda^4}$$

In terms of frequency ν this formula becomes

19.20
$$n(\nu) \, d\nu = \frac{4\pi\nu^2 \, d\nu}{v^3}$$

where v is the wave speed. There are two kinds of elastic waves that can occur in a solid, longitudinal waves whose speed is v_l and transverse waves whose speed is v_t. Furthermore, there are two perpendicular directions of polarization for a transverse wave. Hence the total number of possible standing elastic waves with frequencies between ν and $\nu + d\nu$ is

19.21
$$n(\nu) \, d\nu = 4\pi \left(\frac{1}{v_l{}^3} + \frac{2}{v_t{}^3} \right) \nu^2 \, d\nu$$

Again by analogy with black-body radiation, the average energy $\overline{E}$ per standing wave is assumed to be that of Eq. 16.23,

19.22
$$\overline{E} = \frac{h\nu}{e^{h\nu/kT} - 1}$$

(In other words, Debye asserted that a phonon gas has the same statistical behavior as a photon gas.) The energy contained in the waves whose frequencies range from ν to $\nu + d\nu$ is thus $n(\nu)\overline{E} \, d\nu$. If the volume of a kmole of a particular solid is V_0, its total internal energy U is

$$U = V_0 \int_0^{\nu_m} n(\nu) \, \overline{E} d\nu$$

19.23
$$= 4\pi V_0 \left(\frac{1}{v_l{}^3} + \frac{2}{v_t{}^3} \right) \int_0^{\nu_m} \frac{h\nu^3 \, d\nu}{e^{h\nu/kT} - 1}$$

The upper limit of integration is some maximum frequency ν_m, which reflects the fact that there cannot be an infinite number of standing waves or the solid would have an infinite internal energy. (Another way to look at this limit is that it is meaningless to consider the propagation of a mechanical wave in a medium when the wavelength is smaller than the interatomic distance.) Debye assumed that the total number of different standing waves is the same as the $3N_0$ degrees of freedom of a kmole of a solid, so that

$$3N_0 = V_0 \int_0^{\nu_m} n(\nu)\,d\nu$$

$$= 4\pi V_0 \left(\frac{1}{v_l^3} + \frac{2}{v_t^3}\right)\int_0^{\nu_m} \nu^2\,d\nu$$

19.24
$$= \frac{4}{3}\pi V_0 \left(\frac{1}{v_l^3} + \frac{2}{v_t^3}\right)\nu_m^3$$

and

19.25
$$\nu_m = \left[\frac{9N_0}{4\pi V_0 \left(\dfrac{1}{v_l^3} + \dfrac{2}{v_t^3}\right)}\right]^{1/3}$$

Equation 19.23 can be rewritten in terms of ν_m as

19.26
$$U = \frac{9N_0}{\nu_m^3}\int_0^{\nu_m} \frac{h\nu^3\,d\nu}{e^{h\nu/kT} - 1}$$

It is convenient at this point to change the variable in Eq. 19.26 from ν to the dimensionless quantity x, where

19.27
$$x = \frac{h\nu}{kT}$$

Hence

$$\nu = \frac{kT}{h}x$$

$$d\nu = \frac{kT}{h}dx$$

In addition, we define the *Debye characteristic temperature* θ as

19.28
$$\theta = \frac{h\nu_m}{k}$$

Debye temperature

In terms of x and θ, Eq. 19.26 becomes

$$U = 9N_0 k \frac{T^4}{\theta^3}\int_0^{\theta/T} \frac{x^3\,dx}{e^x - 1}$$

19.29
$$= 9R \frac{T^4}{\theta^3}\int_0^{\theta/T} \frac{x^3\,dx}{e^x - 1}$$

since the universal gas constant R is equal to $N_0 k$. The Debye result for the molar specific heat of a solid at constant volume is therefore

$$c_V = \left(\frac{\partial U}{\partial T}\right)_V$$

19.30
$$= 9R\left[4\left(\frac{T}{\theta}\right)^3 \int_0^{\theta/T} \frac{x^3\,dx}{e^x - 1} - \left(\frac{\theta}{T}\right)\frac{1}{e^{\theta/T} - 1}\right]$$ **Debye specific heat formula**

Equation 19.30 gives the specific heat as a function of T/θ, the ratio between the absolute temperature and the Debye temperature θ of the material. Table 19.1 contains values of θ for a number of substances.

Let us examine the behavior of Debye's expression for c_V at both extremes of temperature to see whether it agrees with the observation that $c_V \rightarrow 3R$ when T is large and that $c_V \propto T^3$ when T is small. When T is large, θ/T is very small, and $e^{\theta/T} \approx 1 + \theta/T$. Hence

$$\left(\frac{\theta}{T}\right)\frac{1}{e^{\theta/T} - 1} \approx \left(\frac{\theta}{T}\right)\frac{1}{1 + \theta/T - 1} \approx 1$$

Furthermore, since $x = h\nu/kT$, x is also small when T is large, and $e^x \approx 1 + x$. Hence

$$\frac{x^3}{e^x - 1} \approx \frac{x^3}{1 + x - 1} \approx x^2$$

and

$$\int_0^{\theta/T} \frac{x^3\,dx}{e^x - 1} \approx \int_0^{\theta/T} x^2\,dx \approx \frac{1}{3}\left(\frac{\theta}{T}\right)^3$$

We therefore have

$$c_V \approx 9R\left[4\left(\frac{T}{\theta}\right)^3 \times \frac{1}{3}\left(\frac{\theta}{T}\right)^3 - 1\right] \approx 9R\left[\frac{4}{3} - 1\right]$$

$$\approx 3R \qquad T \gg \theta$$

which is the Dulong-Petit value.

At very low temperatures $\theta/T \rightarrow \infty$, so that the second term in the brackets of Eq. 19.30 becomes negligible and we can replace the upper limit of integration in the first term by ∞. Since

$$\int_0^\infty \frac{x^3\,dx}{e^x - 1} = \frac{\pi^4}{15}$$

we have

$$c_V \approx 9R\left[4\left(\frac{T}{\theta}\right)^3 \frac{\pi^4}{15}\right]$$

19.31
$$\approx \frac{12}{5}\frac{\pi^4 R}{\theta^3} T^3 \qquad T \ll \theta$$

Thus the Debye theory accounts for the observed dependence of c_V on T^3 at low temperatures.

At intermediate temperatures Debye's formula for the molar specific heat must be evaluated numerically. The resulting curve for c_V is shown in Fig. 19.5; the agreement with experiment is excellent for many substances. The curve levels out at $T/\theta \approx 1$, when $c_V = 5.676$ kcal/kmole °K, which is 95 percent of its maximum value of 5.961 when $T/\theta = \infty$. Hence when $T > \theta$, a solid behaves classically and has a specific heat of $c_V \approx 3R$, the Dulong-Petit value, while when $T < \theta$, its behavior is dominated by quantum effects and c_V decreases with decreasing temperature. The reason for the change from classical to quantum behavior is the same as that mentioned in connection with Einstein's theory of specific heats: at high temperatures the spacing $h\nu$ between possible energies is small relative to kT, so the spectrum is effectively continuous, while at low temperatures the spacing is large relative to kT and inhibits the possession of energies above the zero-point energy.

For all its merits, the Debye model of a solid contains simplifications which limit its range of validity unless modified. For example, the actual spectrum of elastic waves in a solid depends upon its crystal structure, and in general is somewhat different from the universal spectrum of Eq. 19.21 which has a sharp cutoff at ν_m. Also, if the lattice points of

TABLE 19.1. SOME DEBYE TEMPERATURES

Substance	θ, °K
Hg	75
Pb	95
H	100
Na	160
Sn (white)*	200
Sn (gray)*	260
Cu	340
Fe	360
Al	375
Be	1200
C (diamond)	1850
AgBr	150
NaCl	280
CaF$_2$	474

* See Sec. 17.6.

The Debye Theory

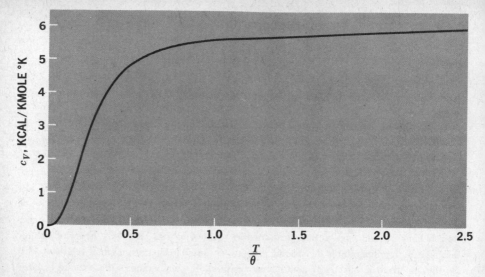

FIGURE 19.5   The variation of c_v with T/θ according to the Debye theory, Eq. 19.30.

the solid are occupied by groups of atoms instead of by individual atoms, as in the case of molecular crystals and ionic crystals involving such ions as NH_4^+, the vibrations of the atoms within the group must be taken into account. Since the internal vibrations of each atom group are independent of those of the other groups, a factor of the Einstein type (Eq. 19.15) must be included for such vibrations. Still other effects influence the specific heats of many solids.

19.6   The Fermi Energy

The various theories of the specific heats of solids that have been considered here are as successful for metals as for nonmetals, which is strange because they completely ignore the presence of free electrons in metals. In a typical metal each atom contributes one electron to the common "electron gas," so in 1 kmole of the metal there are N_0 free electrons. If these electrons behave like molecules in an ideal gas, each would have $\frac{3}{2}kT$ of kinetic energy on the average, and the metal would have

$$U_e = \frac{3}{2}N_0kT = \frac{3}{2}RT$$

of electron energy per kmole. The molar specific heat of the electrons should therefore be

$$c_{Ve} = \left(\frac{\partial U_e}{\partial T}\right)_V = \frac{3}{2}R$$

and the total specific heat of the metal at temperatures above the Debye temperature θ should be

$$c_V \approx 3R + \tfrac{1}{2}R \approx \tfrac{7}{2}R$$

Actually, of course, $c_V \approx 3R$ at high temperatures, from which we conclude that the free electrons do not in fact contribute to the specific heat. Why not?

In order to account for the unexpectedly low specific heats of metals, we must look more closely into the statistical behavior of the electron gas. We might suspect from the discussion of Chap. 16 that such a gas behaves like a system of Fermi particles and hence obeys Fermi-Dirac statistics, and this is indeed the case. The Fermi-Dirac distribution law for the number of electrons n_i with the energy u_i is

$$\textbf{19.32} \qquad\qquad n_i = \frac{g_i}{e^\alpha e^{u_i/kT} + 1}$$

It is more convenient to consider a continuous distribution of electron energies than the discrete distribution of Eq. 19.32, so that the distribution law becomes

$$\textbf{19.33} \qquad\qquad n(u)\ du = \frac{g(u)\ du}{e^\alpha e^{u/kT} + 1}$$

To find $g(u)\ du$, the number of quantum states available to electrons with energies between u and $u + du$, we use the same reasoning as for the photon gas involved in black-body radiation. The correspondence is exact because there are two possible spin states, $m_s = +\tfrac{1}{2}$ and $m_s = -\tfrac{1}{2}$, for electrons, thus doubling the number of available phase-space cells just as the existence of two possible directions of polarization for otherwise identical photons doubles the number of cells for a photon gas. In terms of momentum we found in Sec. 16.4 that

$$g(p)\ dp = \frac{8\pi V p^2\ dp}{h^3}$$

For nonrelativistic electrons

$$p^2\ dp = (2m^3 u)^{1/2}\ du$$

with the result that

$$\textbf{19.34} \qquad\qquad g(u)\ du = \frac{8\ \sqrt{2}\pi V m^{3/2}}{h^3}\ u^{1/2}\ du$$

The next step is to evaluate the parameter α. In order to do this, we consider the condition of the electron gas at low temperatures. As observed in Sec. 16.6, the occupation index when T is small is 1 from $u = 0$ until near the Fermi energy u_F, where it drops rapidly to 0. This situation reflects the effect of the exclusion principle: no states can contain more than one electron, and so the minimum energy configuration of an electron gas is one in which the lowest states are filled and the remaining ones are empty. If we set

19.35
$$\alpha = -\frac{u_F}{kT}$$

the occupation index becomes

19.36
$$f(u) = \frac{n(u)}{g(u)} = \frac{1}{e^{(u-u_F)/kT} + 1}$$

The formula is in accord with the exclusion principle. At $T = 0°K$,

$$f(u) = 1 \text{ when } u < u_F$$
$$= 0 \text{ when } u > u_F$$

As the temperature increases, the occupation index changes from 1 to 0 more and more gradually, as in Fig. 19.6. At all temperatures

$$f(u) = \frac{1}{2} \text{ when } u = u_F$$

If a particular metal sample contains N free electrons, we can calculate its Fermi energy u_F by filling up its energy states with these electrons in order of increasing energy starting from $u = 0$. The highest energy state to be filled will then have the energy $u = u_F$ by definition. The number of electrons that can have the same energy u is equal to the number of states $g(u)$ that have this energy, since each state is limited to one electron. Hence

19.37
$$\int_0^{u_F} g(u) \, du = N$$

Substituting Eq. 19.34 for $g(u) \, du$ yields

$$N = \frac{8\sqrt{2}\pi V m^{3/2}}{h^3} \int_0^{u_F} u_1^{1/2}/du$$
$$= \frac{16\sqrt{2}\pi V m^{3/2}}{3h^3} u_F^{3/2}$$

and

19.38
$$u_F = \frac{h^2}{2m}\left(\frac{3N}{8\pi V}\right)^{2/3}$$ **Fermi energy**

The quantity N/V is the density of free electrons; hence u_F is independent of the dimensions of whatever metal sample is being considered.

Let us use Eq. 19.38 to calculate the Fermi energy in copper. The electron configuration of the ground state of the copper atom is $1s^22s^22p^63s^23p^63d^{10}4s$; that is, each atom has

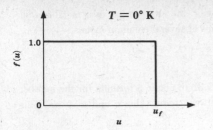
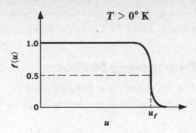

FIGURE 19.6 The occupation index for a Fermi-Dirac distribution at absolute zero and at a higher temperature.

a single $4s$ electron outside closed inner shells. It is therefore reasonable to assume that each copper atom contributes one free electron to the electron gas. The electron density $\eta = N/V$ is accordingly equal to the number of copper atoms per unit volume, which is given by

$$\frac{\text{Atoms}}{\text{Volume}} = \frac{(\text{atoms/kmole}) \times (\text{mass/volume})}{\text{mass/kmole}}$$

19.39

$$= \frac{N_0 \rho}{w}$$

Here

$$N_0 = \text{Avogadro's number} = 6.02 \times 10^{26} \text{ atoms/kmole}$$
$$\rho = \text{density of copper} = 8.94 \times 10^3 \text{ kg/m}^3$$
$$w = \text{atomic mass of copper} = 63.5 \text{ kg/kmole}$$

so that

$$\eta = \frac{6.02 \times 10^{26} \text{ atoms/kmole} \times 8.94 \times 10^3 \text{ kg/m}^3}{63.5 \text{ kg/kmole}}$$

$$= 8.5 \times 10^{28} \text{ atoms/m}^3$$
$$= 8.5 \times 10^{28} \text{ electrons/m}^3$$

The corresponding Fermi energy is, from Eq. 19.38,

$$u_F = \frac{(6.63 \times 10^{-34} \text{ joule-sec})^2}{2 \times 9.11 \times 10^{-31} \text{ kg/electron}} \left(\frac{3 \times 8.5 \times 10^{28} \text{ electrons/m}^3}{8\pi} \right)^{2/3}$$

$$= 1.13 \times 10^{-18} \text{ joule}$$
$$= 7.04 \text{ ev}$$

At absolute zero, $T = 0°\text{K}$, there would be electrons with energies of up to 7.04 ev in copper. By contrast, *all* of the molecules in an ideal gas at absolute zero would have zero

The Fermi Energy

energy. Because of its decidedly nonclassical behavior, the electron gas in a metal is said to be *degenerate*. Table 19.2 gives the Fermi energies of several common metals.

19.7 Electron-energy Distribution

We may now substitute for α and $g(u)\,du$ in Eq. 19.33 to obtain a formula for the number of electrons in an electron gas having energies between some value u and $u + du$. This formula is

19.40
$$n(u)\,du = \frac{(8\sqrt{2}\pi\,Vm^{3/2}/h^3)u^{1/2}\,du}{e^{(u-u_F)/kT} + 1}$$

If we express the numerator of Eq. 19.40 in terms of the Fermi energy u_F, we obtain

19.41
$$n(u)\,du = \frac{(3N/2)u_F^{-3/2}u^{1/2}\,du}{e^{(u-u_F)/kT} + 1}$$

Equation 19.41 is plotted in Fig. 19.7 for the temperatures $T = 0$, 300, and 1200°K.

It is interesting to determine the average electron energy at absolute zero. To do this, we first obtain the total energy U_0 at 0°K, which is

$$U_0 = \int_0^{u_F} un(u)\,du$$

Since at $T = 0°$K all of the electrons have energies less than or equal to the Fermi energy u_F, we may let

$$e^{(u-u_F)/kT} = 0$$

and

$$U_0 = \frac{3N}{2}u_F^{-3/2}\int_0^{u_F} u^{3/2}\,du$$

19.42
$$= \frac{3}{5}Nu_F$$

The average electron energy $\bar{u}_0$ is this total energy divided by the number of electrons present N, which yields

19.43
$$\bar{u}_0 = \frac{3}{5}u_F$$

Since Fermi energies for metals are usually several electron volts, the average electron energy in them at 0°K will also be of this order of magnitude. The temperature of an ideal gas whose molecules have an average kinetic energy of 1 ev is 11,600°K; this means that, if free electrons behaved classically, a sample of copper would have to be at a temperature

TABLE 19.2. SOME FERMI ENERGIES

Metal	Fermi energy, ev
Lithium (Li)	4.72
Sodium (Na)	3.12
Aluminum (Al)	11.8
Potassium (K)	2.14
Cesium (Cs)	1.53
Copper (Cu)	7.04
Zinc (Zn)	11.0
Silver (Ag)	5.51
Gold (Au)	5.54

of about $50,000°K$ for its electrons to have the same average energy they actually have at $0°K$.

The considerable amount of kinetic energy possessed by the valence electrons in the electron gas of a metal represents a repulsive influence, as noted in Chap. 17. The act of assembling a group of metal atoms into a solid requires that additional energy be given to the valence electrons in order to elevate them to the higher energy states required by the exclusion principle. The atoms in a metallic solid, however, are closer together because of their bonds than they would be otherwise. As a result the valence electrons are, on the

FIGURE 19.7 Distribution of electron energies in a metal at various temperatures.

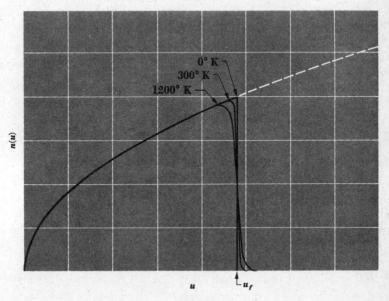

average, closer to an atomic nucleus in a metallic solid than they are in an isolated metal atom. These electrons accordingly have lower potential energies in the former case than in the latter, sufficiently lower to lead to a net cohesive force even when the added electron kinetic energy is taken into account. We shall explore this point further in the following chapter.

19.8 Electronic Specific Heat

The failure of the free electrons in a metal to contribute appreciably to its specific heat follows directly from the nature of the electron-energy distribution. When a metal is heated, only those electrons near the very top of the energy distribution—within about kT of the Fermi energy—are excited to higher energy states. The less energetic electrons cannot absorb more energy because the states above them are already filled, and it is unlikely that an electron with, say, an energy u that is 0.5 ev below u_F can leapfrog the filled states above it to the nearest empty state when kT at room temperature is 0.025 ev and even at 500°K is only 0.043 ev.

As a solid is heated, then, only those electrons within about kT of the Fermi level acquire additional energy. We can estimate the number of such electrons by assuming that $n(u)$ is directly proportional to u near the top of the energy distribution. In this case the number N of electrons that contribute to the specific heat of a kmole of a univalent metal at the temperature T is

$$N \approx N_0 \frac{\Delta u}{u_F} \approx N_0 \left(\frac{kT}{u_F} \right)$$

where u_F is the Fermi energy and N_0 (Avogadro's number) is the number of free electrons per kmole. Each of the N electrons within kT of u_F has, by hypothesis, acquired $\frac{3}{2}kT$ of energy, and so the total electronic contribution to the internal energy of the metal is

$$U_e \approx N_0 \left(\frac{kT}{u_F} \right) \times \frac{3}{2}kT$$

$$\approx \frac{3}{2} \left(\frac{kT}{u_F} \right) RT$$

since $N_0 k = R$. The resulting electronic specific heat is

$$c_{Ve} = \left(\frac{\partial U_e}{\partial T} \right)_V \approx \frac{3}{2} \left(\frac{kT}{u_F} \right) R$$

and not the $\frac{3}{2}R$ that the classical theory of an electron gas predicts. A more detailed calculation shows that

19.44
$$c_{Ve} = \frac{\pi^2}{2} \left(\frac{kT}{u_F} \right) R$$
Electronic specific heat

At room temperature, kT/u_F ranges from 0.016 for cesium to 0.0021 for aluminum for the metals listed in Table 19.2, so the coefficient of R is very much smaller than the classical figure of ½. The dominance of the lattice specific heat c_V over the electronic specific heat is pronounced over a wide temperature range. However, at very low temperatures c_{Ve} becomes important since c_V is proportional to T^3 when $T \ll \theta$ whereas c_{Ve} is proportional to T. At very high temperatures c_V levels out at $\approx 3R$, but c_{Ve} continues to increase, and the contribution of c_{Ve} to the total specific heat is then detectable although melting occurs long before $c_{Ve} > c_V$.

Problems

1. Find the average amplitude of the vibrations of aluminum atoms at the melting point of aluminum, 660°C. What percentage is this of the interatomic spacing?

2. In solid iron the interatomic spacing is 2.48 A and Young's modulus is 9.1×10^{10} n/m². Iron atoms have an average mass of 9.3×10^{-26} kg. (a) Find the frequency of vibration of the atoms in solid iron. (b) Find the average amplitude of these vibrations at the melting point of iron, 1539°C. What percentage is this of the interatomic spacing?

3. Lead has an fcc structure with a lattice constant of 4.94 A. Young's modulus for lead is 1.6×10^{10} n/m². If lead melts when the average amplitude of its atomic vibrations is 15.8 percent of the interatomic spacing, find the melting point of lead.

4. In copper, $v_l = 4.56 \times 10^3$ m/sec and $v_t = 2.25 \times 10^3$ m/sec. The density of copper is 8.96 g/cm³ and its atomic weight is 63.54. (a) Calculate v_m in copper from these data. (b) The Debye temperature for copper, as obtained from specific heat measurements, is 340°K. Find v_m from this figure and compare it with the result obtained in (a).

5. In aluminum, $v_l = 6.32 \times 10^3$ m/sec and $v_t = 3.10 \times 10^3$ m/sec. The density of aluminum is 2.70 g/cm³ and its atomic weight is 26.97. (a) Calculate v_m in aluminum from these data. (b) The Debye temperature for aluminum, as obtained from specific heat measurements, is 375°K. Find v_m from this figure and compare it with the result obtained in (a). (c) Are the preceding values of v_m consistent with the vibrational frequency of aluminum atoms in solid aluminum?

6. What is the connection between the fact that the free electrons in a metal obey Fermi statistics and the fact that the photoelectric effect is virtually temperature-independent?

7. The Fermi energy in silver is 5.51 ev. (a) What is the average energy of the free electrons in silver at 0°K? (b) What temperature is necessary for the average molecular energy in an ideal gas to have this value? (c) What is the speed of an electron with this energy?

8. The Fermi energy in copper is 7.04 ev. (a) Approximately what percentage of the free electrons in copper are in excited states at room temperature? (b) At the melting point of copper, 1083°C?

9. The density of zinc is 7.13 g/cm³ and its atomic weight is 65.4. The electronic structure of zinc is given in Table 10.2, and the effective mass (see Sec. 20.7) of an electron in zinc is 0.85 m_e. Calculate the Fermi energy in zinc.

10. The density of aluminum is 2.70 g/cm³ and its atomic weight is 26.97. The electronic structure of aluminum is given in Table 10.2 (note: the energy difference between 3s and 3p electrons is very small), and the effective mass (see Sec. 20.7) of an electron in aluminum is 0.97 m_e. Calculate the Fermi energy in aluminum.

11. The electronic specific heat of zinc is $\sim 1.3 \times 10^{-4}\, T$ kcal/kg °K. Find the Fermi energy in zinc from this figure.

12. Find the electronic specific heat of silver at room temperature and at the melting point of silver, 960°C. What percentage of the lattice specific heat—assumed given by the Dulong-Petit law—are these figures?

13. Why is the electronic specific heat in nonmetals a negligible quantity?

14. How would the degree of occupancy of each quantum state change if the electrons in a metal were to have a Maxwell-Boltzmann energy distribution instead of a Fermi-Dirac one?

15. Derive Eq. 19.38 for the Fermi energy u_F in the following way. Imagine N electrons contained in a cubical box L long on a side. Obtain a formula for the number of quantum states in the box for which $n\ (= \sqrt{n_x^2 + n_y^2 + n_z^2})$ is less than some value n_F; this can be done most easily with the help of $\mathbf{n}$ space. Multiply the number of quantum states by the number of electrons that can occupy each state, and set the result equal to N, the number of electrons in the box. Solve for n_F^2 and find u_F with the help of Eq. 8.47.

Chapter 20
Band Theory
of Solids

The atoms in almost every crystalline solid, whether a metal or not, are so close together that their valence electrons constitute a single system of electrons common to the entire crystal. The exclusion principle is obeyed by such an electron system because the energy states of the outer electron shells of the atoms are all altered somewhat by their mutual interactions. In place of each precisely defined characteristic energy level of an individual atom, the entire crystal possesses an *energy band* composed of myriad separate levels very close together. Since there are as many of these separate levels as there are atoms in the crystal, the band cannot be distinguished from a continuous spread of permitted energies. As we shall see, the presence of energy bands, the gaps that may occur between them, and the extent to which they are filled by electrons not only determine the electrical behavior of a solid but also have important bearing on other of its properties.

20.1 Energy Bands

There are two ways to consider the origin of energy bands. The simplest is to look into what happens to the energy levels of isolated atoms as they are brought closer and closer together to form a solid. We shall introduce the subject in this way, and then examine some of the consequences of the notion of energy bands. Later in the chapter we shall analyze energy bands in terms of the restrictions imposed by the periodicity of a crystal lattice on the motion of electrons, a more powerful approach that provides the basis of much of the modern theory of solids.

Figure 20.1 shows the energy levels in sodium plotted versus internuclear distance. The $3s$ level is the first occupied level in the sodium atom to broaden into a band; the $2p$ level does not begin to spread out until a quite small internuclear separation. This behavior reflects the order in which the electron subshells of sodium atoms interact as the atoms are brought together. The average energies in the $3p$ and $3s$ bands drop at first, implying attractive forces. The actual internuclear distance in solid sodium is indicated, and it corresponds, as it should, to a situation of minimum average energy.

The energy bands in a solid correspond to the energy levels in an atom, and an electron in a solid can possess only those energies that fall within these energy bands. The

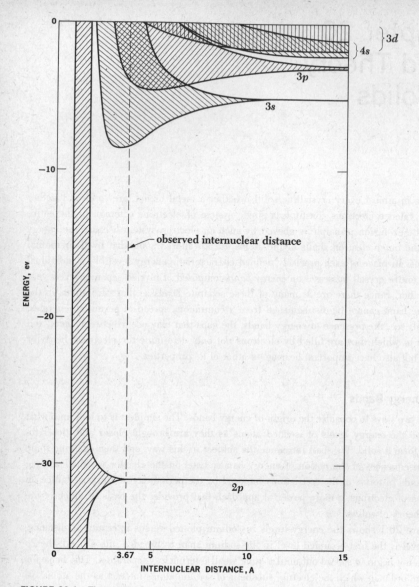

FIGURE 20.1 The energy levels of sodium atoms become bands as their internuclear distance decreases. The observed internuclear distance in solid sodium is 3.67 A.

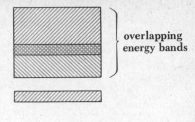

overlapping energy bands

FIGURE 20.2 The energy bands in a solid may overlap. _____

various energy bands in a solid may overlap, as in Fig. 20.2, in which case its electrons have a continuous distribution of permitted energies. In other solids the bands may *not* overlap (Fig. 20.3), and the intervals between them represent energies which their electrons can not possess. Such intervals are called *forbidden bands*. The electrical behavior of a crystalline solid is determined both by its energy-band structure and by how these bands are normally filled by electrons.

Figure 20.4 is a simplified diagram of the energy levels of a sodium atom and the energy bands of solid sodium. A sodium atom has a single 3s electron in its outer shell. This means that the lowest energy band in a sodium crystal is only half occupied, since each level in the band, like each level in the atom, is able to contain *two* electrons. When an electric field is set up across a piece of solid sodium, electrons easily acquire additional energy while remaining in their original energy band. The additional energy is in the form of kinetic energy, and the moving electrons contribute to an electric current by the mechanism to be discussed in Sec. 20.3. Sodium is therefore a good conductor of electricity, as are other crystalline solids with energy bands that are only partially filled.

Figure 20.5 is a simplified diagram of the energy bands of diamond. The two lower energy bands are completely filled with electrons, and there is a gap of 6 ev between the top of the higher of these bands and the empty band above it. This means that at least 6 ev of additional energy must be provided to an electron in a diamond crystal if it is to have any kinetic energy, since it cannot have an energy lying in the forbidden band. An energy

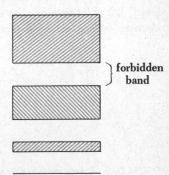

} forbidden band

FIGURE 20.3 A forbidden band separates nonoverlapping energy bands. _____

Energy Bands

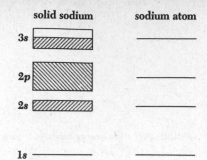

FIGURE 20.4 Energy levels in the sodium atom and the corresponding bands in solid sodium (not to scale).

increment of this magnitude cannot readily be given to an electron in a crystal by an electric field. An electron moving through a crystal undergoes a collision an average of every $\sim 10^{-8}$ m, and it loses much of the energy it gained from any electric field in the collision. (We shall later examine the nature of the collisions.) An electric-field intensity of 6×10^8 volts/m is necessary if an electron is to gain 6 ev in a path length of 10^{-8} m, well over 10^{10} times greater than the electric intensity needed to cause a current to flow in sodium. Diamond is therefore a very poor conductor of electricity and is accordingly classed as an insulator.

Silicon has a crystal structure resembling that of diamond, and, as in diamond, a gap separates the top of a filled energy band from a vacant higher band. The forbidden band in silicon, however, is only 1.1 ev wide. At low temperatures silicon is little better than diamond as a conductor, but at room temperature a small proportion of its electrons have sufficient kinetic energy of thermal origin to jump the forbidden band and enter the energy band above it. These electrons are sufficient to permit a limited amount of current to flow when an electric field is applied. Thus silicon has an electrical resistivity intermediate between those of conductors and those of insulators, and it is termed a *semiconductor*.

20.2 Impurity Semiconductors

The resistivity of semiconductors can be significantly affected by small amounts of impurity. Let us incorporate a few arsenic atoms in a silicon crystal. Arsenic atoms have five electrons in their outermost shells, while silicon atoms have four. (These shells have the

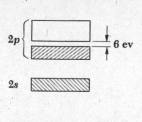

FIGURE 20.5   Energy bands in diamond (not to scale).

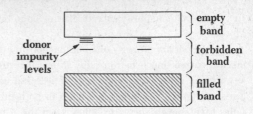

FIGURE 20.6   A trace of arsenic in a silicon crystal provides donor levels in the normally forbidden band, producing an *n*-type semiconductor.

configurations $4s^2 4p^3$ and $3s^2 3p^2$ respectively.) When an arsenic atom replaces a silicon atom in a silicon crystal, four of its electrons are incorporated in covalent bonds with its nearest neighbors. The fifth electron requires little energy to be detached and move about in the crystal (see Problems 7 and 8). As shown in Fig. 20.6, the presence of arsenic as an impurity provides energy levels just below the band which electrons must occupy for conduction to take place. Such levels are termed *donor levels*, and the substance is called an *n-type* semiconductor because electric current in it is carried by negative charges.

If we alternatively incorporate gallium atoms in a silicon crystal, a different effect occurs. Gallium atoms have only three electrons in their outer shells, whose configuration is $4s^2 4p$, and their presence leaves vacancies called *holes* in the electron structure of the crystal. An electron needs relatively little energy to enter a hole, but as it does so, it leaves a new hole in its former location. When an electric field is applied across a silicon crystal containing a trace of gallium, electrons move toward the anode by successively filling holes. The flow of current here is conveniently described with reference to the holes, whose behavior is like that of positive charges since they move toward the negative electrode. A substance of this kind is called a *p-type* semiconductor. (As we shall see later, certain metals, such as zinc, conduct current primarily by the motion of holes.) In the energy-band diagram of Fig. 20.7 we see that the presence of gallium provides energy levels, termed *acceptor levels*, just above the highest filled band. Any electrons that occupy these levels leave behind them, in the formerly filled band, vacancies which permit electric current to flow.

The addition of impurities to germanium and silicon crystals can be so controlled that part of a crystal is *n*-type and the rest *p*-type, with only a thin intermediate region. The simplest of several methods of producing such crystals is to gradually pull out a crystal that is growing in, say, molten germanium that contains a donor impurity, and quickly add an acceptor impurity to the melt in the middle of the process. The first portion of the resulting crystal to solidify will be *n*-type and the remainder will be *p*-type if the propor-

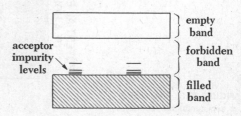

FIGURE 20.7   A trace of gallium in a silicon crystal provides acceptor levels in the normally forbidden band, producing a *p*-type semiconductor.

Impurity Semiconductors

tion of acceptor impurity exceeds that of donor impurity in the last stage of the crystal growth.

An important property of a *p-n* junction in a semiconductor crystal is that it conducts electric current much more readily in one direction than in the other. In the absence of any applied voltage across the ends of a crystal containing a *p-n* junction, there normally are small flows of electrons and holes in both directions across the junction. Thus some holes have enough energy to diffuse from the *p* region into the *n* region, where they recombine with conduction electrons, while at the same time thermal fluctuations lead to

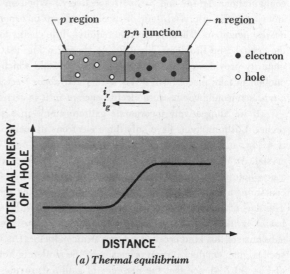

(a) Thermal equilibrium

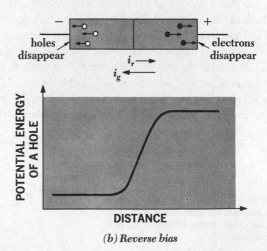

FIGURE 20.8   Operation of a semiconductor rectifier.

(b) Reverse bias

the spontaneous generation of holes in the n region that diffuse across the junction into the p region. At equilibrium the two currents, i_r and i_g respectively, are equal, as is shown in Fig. 20.8a. The other sketch shows how the potential energy of a hole varies through the crystal. At (b) a voltage is applied across the crystal with the p end negative and the n end positive. In this situation, called *reverse bias*, the potential-energy difference across the junction for a hole is increased over that in (a), which impedes the recombination current i_r without affecting i_g appreciably. Hence the limit of the reverse hole current is i_g. At (c) a forward bias is applied, with the p end of the crystal positive and the n end negative. Now the diffusion of holes across the junction into the n region of the crystal is enhanced, and i_r increases exponentially with the applied voltage. The behavior of the electron current corresponds exactly to that of the hole current except for a change in sign, so that under reverse bias few electrons flow from the n to the p region, while under forward bias a high density of electrons can surmount the potential barrier to complement the hole current in the opposite direction, yielding a high resultant electric current. Thus a p-n junction acts as a rectifier. More complex semiconductor devices, notably transistors, are in wide use in modern electronic circuits.

20.3 Ohm's Law

In the past few chapters the point has been made that the valence electrons in a metal constitute a gas of free electrons analogous in some respects to an ideal gas. More precisely, the electrons in the upper part of the energy distribution (within about kT of the Fermi energy) behave in this way, since they are close enough in energy to the vacant

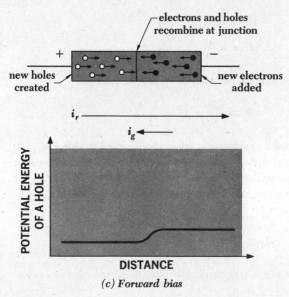

(c) Forward bias

upper part of their energy band to be capable of acquiring additional energy either of thermal origin or from an applied electric field. Before we go on to explain how and why the electron gas in a metal deviates from even this modified picture, let us see how the simple fact of the presence of an electron gas in a metal leads to a theory of electrical conduction that agrees with experience.

The current i that flows through a metallic conductor when the potential difference V is established across its ends is, within wide limits, directly proportional to V. This empirical relationship, called *Ohm's law*, is usually expressed as

20.1
$$i = \frac{V}{R}$$

where R, the *resistance* of the conductor, is a function of its dimensions, composition, and temperature but is independent of V. To derive Ohm's law, we begin by assuming that the free electrons in a metal, like the molecules in a gas, move in random directions and constantly undergo collisions. The collisions here are not with other electrons but with lattice irregularities, either structural defects or atoms momentarily out of place as they vibrate, as we shall learn later in this chapter; the precise nature of the collisions is unimportant to the present analysis so long as they are random.

Between collisions there is some average time interval τ which depends upon the detailed characteristics of the electron gas. When we apply a potential difference across the ends of a conductor, there will be an electric field E within it. The force this field exerts on an electron is eE and is opposite in direction to E, since electrons are negatively charged. The second law of motion yields

$$F = ma$$

$$eE = m\frac{\Delta v}{\tau}$$

where Δv is the change in speed the electron experiences in the time τ. Hence

$$\Delta v = \frac{eE\tau}{m}$$

Each time an electron undergoes a collision, it rebounds in a random direction and, on the average, no longer has any component of motion parallel to the field. At the start of each time interval τ we may imagine the electron at rest in the field direction, and at the end of the interval to have the speed Δv. The average drift speed v_d through the conductor is therefore

$$v_d = \frac{\Delta v}{2} = \frac{eE\tau}{2m}$$

The name "drift speed" is apt since v_d is very small relative to the average speed $\bar{v}$ of the

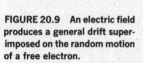

FIGURE 20.9 An electric field produces a general drift super-imposed on the random motion of a free electron.

random electron motion. The effect of imposing the electric field E on the free-electron gas in a metal, then, is to superimpose a general drift on the more pronounced but random motions of the electrons (Fig. 20.9).

We can now compute the total current i flowing in a wire of length L and cross-sectional area A that contains η free electrons per unit volume (Fig. 20.10). The current is

$$\text{Total current} = \text{free-electron density} \times \text{cross-sectional area}$$
$$\times \text{charge per electron} \times \text{electron drift speed}$$

$$i = \eta A e v_d$$
$$= \frac{\eta A e^2 E \tau}{2m}$$

Since the electric-field intensity E in the wire is just

$$E = \frac{V}{L}$$

the current is

20.2
$$i = \frac{\eta e^2 \tau}{2m} \frac{A}{L} V$$

Equation 20.2 becomes Ohm's law if we set

20.3
$$R = \left(\frac{2m}{\eta e^2 \tau}\right) \frac{L}{A}$$

This formula for the resistance of a conductor indicates that R should be proportional to

FIGURE 20.10 A model for deriving Ohm's law.

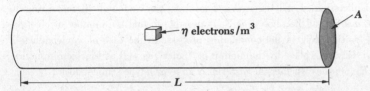

η electrons/m^3

A

L

Ohm's Law   469

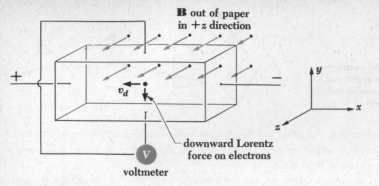

FIGURE 20.11 The Hall effect.

the length of the wire and inversely proportional to its cross-sectional area, which agrees with experience. At a given temperature the quantity in parentheses in Eq. 20.3, the *resistivity* of the metal, is a constant that depends solely upon the temperature and the properties of the metal. Our hypothesis that electrical conduction in metals occurs because of the presence of a gas of free electrons in them is evidently on the right track.

The Hall effect makes it possible to determine experimentally several of the microscopic properties of conductors and semiconductors. As in Fig. 20.11, a block of a solid is placed in perpendicular electric and magnetic fields. If the electric field is in the $+x$ direction, the electrons in the solid experience the force $-eE_x$ and drift in the $-x$ direction with the speed v_d. The magnetic field exerts the Lorentz force $\mathbf{F} = -e\mathbf{v} \times \mathbf{B}$ on moving electrons; this force averages out to zero on the random motions of the electrons, but it has a resultant component in the $-y$ direction because of the drift v_d. This component has the magnitude $F_y = -ev_dB$, and it causes the electrons to collect at the lower face of the block until there is built up between the upper and lower faces an electric field E_y that opposes the further transverse migration of electrons. The condition for force balance on an electron is thus

$$F_y = eE_y$$
$$-ev_dB = eE_y$$
$$E_y = -v_dB$$

The potential difference between the upper and lower faces of the block can be measured; from it E_y can be determined and, since B is known, v_d can be found. By measuring the resistivity of the material at the same time, it is possible to find the number of electrons per unit volume participating in the conduction process. In the case of semiconductors, where conduction may take place by the motion of "holes" as well as by electron motion, the direction of E_y indicates whether the material is an n- or p-type semiconductor.

20.4 Brillouin Zones

We now turn to a more detailed examination of how allowed and forbidden bands in a solid originate. The fundamental idea is that an electron in a crystal moves in a region of periodically varying potential (Fig. 20.12) rather than one of constant potential, and as a result diffraction effects occur that limit the electron to certain ranges of momenta that correspond to the allowed energy bands. In this way of thinking the interactions among the atoms influence valence-electron behavior indirectly through the lattice of the crystal these interactions bring about, instead of directly as in the approach described at the beginning of this chapter.

The de Broglie wavelength of a free electron of momentum p is

20.4
$$\lambda = \frac{h}{p}$$
Free electron

Unbound low-energy electrons can travel freely through a crystal since their wavelengths are long relative to the lattice spacing a. More energetic electrons, such as those with the Fermi energy in a metal, have wavelengths comparable with a, and such electrons are diffracted in precisely the same way as X rays (Sec. 3.4) or electrons in a beam (Sec. 4.5) directed at the crystal from the outside. (When λ is near a, $2a$, $3a$, ... in value, Eq. 20.4 no longer holds, as discussed in Sec. 20.7.) An electron of wavelength λ undergoes Bragg "reflection" from one of the atomic planes in a crystal when it approaches the plane at the angle θ, where from Eq. 3.8

20.5
$$n\lambda = 2a \sin \theta \qquad n = 1, 2, 3, \ldots$$

It is customary to treat the situation of electron waves in a crystal by replacing λ by the propagation constant k introduced in Sec. 4.3, where

FIGURE 20.12 The potential energy of an electron in a periodic array of positive ions.

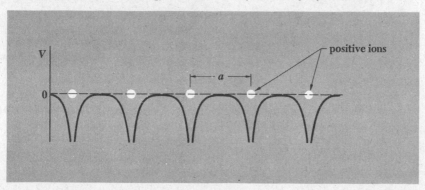

$$k = \frac{2\pi}{\lambda}$$

The propagation constant is equal to the number of radians per meter in the wave train it describes. Since the wave train moves in the same direction as the particle, we can describe the wave train by means of a vector **k**. Bragg's formula in terms of k is

$$k = \frac{n\pi}{a \sin \theta}$$

Figure 20.13 shows Bragg reflection in a two-dimensional square lattice. Evidently we can express the Bragg condition by saying that reflection from the vertical rows of ions occurs when the component of **k** in the x direction, k_x, is equal to $n\pi/a$. Similarly, reflection from the horizontal rows occurs when $k_y = n\pi/a$.

Let us consider first those electrons whose propagation constants are sufficiently small for them to avoid diffraction. If k is less than π/a, the electron can move freely through the lattice in any direction. When $k = \pi/a$, they are prevented from moving in the x or y directions by diffraction. The more k exceeds π/a, the more limited the possible directions

FIGURE 20.13 Bragg reflection from the vertical rows of ions occurs when $k_z = n\pi/a$.

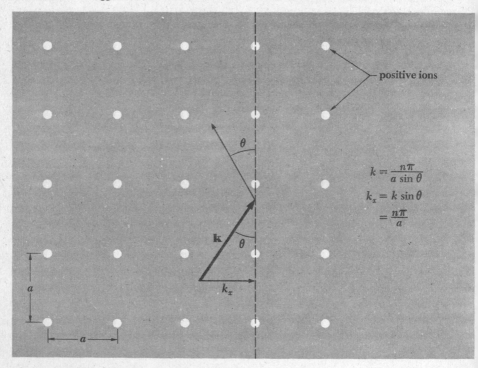

positive ions

$$k = \frac{n\pi}{a \sin \theta}$$
$$k_x = k \sin \theta$$
$$= \frac{n\pi}{a}$$

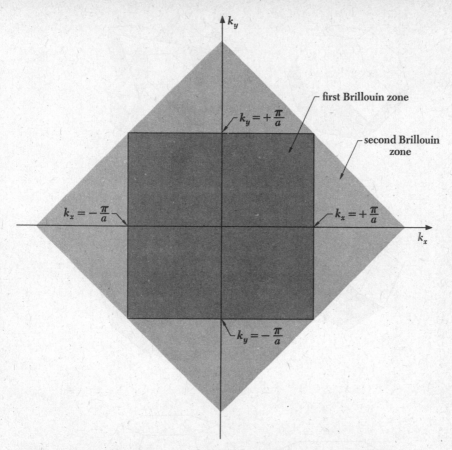

FIGURE 20.14   The first and second Brillouin zones of a two-dimensional square lattice.

of motion, until when $k = \pi/a \sin 45° = \sqrt{2}\,\pi/a$ the electrons are diffracted even when they move diagonally through the lattice.

The region in k-space that low-k electrons can occupy without being diffracted is called the *first Brillouin zone* and is shown in Fig. 20.14. The second Brillouin zone is also shown; it contains electrons with $k > \pi/a$ that do not fit into the first zone yet which have sufficiently small propagation constants to avoid diffraction by the diagonal sets of atomic planes in Fig. 20.13. The second zone contains electrons with k values from π/a to $2\pi/a$ for electrons moving in the $\pm x$ and $\pm y$ directions, with the possible range of k values narrowing as the diagonal directions are approached. Further Brillouin zones can be constructed in the same manner. The extension of this analysis to actual three-dimensional structures leads to the Brillouin zones shown in Fig. 20.15.

Brillouin Zones

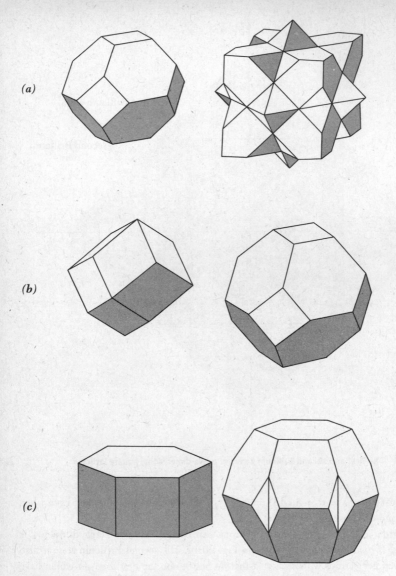

FIGURE 20.15 First and second Brillouin zones in (a) face-centered cubic structure, (b) body-centered cubic structure, and (c) hexagonal close-packed structure.

20.5 Origin of Forbidden Bands

The significance of the Brillouin zones becomes apparent when we examine the energies of the electrons in each zone. The energy of a free electron is related to its momentum p by

$$E = \frac{p^2}{2m}$$

and hence to its propagation constant k by

20.8
$$E = \frac{\hbar^2 k^2}{2m}$$

In the case of an electron in a crystal for which $k \ll \pi/a$, there is practically no interaction with the lattice, and Eq. 20.8 is valid. Since the energy of such an electron depends upon k^2, the contour lines of constant energy in a two-dimensional k-space are simply circles of constant k, as in Fig. 20.16. With increasing k the constant-energy contour lines become

FIGURE 20.16 Energy contours in electron volts in the first and second Brillouin zones of a hypothetical square lattice.

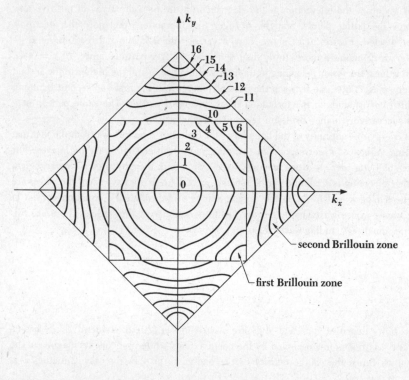

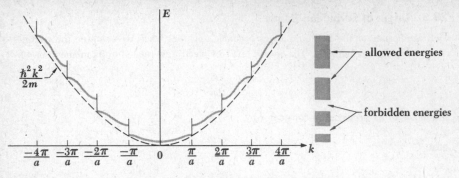

FIGURE 20.17 Electron energy E versus propagation constant k in the k_x direction. The dashed line shows how E varies with k for a free electron.

progressively closer together and also more and more distorted. The reason for the first effect is merely that E varies with k^2. The reason for the second is almost equally straightforward. The closer an electron is to the boundary of a Brillouin zone in k-space, the closer it is to being diffracted by the actual crystal lattice. But in particle terms the diffraction occurs by virtue of the interaction of the electron with the periodic array of positive ions that occupy the lattice points, and the stronger the interaction, the more the electron's energy is affected. Figure 20.17 shows how E varies with k in the x direction. As k approaches π/a, E increases more slowly than $\hbar^2 k^2/2m$, the free-particle figure. At $k = \pi/a$, E has two values, the lower belonging to the first Brillouin zone and the higher to the second zone. There is a definite gap between the possible energies in the first and second Brillouin zones which corresponds to the forbidden band spoken of earlier. The same pattern continues as successively higher Brillouin zones are reached.

The energy discontinuity at the boundary of a Brillouin zone follows from the fact that the limiting values of k correspond to standing waves rather than traveling waves. For clarity we shall consider electrons moving in the x direction; the extension of the argument to any other direction is straightforward. When $k = \pm\pi/a$, as we have seen, the waves are Bragg-reflected back and forth, so the only solutions of Schrödinger's equation consist of standing waves whose wavelength is equal to the periodicity of the lattice. There are two possibilities for these standing waves for $n = 1$, namely

20.9
$$\psi_1 = A \sin \frac{\pi x}{a}$$

20.10
$$\psi_2 = A \cos \frac{\pi x}{a}$$

The probability densities $|\psi_1|^2$ and $|\psi_2|^2$ are plotted in Fig. 20.18. Evidently $|\psi_1|^2$ has its minima at the lattice points occupied by the positive ions, while $|\psi_2|^2$ has its maxima at the lattice points. Since the charge density corresponding to an electron wave function ψ is

$e|\psi|^2$, the charge density in the case of ψ_1 is concentrated *between* the positive ions while in the case of ψ_2 it is concentrated *at* the positive ions. The potential energy of an electron in a lattice of positive ions is greatest midway between each pair of ions and least at the ions themselves, so the electron energies E_1 and E_2 associated with the standing waves ψ_1 and ψ_2 are different. No other solutions are possible when $k = \pm\pi/a$, and accordingly no electron can have an energy between E_1 and E_2.

Figure 20.19 shows the distribution of electron energies that corresponds to the Brillouin zones pictured in Fig. 20.16. At low energies (in this hypothetical situation for $E <$ ~2 ev) the curve is almost exactly the same as that of Fig. 19.7 based on the free-electron theory, which is not surprising since at low energies k is small and the electrons in a periodic lattice then *do* behave like free electrons. With increasing energy, however, the number of available energy states goes beyond that of the free-electron theory owing to the distortion of the energy contours by the lattice: there are more different k values for each energy. Then, when $k = \pm\pi/a$, the energy contours reach the boundaries of the first zone, and energies higher than about 4 ev (in this particular model) are forbidden for electrons in the k_x and k_y directions although permitted in other directions. As the energy goes farther and farther beyond 4 ev, the available energy states become restricted more and

FIGURE 20.18   Distributions of the probability densities $|\psi_1|^2$ and $|\psi_2|^2$.

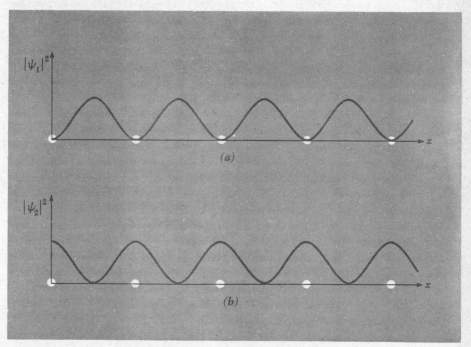

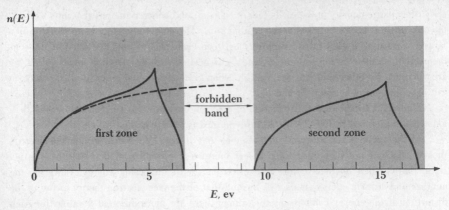

FIGURE 20.19 The distributions of electron energies in the Brillouin zones of Fig. 20.16. The dashed line is the distribution predicted by the free-electron theory.

more to the corners of the zone, and $n(E)$ falls. Finally, at approximately 6½ ev, there are no more states and $n(E) = 0$. The lowest possible energy in the second zone is somewhat less than 10 ev, and another curve similar in shape to the first begins. Here the gap between the possible energies in the two zones is about 3 ev, so the forbidden band is about 3 ev wide.

Although there must be an energy gap between successive Brillouin zones in any given direction, the various gaps may overlap permitted energies in other directions so that there is no forbidden band in the crystal as a whole. Figure 20.20 contains graphs of E versus k for three directions in a crystal that has a forbidden band and in a crystal whose allowed bands overlap sufficiently to avoid having a forbidden band.

The electrical behavior of a solid depends upon the degree of occupancy of its energy bands as well as upon the nature of the band structure, as noted in Sec. 20.1. There are two available energy states (one for each spin) in each band for each structural unit in the crystal. (By "structural unit" in this context is meant an atom in a metal or covalent elemental solid such as diamond, a molecule in a molecular solid, and an ion pair in an ionic solid.) A solid will be an insulator if two conditions are met: (1) It must have an even number of valence electrons per structural unit, and (2) the band that contains the highest-energy electrons must be separated from the allowed band above it by an energy gap large compared with kT. The reason for condition (1) is that it ensures that the highest-energy band be completely filled, and the reason for (2) is that none of the electrons be able to cross the gap to reach unfilled states. Thus diamond, with four electrons per atom, solid hydrogen, with two electrons per H_2 molecule, and NaCl, with eight electrons per Na^+—Cl^- ion pair, all have wide forbidden bands in addition and are insulators. Figure 20.21a shows the energy contours of a hypothetical insulator.

A conductor is characterized by its violation of either (or both) of the above conditions.

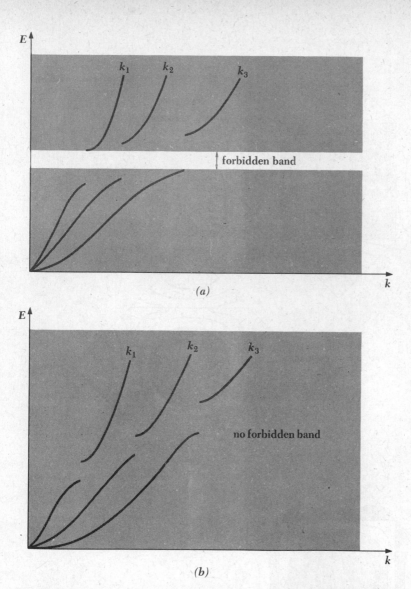

FIGURE 20.20 *E* versus *k* curves for three directions in two crystals. In (a) there is a forbidden band, while in (b) the allowed energy bands overlap and there is no forbidden band.

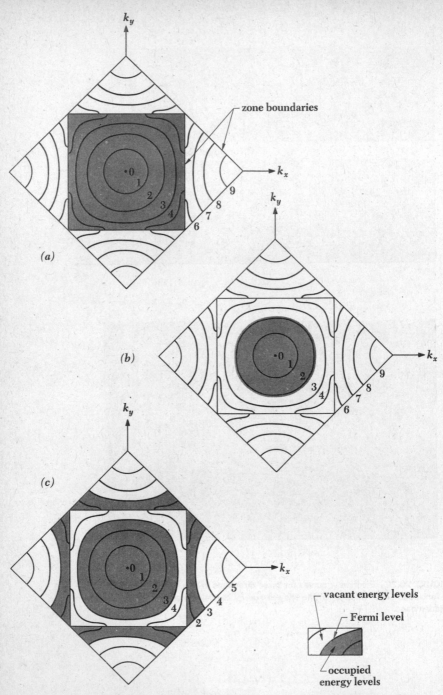

FIGURE 20.21   Electron energy contours and Fermi levels in three types of solid: (a) insulator; (b) monovalent metal; (c) divalent metal. Energies are in electron volts.

Thus the alkali metals, with an odd number of valence electrons per structural unit (namely one per atom), are conductors, as are such divalent metals as magnesium and zinc which have overlapping energy bands. Figure 20.21b and c respectively show the energy contours of these two types of metal. When the forbidden band in an insulator is narrow or the amount of overlap in a metal is small, the electrical conductivity falls in the semiconductor region, and it is not really correct to speak of the substance as either a metal or a nonmetal. The boundary between filled and empty electron energy states in three-dimensional k-space is called the *Fermi surface*.

Hall effect studies indicate that the conductivity of the divalent metals beryllium, zinc, and cadmium is largely due to positive charge carriers, not to electrons. This unexpected finding is readily accounted for on the basis of the band picture by assuming that the overlap of the Fermi surface into the highest band is small, leaving vacant states—which are holes —in the band below it. The holes in the lower band carry the bulk of the current while the electrons in the upper band play a minor role.

20.6   Origin of Resistivity

Electron waves in a perfect metal crystal would undergo neither scattering nor attenuation, and such a metal would have infinite conductivity. No real crystal is ever perfect, of course, since thermal vibrations occur at all temperatures even if no structural defects such as interstitials, vacancies, or impurity atoms are present. The scattering cross section of a positive ion in a metal crystal, which is a measure of its effectiveness in scattering an incident electron wave, is proportional to $\overline{A}^2$, the square of its average amplitude of vibration. Since $\overline{A}^2$ is itself proportional to the absolute temperature T (see Eq. 19.10) at temperatures above the Debye temperature θ, where classical theory holds, it is reasonable to expect that the resistivity ρ of a metal should be proportional to T when $T > \theta$. This prediction is obeyed by pure metals at such temperatures; when $T < \theta$, ρ varies more rapidly with T, just as the specific heat does.

Metals with structural imperfections are another matter. The irregularities in their lattices are not temperature dependent to any great extent, and such metals may have a considerable residual resistivity even at low temperatures when the resistivity due to thermal vibration is small. This observation is summarized in Mathiessen's rule, which states that the various contributions to the resistivity of a metal are additive:

$$\rho = \rho_{\text{thermal}} + \rho_{\text{structural}}$$

The resistivities of cold-worked metals (for instance, "hard-drawn" wires) are lowered when they are annealed because the number of defects are thereby decreased. Most alloys have sufficiently irregular structures to considerably interfere with the propagation of electron waves; for example, some steels have resistivities 10 times that of pure iron. As a general rule, the purer a metal, the less its resistivity.

20.7 Effective Mass

An electron in a crystal interacts with the crystal lattice, and because of this interaction its response to external forces is not, in general, the same as that of a free electron. There is nothing unusual about this phenomenon—no particle subject to constraints behaves like a free particle. What *is* unusual is that the deviations of an electron in a crystal from free-electron behavior under the influence of external forces can all be incorporated into the simple statement that the *effective mass* of such an electron is not the same as its actual mass.

The inertial mass of a particle in classical mechanics is defined by Newton's second law of motion $\mathbf{F} = m\mathbf{a}$ in terms of its acceleration $\mathbf{a}$ when a resultant force $\mathbf{F}$ acts on it. If we consider a free electron moving in the x direction in an electric field of intensity $\mathcal{E}$, we have

20.11
$$e\mathcal{E} = ma = m\frac{dv}{dt}$$

Let us now look at what happens when the same electric field is applied to a crystal that contains an electron whose motion is described in terms of a wave group whose propagation constants are centered on some value k (Sec. 4.4). The velocity of the group, which is the wave equivalent of the velocity of the particle it represents, is

20.12
$$v_g = \frac{d\omega}{dk}$$

Here ω is the angular frequency associated with the energy E, where

$$\omega = 2\pi\nu = \frac{2\pi E}{h} = \frac{E}{\hbar}$$

so that

20.13
$$v_g = \frac{d\omega}{dk} = \frac{1}{\hbar}\frac{dE}{dk} \qquad \text{Group velocity}$$

Figure 20.22 shows how v_g varies with k in the x direction of a simple square lattice. The dashed line is a plot of the same relationship for a free electron, where

$$k = \frac{2\pi}{\lambda} = \frac{2\pi p}{h} = \frac{mv'_g}{\hbar}$$

and

20.14
$$v'_g = \frac{\hbar k}{m} \qquad \text{Free-electron group velocity}$$

The deviations of v_g from v'_g in the vicinity of $k = \pm n\pi/a$ correspond to the deviations of the electron energy in the same lattice from the free-electron value of $\hbar^2 k^2/2m$ so conspicu-

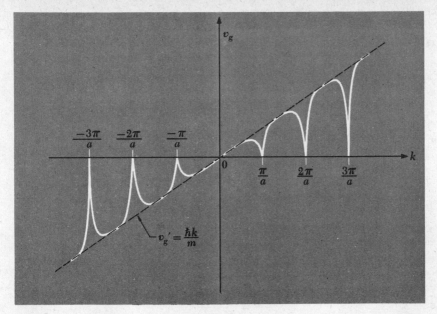

FIGURE 20.22 The group velocity v_g as a function of propagation vector k in the x direction of the lattice shown in Fig. 20.13. The dashed line shows how v_g varies with k for a free particle.

ous in Fig. 20.17; both result from the diffraction of the electron waves by the periodic lattice. The result that $v_g = 0$ at $k = \pm n\pi/a$, the Brillouin zone boundaries, comes about because the electron waves are then standing waves, not traveling waves, and the electron they represent is not going anywhere.

The external electric field $\mathcal{E}$ does an amount of work

$$\Delta E = F \, \Delta x = e\mathcal{E}v_g \, \Delta t$$

on the electron in the time Δt, since the electron moves through the distance $\Delta x = v_g \, \Delta t$ in this time interval. Now we can also express ΔE by

$$\Delta E = \frac{dE}{dk} \, \Delta k = \hbar v_g \, \Delta k$$

where Δk is the change in k that corresponds to the change in energy. Equating these formulas for ΔE,

$$e\mathcal{E}v_g \, \Delta t = \hbar v_g \, \Delta k$$

$$e\mathcal{E} = \hbar \frac{\Delta k}{\Delta t}$$

Effective Mass

which becomes in the limit of $\Delta t \to 0$

20.15
$$e\mathcal{E} = \hbar \frac{dk}{dt}$$

The acceleration dv_g/dt of the wave group is

$$\frac{dv_g}{dt} = \frac{\partial v_g}{\partial k} \frac{dk}{dt}$$

From Eq. 20.13,

$$\frac{\partial v_g}{\partial k} = \frac{1}{\hbar} \frac{\partial^2 E}{\partial k^2}$$

and from Eq. 20.15,

$$\frac{dk}{dt} = \frac{e\mathcal{E}}{\hbar}$$

Hence

20.16
$$\frac{dv_g}{dt} = \frac{e\mathcal{E}}{\hbar^2} \frac{\partial^2 E}{\partial k^2}$$

When we compare this result with Eq. 20.11, $e\mathcal{E} = m \, dv/dt$, we see that the two are the same if the quantity

20.17
$$m^* = \frac{\hbar^2}{\partial^2 E/\partial k^2}$$
Effective mass

is taken to represent the mass of the electron in the lattice. In other words, an electron in a crystal responds to an external force precisely as a free electron would if its mass were m^*.

Figure 20.23 shows how the effective mass m^* of an electron moving in the x direction of a square lattice varies with k. What can it mean physically for an electron to have an effective mass of 0? Or a *negative* effective mass? To answer the first question, we note that $m^* = 0$ right on the boundaries of the Brillouin zones. An electron with a propagation constant that places it on such a boundary has an effective velocity of 0 even though it may possess considerable energy. If the electron is given a slight push, so to speak, its total energy changes by only a small amount, but because it is then able to move through the lattice its *change in momentum* is very great in relation to the impulse provided. Hence an electron in the vicinity of a zone boundary responds to an applied force as though its mass were very small, in fact with $m^* = 0$ right at $k = \pm n\pi/a$.

A negative effective mass occurs when an electron has a propagation constant that places it near the outer (high-k) boundary of a Brillouin zone. Then an increase in the electron's energy means an increase in the momentum transferred to the lattice ions; in wave

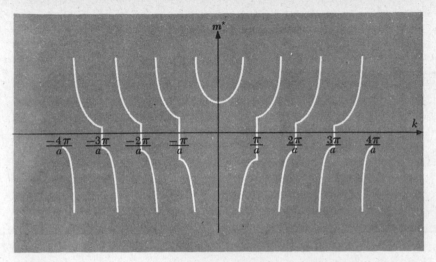

FIGURE 20.23 Effective mass m^* for an electron moving in a square lattice versus propagation constant k.

terms, the wave group is diffracted to a greater extent, even though not completely Bragg-reflected as it is at the zone boundaries. Hence the electron responds to an applied force with a *decrease* in its momentum, which is nicely summarized in the notion of a negative effective mass.

To express the preceding results in a band rather than zone framework, the effective mass of an electron is positive in the lower part of an energy band, where an increase in energy is accompanied by an increase in momentum, while m^* is negative in the upper part of a band, where an increase in energy is accompanied by a decrease in momentum. But now a paradox seems to arise, because the highest-energy electrons in an almost-filled upper band have negative effective masses, yet solids with this type of band are nevertheless able to conduct electric current: in such solids an increase in applied voltage leads to an increase in the current, not to a decrease. The paradox is resolved when we recall that the current carriers under these circumstances are holes—vacant electron states—and not electrons. A hole acts as if it has a charge of $+e$, opposite in sign to the electron charge. Figure 20.24 is a plot of E versus k near the top of a band whose uppermost state is vacant. If an electric field is applied, the electron with the highest energy may acquire enough additional energy to occupy the vacancy, leaving behind it a new vacancy lower in energy. Applying the field has thus increased the propagation constant k of the electron but decreased that of the hole. The electron is now at the top of its energy band and hence is at the Brillouin zone boundary in k-space. It cannot carry current since it is in a stationary state, but the hole has moved *away* from the boundary and therefore can act as a current carrier. Thus in a situation in

Effective Mass

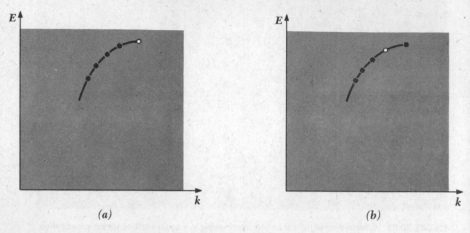

FIGURE 20.24 (a) Before field is applied. (b) Just after field is applied.

which an electron exhibits a negative effective mass, in that increasing its energy decreases its momentum, exactly the opposite happens to a hole and it exhibits a positive effective mass. That this is a general conclusion is evident from Eq. 20.15: changing the sign of e changes that of dk/dt. If we denote the effective mass of a hole by m_h,

20.18
$$m_h^* = -m^* = -\frac{\hbar^2}{\partial^2 E/\partial k^2}$$

Thus electrical conduction in a solid with an almost-filled band takes place by the motion of holes which behave like positive charges with positive effective masses. In a solid with an almost-empty band, conduction takes place by the motion of negatively charged electrons with positive effective masses, since here m_h^* is negative.

The most important results of the free-electron theory of metals discussed in the previous chapter can be incorporated in the more realistic band theory merely by replacing the electron mass m by the average effective mass m^* at the Fermi surface. Thus the Fermi energy in a metal is given by

20.19
$$u_F = \frac{h^2}{2m^*}\left(\frac{3N}{8\pi V}\right)^{2/3}$$
Fermi energy

where N/V is the density of valence electrons. Table 20.1 is a list of effective mass ratios m^*/m in metals. The electronic specific heat is, from Eq. 19.44, inversely proportional to u_F and therefore directly proportional to m^*. Metals such as cobalt, nickel, and platinum, which have high m^*/m ratios, accordingly have high electronic specific heats.

TABLE 20.1. EFFECTIVE MASS RATIOS m^*/m
AT THE FERMI SURFACE IN SOME METALS

Metal	m^*/m
Lithium (Li)	1.2
Beryllium (Be)	1.6
Sodium (Na)	1.2
Aluminum (Al)	0.97
Cobalt (Co)	14
Nickel (Ni)	28
Copper (Cu)	1.01
Zinc (Zn)	0.85
Silver (Ag)	0.99
Platinum (Pt)	13

Problems

1. A small proportion of indium is incorporated in a germanium crystal. Is the crystal an n-type or a p-type semiconductor?

2. A small proportion of antimony is incorporated in a germanium crystal. Is the crystal an n-type or a p-type semiconductor?

3. Use the notion of energy bands to explain the following optical properties of solids:

(a) All metals are opaque to light of all wavelengths.
(b) Semiconductors are transparent to infrared light although opaque to visible light.
(c) Most insulators are transparent to visible light.

4. The energy gap in silicon is 1.1 ev and in diamond it is 6 ev. Discuss the transparency of these substances to visible light.

5. Find the ratio between the kinetic energies of an electron in a two-dimensional square lattice which has $k_x = k_y = \pi/a$ and an electron which has $k_x = \pi/a, k_y = 0$.

6. Draw the third Brillouin zone of the two-dimensional square lattice whose first two Brillouin zones are shown in Fig. 20.14.

7. Phosphorus is present in a germanium sample. Assume that one of its five valence electrons revolves in a Bohr orbit around each P^+ ion in the germanium lattice. (a) If the effective mass of the electron is $0.17\ m_e$ and the dielectric constant of germanium is 16, find the radius of the first Bohr orbit of the electron. (b) The energy gap between the valence and

conduction bands in germanium is 0.65 ev. How does the ionization energy of the above electron compare with this energy and with kT at room temperature?

8. Repeat Prob. 7 for a silicon sample that contains arsenic. The effective mass of an electron in silicon is about 0.31 m_e, the dielectric constant of silicon is 12, and the energy gap in silicon is 1.1 ev.

9. The calculation of the Fermi energy in copper made in Sec. 19.6 did not take into account the difference between m_e and m_e^*, and yet the u_F value obtained was approximately correct. Why?

10. The effective mass m^* of a current carrier in a semiconductor can be directly determined by means of a *cyclotron resonance* experiment in which the carriers (whether electrons or holes) move in spiral orbits about the direction of an externally applied magnetic field **B**. An alternating electric field is applied perpendicular to **B**, and resonant absorption of energy from this field occurs when its frequency ν is equal to the frequency of revolution ν_c of the carrier. (a) Derive an equation for ν_c in terms of m^*, e, and B. (b) In a certain experiment, $B = 0.1$ weber/m² and maximum absorption is found to occur at $\nu = 1.4 \times 10^{10}$ sec^{-1}. Find m^*. (c) Find the maximum orbital radius of a charge carrier in this experiment whose speed is 3×10^4 m/sec.

Chapter 21
The
Atomic
Nucleus

Thus far we have regarded the atomic nucleus solely as a point mass that possesses positive charge. The behavior of atomic electrons is responsible for the chief properties (except mass) of atoms, molecules, and solids, not the behavior of atomic nuclei. But the nucleus itself is far from insignificance in the grand scheme of things. For instance, the elements exist because of the ability of nuclei to possess multiple electric charges, and explaining this ability is the central problem of nuclear physics. Furthermore, the energy that powers the continuing evolution of the universe apparently can all be traced to nuclear reactions and transformations. And, of course, the mundane applications of nuclear energy are familiar enough.

21.1 Atomic Masses

The nucleus of an atom contains nearly all of its mass, and a good deal of information on nuclear properties can be inferred from a knowledge of atomic masses. An instrument used to measure atomic masses is called a *mass spectrometer*; modern spectrometers and techniques are capable of precisions of better than 1 part in 10^6.

A variety of mass spectrometers have been devised, an especially simple one of which is shown schematically in Fig. 21.1. The first step in its operation is the production of ions of the element or compound under study. If the substance is a gas, ions can readily be formed in an electrical discharge or by electron bombardment, while if it is a solid, it may be deposited on a filament which is then strongly heated or incorporated into an electrode used as one terminal of an arc discharge. The ions emerge from their source through a slit with the charge $+e$ and are then accelerated by an electric field. (The presence of ions with other charges is easily taken into account; we consider only those with single charges for simplicity.) When they enter the spectrometer itself the ions as a rule are traveling in slightly different directions with slightly different speeds. A pair of slits serves to collimate the ion beam, which then passes through a *velocity selector*. The velocity selector consists of uniform electric and magnetic fields that are perpendicular to each

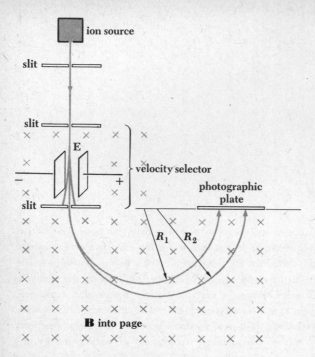

FIGURE 21.1 A simple mass spectrometer.

other and to the beam. The electric field exerts the force

$$F_e = eE$$

on the ions, while the magnetic field exerts the force

$$F_m = Bev$$

on them in the opposite direction. In order for an ion to reach the slit at the far end of the velocity selector, it must suffer no deflection within the selector, which means that only those ions escape for which

$$F_e = F_m$$

$$v = \frac{E}{B}$$

The ions in the beam are now all moving in the same direction with the same velocity. Once past the velocity selector, they enter a uniform magnetic field and follow circular paths whose radii R may be found by equating the magnetic force Bev on them with the centripetal force mv^2/R:

$$Bev = \frac{mv^2}{R}$$

$$R = \frac{mv}{eB}$$

Since v, e, and B are known, a measurement of R yields a value for the ion mass m. In early spectrometers the ions fell upon a photographic plate, permitting R to be determined by the position of their image, while in modern ones B is varied to bring the ion beam to a fixed detector for which R is known.

Atomic masses refer to the masses of neutral atoms, not of stripped nuclei. Thus the masses of the orbital electrons and the mass equivalent of their binding energies are incorporated in the figures. Atomic masses are conventionally expressed in *mass units* (u) such that the mass of the most abundant type of carbon atom is, by definition, exactly 12.000 . . . u. The value of a mass unit is, to five significant figures,

$$1\,u = 1.6604 \times 10^{-27}\,kg$$

and its energy equivalent is 931.48 Mev.

Not long after the development of methods for determining atomic masses early in this century, it was discovered that not all of the atoms of a particular element have the same mass. The different varieties of the same element are called its *isotopes*. Another widely used term, *nuclide*, refers to a particular species of nucleus; thus each isotope of an element is a nuclide.

The atomic masses listed in Table 10.1 refer to the *average* atomic mass of each element, which is the quantity of primary interest to chemists. (The figures in Table 10.1 are expressed in u.) Table 21.1 contains the atomic masses and relative abundances of the five stable isotopes of zinc. The individual masses range from 63.92914 u to 69.92535 u, and the relative abundances range from 0.62 percent to 48.89 percent. The average mass is 65.38 u, which is accordingly the atomic mass of zinc. Twenty elements do not have isotopes, but are composed of a single nuclide each; beryllium, fluorine, sodium, and aluminum are examples.

Even hydrogen is found to have isotopes, though the two heavier ones make up only

TABLE 21.1. PROPERTIES OF THE STABLE ISOTOPES OF ZINC, $Z = 30$

Mass number of isotope	Atomic mass, u	Relative abundance, %
64	63.92914	48.89
66	65.92605	27.81
67	66.92715	4.11
68	67.92486	18.56
70	69.92535	0.62

about 0.015 percent of natural hydrogen. Their atomic masses are 1.007825, 2.014102, and 3.01605 u; the heavier isotopes are known as *deuterium* and *tritium* respectively. (Tritium nuclei, called *tritons*, are unstable and decay radioactively into an isotope of helium.) The nucleus of the lightest isotope is the *proton*, whose mass of

$$m_p = 1.0072766 \, \text{u}$$
$$= 1.6725 \times 10^{-27} \, \text{kg}$$

is, within experimental error, equal to the mass of the entire atom minus the mass of the electron it also contains. The proton, like the electron, is an elementary particle rather than a composite of other particles. (We shall consider the notion of elementary particle in some detail in Chap. 25.)

In tables of nuclear properties, it is customary to list the *mass excess* of an isotope instead of its actual atomic mass. The mass excess is defined as

$$\text{Mass excess} = M - A$$

where M is the atomic mass in u and A is the mass number (see Sec. 21.3) of the isotope. The mass excess is a positive quantity for very light and very heavy nuclei and a negative one for the rest. For example, the mass excess of $_{40}Zn^{64}$ is

$$\text{Mass excess} = 63.92914 \, \text{u} - 64 \, \text{u}$$
$$= -0.07086 \, \text{u}$$

21.2 Nuclear Electrons

An interesting regularity is apparent in listings of nuclide masses: the values are always very close to being integral multiples of the mass of the hydrogen atom, 1.007825 u. For example, the deuterium atom is approximately twice as massive as the hydrogen atom, and the tritium atom approximately three times as massive. The masses of the zinc isotopes listed in Table 21.1 further illustrate this pattern, being quite near to 64, 66, 67, 68, and 70 times the hydrogen-atom mass. It is therefore tempting to regard all nuclei as consisting of protons—hydrogen nuclei—somehow bound together. However, a closer look rules out this notion, since a nuclide mass is invariably greater than the mass of a number of hydrogen atoms equal to its atomic number Z—and the nuclear charge of an atom is $+Ze$. The atomic number of zinc is 30, but its isotopes all have masses more than double that of 30 hydrogen atoms.

Another possibility comes to mind. Perhaps electrons may be present in nuclei which neutralize the positive charge of some of the protons. Thus the helium nucleus, whose mass is four times that of the proton though its charge is only $+2e$, would be regarded as being composed of four protons and two electrons. This explanation is buttressed by the fact that certain radioactive nuclei spontaneously emit electrons, a phenomenon called *beta decay*, whose occurence is easy to account for if electrons are present in nuclei.

Despite the superficial attraction of the hypothesis of nuclear electrons, however, there are a number of strong arguments against it:

1. *Nuclear size.* Nuclei are only $\sim 10^{-14}$ m across. To confine an electron to so small a region requires, by the uncertainty principle, an uncertainty in its momentum of $\Delta p \geq 1.1 \times 10^{-20}$ kg-m/sec, as was calculated in Sec. 4.7. The electron momentum must be at least as large as the minimum value of Δp. The electron kinetic energy that corresponds to a momentum of 1.1×10^{-20} kg-m/sec is 21 Mev. (This figure may also be obtained by calculating the lowest energy level of an electron in a box of nuclear dimensions; since $T \gg m_0 c^2$, the latter calculation must be made relativistically.) However, the electrons emitted during beta decay have energies of only 2 or 3 Mev—an order of magnitude smaller than the energies they must have had within the nucleus if they were to have existed there.

We might remark that the uncertainty principle yields a very different result when applied to protons within a nucleus. For a proton with a momentum of 1.1×10^{-20} kg-m/sec, $T \ll m_0 c^2$, and its kinetic energy can be calculated classically. We have

$$T = \frac{p^2}{2m}$$

$$= \frac{(1.1 \times 10^{-20} \text{ kg-m/sec})^2}{2 \times 1.67 \times 10^{-27} \text{ kg}}$$

$$= 3.6 \times 10^{-14} \text{ joule}$$

$$= 0.23 \text{ Mev}$$

The presence of protons with such kinetic energies in a nucleus is entirely plausible.

2. *Nuclear spin.* Protons and electrons are Fermi particles with spins of ½, that is, angular momenta of $\frac{1}{2}\hbar$. Thus nuclei with an even number of protons plus electrons should have integral spins, while those with an odd number of protons plus electrons should have half-integral spins. This prediction is not obeyed. The fact that the deuteron, which is the nucleus of an isotope of hydrogen, has an atomic number of 1 and a mass number of 2, would be interpreted as implying the presence of two protons and one electron. Depending upon the orientations of the particles, the nuclear spin of $_1H^2$ should therefore be ¾, ½, −½, or −¾. However, the observed spin of the deuteron is 1, something that cannot be reconciled with the hypothesis of nuclear electrons.

3. *Magnetic moment.* The proton has a magnetic moment only about 0.15 percent that of the electron, so nuclear magnetic moments ought to be of the same order of magnitude as that of the electron if electrons are present in nuclei. However, the observed magnetic moments of nuclei are comparable with that of the proton, not with that of the electron, a discrepancy that cannot be understood if electrons are nuclear constituents.

Nuclear Electrons   **493**

4. *Electron-nuclear interaction.* It is observed that the forces that act between nuclear particles lead to binding energies of the order of 8 Mev per particle. It is therefore hard to see why, if electrons can interact strongly enough with protons to form nuclei, the orbital electrons in an atom interact only electrostatically with its nucleus. That is, how can half the electrons in an atom escape the strong binding of the other half? Furthermore, when fast electrons are scattered by nuclei, they behave as though acted upon solely by electrostatic forces, while the nuclear scattering of fast protons reveals departures from electrostatic influences that can be ascribed only to a specifically nuclear force.

The difficulties of the nuclear electron hypothesis were known for some time before the correct explanation for nuclear masses came to light, but there seemed to be no serious alternative. The problem of the mysterious ingredient besides the proton in atomic nuclei was not solved until 1932.

21.3 The Neutron

The composition of atomic nuclei was finally understood in 1932. Two years earlier the German physicists W. Bothe and H. Becker had bombarded beryllium with alpha particles from a sample of polonium and found that radiation was emitted which was able to penetrate matter readily. Bothe and Becker ascertained that the radiation did not consist of charged particles and assumed, quite naturally, that it consisted of gamma rays. (Gamma rays are electromagnetic waves of extremely short wavelength.) The ability of the radiation to pass through as much as several centimeters of lead without being absorbed suggested gamma rays of unprecedentedly short wavelength. Other physicists became interested in this radiation, and a number of experiments were performed to determine its properties in detail. In one such experiment Irène Curie and F. Joliot observed that when the radiation struck a slab of paraffin, a hydrogen-rich substance, protons were knocked out. At first glance this is not very surprising: X rays can give energy to electrons in Compton collisions, and there is no reason why shorter-wavelength gamma rays cannot give energy to protons in similar processes.

Curie and Joliot found proton recoil energies of up to about 5.3 Mev. From Eq. 3.15 for the Compton effect we can calculate the minimum gamma-ray photon energy $E = h\nu$ needed to transfer the kinetic energy T to a proton. Equation 3.15 states that

$$2m_0c^2(h\nu - h\nu') = 2(h\nu)(h\nu')(1 - cos\ \phi)$$

where ϕ is the angle through which the photon is scattered. Maximum energy transfer occurs when $\phi = 180°$, corresponding to a head-on collision, so that $(1 - cos\ \phi) = 2$. Here m_0 is the proton rest mass and $T = h\nu - h\nu'$. Hence

$$m_0c^2T = 2E(E - T)$$

and

21.1
$$E = \frac{1}{2}[T + (T^2 + 2m_0c^2T)^{1/2}]$$

The proton rest energy m_0c^2 is 938 Mev and the observed maximum proton recoil energy T is 5.3 Mev. Substituting these values in Eq. 21.1 yields a minimum initial gamma-ray photon energy of 53 Mev.

This result seemed peculiar because no nuclear radiation known at the time had more than a small fraction of this considerable energy. The peculiarity became even more striking when it was calculated that the presumed reaction of an alpha particle and a beryllium nucleus to yield a carbon nucleus would result in a mass decrease of 0.01144 u, which is equivalent to only 10.7 Mev—one-fifth the energy needed by a gamma-ray photon if it is to knock 5.3 Mev protons out of paraffin.

In 1932 James Chadwick, an associate of Rutherford, proposed an alternative hypothesis for the now-mysterious radiation emitted by beryllium when bombarded by alpha particles. He assumed that the radiation consisted of neutral *particles* whose mass is approximately the same as that of the proton. The electrical neutrality of these particles, which were called *neutrons*, accounted for their ability to penetrate matter readily. Their mass accounted nicely for the observed proton recoil energies: a moving particle colliding head-on with one at rest whose mass is the same can transfer all of its kinetic energy to the latter. A maximum proton energy of 5.3 Mev thus implies a neutron energy of 5.3 Mev, not the 53 Mev required by a gamma ray to cause the same effect. Other experiments had shown that such light nuclei as those of helium, carbon, and nitrogen could also be knocked out of appropriate absorbers by the beryllium radiation, and the measurements made of the energies of these nuclei fit in well with the neutron hypothesis. In fact, Chadwick arrived at the neutron-mass figure of $m_n \approx m_p$ from an analysis of observed proton and nitrogen nuclei recoil energies; no other mass gave as good agreement with the experimental data.

Before we consider the role of the neutron in nuclear structure, we should note that it is not a stable particle outside nuclei. The free neutron decays radioactively into a proton, an electron, and an antineutrino; the half life of the free neutron is 10.8 min according to the most recent measurements.

Immediately after its discovery the neutron was recognized as the missing ingredient in atomic nuclei. Its mass of

$$m_n = 1.0086654 \text{ u}$$
$$= 1.6748 \times 10^{-27} \text{ kg}$$

which is slightly more than that of the proton, its electrical neutrality, and its spin of ½ all fit in perfectly with the observed properties of nuclei when it is assumed that nuclei are composed solely of neutrons and protons.

The following terms and symbols are widely used to describe a nucleus:

Z = atomic number = number of protons

N = neutron number = number of neutrons

$A = Z + N$ = mass number = total number of neutrons and protons

The term *nucleon* refers to both protons and neutrons, so that the mass number A is the number of nucleons in a particular nucleus. Nuclides are identified according to the scheme

$$_Z X^A$$

where X is the chemical symbol of the species. Thus the arsenic isotope of mass number 75 is denoted by

$$_{33}As^{75}$$

since the atomic number of arsenic is 33. Similarly a nucleus of ordinary hydrogen, which is a proton, is denoted by

$$_1H^1$$

Here the atomic and mass numbers are the same because no neutrons are present.

The fact that nuclei are composed of neutrons as well as protons immediately explains the existence of isotopes: the isotopes of an element all contain the same numbers of protons but have different numbers of neutrons. Since its nuclear charge is what is ultimately responsible for the characteristic properties of an atom, the isotopes of an element all have identical chemical behavior and differ physically only in mass.

21.4   Stable Nuclei

Not all combinations of neutrons and protons form stable nuclei. In general, light nuclei ($A < 20$) contain approximately equal numbers of neutrons and protons, while in heavier nuclei the proportion of neutrons becomes progressively greater. This is evident from Fig. 21.2, which is a plot of N versus Z for stable nuclei. The tendency for N to equal Z follows from the existence of nuclear energy levels, whose origin and properties we shall examine in Chap. 22. Nucleons, which have spins of ½, obey the exclusion principle. As a result, each nuclear energy level can contain two neutrons of opposite spins and two protons of opposite spins. Energy levels in nuclei are filled in sequence, just as energy levels in atoms are, to achieve configurations of minimum energy and therefore maximum stability. A nucleus with, say, three neutrons and one proton outside filled inner levels will have more energy than one with two neutrons and two protons in the same situation, since in the former case one of the neutrons must go into a higher level while in the latter case all four nucleons fit into the lowest available level. Figure 21.3 shows how this notion accounts for the absence of a stable $_5B^{12}$ isotope while permitting $_6C^{12}$ to exist.

The preceding argument is only part of the story. Protons are positively charged and repel one another electrostatically. This repulsion becomes so great in nuclei with more than 10 protons or so that an excess of neutrons, which produce only attractive forces, is required for stability; thus the curve of Fig. 21.2 departs more and more from the $N = Z$ line as Z increases. Even in light nuclei N may exceed Z, but is never smaller; $_5B^{11}$ is stable, for instance, but not $_6C^{11}$.

Nuclear forces are limited in range, and as a result nucleons interact strongly only

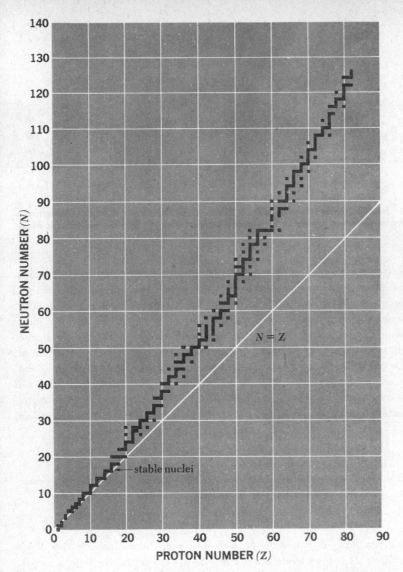

FIGURE 21.2 Neutron-proton diagram for stable nuclides. There are no stable nuclides with $Z = 43$ or 61, with $N = 19, 35, 39, 45, 61, 89, 115,$ or 126, or with $A = Z + N = 5$ or 8. All nuclides with $Z > 83$, $N > 126$, and $A > 209$ are unstable.

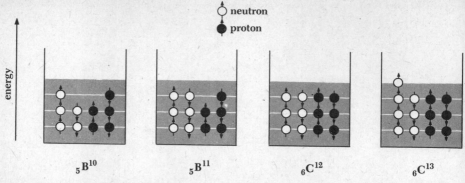

FIGURE 21.3 Simplified energy-level diagrams of stable boron and carbon isotopes. The exclusion principle limits the occupancy of each level to two neutrons of opposite spin and two protons of opposite spin.

with their nearest neighbors. This effect is referred to as the *saturation* of nuclear forces. Because the coulomb repulsion of the protons is appreciable throughout the entire nucleus, there is a limit to the ability of neutrons to prevent the disruption of a large nucleus. This limit is represented by the bismuth isotope $_{83}Bi^{209}$, which is the heaviest stable nuclide. All nuclei with $Z > 83$ and $A > 209$ spontaneously transform themselves into lighter ones through the emission of one or more alpha particles, which are $_2He^4$ nuclei. Since an alpha particle consists of two protons and two neutrons, an alpha decay reduces the Z and the N of the original nucleus by two each. If the resulting daughter nucleus has either too small or too large a neutron/proton ratio for stability, it may beta decay to a more appropriate configuration. In negative beta decay, a neutron is transformed into a proton and an electron:

$$n \rightarrow p + e^-$$

The electron leaves the nucleus and is observed as a "beta particle." In positive beta decay, a proton becomes a neutron and a positron is emitted:

$$p \rightarrow n + e^+$$

Thus negative beta decay decreases the proportion of neutrons and positive beta decay increases it. Figure 21.4 shows how alpha and beta decays enable stability to be achieved. Radioactivity is considered in more detail in Chap. 23.

21.5 Nuclear Sizes

The Rutherford scattering experiment provided the first evidence that nuclei are of finite size. In that experiment, as we saw in Chap. 5, an incident alpha particle is deflected by a target nucleus in a manner consistent with Coulomb's law provided the distance between them exceeds about 10^{-14} m. For smaller separations the predictions of Coulomb's law are not obeyed because the nucleus no longer appears as a point charge to the alpha particle.

Since Rutherford's time a variety of experiments have been performed to determine nuclear dimensions, with particle scattering still a favored technique. Fast electrons and neutrons are ideal for this purpose, since an electron interacts with a nucleus only through electrical forces while a neutron interacts only through specifically nuclear forces. Thus electron scattering provides information on the distribution of charge in a nucleus and neutron scattering provides information on the distribution of nuclear matter. In both cases the de Broglie wavelength of the particle must be smaller than the radius of the nucleus under study. A particle with a wavelength λ of 10^{-14} m has a momentum of

$$p = \frac{h}{\lambda}$$

$$= \frac{6.63 \times 10^{-34} \text{ joule-sec}}{10^{-14} \text{ m}}$$

$$= 6.63 \times 10^{-20} \text{ kg-m/sec}$$

FIGURE 21.4 Alpha and beta decays permit an unstable nucleus to reach a stable configuration.

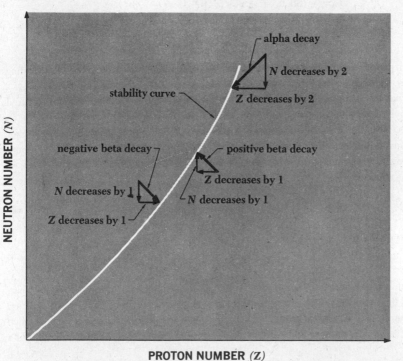

For an electron, where $E \gg m_0c^2$ here, the corresponding kinetic energy is

$$T = pc$$
$$= 6.63 \times 10^{-20} \text{ kg-m/sec} \times 3.00 \times 10^8 \text{ m/sec}$$
$$= 1.99 \times 10^{-11} \text{ joule}$$
$$= 124 \text{ Mev}$$

For a neutron of the above momentum, where m_0c^2 (939 Mev) is comparable with the total energy E,

$$T = E - m_0c^2$$
$$= \sqrt{m_0{}^2c^4 + p^2c^2} - m_0c^2$$

Here, since

$$m_0{}^2c^4 = (939 \text{ Mev})^2 = 88.17 \times 10^4 \text{ Mev}^2$$
$$p^2c^4 = (124 \text{ Mev})^2 = 1.54 \times 10^4 \text{ Mev}^2$$

we have

$$T = \sqrt{(88.17 + 1.54) \times 10^4 \text{ Mev}^2} - 939 \text{ Mev}$$
$$= 947 \text{ Mev} - 939 \text{ Mev}$$
$$= 8 \text{ Mev}$$

Hence electrons with energies exceeding 124 Mev and neutrons with energies exceeding 8 Mev are suitable for exploring nuclear dimensions.

Actual experiments on the sizes of nuclei have employed electrons of several hundred Mev to over 1 Gev (1 Gev = 1,000 Mev = 10^9 ev) and neutrons of 20 Mev and up. In every case it is found that the volume of a nucleus is directly proportional to the number of nucleons it contains, which is its mass number A. If a nuclear radius is R, the corresponding volume is $\frac{4}{3}\pi R^3$ and so R^3 is proportional to A. This relationship is usually expressed in inverse form as

21.2 $$R = R_0 A^{1/3}$$ **Nuclear radii**

The value of R_0 is

$$R_0 \approx 1.3 \times 10^{-15} \text{ m}$$

The indefiniteness in R_0 is a consequence, not just of experimental error, but of the characters of the various experiments: electrons and neutrons interact differently with nuclei. The value of R_0 is slightly smaller when it is deduced from electron scattering, which implies that nuclear matter and nuclear charge are not identically distributed throughout a nucleus.

As we saw in the earlier part of this book, the angstrom unit ($1 \text{ A} = 10^{-10}$ m) is a convenient unit of length for expressing distances in the atomic realm. For example, the radius of the hydrogen atom is 0.53 A, the C and O atoms in a CO molecule are 1.13 A apart, and the Na^+ and Cl^- ions in crystalline NaCl are 2.81 A apart. Nuclei are so small that the *fermi* (f), only 10^{-5} the size of the angstrom, is an appropriate unit of length:

$$1 \text{ fermi} = 1 \text{ f} = 10^{-15} \text{ m}$$

Hence we can write

21.3 $$R \approx 1.3 \, A^{1/3} \text{ f}$$

for nuclear radii. From this formula we find that the radius of the $_6C^{12}$ nucleus is

$$R \approx 1.3 \times (12)^{1/3} \text{ f} \approx 3.0 \text{ f}$$

Similarly, the radius of the $_{47}Ag^{107}$ nucleus is 6.2 f and that of the $_{92}U^{238}$ nucleus is 8.1 f.

Now that we know nuclear sizes as well as masses, we can compute the density of nuclear matter. In the case of $_6C^{12}$, whose atomic mass is 12.0 u, we have for the nuclear density (the masses and binding energies of the six electrons may be neglected here)

$$\rho = \frac{m}{\frac{4}{3}\pi R^3}$$

$$= \frac{12.0 \text{ u} \times 1.66 \times 10^{-27} \text{ kg/u}}{\frac{4}{3}\pi \times (3.0 \times 10^{-15} \text{ m})^3}$$

$$= 2 \times 10^{17} \text{ kg/m}^3$$

This figure—equivalent to 3 billion tons per cubic inch!—is essentially the same for all nuclei. Certain stars, the "white dwarfs" mentioned in Chap. 3, are composed of atoms whose electron shells have collapsed owing to enormous pressure, and the densities of such stars approach that of pure nuclear matter.

We have been assuming that nuclei are spherical. How can nuclear shapes be determined? If the distribution of charge in a nucleus is not spherically symmetric, the nucleus will have an electric quadrupole moment. A nuclear quadrupole moment will interact with the orbital electrons of the atom, and the consequent shifts in atomic energy levels will lead to hyperfine splitting of the spectral lines. Of course, this source of hyperfine structure must be distinguished from that due to the magnetic moment of the nucleus, but when this is done it is found that deviations from sphericity actually do occur in nuclei whose spin quantum numbers are 1 or more. Such nuclei may be prolate or oblate spheroids, but the difference between major and minor axes never exceeds ~20 percent and is usually much less. For almost all purposes it is sufficient to regard nuclei as being spherical, but in the next chapter we shall see that the departures from sphericity, small as they are, furnish valuable information on nuclear structure.

21.6 Binding Energy

A stable atom invariably has a smaller mass than the combined masses of its constituent particles. The deuterium atom $_1H^2$, for example, has a mass of 2.014102 u, while the mass of a hydrogen atom ($_1H^1$) plus that of a neutron is

$$m_{\text{hydrogen}} + m_n = 1.007825 \text{ u} + 1.008665 \text{ u}$$
$$= 2.016490 \text{ u}$$

which is 0.002388 u greater. Since a deuterium nucleus—called a *deuteron*—is composed of a proton and a neutron, and both $_1H^1$ and $_1H^2$ have single orbital electrons, it is evident that the *mass defect* of 0.002388 u is related to the binding of a proton and a neutron to form a deuteron. A mass of 0.002388 u is equal to

$$0.002388 \text{ u} \times 931 \text{ Mev/u} = 2.23 \text{ Mev}$$

When a deuteron is formed from a free proton and neutron, then, 2.23 Mev of energy is liberated. Conversely, 2.23 Mev must be supplied from an external source to break a deuteron up into a proton and a neutron. This inference is supported by experiments on the photodisintegration of the deuteron, which show that a gamma-ray photon must have an energy of at least 2.23 Mev to disrupt a deuteron (Fig. 21.5).

The energy equivalent of the mass defect in a nucleus is called its *binding energy* and is a measure of the stability of the nucleus. Binding energies arise from the action of the forces that hold nucleons together to form nuclei, just as ionization energies of atoms, which must be provided to remove electrons from them, arise from the action of electrostatic forces. Binding energies range from 2.23 Mev for the deuteron, which is the smallest compound nucleus, up to 1,640 Mev for $_{83}Bi^{209}$, the heaviest stable nucleus.

The binding energy per nucleon, arrived at by dividing the total binding energy of a

FIGURE 21.5 The binding energy of the deuteron is 2.23 Mev, which is confirmed by experiments that show that a gamma-ray photon with a minimum energy of 2.23 Mev can split a deuteron into a free neutron and a free proton.

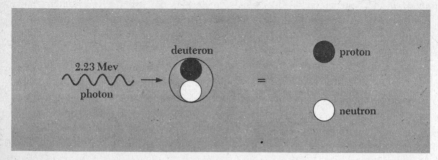

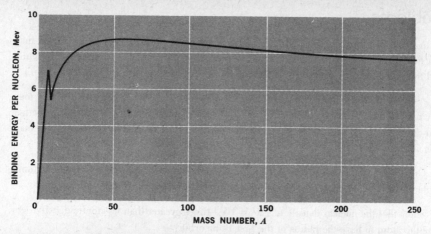

FIGURE 21.6 Binding energy per nucleon as a function of mass number. The peak at $A = 4$ corresponds to the $_2\text{He}^4$ nucleus.

nucleus by the number of nucleons it contains, is a most interesting quantity. The binding energy per nucleon is plotted as a function of mass number A in Fig. 21.6. The curve rises steeply at first and then more gradually until it reaches a maximum of 8.79 Mev at $A = 56$, corresponding to the iron nucleus $_{26}\text{Fe}^{56}$, and then drops slowly to about 7.6 Mev at the highest mass numbers. Evidently nuclei of intermediate mass are the most stable, since the greatest amount of energy must be supplied to liberate each of their nucleons. This fact suggests that energy will be evolved if heavy nuclei can somehow be split into lighter ones or if light nuclei can somehow be joined to form heavier ones. The former process is known as *nuclear fission* and the latter as *nuclear fusion*, and both indeed occur under proper circumstances and do evolve energy as predicted.

Nuclear binding energies are strikingly large. To appreciate their magnitude, it is helpful to convert the figures from Mev/nucleon to more familiar units, say kcal/kg. Since 1 ev $= 1.60 \times 10^{-19}$ joule and 1 joule $= 2.39 \times 10^{-4}$ kcal, we find that 1 Mev $= 3.83 \times 10^{-17}$ kcal. One mass unit is equal to 1.66×10^{-27} kg, and each nucleon in a nucleus has a mass of very nearly 1 u. Hence

$$1 \frac{\text{Mev}}{\text{nucleon}} = \frac{3.83 \times 10^{-17} \text{ kcal}}{1.66 \times 10^{-27} \text{ kg}} = 2.31 \times 10^{10} \frac{\text{kcal}}{\text{kg}}$$

A binding energy of 8 Mev/nucleon, a typical value, is therefore equivalent to 1.85×10^{11} kcal/kg. By contrast, the heat of vaporization of water is a mere 540 kcal/kg, and even the heat of combustion of gasoline, 1.13×10^4 kcal/kg, is 10 million times smaller.

Binding Energy

Problems

1. A beam of singly charged ions of $_3Li^6$ with energies of 400 ev enters a uniform magnetic field of flux density 0.08 weber/m^2. The ions move perpendicular to the field direction. Find the radius of their path in the magnetic field. (The $_3Li^6$ atomic mass is 6.01513 u.)

2. A beam of singly charged boron ions with energies of 1,000 ev enters a uniform magnetic field of flux density 0.2 weber/m^2. The ions move perpendicular to the field direction. Find the radii of the path of the $_5B^{10}$ (10.013 u) and $_5B^{11}$ (11.009 u) isotopes in the magnetic field.

3. Ordinary boron is a mixture of the $_5B^{10}$ and $_5B^{11}$ isotopes and has a composite atomic weight of 10.82 u. What percentage of each isotope is present in ordinary boron?

4. Show that the nuclear density of $_1H^1$ is 10^{14} times greater than its atomic density. (Assume the atom to have the radius of the first Bohr orbit.)

5. The binding energy of $_{17}Cl^{35}$ is 298 Mev. Find its mass in u.

6. The mass of $_{10}Ne^{20}$ is 19.9924 u. Find its binding energy in Mev.

7. Find the average binding energy per nucleon in $_8O^{16}$ (the mass of the neutral $_8O^{16}$ atom is 15.9949 u).

8. How much energy is required to remove one proton from $_8O^{16}$? (The mass of the neutral $_7N^{15}$ atom is 15.0001 u; that of the neutral $_8O^{15}$ atom is. 15.0030 u.)

9. How much energy is required to remove one neutron from $_8O^{16}$?

10. Compare the minimum energy a gamma-ray photon must possess if it is to disintegrate an alpha particle into a triton and a proton with that it must possess if it is to disintegrate an alpha particle into a $_2He^3$ nucleus and a neutron. (The atomic masses of $_1H^3$ and $_2He^3$ are respectively 3.01605 u and 3.01603 u.)

Chapter 22
Nuclear Forces
and Models

The unique short-range forces that bind nucleons so securely into nuclei constitute by far the strongest class of forces known. Unfortunately nuclear forces are nowhere near as well understood as electrical forces, and in consequence the theory of nuclear structure is still primitive as compared with the theory of atomic structure. However, even without a satisfactory understanding of nuclear forces, considerable progress has been made in recent years in interpreting the properties and behavior of nuclei in terms of detailed models, and we shall examine some of the concepts embodied in these models in this chapter.

22.1 The Deuteron

The simplest nucleus containing more than one nucleon is the *deuteron,* which consists of a proton and a neutron. The deuteron binding energy is 2.23 Mev, a figure that can be obtained either from the discrepancy in mass between m_{deuteron} and $m_p + m_n$ or from photodisintegration experiments which show that only gamma rays with $h\nu \geqslant 2.23$ Mev can disrupt deuterons into their constituent nucleons. In Chap. 9 we analyzed another two-body system, the hydrogen atom, with the help of quantum mechanics, but in that case the precise nature of the force between the proton and the electron was known. If a force law is known for an interaction, the corresponding potential energy function V can be found and substituted into Schrödinger's equation. Our understanding of nuclear forces is less complete than our understanding of coulomb forces, however, and so it is not possible to discuss the deuteron in as much quantitative detail as the hydrogen atom.

The actual potential energy V of the deuteron, that is, the potential energy of either nucleon with respect to the other, depends upon the distance r between the centers of the neutron and proton more or less as shown by the solid line in Fig. 22.1. (The repulsive "core" perhaps 0.4×10^{-15} m in radius reflects the inability of nucleons to mesh together more than a certain amount.) We shall approximate this $V(r)$ by the "square well" shown as a dashed line in the figure. This approximation means that we consider the nuclear force between neutron and proton to be zero when they are more than r_0 apart, and to have a constant magnitude, leading to the constant potential energy $- V_0$, when they are closer together than r_0. Thus the parameters V_0 and r_0 are representative of the strength and range,

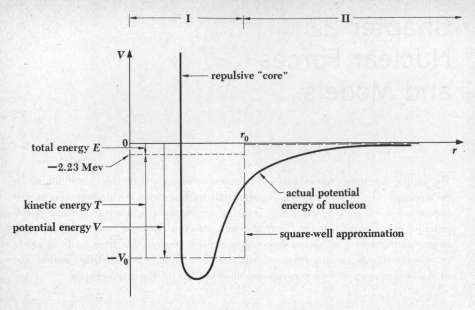

FIGURE 22.1 The actual potential energy of either proton or neutron in a deuteron and the square-well approximation to this potential energy as functions of the distance between proton and neutron.

respectively, of the interaction holding the deuteron together, and the square-well potential itself is representative of the short-range character of the interaction.

A square-well potential means that V is a function of r alone, and therefore, as in the case of other central-force potentials, it is easiest to examine the problem in a spherical polar-coordinate system (see Fig. 9.1). In spherical polar coordinates Schrödinger's equation for a particle of mass m is, with $\hbar = h/2\pi$,

22.1 $\quad \dfrac{1}{r^2}\dfrac{\partial}{\partial r}\left(r^2\dfrac{\partial\psi}{\partial r}\right) + \dfrac{1}{r^2\sin\theta}\dfrac{\partial}{\partial\theta}\left(\sin\theta\dfrac{\partial\psi}{\partial\theta}\right) + \dfrac{1}{r^2\sin^2\theta}\dfrac{\partial^2\psi}{\partial\phi^2} + \dfrac{2m}{\hbar^2}(E-V)\psi = 0$

Let us choose the particle in question to be the neutron, so that we imagine it moving in the force field of the proton. (The opposite choice would, of course, yield identical results.) We note from Fig. 22.1 that E, the total energy of the neutron, is negative and is the same as the binding energy of the deuteron.

In analyzing the hydrogen atom, where one particle is much heavier than the other, it is still necessary to consider the effects of nuclear motion, and we did this in Chap. 9 by replacing the electron mass m_e by its reduced mass m'. In this way the problem of a proton and an electron moving about a common center of mass is replaced by the problem of a single particle of mass m' moving about a fixed point. A similar procedure is even more appropriate here, since neutron and proton masses are almost the same. According to Eq. 6.30, the reduced mass of a neutron-proton system is

22.2
$$m' = \frac{m_n m_p}{m_n + m_p}$$

and so we replace the m of Eq. 22.1 with m' as given above.

We now assume that the solution of Eq. 22.1 can be written as the product of radial and angular functions,

22.3
$$\psi(r, \theta, \phi) = R(r)\Theta(\theta)\Phi(\phi)$$

As before, the function $R(r)$ describes how the wave function ψ varies along a radius vector from the nucleus, with θ and ϕ constant; the function $\Theta(\theta)$ describes how ψ varies with zenith angle θ along a meridian on a sphere centered at $r = 0$, with r and ϕ constant; and the function $\Phi(\phi)$ describes how ψ varies with azimuth angle ϕ along a parallel on a sphere centered at $r = 0$, with r and θ constant.

Although angular motion can occur in a square-well potential, our interest for the moment is in radial motion, that is, in oscillations of the neutron and proton about their center of mass. If there is no angular motion, Θ and Φ are both constant and their derivatives are zero. With $\partial\psi/\partial\theta = \partial^2\psi/\partial\phi^2 = 0$, Eq. 22.1 becomes

22.4
$$\frac{1}{r^2}\frac{d}{dr}\left(r^2\frac{dR}{dr}\right) + \frac{2m'}{\hbar^2}(E - V)R = 0$$

A further simplification can be made by letting

22.5
$$u(r) = rR(r)$$

In terms of the new function u the wave equation becomes

22.6
$$\frac{d^2u}{dr^2} + \frac{2m'}{\hbar^2}(E - V)u = 0$$

Because V has two different values, $V = -V_0$ inside the well and $V = 0$ outside it, there are two different solutions to Eq. 22.6, u_I for $r \leqslant r_0$ and u_{II} for $r \geqslant r_0$. Inside the well the wave equation is

$$\frac{d^2u_I}{dr^2} + \frac{2m'}{\hbar^2}(E + V_0)u_I = 0$$

or, if we let

22.7
$$a^2 = \frac{2m'}{\hbar^2}(E + V_0)$$

The Deuteron

it becomes simply

22.8
$$\frac{d^2u_I}{dr^2} + a^2u_I = 0$$

(We note from Fig. 22.1 that, since $|V_0| > |E|$, the quantities $E + V_0$ and hence a^2 are positive.) Equation 22.8 is the same as the wave equation for a particle in a box of Sec. 8.1, and again the solution is

22.9
$$u_I = A \cos ar + B \sin ar$$

We recall that the radial wave function R is given by $R = u/r$, which means that the cosine solution must be discarded if R is not to be infinite at $r = 0$. Hence $A = 0$ and u_I inside the well is just

22.10
$$u_I = B \sin ar$$

Outside the well $V = 0$ and so

22.11
$$\frac{d^2u_{II}}{dr^2} + \frac{2m'}{\hbar^2} Eu_{II} = 0$$

The total energy E of the neutron is a negative quantity since it is bound to the proton. Therefore

22.12
$$b^2 = \frac{2m'}{\hbar^2} (-E)$$

is a positive quantity, and we can write

22.13
$$\frac{d^2u_{II}}{dr^2} - b^2u_{II} = 0$$

The solution of Eq. 22.13 is

22.14
$$u_{II} = Ce^{-br} + De^{br}$$

Because it must be true that $u \to 0$ as $r \to \infty$, we conclude that $D = 0$. Hence outside the well

22.15
$$u_{II} = Ce^{-br}$$

22.2 Ground State of the Deuteron

We now have expressions for u (and hence for ψ) both inside and outside the well, and it remains to match these expressions and their first derivatives at the well boundary since it is necessary that both u and du/dr be continuous everywhere in the region for which they are defined. At $r = r_0$, then,

$$u_{\text{I}} = u_{\text{II}}$$

22.16
$$B \sin a r_0 = C e^{-b r_0}$$

and

$$\frac{d u_{\text{I}}}{dr} = \frac{d u_{\text{II}}}{dr}$$

22.17
$$a B \cos a r_0 = -b C e^{-b r_0}$$

By dividing Eq. 22.16 by Eq. 22.17 we eliminate the coefficients B and C and obtain the transcendental equation

22.18
$$\tan a r_0 = -\frac{a}{b}$$

Equation 22.18 cannot be solved analytically, but it can be solved either graphically or numerically to any desired degree of accuracy. We note that

22.19
$$\frac{a}{b} = \frac{\sqrt{2m'(E + V_0)/\hbar}}{\sqrt{2m'(-E)/\hbar}} = \sqrt{\frac{E + V_0}{-E}}$$

where $-E$ is the binding energy of the deuteron and V_0 is the depth of the potential well. Since $|V_0| > |E|$, in order to obtain a first crude approximation we might assume that a/b is so large that

$$\tan a r_0 \approx \infty$$

Since $\tan \theta$ becomes infinite at $\theta = \pi/2, \pi, 3\pi/2, \ldots, n\pi/2$, in this approximation the ground state of the deuteron, for which $n = 1$, corresponds to

$$a r_0 \approx \frac{\pi}{2}$$

(In fact, this is the only bound state of the deuteron.) Hence

$$\frac{\sqrt{2m'(E + V_0)}}{\hbar} r_0 \approx \frac{\sqrt{2m' V_0}}{\hbar} r_0 \approx \frac{\pi}{2}$$

since we are assuming that E is negligible compared with V_0, and

22.20
$$V_0 \approx \frac{\pi^2 \hbar^2}{8 m' r_0{}^2}$$

The above approximation is equivalent to assuming that the function u_{I} inside the well is at its maximum (corresponding to $ar = 90°$) at the boundary of the well. Actually, u_{I}

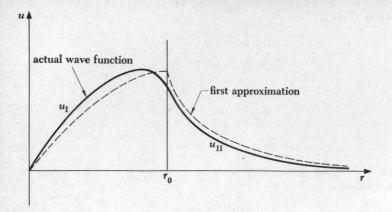

FIGURE 22.2 The wave function $u(r)$ of either proton or neutron in a deuteron.

must be somewhat past its maximum there in order to join smoothly with the function u_{II} outside the well, as shown in Fig. 22.2; a more detailed calculation shows that $ar \approx 116°$ at $r = r_0$. The difference between the two results is due to our neglect of the binding energy $-E$ relative to V_0 in obtaining Eq. 22.20, and when this neglect is remedied, the better approximation

22.21
$$V_0 \approx \frac{\pi^2 \hbar^2}{8m' r_0^2} + \frac{2\hbar}{r_0} \sqrt{\frac{E}{2m'}}$$

is obtained.

Equation 22.21 is a relationship among r_0, the radius of the potential well and therefore representative of the range of nuclear forces, V_0, the depth of the well and therefore representative of the interaction potential energy between two nucleons due to nuclear forces, and $-E$, the binding energy of the deuteron. What we now ask is whether this relationship corresponds to reality in the sense that a reasonable choice of r_0 leads to a reasonable value for V_0. (There is nothing in our simple model that can point to a unique value for either r_0 or V_0.) Such a choice for r_0 might be 2 f. Substituting for the known quantities in Eq. 22.21 and expressing energies in Mev yields

$$V_0 \approx \frac{1.0 \times 10^{-28} \text{ Mev-m}^2}{r_0^2} + \frac{1.9 \times 10^{-14} \text{ Mev-m}}{r_0}$$

and so, for $r_0 = {}_{-} {}_{\perp} = 2 \times 10^{-15}$ m,

$$V_0 \approx 35 \text{ Mev}$$

This is a plausible figure for V_0, from which we conclude that the basic features of our model—nucleons that maintain their identities in the nucleus instead of fusing together, strong nuclear forces with a short range and relatively minor angular dependence—are valid.

22.3 Triplet and Singlet States

Equation 22.21 contains all the information about the neutron-proton force that can be obtained from the observation that the deuteron is a stable system with a binding energy of $-E$. To go further we require additional experimental data. Among the most significant such data is that concerning angular momentum, which was ignored in the preceding analysis of the deuteron when it was assumed that the neutron-proton potential is a function of r alone. The latter is not correct; in general, angular momentum *does* play a significant role in nuclear structure, although certain aspects of this structure are virtually independent of it. In the case of the deuteron, for example, the proton and neutron interact in such a manner that binding occurs only when their spins are parallel to produce a *triplet state*, not when their spins are antiparallel to produce a *singlet state*. Evidently the neutron-proton force depends on the relative orientation of the spins and is weaker when the spins are antiparallel.

The difference between the triplet and singlet potentials, together with the Pauli exclusion principle, makes it possible to see why diprotons and dineutrons do not exist despite the observed stability of the deuteron and the charge-independence of nuclear forces. The exclusion principle prevents a diproton or a dineutron from occurring in a triplet state, since with parallel spins both nucleons in each system would be in identical quantum states. No such restriction applies to the deuteron, since the neutron and proton of which it consists are distinguishable particles even with parallel spins. However, while a diproton or dineutron could occur in principle in a singlet state, the singlet nuclear force is not sufficiently strong to produce binding—and, indeed, diprotons and dineutrons have never been found.

22.4 Meson Theory of Nuclear Forces

If nuclear forces were exclusively attractive, a nucleus would be stable only if its volume were so small (about 2 f in radius) that each nucleon interacted with all the others. The binding energy per nucleon would then be proportional to A, the number of nucleons present. In fact, nuclear volumes are found to be proportional to A and the binding energy per nucleon is roughly the same for all nuclei: each nucleon interacts only with a small number of its nearest neighbors. Thus there must be a repulsive component in nuclear forces that keeps nuclei from collapsing, as indicated in Fig. 22.1, which means that these forces are not analogous to the "ordinary" gravitational and electrical forces.

We encountered a similar situation in Sec. 12.4, where the coulomb forces present in the $H_2{}^+$ molecular ion turned out to be supplemented by an *exchange force* which arose because of the possibility that the electron can shift from one of the protons to the other. Depending on whether the wave function of the system is symmetric or antisymmetric for the particle exchange, the exchange force is either attractive or repulsive. It is tempting to consider the interaction between nucleons to be, at least in part, a consequence of some kind of exchange force as well. For instance, exchange forces provide an explanation for the stability of the triplet state of the deuteron, which is described by a symmetric wave function since the spins are parallel, and the instability of the singlet state, which is described by an antisymmetric wave function. Since the nucleons in a nucleus are all in different quantum states (by the exclusion principle), both attractive and repulsive exchange forces would occur, and a mixture of an "ordinary" attractive nuclear force and such exchange forces is able to account in a general way for a great many nuclear properties.

The next question is, what kinds of particles are exchanged between nearby nucleons? In 1932, Heisenberg suggested that electrons and positrons shift back and forth between nucleons; for instance, a neutron might emit an electron and become a proton, while a proton absorbing the electron would then become a neutron. However, calculations based on beta-decay data showed that the forces resulting from electron and positron exchange by nucleons are too small by the huge factor of 10^{14} to be significant in nuclear structure. Then, in 1935, the Japanese physicist Hideki Yukawa proposed that particles called *mesons*, heavier than electrons, are involved in nuclear forces, and he was able to show that the interactions they produce between nucleons are of the correct order of magnitude.

According to the meson theory of nuclear forces, all nucleons consist of identical cores surrounded by a "cloud" of one or more mesons. Mesons may be neutral or carry either charge, and the sole difference between neutrons and protons is supposed to lie in the composition of their respective meson clouds. The forces that act between one neutron and another and between one proton and another are the result of the exchange of neutral mesons (designated π^0) between them. The force between a neutron and a proton is the result of the exchange of charged mesons (π^+ and π^-) between them. Thus a neutron emits a π^- meson and is converted into a proton:

$$n \rightarrow p + \pi^-$$

while the absorption of the π^- by the proton the neutron was interacting with converts it into a neutron:

$$p + \pi^- \rightarrow n$$

In the reverse process, a proton emits a π^+ meson whose absorption by a neutron converts it into a proton:

$$p \rightarrow n + \pi^+$$
$$n + \pi^+ \rightarrow p$$

While there is no simple mathematical way of demonstrating how the exchange of particles between two bodies can lead to attractive and repulsive forces, a rough analogy may make the process intuitively meaningful. Let us imagine two boys exchanging basket-

repulsive force due to particle exchange

FIGURE 22.3 Attractive and repulsive forces can both arise from particle exchange.

attractive force due to particle exchange

balls (Fig. 22.3). If they throw the balls at each other, they each move backward, and when they catch the balls thrown at them, their backward momentum increases. Thus this method of exchanging the basketballs yields the same effect as a repulsive force between the boys. If the boys snatch the basketballs from each other's hands, however, the result will be equivalent to an attractive force acting between them.

A fundamental problem presents itself at this point. If nucleons constantly emit and absorb mesons, why are neutrons or protons never found with other than their usual masses? The answer is based upon the uncertainty principle. The laws of physics refer exclusively to experimentally measurable quantities, and the uncertainty principle limits the accuracy with which certain combinations of measurements can be made. The emission of a meson by a nucleon which does not change in mass—a clear violation of the law of conservation of energy—can occur provided that the nucleon absorbs a meson emitted by the neighboring nucleon it is interacting with so soon afterward that *even in principle* it is impossible to determine whether or not any mass change actually has been involved. Since the uncertainty principle may be written

22.22
$$\Delta E\, \Delta t \geq \hbar$$

an event in which an amount of energy ΔE is not conserved is not prohibited so long as the duration of the event does not exceed approximately $\hbar/\Delta E$.

We know that nuclear forces have a maximum range R of about 1.7 f, so that if we assume a meson travels between nuclei at approximately the speed of light c, the time interval Δt during which it is in flight is

$$22.23 \qquad\qquad \Delta t = \frac{R}{c}$$

The emission of a meson of mass m_π represents the nonconservation of

$$22.24 \qquad\qquad \Delta E = m_\pi c^2$$

of energy. According to Eq. 22.22, this can occur if $\Delta E \, \Delta t \geq \hbar$; that is, if

$$(m_\pi c^2) \left(\frac{R}{c} \right) \geq \hbar$$

Hence the minimum meson mass is specified by

$$22.25 \qquad\qquad m_\pi \geq \frac{\hbar}{Rc}$$

$$\geq 1.9 \times 10^{-28} \text{ kg}$$

which is about 200 m_e, that is, 200 electron masses.

Is the meson theory of nuclear forces correct? We are entitled to expect of any hypothesis that it make quantitative predictions in agreement with experiment; yet by and large the meson theory cannot account for nuclear properties in the detailed manner that quantum theory can account for atomic properties. There are exceptions to this disappointing picture, to be sure, for instance the problem of the magnetic moments of the proton and neutron. By analogy with the magnetic moment of the electron (Sec. 10.1), we would expect that of the proton to be $e\hbar/2m_p$ and that of the neutron, which is uncharged, to be 0. Actually, the magnetic moment of the proton is about 2.8 $e\hbar/2m_p$ and that of the neutron about 1.9 $e\hbar/2m_p$. A plausible explanation for these measurements is that both nucleons have structures that consist of "bare" cores of nuclear matter surrounded by clouds of charged mesons that contribute to the observed magnetic moments. Furthermore, particles *have* been discovered whose mass and behavior are in accord with meson theory, as we shall learn in Chap. 25. Hence there can be little doubt that Yukawa was on the right track, and it is possible, perhaps likely, that an elaboration of the meson theory will prove successful.

22.5 The Liquid-drop Model

While the attractive forces that nucleons exert upon one another are very strong, their range is so small that each particle in a nucleus interacts solely with its nearest neighbors. This situation is the same as that of atoms in a solid, which ideally vibrate about fixed positions in a crystal lattice, or that of molecules in a liquid, which ideally are free to move about while maintaining a fixed intermolecular distance. The analogy with a solid cannot be pursued because a calculation shows that the vibrations of the nucleons about their average positions would be too great for the nucleus to be stable. The analogy with a liquid, on the other hand, turns out to be extremely useful in understanding certain aspects of nuclear behavior.

Let us see how the picture of a nucleus as a drop of liquid accounts for the observed variation of binding energy per nucleon with mass number. We shall start by assuming that the energy associated with each nucleon-nucleon bond has some value U; this energy is really negative, since attractive forces are involved, but is usually written as positive because binding energy is considered a positive quantity for convenience. Because each bond energy U is shared by two nucleons, each has a binding energy of $\frac{1}{2}U$. When an assembly of spheres of the same size is packed together into the smallest volume, as we suppose is the case of nucleons within a nucleus, each interior sphere has 12 other spheres in contact with it (Fig. 22.4). Hence each interior nucleon in a nucleus has a binding energy of $12 \times \frac{1}{2}U$ or $6U$. If all A nucleons in a nucleus were in its interior, the total binding energy of the nucleus would be

22.26
$$E_v = 6AU$$

Equation 22.26 is often written simply as

22.27
$$E_v = a_1A$$

The energy E_v is called the *volume energy* of a nucleus and is directly proportional to A.

Actually, of course, some nucleons are on the surface of every nucleus and therefore have fewer than 12 neighbors. The number of such nucleons depends upon the surface area of the nucleus in question. A nucleus of radius R has an area of

$$4\pi R^2 = 4\pi R_0^2 A^{2/3}$$

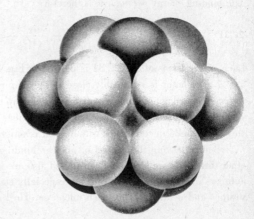

FIGURE 22.4 In a tightly packed assembly of identical spheres, each interior sphere is in contact with twelve others.

The Liquid-drop Model

Hence the number of nucleons with fewer than the maximum number of bonds is proportional to $A^{2/3}$, reducing the total binding energy by

22.28
$$E_s = -a_2 A^{2/3}$$

The negative energy E_s is called the *surface energy* of a nucleus; it is most significant for the lighter nuclei since a greater fraction of their nucleons are on the surface. Because natural systems always tend to evolve toward configurations of minimum potential energy, nuclei tend toward configurations of maximum binding energy. (We recall that binding energy is the mass-energy difference between a nucleus and the same numbers of free neutrons and protons.) Hence a nucleus should exhibit the same surface-tension effects as a liquid drop, and in the absence of external forces it should be spherical since a sphere has the least surface area for a given volume.

The electrostatic repulsion between each pair of protons in a nucleus also contributes toward decreasing its binding energy. The *coulomb energy* E_c of a nucleus is the work that must be done to bring together Z protons from infinity into a volume equal to that of the nucleus. Hence E_c is proportional to $Z(Z-1)/2$, the number of proton pairs in a nucleus containing Z protons, and inversely proportional to the nuclear radius $R = R_0 A^{1/3}$:

22.29
$$E_c = -a_3 \frac{Z(Z-1)}{A^{1/3}}$$

The coulomb energy is negative because it arises from a force that opposes nuclear stability.

The total *binding energy* E_b of a nucleus is the sum of its volume, surface, and coulomb energies:

$$E_b = E_v + E_s + E_c$$

22.30
$$= a_1 A - a_2 A^{2/3} - a_3 \frac{Z(Z-1)}{A^{1/3}}$$

The binding energy *per nucleon* is therefore

22.31
$$\frac{E_b}{A} = a_1 - \frac{a_2}{A^{1/3}} - a_3 \frac{Z(Z-1)}{A^{4/3}}$$

Each of the terms of Eq. 22.31 is plotted in Fig. 22.5 versus A, together with their sum, E_b/A. The latter is reasonably close to the empirical curve of E_b/A shown in Fig. 21.6. Hence the analogy of a nucleus with a liquid drop has some validity at least, and we may be encouraged to see what further aspects of nuclear behavior it can illuminate. We shall do this in Chap. 24 in connection with nuclear reactions.

Before leaving the subject of nuclear binding energy, it should be noted that effects other than those we have here considered also are involved. For instance, nuclei with equal numbers of protons and neutrons are especially stable, as are nuclei with even numbers of protons and neutrons. Thus such nuclei as $_2He^4$, $_6C^{12}$, and $_8O^{16}$ appear as peaks on the

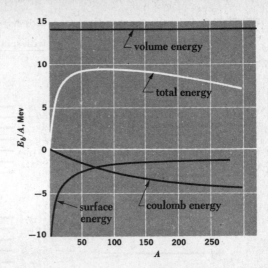

FIGURE 22.5 The binding energy per nucleon is the sum of the volume, surface, and coulomb energies.

empirical binding energy per nucleon curve. These peaks imply that the energy states of neutrons and protons in a nucleus are almost identical and that each state can be occupied by two particles of opposite spin, as discussed in Sec. 21.4.

22.6   The Shell Model

The basic assumption of the liquid-drop model is that the constituents of a nucleus interact only with their nearest neighbors, like the molecules of a liquid. There is a good deal of empirical support for this assumption. There is also, however, extensive experimental evidence for the contrary hypothesis that the nucleons in a nucleus interact primarily with a general force field rather than directly with one another. This latter situation resembles that of electrons in an atom, where only certain quantum states are permitted and no more than two electrons, which are Fermi particles, can occupy each state. Nucleons are also Fermi particles, and several nuclear properties vary periodically with Z and N in a manner reminiscent of the periodic variation of atomic properties with Z.

The electrons in an atom may be thought of as occupying positions in "shells" designated by the various principal quantum numbers, and the degree of occupancy of the outermost shell is what determines certain important aspects of an atom's behavior. For instance, atoms with 2, 10, 18, 36, 54, and 86 electrons have all their electron shells completely filled. Such electron structures are stable, thereby accounting for the chemical inertness of the rare gases. The same kind of situation is observed with respect to nuclei; nuclei having 2, 8, 20, 28, 50, 82, and 126 neutrons or protons are more abundant than other nuclei of similar mass numbers, suggesting that their structures are more stable. Since complex nuclei arose from reactions among lighter ones, the evolution of heavier

The Shell Model   **517**

FIGURE 22.6 Sequence of nucleon energy levels according to the shell model (not to scale).

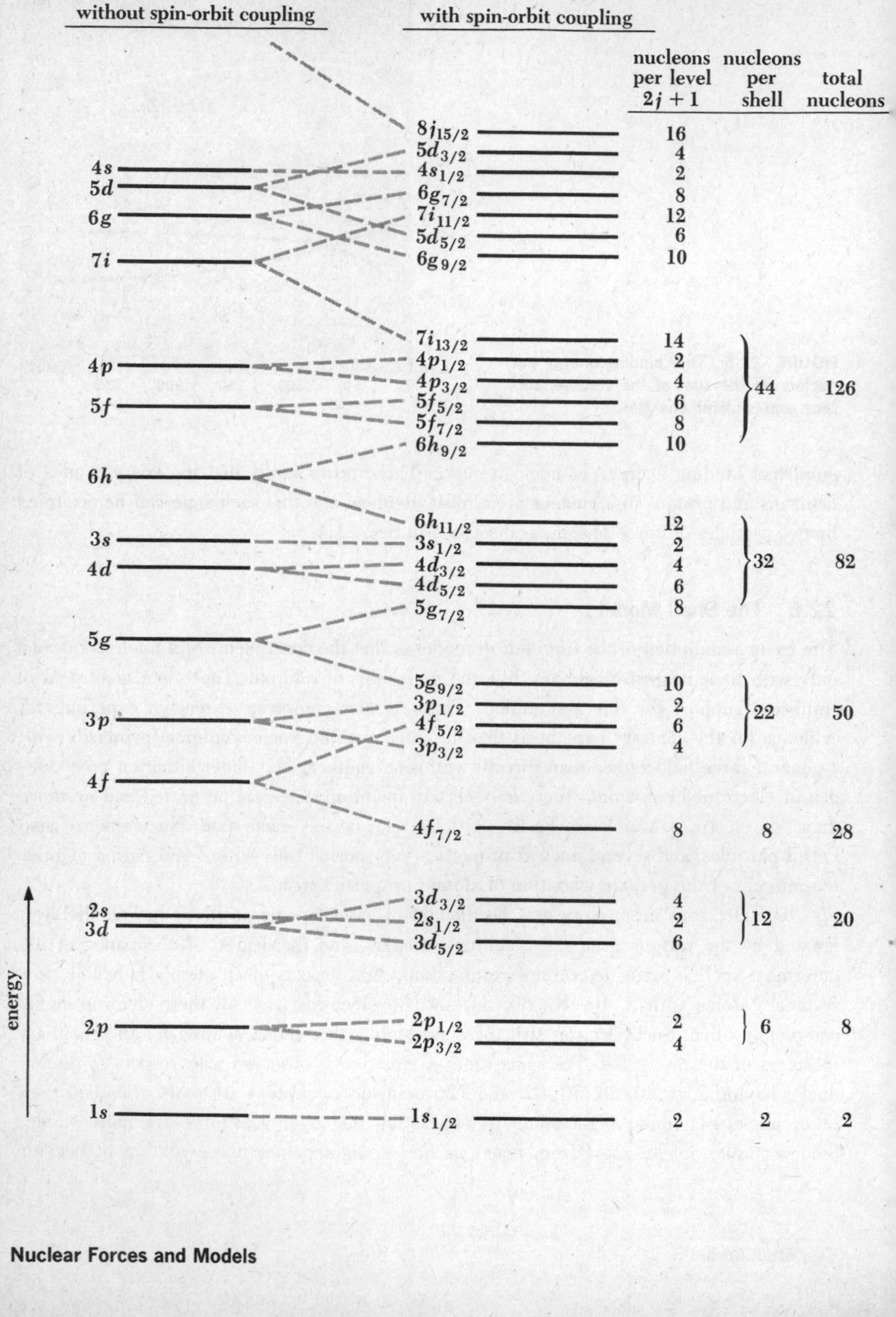

		nucleons per level $2j + 1$	nucleons per shell	total nucleons
without spin-orbit coupling	with spin-orbit coupling			

$8j_{15/2}$ — 16
$5d_{3/2}$ — 4
$4s$ —— $4s_{1/2}$ — 2
$5d$ —— $6g_{7/2}$ — 8
$6g$ —— $7i_{11/2}$ — 12
$5d_{5/2}$ — 6
$7i$ —— $6g_{9/2}$ — 10

$7i_{13/2}$ — 14
$4p$ —— $4p_{1/2}$ — 2
$4p_{3/2}$ — 4
$5f$ —— $5f_{5/2}$ — 6 } 44   126
$5f_{7/2}$ — 8
$6h_{9/2}$ — 10
$6h$ ——

$6h_{11/2}$ — 12
$3s$ —— $3s_{1/2}$ — 2
$4d$ —— $4d_{3/2}$ — 4 } 32   82
$4d_{5/2}$ — 6
$5g_{7/2}$ — 8
$5g$ ——

$5g_{9/2}$ — 10
$3p$ —— $3p_{1/2}$ — 2
$4f_{5/2}$ — 6 } 22   50
$3p_{3/2}$ — 4
$4f$ ——

$4f_{7/2}$ — 8   8   28

$3d_{3/2}$ — 4
$2s$ —— $2s_{1/2}$ — 2 } 12   20
$3d$ —— $3d_{5/2}$ — 6

$2p$ —— $2p_{1/2}$ — 2 } 6   8
$2p_{3/2}$ — 4

energy

$1s$ —— $1s_{1/2}$ — 2   2   2

and heavier nuclei became retarded when each relatively inert nucleus was formed; this accounts for their abundance.

Other evidence also points up the significance of the numbers 2, 8, 20, 28, 50, 82, and 126, which have become known as *magic numbers*, in nuclear structure. An example is the observed pattern of nuclear electric quadrupole moments, which are measures of the departures of nuclear charge distributions from sphericity. A spherical nucleus has no quadrupole moment, while one shaped like a football has a positive moment and one shaped like a pumpkin has a negative moment. Nuclei of magic N and Z are found to have zero quadrupole moments and hence are spherical, while other nuclei are distorted in shape.

The *shell model* of the nucleus is an attempt to account for the existence of magic numbers and certain other nuclear properties in terms of interactions between an individual nucleon and a force field produced by all the other nucleons. A potential energy function is used that corresponds to a square well about 50 Mev deep with rounded corners, so that there is a more realistic gradual change from $V = V_0$ to $V = 0$ than the sudden change of the pure square-well potential we used in treating the deuteron. Schrödinger's equation for a particle in a potential well of this kind is then solved, and it is found that stationary states of the system occur characterized by quantum numbers n, l, and m_l whose significance is the same as in the analogous case of stationary states of atomic electrons. Neutrons and protons occupy separate sets of states in a nucleus since the latter interact electrically as well as through the specifically nuclear charge.

In order to obtain a series of energy levels that leads to the observed magic numbers, it is merely necessary to assume a spin-orbit interaction whose magnitude is such that the consequent splitting of energy levels into sublevels is large for large l, that is, for large orbital angular momenta. It is assumed that LS coupling holds only for the very lightest nuclei, in which the l values are necessarily small in their normal configurations. In this scheme, as we saw in Chap. 10, the intrinsic spin angular momenta $\mathbf{S}_i$ of the particles concerned (the neutrons form one group and the protons another) are coupled together into a total spin momentum $\mathbf{S}$, and the orbital angular momenta $\mathbf{L}_i$ are separately coupled together into a total orbital momentum $\mathbf{L}$; $\mathbf{S}$ and $\mathbf{L}$ are then coupled to form a total angular momentum $\mathbf{J}$ of magnitude $\sqrt{J(J+1)}\hbar$. After a transition region in which an intermediate coupling scheme holds, the heavier nuclei exhibit jj coupling. In this case the $\mathbf{S}_i$ and $\mathbf{L}_i$ of each particle are first coupled to form a $\mathbf{J}_i$ for that particle of magnitude $\sqrt{j(j+1)}\,\hbar$, and the various $\mathbf{J}_i$ then couple together to form the total angular momentum $\mathbf{J}$. The jj coupling scheme holds for the great majority of nuclei.

When an appropriate strength is assumed for the spin-orbit interaction, the energy levels of either class of nucleon fall into the sequence shown in Fig. 22.6. The levels are designated by a prefix equal to the total quantum number n, a letter that indicates l for each particle in that level according to the usual pattern ($s, p, d, f, g, \ldots$ correspond respectively to $l = 0, 1, 2, 3, 4, \ldots$), and a subscript equal to j. The spin-orbit interaction splits each state of given j into $2j + 1$ substates, since there are $2j + 1$ allowed orientations of $\mathbf{J}_i$. Large energy gaps appear in the spacing of the levels at intervals that are consistent with the notion of separate shells. The number of available nuclear states in each nuclear shell is, in ascending order of energy, 2, 6, 12, 8, 22, 32, and 44; hence shells are filled when there are 2, 8, 20, 28, 50, 82, or 126 neutrons or protons in a nucleus.

The shell model is able to account for several nuclear phenomena in addition to magic numbers. Since each energy sublevel can contain two particles (spin up and spin down), only filled sublevels are present when there are even numbers of neutrons and protons in a nucleus ("even-even" nucleus). At the other extreme, a nucleus with odd numbers of neutrons and protons ("odd-odd" nucleus) contains unfilled sublevels for both kinds of particle. The stability we expect to be conferred by filled sublevels is borne out by the fact that 160 stable even-even nuclides are known, as against only four stable odd-odd nuclides: $_1H^2$, $_3Li^6$, $_5B^{10}$, and $_7N^{14}$.

Further evidence in favor of the shell model is its ability to predict total nuclear angular momenta. In even-even nuclei, all of the protons and neutrons should pair off so as to cancel out one another's spin and orbital angular momenta. Thus even-even nuclei ought to have zero nuclear angular momenta, as observed. In even-odd (even Z, odd N) and odd-even (odd Z, even N) nuclei, the half-integral spin of the single "extra" nucleon should be combined with the integral angular momentum of the rest of the nucleus for a half-integral total angular momentum, and odd-odd nuclei each have an extra neutron and an extra proton whose half-integral spins should yield integral total angular momenta. Both of these predictions are experimentally confirmed.

If the nucleons in a nucleus are so close together and interact so strongly that the nucleus can be considered as analogous to a liquid drop, how can these same nucleons be regarded as moving independently of each other in a common force field as required by the shell model? It would seem that the points of view are mutually exclusive, since a nucleon moving about in a liquid-drop nucleus must surely undergo frequent collisions with other nucleons. A closer look shows that there is no contradiction. In the ground state of a nucleus, the neutrons and protons fill the energy levels available to them in order of increasing energy in such a way as to obey the exclusion principle (see Fig. 21.3). In a collision, energy is transferred from one nucleon to another, leaving the former in a state of reduced energy and the latter in one of increased energy. But all the available levels of lower energy are already filled, so such an energy transfer can take place only if the exclusion principle is violated. Of course, it is possible for two nucleons of the same kind to merely exchange their respective energies, but such a collision is hardly significant since the system remains in exactly the same state it was in initially. In essence, then, the exclusion principle prevents nucleon-nucleon collisions even in a tightly packed nucleus, and thereby justifies the independent-particle approach to nuclear structure.

Both the liquid-drop and shell models of the nucleus are, in their very different ways, able to account for much that is known of nuclear behavior. Recently attempts have been made to devise theories that combine the best features of each of these models in a consistent scheme, and some success has been achieved in the endeavor. The resulting *collective model* includes the possibility of a nucleus vibrating and rotating as a whole. The situation is complicated by the nonspherical shape of all but even-even nuclei and the centrifugal distortion experienced by a rotating nucleus; the detailed theory is consistent with the spacing of excited nuclear levels inferred from the gamma-ray spectra of nuclei and in other ways.

Problems

1. Show that the electrostatic potential energy of two protons 1.7×10^{-15} m apart is of the correct order of magnitude to account for the difference in binding energy between $_1H^3$ and $_2He^3$. How does this result bear upon the problem of the charge-independence of nuclear forces? (The masses of the neutral atoms are, respectively, 3.016049 and 3.016029 u.)

2. Protons, neutrons, and electrons all have spins of ½. Why do $_2He^4$ atoms obey Bose-Einstein statistics while $_2He^3$ atoms obey Fermi-Dirac statistics?

3. Show that the results of Sec. 22.2—a potential well for the deuteron that is about 35 Mev deep and 2 f in radius— are consistent with the uncertainty principle.

4. Van der Waals forces are limited to very short ranges and do not have an inverse-square dependence on distance, yet nobody suggests that the exchange of a special meson-like particle is responsible for such forces. Discuss the differences and similarities between nuclear forces and van der Waals forces.

5. Calculate the approximate value of a_3 in Eq. 22.29 using whatever assumptions seem appropriate.

6. The *Klein-Gordon equation,*

$$\nabla^2 \psi - \frac{1}{c^2} \frac{\partial^2 \psi}{\partial t^2} = \frac{m_\pi^2 c^2}{\hbar^2} \psi$$

is a relativistic wave equation that describes the π meson. (a) Derive the Klein-Gordon equation by starting from the relativistic formula $E^2 = m_\pi^2 c^4 + p^2 c^2$ and substituting the appropriate operation for E^2 and for p^2. (b) Verify that

$$\psi_\pi = -\frac{a^2}{r} e^{-r/R} \qquad R = \frac{\hbar}{m_\pi c}$$

is a solution of the Klein-Gordon equation for $r > 0$. How is this result related to the discussion of the meson mass m_π in Sec. 22.4? (c) Verify that

$$\psi_0 = -\frac{b^2}{r}$$

is a solution of the Klein-Gordon equation for $r > 0$ when $m_\pi = 0$. This result corresponds to the electromagnetic field, whose "mesons" are massless quanta. (d) What are the characteristics of the interaction potentials suggested by ψ_π and ψ_0?

7. According to the *Fermi gas model* of the nucleus, its protons and neutrons exist in a box of nuclear dimensions and fill the lowest available quantum states to the extent permitted by the exclusion principle. Since both neutrons and protons have spins of ½, they are Fermi particles and must obey Fermi-Dirac statistics. (a) Starting from Eq. 19.36, derive an equation for the Fermi energy in a nucleus under the assumption that equal numbers of neutrons and protons are present. (b) Find the Fermi energy for such a nucleus in which $R_0 = 1.3$ f.

Chapter 23
Radioactivity

Perhaps no single phenomenon has played so significant a role in the development of both atomic and nuclear physics as *radioactivity*. A nucleus undergoing radioactive decay spontaneously emits a $_2\text{He}^4$ nucleus (alpha particle), an electron (beta particle), or a photon (gamma ray), thereby either ridding itself of nuclear excitation energy or achieving a configuration that is or will lead to one of greater stability. In this chapter we are concerned with the various physical processes involved in radioactive decay itself, and not with the many applications of radioactivity in the past and present.

23.1 Statistics of Radioactive Decay

The *activity* of a sample of any radioactive material is the rate at which the nuclei of its constituent atoms decay. If N is the number of nuclei present at a certain time in the sample, its activity R is given by

23.1
$$R = -\frac{dN}{dt}$$

The minus sign is inserted to make R a positive quantity, since dN/dt is, of course, intrinsically negative. While the natural units for activity are disintegrations per second, it is customary to express R in terms of the *curie* and its submultiples, the *millicurie* (mc) and *microcurie* (μc). By definition,

$$1 \text{ curie} = 3.70 \times 10^{10} \text{ disintegrations/sec}$$
$$1 \text{ mc} = 10^{-3} \text{ curie} = 3.70 \times 10^{7} \text{ disintegrations/sec}$$
$$1 \text{ μc} = 10^{-6} \text{ curie} = 3.70 \times 10^{4} \text{ disintegrations/sec}$$

Experimental measurements on the activities of radioactive samples indicate that, in every case, they fall off exponentially with time. Figure 23.1 is a graph of R versus t for a typical radioisotope. We note that in every 5-hr period, regardless of when the period starts, the activity drops to half of what it was at the start of the period. Accordingly the *half life* $T_{1/2}$ of the isotope is 5 hr. Every radioisotope has a characteristic half life; some have half lives of a millionth of a second, others have half lives that range up to billions of years. When the observations plotted in Fig. 23.1 began, the activity of the sample was R_0. Five

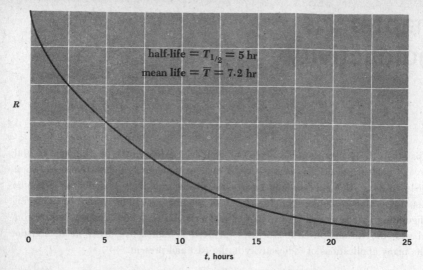

FIGURE 23.1   The activity of a radioisotope decreases exponentially with time.

hr later it decreased to $0.5R_0$. After another 5 hr, R again decreased by a factor of 2 to $0.25R_0$. That is, the activity of the sample was only 0.25 its initial value after an interval of $2T_{1/2}$. With the lapse of another half life of 5 hr, corresponding to a total interval of $3T$, R became $\frac{1}{2}(0.25R_0)$, or $0.125R_0$.

The behavior illustrated in Fig. 23.1 indicates that we can express our empirical information about the time variation of activity in the form

23.2
$$R = R_0e^{-\lambda t}$$

where λ, called the *decay constant*, has a different value for each radioisotope. The connection between decay constant λ and half life $T_{1/2}$ is easy to establish. After a half life has elapsed, that is, when $t = T_{1/2}$, the activity R drops to $\frac{1}{2}R_0$ by definition. Hence

$$R = R_0e^{-\lambda t}$$
$$\tfrac{1}{2}R_0 = R_0e^{-\lambda T_{1/2}}$$
$$e^{\lambda T_{1/2}} = 2$$

Taking natural logarithms of both sides of this equation,

$$\lambda T_{1/2} = \ln 2$$

23.3
$$T_{1/2} = \frac{\ln 2}{\lambda} = \frac{0.693}{\lambda}$$

Half life

The decay constant of the radioisotope whose half life is 5 hr is therefore

$$\lambda = \frac{0.693}{T_{1/2}}$$

$$= \frac{0.693}{5 \text{ hr} \times 3,600 \text{ sec/hr}}$$

$$= 3.85 \times 10^{-5} \text{ sec}^{-1}$$

The fact that radioactive decay follows the exponential law of Eq. 23.2 is strong evidence that this phenomenon is statistical in nature: every nucleus in a sample of radioactive material has a certain probability of decaying, but there is no way of knowing in advance *which* nuclei will actually decay in a particular time span. If the sample is large enough—that is, if many nuclei are present—the actual fraction of it that decays in a certain time span will be very close to the a priori probability for any individual nucleus to decay. The statement that a certain radioisotope has a half life of 5 hr, then, signifies that every nucleus of this isotope has a 50 percent chance of decaying in any 5-hr period. This does *not* mean a 100 percent probability of decaying in 10 hr; a nucleus does not have a memory, and its decay probability per unit time is constant until it actually does decay. A half life of 5 hr implies a 75 percent probability of decay in 10 hr, which increases to 87.5 percent in 15 hr, to 93.75 percent in 20 hr, and so on, because in every 5-hr interval the probability is 50 percent.

The empirical activity law of Eq. 23.2 follows directly from the assumption of a constant probability λ per unit time for the decay of each nucleus of a given isotope. Since λ is the probability per unit time, $\lambda\, dt$ is the probability that any nucleus will undergo decay in a time interval dt. If a sample contains N undecayed nuclei, the number dN that decay in a time dt is the product of the number of nuclei N and the probability $\lambda\, dt$ that each will decay in dt. That is,

23.4
$$dN = -N\lambda\, dt$$

where the minus sign is required because N decreases with increasing t. Equation 23.4 can be rewritten

$$\frac{dN}{N} = -N\lambda\, dt$$

and each side can now be integrated:

$$\int_{N_0}^{N} \frac{dN}{N} = -\lambda \int_{0}^{t} dt$$

$$\ln N - \ln N_0 = -\lambda t$$

23.5
$$N = N_0 e^{-\lambda t}$$

Equation 23.5 is a formula that gives the number N of undecayed nuclei at the time t in terms of the decay probability per unit time λ of the isotope involved and the number N_0 of undecayed nuclei at $t = 0$.

Since the activity of a radioactive sample is defined as

$$R = -\frac{dN}{dt}$$

we see that, from Eq. 23.5,

$$R = \lambda N_0 e^{-\lambda t}$$

This agrees with the empirical activity law if

$$R_0 = \lambda N_0$$

or, in general, if

23.6
$$R = \lambda N$$

Evidently the decay constant λ of a radioisotope is the same as the probability per unit time for the decay of a nucleus of that isotope.

Equation 23.6 permits us to calculate the activity of a radioisotope sample if we know its mass, atomic mass, and decay constant. As an example, let us determine the activity of a 1-g sample of $_{38}Sr^{90}$, whose half life against beta decay is 28 years. The decay constant of $_{38}Sr^{90}$ is

$$\lambda = \frac{0.693}{T_{1/2}}$$

$$= \frac{0.693}{28 \text{ years} \times 3.16 \times 10^7 \text{ sec/year}}$$

$$= 7.83 \times 10^{-10} \text{ sec}^{-1}$$

A kmole of an isotope has a mass very nearly equal to the mass number of that isotope expressed in kilograms. Hence 1 g of $_{38}Sr^{90}$ contains

$$\frac{10^{-3} \text{ kg}}{90 \text{ kg/kmole}} = 1.11 \times 10^{-5} \text{ kmole}$$

One kmole of any isotope contains Avogadro's number of atoms, and so 1 g of $_{38}Sr^{90}$ contains

$$1.11 \times 10^{-5} \text{ kmole} \times 6.02 \times 10^{26} \text{ atoms/kmole} = 6.68 \times 10^{21} \text{ atoms}$$

Thus the activity of the sample is

$$R = \lambda N$$
$$= 7.83 \times 10^{-10} \times 6.68 \times 10^{21} \text{ sec}^{-1}$$
$$= 5.23 \times 10^{12} \text{ sec}^{-1}$$
$$= 141 \text{ curies}$$

It is worth keeping in mind that the half life of a radioisotope is not the same as its *mean lifetime* $\bar{T}$. The mean lifetime of an isotope is the reciprocal of its decay probability per unit time:

23.7
$$\bar{T} = \frac{1}{\lambda}$$

Hence

23.8
$$\bar{T} = \frac{1}{\lambda} = \frac{T_{1/2}}{0.693} = 1.44 T_{1/2} \qquad \textbf{Mean lifetime}$$

$\bar{T}$ is nearly half again more than $T_{1/2}$. The mean lifetime of an isotope whose half life is 5 hr is 7.2 hr.

23.2 Radioactive Series

Most of the radioactive elements found in nature are members of four *radioactive series,* with each series consisting of a succession of daughter products all ultimately derived from a single parent nuclide. The reason that there are exactly four such series follows from the fact that alpha decay reduces the mass number of a nucleus by 4. Thus the nuclides whose mass numbers are all given by

23.9
$$A = 4n$$

where n is an integer, can decay into one another in descending order of mass number. Radioactive nuclides whose mass numbers obey Eq. 23.9 are said to be members of the $4n$ series. The members of the $4n + 1$ series have mass numbers specified by

23.10
$$A = 4n + 1$$

and members of the $4n + 2$ and $4n + 3$ series have mass numbers specified respectively by

23.11
$$A = 4n + 2$$

23.12
$$A = 4n + 3$$

The members of each of these series, too, can decay into one another in descending order of mass number.

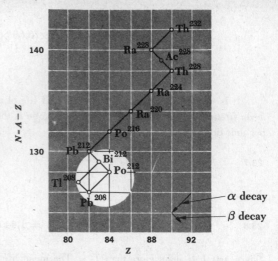

FIGURE 23.2 The thorium decay series ($A = 4n$). The decay of $_{83}Bi^{212}$ may proceed either by alpha emission and then beta emission or in the reverse order.

Table 23.1 is a list of the names of four important radioactive series, their parent nuclides and the half lives of these parents, and the stable daughters which are end products of the series. The half life of neptunium is so short compared with the estimated age ($\sim 10^{10}$ years) of the universe that the members of this series are not found in nature today. They have, however, been produced in the laboratory by the neutron bombardment of other heavy nuclei, as we shall describe in the next chapter. The sequences of alpha and beta decays that lead from parent to stable end product in each series are shown in Figs. 23.2 to 23.5. Some nuclides may decay either by beta or alpha emission, so that the decay chain *branches* at

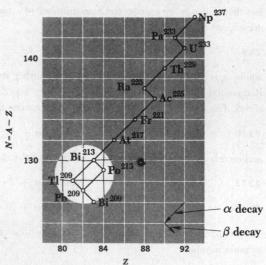

FIGURE 23.3 The neptunium decay series ($A = 4n + 1$). The decay of $_{83}Bi^{213}$ may proceed either by alpha emission and then beta emission or in the reverse order.

TABLE 23.1. FOUR RADIOACTIVE SERIES

Mass numbers	Series	Parent	Half life, years	Stable end product
$4n$	Thorium	$_{90}\text{Th}^{232}$	1.39×10^{10}	$_{82}\text{Pb}^{208}$
$4n + 1$	Neptunium	$_{93}\text{Np}^{237}$	2.25×10^{6}	$_{83}\text{Bi}^{209}$
$4n + 2$	Uranium	$_{92}\text{U}^{238}$	4.51×10^{9}	$_{82}\text{Pb}^{206}$
$4n + 3$	Actinium	$_{92}\text{U}^{235}$	7.07×10^{8}	$_{82}\text{Pb}^{207}$

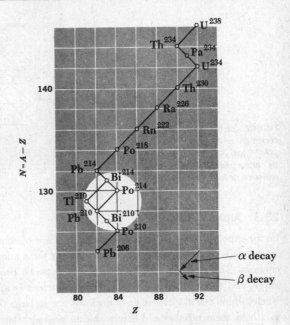

FIGURE 23.4 The uranium decay series ($A = 4n + 2$). The decay of $_{83}\text{Bi}^{214}$ may proceed either by alpha emission and then beta emission or in the reverse order.

them. Thus $_{83}\text{Bi}^{212}$, a member of the thorium series, has a 66.3 percent chance of beta decaying into $_{84}\text{Po}^{212}$ and a 33.7 percent chance of alpha decaying into $_{81}\text{Tl}^{208}$. The beta decay is followed by a subsequent alpha decay and the alpha decay is followed by a subsequent beta decay, so that both branches lead to $_{82}\text{Pb}^{208}$.

Several alpha-radioactive nuclides whose atomic numbers are less than 82 are found in nature, though they are not very abundant. These nuclides are listed in Table 23.2.

TABLE 23.2. LIGHT ALPHA EMITTERS

Nuclide	Half life, years
$_{60}\text{Nd}^{144}$	1×10^{15}
$_{62}\text{Sm}^{147}$	1.4×10^{11}
$_{64}\text{Gd}^{152}$	1.1×10^{14}
$_{72}\text{Hf}^{174}$	2×10^{15}
$_{78}\text{Pt}^{190}$	6×10^{11}

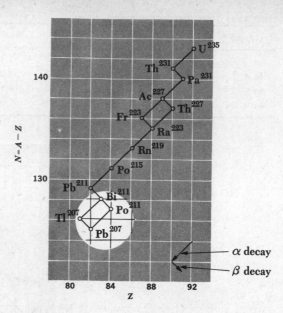

FIGURE 23.5 The actinium decay series ($A = 4n + 3$). The decays of $_{89}Ac^{227}$ and $_{83}Bi^{211}$ may proceed either by alpha emission and then beta emission or in the reverse order.

The figure axes are labeled $N = A - Z$ (vertical) with values 130 and 140 marked, and Z (horizontal) with values 80, 84, 88, 92 marked. The decay series shows: U^{235}, Th^{231}, Pa^{231}, Ac^{227}, Th^{227}, Fr^{223}, Ra^{223}, Rn^{219}, Po^{215}, Pb^{211}, Bi^{211}, Po^{211}, Tl^{207}, Pb^{207}. Arrows indicate α decay and β decay.

23.3 Alpha Decay

Because the attractive forces between nucleons are of short range, the total binding energy in a nucleus is approximately proportional to its mass number A, the number of nucleons it contains. The repulsive electrostatic forces between protons, however, are of unlimited range, and the total disruptive energy in a nucleus is approximately proportional to Z^2. Nuclei which contain 210 or more nucleons are so large that the short-range nuclear forces that hold them together are barely able to counterbalance the mutual repulsion of their protons. Alpha decay occurs in such nuclei as a means of increasing their stability by reducing their size.

Why are alpha particles invariably emitted rather than, say, individual protons or $_2He^3$ nuclei? The answer follows from the high binding energy of the alpha particle. To escape from a nucleus, a particle must have kinetic energy, and the alpha-particle mass is sufficiently smaller than that of its constituent nucleons for such energy to be available. To illustrate this point, we can compute, from the known masses of each particle and the parent and daughter nuclei, the kinetic energy Q released when various particles are emitted by a heavy nucleus. This is given by

$$Q = (m_i - m_f - m)c^2$$

where m_i is the mass of the initial nucleus, m_f the mass of the final nucleus, and m the particle mass. The result is that *only* the emission of an alpha particle is energetically possible; other decay modes would require energy to be supplied from outside the nucleus

Thus alpha decay in $_{92}U^{232}$ is accompanied by the release of 5.4 Mev, while 6.1 Mev would somehow have to be furnished from an external source if a proton is to be emitted and 9.6 Mev if a $_2He^3$ nucleus is to be emitted. The observed disintegration energies in alpha decay agree with the corresponding predicted values based upon the nuclear masses involved.

The kinetic energy T_α of the emitted alpha particle is never quite equal to the disintegration energy Q because, since momentum must be conserved, the nucleus recoils with a small amount of kinetic energy when the alpha particle emerges. It is easy to show that, as a consequence of momentum and energy conservation, T_α is related to Q and the mass number A of the original nucleus by

$$T_\alpha \approx \frac{A-4}{A} Q$$

The mass numbers of nearly all alpha emitters exceed 210, and so most of the disintegration energy appears as the kinetic energy of the alpha particle. In the decay of $_{86}Rn^{222}$, $Q = 5.587$ Mev while $T_\alpha = 5.486$ Mev.

While a heavy nucleus can, in principle, spontaneously reduce its bulk by alpha decay, there remains the problem of *how* an alpha particle can actually escape from the nucleus. Figure 23.6 is a plot of the potential energy V of an alpha particle as a function of its distance r from the center of a certain heavy nucleus. The height of the potential barrier is about 25 Mev, which is equal to the work that must be done against the repulsive electrostatic force to bring an alpha particle from infinity to a position adjacent to the nucleus but

FIGURE 23.6   The potential energy of an alpha particle as a function of its distance from the center of a nucleus.

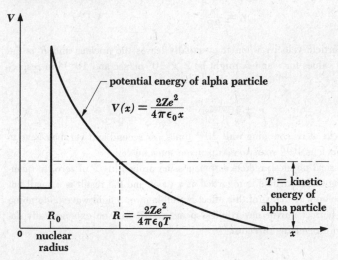

just outside the range of its attractive forces. We may therefore regard an alpha particle in such a nucleus as being inside a box whose walls require an energy of 25 Mev to be surmounted. However, decay alpha particles have energies that range from 4 to 9 Mev, depending upon the particular nuclide involved—16 to 21 Mev short of the energy needed for escape.

While alpha decay is inexplicable on the basis of classical arguments, quantum mechanics provides a straightforward explanation. In fact, the theory of alpha decay developed independently in 1928 by Gamow and by Gurney and Condon was greeted as an especially striking confirmation of quantum mechanics. In the following two sections we shall show how even a simplified treatment of the problem of the escape of an alpha particle from a nucleus gives results in agreement with experiment.

The basic notions of this theory are:

1. An alpha particle may exist as an entity within a heavy nucleus;

2. Such a particle is in constant motion and is contained in the nucleus by the surrounding potential barrier;

3. There is a small—but definite—likelihood that the particle may pass through the barrier (despite its height) each time a collision with it occurs.

Thus the decay probability per unit time λ can be expressed as

$$\lambda = \nu P$$

where ν is the number of times per second an alpha particle within a nucleus strikes the potential barrier around it and P is the probability that the particle will be transmitted through the barrier. If we suppose that at any moment only one alpha particle exists as such in a nucleus and that it moves back and forth along a nuclear diameter,

$$\nu = \frac{v}{2R}$$

where v is the alpha-particle velocity when it eventually leaves the nucleus and R is the nuclear radius. Typical values for v and R might be 2×10^7 m/sec and 10^{-14} m respectively, so that

$$\nu \approx 10^{21} \text{ sec}^{-1}$$

The alpha particle knocks at its confining wall 10^{21} times per second and yet may have to wait an average of as much as 10^{10} years to escape from some nuclei!

Since $V > E$, classical physics predicts a transmission probability P of zero. In quantum mechanics a moving alpha particle is regarded as a wave, and the result is a small but definite value for P. The optical analog of this effect is well-known: a light wave undergoing reflection from even a perfect mirror nevertheless penetrates it with an exponentially decreasing amplitude before reversing direction.

23.4 Barrier Penetration

Let us consider the case of a beam of particles of kinetic energy T incident from the left on a potential barrier of height V and width L, as in Fig. 23.7. On both sides of the barrier $V = 0$, which means that no forces act upon the particles there. In these regions Schrödinger's equation for the particles is

23.13
$$\frac{\partial^2 \psi_I}{\partial x^2} + \frac{2m}{\hbar^2} E\psi_I = 0$$

and

23.14
$$\frac{\partial^2 \psi_{III}}{\partial x^2} + \frac{2m}{\hbar^2} E\psi_{III} = 0$$

Let us assume that

23.15
$$\psi_I = Ae^{iax} + Be^{-iax}$$

23.16
$$\psi_{III} = Ee^{iax} + Fe^{-iax}$$

are solutions to Eqs. 23.13 and 23.14 respectively. The various terms in these solutions are not hard to interpret. As shown schematically in Fig. 23.8, Ae^{iax} is a wave of amplitude A incident from the left on the barrier. That is,

23.17
$$\psi_{I+} = Ae^{iax}$$

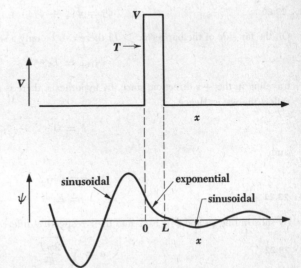

FIGURE 23.7   A beam of particles can "leak" through a finite barrier.

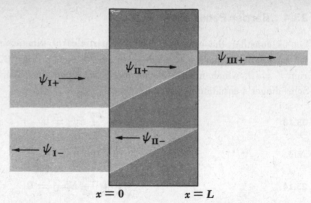

FIGURE 23.8 Schematic representation of barrier penetration.

$x = 0$ $x = L$

This wave corresponds to the incident beam of particles in the sense that $|\psi_{I+}|^2$ is their probability density. If v is the group velocity of the wave, which equals the particle velocity,

23.18
$$|\psi_{I+}|^2 v$$

is the flux of particles that arrive at the barrier. At $x = 0$ the incident wave strikes the barrier and is partially reflected, with

23.19
$$\psi_{I-} = Be^{-iax}$$

representing the reflected wave. Hence

23.20
$$\psi_{I} = \psi_{I+} + \psi_{I-}$$

On the far side of the barrier $(x > L)$ there can be only a wave

$$\psi_{III+} = Ee^{iax}$$

traveling in the $+x$ direction, since, by hypothesis, there is nothing in region III that could reflect the wave. Hence

$$F = 0$$

and

$$\psi_{III} = \psi_{III+}$$

23.21
$$= Ee^{iax}$$

By substituting ψ_I and ψ_{III} back into their respective differential equations, we find that

23.22
$$a = \sqrt{\frac{2mT}{\hbar^2}}$$

It is evident that the transmission probability P for a particle to pass through the barrier is the ratio

23.23
$$P = \frac{|\psi_{III}|^2}{|\psi_I|^2} = \frac{EE^*}{AA^*}$$

between its probability density in region III and its probability density in region I. Classically $P = 0$ because the particle cannot exist inside the barrier; let us see what the quantum-mechanical result is.

In region II Schrödinger's equation for the particles is

23.24
$$\frac{\partial^2 \psi_{II}}{\partial x^2} + \frac{2m}{\hbar^2}(T - V)\psi_{II} = 0$$

Its solution is

23.25
$$\psi_{II} = Ce^{ibx} + De^{-ibx}$$

where

23.26
$$b = \sqrt{\frac{2m(T - V)}{\hbar^2}}$$

Since $V > T$, b is imaginary and we may define a new wave number b' by

23.27
$$\begin{aligned} b' &= -ib \\ &= \sqrt{\frac{2m(V - T)}{\hbar^2}} \end{aligned}$$

Hence

23.28
$$\psi_{II} = Ce^{-b'x} + De^{b'x}$$

The term

23.29
$$\psi_{II+} = Ce^{-b'x}$$

is an exponentially decreasing wave function that corresponds to a nonoscillatory disturbance moving to the right through the barrier. Within the barrier part of the disturbance is reflected, and

23.30
$$\psi_{II-} = De^{b'x}$$

is an exponentially decreasing wave function that corresponds to the reflected disturbance moving to the left.

Even though ψ_{II} does not oscillate, and therefore does not represent a moving particle of positive kinetic energy, the probability density $|\psi_{II}|^2$ is not zero. There is a finite prob-

Barrier Penetration

ability of finding a particle within the barrier. A particle at the far end of the barrier that is not reflected there will emerge into region III with the same kinetic energy T it originally had, and its wave function will be ψ_{III} as it continues moving unimpeded in the $+x$ direction. In the limit of an infinitely thick barrier, $\psi_{III} = 0$, which implies that all of the incident particles are reflected. The reflection process takes place *within* the barrier, however, not at its left-hand wall, and a barrier of finite width therefore permits a fraction P of the initial beam to pass through it.

In order to calculate P, we must apply certain boundary conditions to ψ_I, ψ_{II}, and ψ_{III}. Figure 23.7 is a schematic representation of the wave functions in regions I, II, and III which may help in visualizing the boundary conditions. As discussed in Chap. 7, both ψ and its derivative $\partial\psi/\partial x$ must be continuous everywhere. With reference to Fig. 23.7, these conditions mean that, at each wall of the barrier, the wave functions inside and outside must not only have the same value but also the same slope, so that they match up perfectly. Hence at the left-hand wall of the barrier

23.31a
$$\psi_I = \psi_{II}$$

$$x = 0$$

23.31b
$$\frac{\partial\psi_I}{\partial x} = \frac{\partial\psi_{II}}{\partial x}$$

and at the right-hand wall

23.32a
$$\psi_{II} = \psi_{III}$$

$$x = L$$

23.32b
$$\frac{\partial\psi_{II}}{\partial x} = \frac{\partial\psi_{III}}{\partial x}$$

Substituting ψ_I, ψ_{II}, and ψ_{III} from Eqs. 23.15, 23.28, and 23.21 into the above equations yields

23.33
$$A + B = C + D$$

23.34
$$iaA - iaB = -b'C + b'D$$

23.35
$$b'Ce^{-b'L} + b'\,De^{b'L} = Ee^{iaL}$$

23.36
$$-aCe^{-b'L} + aDe^{b'L} = iaEe^{iaL}$$

Equations 23.33 to 23.36 may be readily solved to yield

23.37
$$\left(\frac{A}{E}\right) = \left[\frac{1}{2} + \frac{i}{4}\left(\frac{b'}{a} - \frac{a}{b'}\right)\right]e^{(ia+b')L} + \left[\frac{1}{2} - \frac{i}{4}\left(\frac{b'}{a} - \frac{a}{b'}\right)\right]e^{(ia-b')L}$$

The complex conjugate of A/E, which we require to compute the transmission probability P, is found by replacing i by $-i$ wherever it occurs in A/E:

23.38
$$\left(\frac{A}{E}\right)^* = \left[\frac{1}{2} - \frac{i}{4}\left(\frac{b'}{a} - \frac{a}{b'}\right)\right]e^{(-ia+b')L} + \left[\frac{1}{2} + \frac{i}{4}\left(\frac{b'}{a} - \frac{a}{b'}\right)\right]e^{(-ia-b')L}$$

Let us assume that the potential barrier is high relative to the kinetic energy of an incident particle; this means that $b' > a$ and

23.39
$$\left(\frac{b'}{a} - \frac{a}{b'}\right) \approx \frac{b'}{a}$$

Let us also assume that the barrier is wide enough for ψ_{II} to be severely attenuated between $x = 0$ and $x = L$; this means that $b'L \gg 1$ and

23.40
$$e^{b'L} \gg e^{-b'L}$$

Hence Eqs. 23.37 and 23.38 may be approximated by

23.41
$$\left(\frac{A}{E}\right) = \left(\frac{1}{2} + \frac{ib'}{4a}\right)e^{(ia+b')L}$$

and

23.42
$$\left(\frac{A}{E}\right)^* = \left(\frac{1}{2} - \frac{ib'}{4a}\right)e^{(-ia+b')L}$$

Multiplying (A/E) and $(A/E)^*$ yields

$$\left(\frac{A}{E}\right)\left(\frac{A}{E}\right)^* = \left(\frac{1}{4} + \frac{b'^2}{16a^2}\right)e^{2b'L}$$

and so the transmission probability P is

$$P = \frac{EE^*}{AA^*} = \left[\left(\frac{A}{E}\right)\left(\frac{A}{E}\right)^*\right]^{-1}$$

23.43
$$= \left[\frac{16}{4 + (b'/a)^2}\right]e^{-2b'L}$$

Since from the definitions of a (Eq. 23.22) and b' (Eq. 23.27)

$$\left(\frac{b'}{a}\right)^2 = \frac{V}{T} - 1$$

the variation in the coefficient of the exponential of Eq. 23.43 with T and V is negligible compared with the variation in the exponential itself. The coefficient, furthermore, is never far from unity, and so

23.44
$$P \approx e^{-2b'L}$$

is a good approximation for the transmission probability. We shall find it convenient to write Eq. 23.44 as

23.45
$$\ln P = -2b'L$$

Barrier Penetration

23.5 Theory of Alpha Decay

Equation 23.45 is derived for a rectangular potential barrier, while an alpha particle inside a nucleus is faced with a barrier of varying height, as in Fig. 23.6. We must therefore replace $\ln P = -2b'L$ by

23.46
$$\ln P = -2 \int_0^L b'(x)\, dx = -2 \int_{R_0}^R b'(x)\, dx$$

where R_0 is the radius of the nucleus and R the distance from its center at which $V = T$. Beyond R the kinetic energy of the alpha particle is positive, and it is able to move freely (Fig. 23.9). Now

$$V(x) = \frac{2Ze^2}{4\pi\varepsilon_0 x}$$

FIGURE 23.9 Alpha decay from the point of view of wave mechanics.

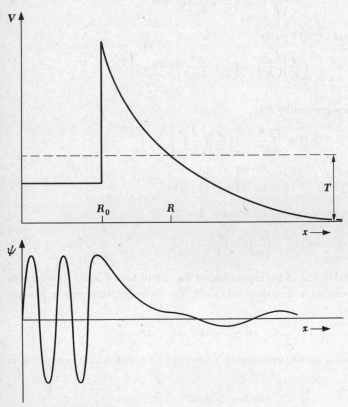

is the electrostatic potential energy of an alpha particle at a distance x from the center of a nucleus of charge Ze (that is, Ze is the nuclear charge *minus* the alpha-particle charge of $2e$). We therefore have

$$b' = \sqrt{\frac{2m(V-T)}{\hbar^2}}$$

$$= \left(\frac{2m}{\hbar^2}\right)^{1/2} \left(\frac{2Ze^2}{4\pi\varepsilon_0 x} - T\right)^{1/2}$$

and, since $T = V$ when $x = R$,

$$b' = \left(\frac{2mT}{\hbar^2}\right)^{1/2} \left(\frac{R}{x} - 1\right)^{1/2}$$

Hence

$$\ln P = -2 \int_{R_0}^{R} b'(x) \, dx$$

$$= -2\left(\frac{2mT}{\hbar^2}\right)^{1/2} \int_{R_0}^{R} \left(\frac{R}{x} - 1\right)^{1/2} dx$$

23.47
$$= -2\left(\frac{2mT}{\hbar^2}\right)^{1/2} R\left[\cos^{-1}\left(\frac{R_0}{R}\right)^{1/2} - \left(\frac{R_0}{R}\right)^{1/2}\left(1 - \frac{R_0}{R}\right)^{1/2}\right]$$

Because the potential barrier is relatively wide, $R \gg R_0$, and

$$\cos^{-1}\left(\frac{R_0}{R}\right)^{1/2} \approx \frac{\pi}{2} - \left(\frac{R_0}{R}\right)^{1/2}$$

$$\left(1 - \frac{R_0}{R}\right)^{1/2} \approx 1$$

with the result that

$$\ln P = -2\left(\frac{2mT}{\hbar^2}\right)^{1/2} R\left[\frac{\pi}{2} - 2\left(\frac{R_0}{R}\right)^{1/2}\right]$$

Replacing R by

$$R = \frac{2Ze^2}{4\pi\varepsilon_0 T}$$

we obtain

23.48
$$\ln P = \frac{4e}{\hbar}\left(\frac{m}{\pi\varepsilon_0}\right)^{1/2} Z^{1/2} R_0^{1/2} - \frac{e^2}{\hbar\varepsilon_0}\left(\frac{m}{2}\right)^{1/2} ZT^{-1/2}$$

Theory of Alpha Decay

The result of evaluating the various constants in Eq. 23.48 is

$$\ln P = 2.97Z^{1/2} R_0^{1/2} - 3.95ZT^{-1/2}$$

where T (the alpha-particle kinetic energy) is expressed in Mev, R_0 (the nuclear radius) is expressed in f($1 \text{ f} = 10^{-15}$ m), and Z is the atomic number of the nucleus minus the alpha particle. The decay constant λ, given by

$$\lambda = \nu P$$

$$= \frac{v}{2R} P$$

may therefore be written

23.49 $$\qquad \ln \lambda = \ln \left(\frac{v}{2R_0}\right) + 2.97Z^{1/2} R_0^{1/2} - 3.95ZT^{-1/2} \qquad \text{**Alpha decay**}$$

To express Eq. 23.49 in terms of common logarithms, we note that

$$\ln A = \frac{\log_{10} A}{\log_{10} e} = \frac{\log_{10} A}{0.4343}$$

and so

$$\log_{10} \lambda = \log_{10} \left(\frac{v}{2R_0}\right) + 0.4343 \, (2.97Z^{1/2} R_0^{1/2} - 3.95ZT^{-1/2})$$

$$= \log_{10} \left(\frac{v}{2R_0}\right) + 1.29Z^{1/2} R_0^{1/2} - 1.72ZT^{-1/2}$$

Figure 23.10 is a plot of $\log_{10}\lambda$ versus $ZT^{-1/2}$ for a number of alpha-radioactive nuclides. The straight line fitted to the experimental data has the -1.72 slope predicted throughout the entire range of decay constants. We can use the position of the line to determine R_0, the nuclear radius. The result is just about what is obtained from nuclear scattering experiments like that of Rutherford, namely, ~ 10 f in such very heavy nuclei. This approach constitutes an independent means for determining nuclear sizes.

The quantum-mechanical analysis of alpha-particle emission, which is in complete accord with the observed data, is significant on two grounds. First, it makes understandable the enormous variation in half life with disintegration energy. The slowest decay is that of $_{90}\text{Th}^{232}$, whose half life is 1.3×10^{10} years, and the fastest decay is that of $_{84}\text{Po}^{212}$, whose half life is 3.0×10^{-7} sec. While its half life is 10^{24} greater, the distintegration energy of $_{90}\text{Th}^{232}$ (4.05 Mev) is only about half that of $_{84}\text{Po}^{212}$ (8.95 Mev)—behavior predicted by Eq. 23.49.

The second significant feature of the theory of alpha decay is its explanation of this phenomenon in terms of the penetration of a potential barrier by a particle which does not have enough energy to surmount the barrier. In classical physics such penetration cannot

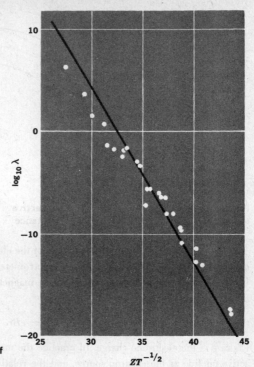

FIGURE 23.10 Experimental verification of the theory of alpha decay.

occur: a baseball thrown against the Great Wall of China has, classically, a 0 percent probability of getting through. In quantum mechanics the probability is not much more than 0 percent, but it is *not* identically equal to 0.

23.6 Beta Decay

Beta decay, like alpha decay, is a means whereby a nucleus can alter its Z/N ratio to achieve greater stability. Beta decay, however, presents a considerably greater problem to the physicist who seeks to understand natural phenomena. The most obvious difficulty is that in beta decay a nucleus emits an electron, while, as we have seen in the previous chapter, there are strong arguments against the presence of electrons in nuclei. Since beta decay is essentially the spontaneous conversion of a nuclear neutron into a proton and electron, this difficulty is disposed of if we simply assume that the electron leaves the nucleus immediately after its creation. A more serious problem is that observations of beta decay reveal that three conservation principles, those of energy, momentum, and angular momentum, are apparently being violated.

Beta Decay

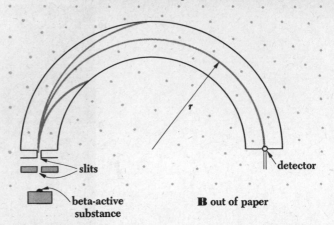

FIGURE 23.11 A beta-ray spectrometer.

slits

beta-active substance

detector

B out of paper

A device for determining the energies of the electrons emitted in beta decay is called a *beta-ray spectrometer*. A simple beta-ray spectrometer is shown schematically in Fig. 23.11. If r is the fixed radius of curvature and B the magnetic flux density, the electron momentum p is given by

$$p = eBr$$

where e is the electronic charge. In practice the magnetic field is varied with a beta-radioactive nuclide as the electron source, and the relative momentum distribution determined. The electrons emitted in beta decay often have kinetic energies comparable with the rest energy of the electron, so the relativistic formula

$$T = \sqrt{m_0{}^2 c^4 + p^2 c^2} - m_0 c^2$$

must be used to convert the observed momenta to kinetic energies.

The electron energies observed in the beta decay of a particular nuclide are found to vary *continuously* from 0 to a maximum value T_{max} characteristic of the nuclide. Figure 23.12 shows the energy spectrum of the electrons emitted in the beta decay of $_{83}Bi^{210}$; here $T_{max} = 1.17$ Mev. In every case the maximum energy

$$E_{max} = m_0 c^2 + T_{max}$$

carried off by the decay electron is equal to the energy equivalent of the mass difference between the parent and daughter nuclei. Only seldom, however, is an emitted electron found with an energy of T_{max}.

It was suspected at one time that the "missing" energy was lost during collisions between the emitted electron and the atomic electrons surrounding the nucleus. An experiment first performed in 1927 showed that this hypothesis is not correct. In the experiment a sample of a beta-radioactive nuclide is placed in a calorimeter, and the heat evolved after a given number of decays is measured. The evolved heat divided by the number of decays gives the average energy per decay. In the case of $_{83}Bi^{210}$ the average evolved energy was found to be 0.35 Mev, which is very close to the 0.39-Mev average of the spectrum in Fig. 23.12 but far away indeed from the T_{max} value of 1.17 Mev. The conclusion is that the observed continuous spectra represent the actual energy distributions of the electrons emitted by beta-radioactive nuclei.

Linear and angular momenta are also found not to be conserved in beta decay. In the beta decay of certain nuclides the directions of the emitted electrons and of the recoiling nuclei can be observed; they are almost never exactly opposite as required for momentum conservation. The nonconservation of angular momentum follows from the known spins of $\frac{1}{2}$ of the electron, proton, and neutron. Beta decay involves the conversion of a nuclear neutron into a proton:

$$n \rightarrow p + e^-$$

Since the spin of each of the particles involved is $\frac{1}{2}$, this reaction cannot take place if spin (and hence angular momentum) is to be conserved.

23.7 The Neutrino

In 1930 Pauli proposed that if an uncharged particle of small or zero mass and spin $\frac{1}{2}$ is emitted in beta decay together with the electron, the energy, momentum, and angular-momentum discrepancies discussed above would be removed. It was supposed that this particle, later christened the *neutrino*, carries off an energy equal to the difference between T_{max} and the actual electron kinetic energy (the recoil nucleus carries away negligible kinetic energy) and, in so doing, has a momentum exactly balancing those of the electron and the recoiling daughter nucleus. Subsequently it was found that there are *two* kinds of neutrino involved in beta decay, the neutrino itself (symbol ν) and the antineutrino (symbol $\bar{\nu}$). We shall discuss the distinction between them in Chap. 25. In ordinary beta decay it is an antineutrino that is emitted:

23.50 $$n \rightarrow p + e^- + \bar{\nu}$$ Beta decay

The neutrino hypothesis has turned out to be completely successful. The neutrino mass was not expected to be more than a small fraction of the electron mass because T_{max} is observed to be equal (within experimental error) to the value calculated from the parent-daughter mass difference; the neutrino mass is now believed to be zero. The reason neutrinos were not experimentally detected until recently is that their interaction with matter is extremely feeble. Lacking charge and mass, and not electromagnetic in nature as is the photon, the neutrino can pass unimpeded through vast amounts of matter. A neutrino would have to pass through over 100 *light-years* of solid iron on the average before interacting! The only interaction with matter a neutrino can experience is through a process called inverse beta decay, which we shall consider shortly.

23.8 Positron Emission and Electron Capture

Positive electrons, usually called *positrons*, were discovered in 1932 and two years later were found to be spontaneously emitted by certain nuclei. The properties of the positron are identical with those of the electron except that it carries a charge of $+e$ instead of $-e$. Positron emission corresponds to the conversion of a nuclear proton into a neutron, a positron, and a neutrino:

23.51 $$p \rightarrow n + e^+ + \nu$$ **Positron emission**

While a neutron outside a nucleus can undergo negative beta decay into a proton because its mass is greater than that of the proton, the lighter proton cannot be transformed into a neutron except within a nucleus.

Positron emission leads to a daughter nucleus of lower atomic number Z while leaving the mass number A uncharged. Thus negative and positive beta decays may be represented as

23.52 $$_ZX^A \rightarrow {}_{Z+1}X^A + e^- + \bar{\nu}$$
23.53 $$_ZX^A \rightarrow {}_{Z-1}X^A + e^+ + \nu$$

Positron emission resembles electron emission in all respects except for the shapes of their respective energy spectra. The difference is most pronounced at the low-energy ends, as can be seen by comparing the spectrum of positrons from $_{29}Cu^{64}$ shown in Fig. 23.13 with the spectrum of electrons from $_{83}Bi^{210}$ shown in Fig. 23.12. There are many low-energy electrons, while there are very few low-energy positrons. The reason for this difference is not hard to find. An electron leaving a nucleus is attracted by the positive nuclear charge, which causes it to slow down somewhat. A positron leaving a nucleus, on the other hand, is repelled by the positive nuclear charge and is accordingly accelerated outward. Thus the average electron energy in beta decay is about $0.3T_{max}$, while the average positron energy is about $0.4T_{max}$.

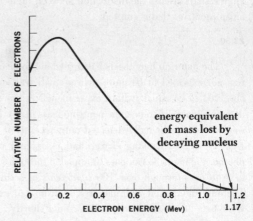

FIGURE 23.12 Energy spectrum of electrons from the beta decay of $_{83}Bi^{212}$.

energy equivalent of mass lost by decaying nucleus

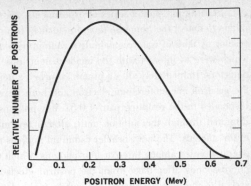

FIGURE 23.13 Energy spectrum of positrons from the beta decay of $_{29}Cu^{64}$.

POSITRON ENERGY (Mev)

Closely connected with positron emission is the phenomenon of *electron capture*. In electron capture a nucleus absorbs one of its inner orbital electrons, with the result that a nuclear proton becomes a neutron and a neutrino is emitted. Thus the fundamental reaction in electron capture is

23.54 $$p + e^- \rightarrow n + \nu$$ Electron capture

Electron capture is competitive with positron emission since both processes lead to the same nuclear transformation. Electron capture occurs more often than positron emission in heavy elements because the electron orbits in such elements have smaller radii; the closer proximity of the electrons promotes their interaction with the nucleus. Since almost the only unstable nuclei found in nature are of high Z, positron emission was not discovered until several decades after electron emission had been established.

23.9 Inverse Beta Decay

The beta decay of a proton within a nucleus follows the scheme

23.55 $$p \rightarrow n + e^+ + \nu$$

Because the absorption of an electron by a nucleus is equivalent to its emission of a positron, the electron capture reaction

23.56 $$p + e^- \rightarrow n + \nu$$

is essentially the same as the beta decay of Eq. 23.55. Similarly, the absorption of an antineutrino is equivalent to the emission of a neutrino, so that the reaction

23.57 $$p + \bar{\nu} \rightarrow n + e^+$$

also involves the same physical process as that of Eq. 23.55. This latter reaction, called *inverse beta decay*, is interesting because it provides a method for establishing the actual existence of neutrinos.

Starting in 1953, a series of experiments were begun by F. Reines, C. L. Cowan, and others to detect the immense flux of neutrinos from the beta decays that occur in a nuclear reactor. A tank of water containing a cadmium compound in solution supplied the protons which were to interact with the incident neutrinos. Surrounding the tank were gamma-ray detectors. Immediately after a proton absorbed a neutrino to yield a positron and a neutron, the positron encountered an electron and both were annihilated. The gamma-ray detectors responded to the resulting pair of 0.51-Mev photons. Meanwhile the newly formed neutron migrated through the solution until, after a few microseconds, it was captured by a cadmium nucleus. The new, heavier cadmium nucleus then released about 8 Mev of excitation energy divided among three or four photons, which were picked up by the detectors several microseconds after those from the positron-electron annihilation. In principle, then, the arrival of the above sequence of photons at the detector is a sure sign that the reaction of Eq. 23.57 has occurred. To avoid any uncertainty, the experiment was performed with the reactor alternately on and off, and the expected variation in the frequency of neutrino-capture events was observed. Thus the existence of the neutrino may be regarded as experimentally established.

Inverse beta decay is the sole known means whereby neutrinos and antineutrinos interact with matter:

$$p + \bar{v} \rightarrow n + e^+$$
$$n + v \rightarrow p + e^-$$

The probability for these reactions is almost vanishingly small; this is why neutrinos are able to traverse freely such vast amounts of matter. Once liberated, neutrinos travel freely through space and matter indefinitely, constituting a kind of independent universe within the universe of other particles.

23.10 Gamma Decay

Nuclei can exist in states of definite energies, just as atoms can. An excited nucleus is denoted by an asterisk after its usual symbol; thus $_{38}Sr^{87*}$ refers to $_{38}Sr^{87}$ in an excited state. Excited nuclei return to their ground states by emitting photons whose energies correspond to the energy differences between the various initial and final states in the transitions involved. The photons emitted by nuclei range in energy up to several Mev, and are traditionally called *gamma rays*.

A simple example of the relationship between energy levels and decay schemes is shown in Fig. 23.14, which pictures the beta decay of $_{12}Mg^{27}$ to $_{13}Al^{27}$. The half life of the decay is 9.5 min, and it may take place to either of the two excited states of $_{13}Al^{27}$. The resulting $_{13}Al^{27*}$ nucleus then undergoes one or two gamma decays to reach the ground state.

As an alternative to gamma-decay, an excited nucleus in some cases may return to its ground state by giving up its excitation energy to one of the orbital electrons around it.

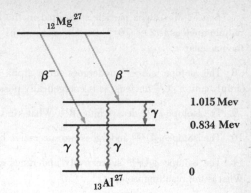

FIGURE 23.14   Successive beta and gamma emission in the decay of $_{12}Mg^{27}$ to $_{13}Al^{27}$.

While we can think of this process, which is known as *internal conversion*, as a kind of photoelectric effect in which a nuclear photon is absorbed by an atomic electron, it is in better accord with experiment to regard internal conversion as representing a direct transfer of excitation energy from a nucleus to an electron. The emitted electron has a kinetic energy equal to the lost nuclear excitation energy minus the binding energy of the electron in the atom.

Most excited nuclei have very short half lives against gamma decay, but a few remain excited for as long as several hours. A long-lived excited nucleus is called an *isomer* of the same nucleus in its ground state. The excited nucleus $_{38}Sr^{87*}$ has a half life of 2.8 hr and is accordingly an isomer of $_{38}Sr^{87}$.

Problems

1. Tritium ($_1H^3$) has a half life of 12.5 years against beta decay. What fraction of a sample of pure tritium will remain undecayed after 25 years?

2. The half life of $_{11}Na^{24}$ is 15 hr. How long does it take for 93.75 percent of a sample of this isotope to decay?

3. One g of radium has an activity of 1 curie. From this fact determine the half life of radium.

4. The mass of a millicurie of $_{82}Pb^{214}$ is 3×10^{-14} kg. From this fact find the decay constant of $_{82}Pb^{214}$. (Assume atomic mass in u equal to mass number in Probs. 4 to 6.)

5. The half life of $_{92}U^{238}$ against alpha decay is 4.5×10^9 years. How many disintegrations per second occur in 1 g of $_{92}U^{238}$?

6. The potassium isotope $_{19}K^{40}$ undergoes beta decay with a half life of 1.83×10^9 years. Find the number of beta decays that occur per second in 1 g of pure $_{19}K^{40}$.

7. A 5.78-Mev alpha particle is emitted in the decay of radium. If the diameter of the radium nucleus is 2×10^{-14} m, how many alpha-particle de Broglie wavelengths fit inside the nucleus?

8. The isotope $_{89}Ac^{225}$ undergoes three alpha decays in succession. Why does it not simply emit a $_6C^{12}$ nucleus, as is energetically possible?

9. The isotope $_6C^{11}$ decays into $_5B^{11}$. What kind of particle is emitted?

10. The isotope $_{92}U^{239}$ undergoes two successive beta decays. What is the resulting isotope?

11. The isotope $_{92}U^{238}$ successively undergoes eight alpha decays and six beta decays. What is the resulting isotope?

12. Identify the nuclei that result from the negative beta decay of $_{79}Au^{198}$, $_{29}Cu^{66}$, and $_{15}P^{32}$.

13. Identify the nuclei that result from the positive beta decay of $_{48}Cd^{107}$, $_{19}K^{38}$, and $_{51}Sb^{120}$.

14. Calculate the maximum energy of the electrons emitted in the beta decay of $_5B^{12}$. The atomic mass of $_5B^{12}$ is 12.0144 u and that of $_6C^{12}$ is 12.0000 u.

15. Why does $_4Be^7$ invariably decay by electron capture instead of by positron emission? The atomic mass of $_4Be^7$ is 7.0169 u and that of $_3Li^7$ is 7.0160 u. Note that $_4Be^7$ contains one more atomic electron than does $_3Li^7$.

16. Determine the ground and lowest excited states of the 39th proton in $_{39}Y^{89}$ with the help of Fig. 22.6. Use this information to explain the isomerism of $_{39}Y^{89}$ together with the fact, noted in Chap. 11, that radiative transitions between states with very different angular momenta are extremely improbable.

Chapter 24
Nuclear Reactions

Nuclear reactions, like chemical reactions, provide both information and means for utilizing this information in a practical way. Our basic knowledge of nuclei has come from the study of their interactions with other nuclei and with such bombarding particles as electrons and neutrons, and fission and fusion reactions are the means whereby the vast reservoir of nuclear energy has been broached. We shall begin our examination of nuclear reactions with descriptions of the center-of-mass coordinate system and the idea of interaction cross section, concepts that are required in order to make sense of the experimental data, and then go on to examine some specific reactions.

24.1   Center-of-mass Coordinate System

Most of what we know about atomic nuclei and their interactions has come from experiments in which energetic bombarding particles collide with stationary target nuclei. The analysis of such collisions is greatly simplified by the use of a coordinate system that is moving with the center of mass of the colliding bodies, rather than the usual coordinate system that is fixed in the laboratory.

We shall restrict ourselves to the case of a particle of mass m_1 moving in the $+x$ direction with the velocity v (where $v \ll c$) which strikes another particle, initially at rest, whose mass is m_2 (Fig. 24.1). This corresponds to the situation usually found in nuclear physics. When particle 1 is at x_1 and particle 2 at x_2, the position X of the center of mass of the two particles is defined by the equation

24.1 $$(m_1 + m_2)X = m_1x_1 + m_2x_2$$ **Center-of-mass position**

FIGURE 24.1   Motion in the laboratory coordinate system before collision, where $m_2 = 2m_1$.

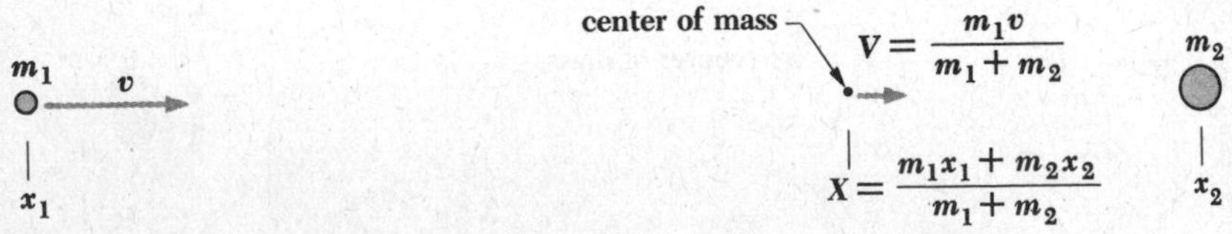

Differentiating with respect to time,

24.2
$$(m_1 + m_2)\frac{dX}{dt} = m_1\frac{dx_1}{dt} + m_2\frac{dx_2}{dt}$$

No variation in mass is involved since only nonrelativistic velocities are being considered. In the present case

$$\frac{dx_1}{dt} = v$$

$$\frac{dx_2}{dt} = 0$$

and the velocity V of the center of mass is

$$V = \frac{dX}{dt}$$

Hence Eq. 24.2 becomes

$$(m_1 + m_2)V = m_1 v$$

and

24.3
$$V = \left(\frac{m_1}{m_1 + m_2}\right)v \qquad \text{Center-of-mass velocity}$$

The center of mass moves in the same direction as v, but its speed is smaller by the factor $m_1/(m_1 + m_2)$.

Let us examine the two particles from the point of view of an observer located at the center of mass and moving with it (Fig. 24.2). This observer sees particle 1 approaching *from the left* at the velocity $v - V$ relative to him, while particle 2 approaches *from the right* at the velocity $-V$. In the center-of-mass coordinate system the total momentum of the two particles is therefore

$$m_1(v - V) - m_2 V = m_1 v - m_1\left(\frac{m_1}{m_1 + m_2}\right)v - m_2\left(\frac{m_1}{m_1 + m_2}\right)v$$

If the total momentum is 0 before the collision, it must be 0 afterward as well. Regardless

FIGURE 24.2 Motion in the center-of-mass coordinate system before collision, where $m_2 = 2m_1$.

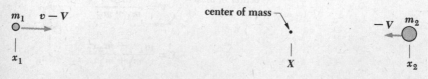

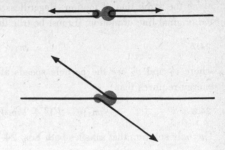

FIGURE 24.3 Head-on (top) and glancing (bottom) collisions as seen in center-of-mass coordinate system.

of the nature of the interaction between the particles during the actual collision, they must move apart (in the center-of-mass system) with equal and opposite momenta. In the event of a glancing rather than head-on collision the particles will no longer move along the x axis, as in Fig. 24.3, but they must nevertheless have equal and opposite momenta for the total momentum to remain 0.

In the center-of-mass system the total kinetic energy T' also is unlike that in the laboratory system. In the center-of-mass system

$$T' = \tfrac{1}{2}m_1(v - V)^2 + \tfrac{1}{2}m_2V^2$$
$$= \tfrac{1}{2}m_1v^2 - \tfrac{1}{2}(m_1 + m_2)V^2$$

while the total kinetic energy T in the laboratory system is

24.4 $$T = \tfrac{1}{2}m_1v^2$$ **Laboratory system**

Therefore

24.5 $$T' = T - \tfrac{1}{2}(m_1 + m_2)V^2$$ **Center-of-mass system**

The total kinetic energy of the particles relative to the center of mass is their total kinetic energy in the laboratory system minus the kinetic energy $\tfrac{1}{2}(m_1 + m_2)V^2$ of the moving center of mass. It is appropriate to regard T' as the kinetic energy of the relative motion of the particles. The ratio between T' and T is

24.6 $$\frac{T'}{T} = \frac{m_2}{m_1 + m_2}$$

An elastic collision between two particles is one in which there is no loss of kinetic energy, although the distribution of kinetic energy between the particles may change. In an elastic collision as viewed from the center-of-mass system, the initial and final total momenta are 0, and the sum of the initial kinetic energies of the particles is the same as the sum of their final kinetic energies. To satisfy the first condition, as we noted above, the particles must recede from the center of mass after impact in exactly opposite directions along the same line of motion. Regardless of whether this line of motion is the same as the original line of motion, it must be true that

24.7
$$m_1 v_1' = m_2 v_2'$$

where v_1' and v_2' are the particle speeds after the collision. The conservation of kinetic energy requires that

24.8
$$\tfrac{1}{2} m_1 (v - V)^2 + \tfrac{1}{2} m_2 V^2 = \tfrac{1}{2} m_1 v_1'^2 + \tfrac{1}{2} m_2 v_2'^2$$

The only solution that satisfies both Eqs. 24.7 and 24.8 is

24.9
$$v_1' = v - V$$

24.10
$$v_2' = V$$

The particles recede from their center of mass after colliding with the *same speeds* with which they approached it before the collision.

In an inelastic collision, part of the initial kinetic energy is dissipated in some manner. In an inelastic collision of macroscopic bodies, the lost kinetic energy usually reappears as heat and sound. When a nucleus experiences an inelastic collision, the lost kinetic energy usually becomes excitation energy that is subsequently given off as gamma rays. If enough energy is involved, nucleons or mesons may be emitted, as we shall learn in the next chapter. Analyzing a collision in center-of-mass coordinates makes it easy to determine the maximum amount of kinetic energy that can be dissipated while still conserving momentum. As shown in Fig. 24.4, a collision in which both particles lose all of their kinetic energy relative to the center of mass can still conserve momentum. Hence the maximum kinetic-energy decrease in a collision is T', the total initial kinetic energy in the center-of-mass system, which is related to the total initial kinetic energy T in the laboratory system by

24.11
$$T' = \left(\frac{m_2}{m_1 + m_2} \right) T$$

FIGURE 24.4 A completely inelastic collision as seen in laboratory and center-of-mass coordinate systems.

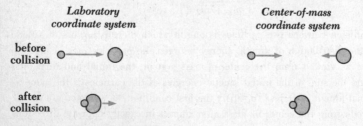

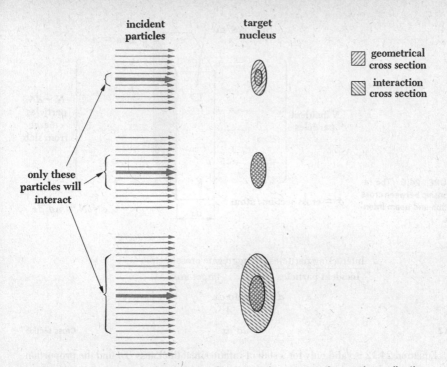

incident particles

target nucleus

geometrical cross section

interaction cross section

only these particles will interact

FIGURE 24.5 The concept of cross section. The interaction cross section may be smaller than, equal to, or larger than the geometrical cross section.

24.2 Cross Section

A very convenient way to express the probability that a bombarding particle will interact in a certain way with a target particle employs the idea of *cross section* that was introduced in Chap. 5 in connection with the Rutherford scattering experiment. What we do is visualize each target particle as presenting a certain area, called its cross section, to the incident particles, as in Fig. 24.5. Any incident particle that is directed at this area interacts with the target particle. Hence the greater the cross section, the greater the likelihood of an interaction. The interaction cross section of a target particle varies with the nature of the process involved and with the energy of the incident particle; it may be greater or less than the geometrical cross section of the particle.

Suppose we have a slab of some material whose area is A and whose thickness is dx (Fig. 24.6). If the material contains n atoms per unit volume, there are a total of $nA\ dx$ nuclei in the slab, since its volume is $A\ dx$. Each nucleus, has a cross section of σ for some particular interaction, so that the aggregate cross section of all of the nuclei in the slab is $nA\sigma\ dx$. If there are N incident particles in a bombarding beam, the number dN that interact with nuclei in the slab is therefore specified by

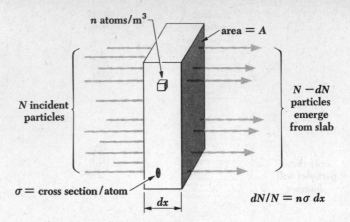

n atoms/m^3

area $= A$

N incident particles

$N - dN$ particles emerge from slab

FIGURE 24.6 The relationship between cross section and beam intensity.

$\sigma =$ cross section / atom

dx

$dN/N = n\sigma\,dx$

$$\frac{\text{Interacting particles}}{\text{Incident particles}} = \frac{\text{aggregate cross section}}{\text{target area}}$$

$$\frac{dN}{N} = \frac{nA\sigma\,dx}{A}$$

24.12 $= n\sigma\,dx$ **Cross section**

Equation 24.12 is valid only for a slab of infinitesimal thickness. To find the proportion of incident particles that interact with nuclei in a slab of finite thickness, we must integrate dN/N. If we assume that each incident particle is capable of only a single interaction, dN particles may be thought of as being removed from the beam in passing through the first dx of the slab. Hence we must introduce a minus sign in Eq. 24.12, which becomes

24.13 $$-\frac{dN}{N} = n\sigma\,dx$$

Denoting the initial number of incident particles by N_0, we have

24.14
$$\int_{N_0}^{N} \frac{dN}{N} = -n\sigma \int_0^x dx$$
$$\ln N - \ln N_0 = -n\sigma x$$
$$N = N_0 e^{-n\sigma x}$$

The number of surviving particles N decreases exponentially with increasing slab thickness x (Fig. 24.7). The number of particles that have undergone interactions in the slab is evidently

24.15 $$N_0 - N = N_0(1 - e^{-n\sigma x})$$

It is easy to show that Eq. 24.15 reduces to Eq. 24.12 as x decreases. If x is very small,

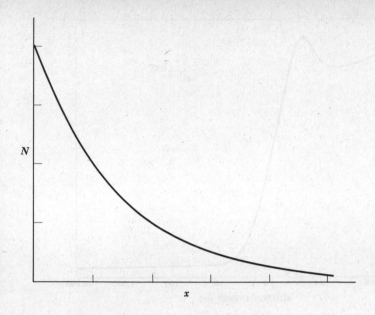

$$e^{-n\sigma x} \approx 1 - n\sigma x$$
$$N_0 - N \approx N_0 - N_0(1 - n\sigma x)$$
$$\approx N_0 n\sigma x$$

and

$$\frac{dN}{N} \approx \frac{N_0 - N}{N_0} \approx n\sigma x$$

The notion of cross section is often applied in situations where the interaction between incident and target particles is such that the former are absorbed. In such situations the quantity $n\sigma$ is called the *absorption coefficient* of the target material. The symbol for absorption coefficient is α:

24.16 $$\alpha = n\sigma$$

The greater the absorption coefficient, the more effective the material in stopping the incident particles.

While cross sections, which are areas, should be expressed in m^2, it is convenient and customary to express them in *barns*, where 1 barn $= 10^{-28}$ m^2. The barn is of the order of magnitude of the geometrical cross section of a nucleus. The cross sections for most nuclear reactions depend upon the energy of the incident particle. Figure 24.8 shows how the neutron-absorption cross section of $_{48}Cd^{113}$ varies with neutron energy; the narrow peak at 0.176 ev is associated with a specific energy level in the resulting $_{48}Cd^{114}$ nucleus.

Cross Section

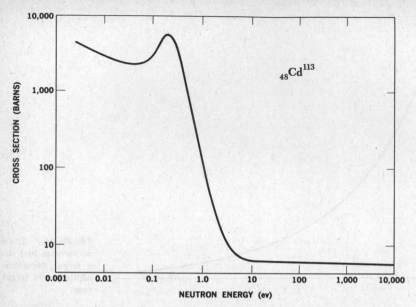

FIGURE 24.8 The cross section for neutron absorption in $_{48}Cd^{113}$ varies with neutron energy.

24.3 Mean Free Path

The *mean free path l* of a particle in a material is the average distance it can travel in the material before interacting with a target nucleus. The probability f that an incident particle will undergo an interaction in a slab Δx thick of the material is

24.17
$$f = n\sigma \, \Delta x$$

The number of times H it must traverse the slab before interacting is therefore

24.18
$$H = \frac{1}{n\sigma \, \Delta x}$$

on the average. To verify this statement, we note that

$$Hf = 1$$

The product of the probability for interacting and the mean number of passages before interaction is 1, as it must be. The average distance the particle travels before interacting is therefore

$$H \, \Delta x = \frac{1}{n\sigma}$$

556   Nuclear Reactions

which is, by definition, the mean free path. Hence

24.19 $$l = \frac{1}{n\sigma}$$ **Mean free path**

a most useful formula.

The cross section for the interaction of a neutrino with matter (Sec. 23.9) has been found to be approximately 10^{-47} m². Let us use Eq. 24.19 to find the mean free path of neutrinos in solid iron. The atomic mass of iron is 55.9, so that the mass of an iron atom is, on the average,

$$m_{Fe} = 55.9 \ \text{u/atom} \times 1.66 \times 10^{-27} \ \text{kg/u}$$
$$= 9.3 \times 10^{-26} \ \text{kg/atom}$$

Since the density of iron is 7.8×10^3 kg/m³, the number of atoms per m³ in iron is

$$n = \frac{7.8 \times 10^3 \ \text{kg/m}^3}{9.3 \times 10^{-26} \ \text{kg/atom}}$$
$$= 8.4 \times 10^{28} \ \text{atoms/m}^3$$

The mean free path for neutrinos in iron is therefore

$$l = \frac{1}{n\sigma} = \frac{1}{8.4 \times 10^{28} \ \text{atoms/m}^3 \times 10^{-47} \ \text{m}^2}$$
$$= 1.2 \times 10^{18} \ \text{m}$$

A light-year (the distance light travels in free space in a year) is equal to 9.46×10^{15} m, and so the mean free path turns out to be

$$l = \frac{1.2 \times 10^{18} \ \text{m}}{9.46 \times 10^{15} \ \text{m/light-year}} = 130 \ \text{light-years}$$

in solid iron! An immense flux of neutrinos is produced in the sun and other stars in the course of the nuclear reactions that occur within them, and this flux moves practically unimpeded through the universe. There are already far more neutrinos than atoms in the universe, and their number continues to increase. The energy these neutrinos carry is—apparently—lost forever in the sense of being unavailable for conversion into other forms.

24.4 The Compound Nucleus

Many nuclear reactions actually involve two separate stages. In the first, an incident particle strikes a target nucleus and the two combine to form a new nucleus, called a *compound nucleus*, whose atomic and mass numbers are respectively the sum of the atomic numbers

TABLE 24.1. NUCLEAR REACTIONS WHOSE PRODUCT IS THE COMPOUND NUCLEUS $_7N^{14*}$ The excitation energies given are calculated from the masses of the particles involved; the kinetic energy of an incident particle will add to the excitation energy by an amount depending upon the dynamics of the reaction in the center-of-mass coordinate system

$$_7N^{13} + _0n^1 \rightarrow _7N^{14*} \text{ (10.5 Mev)}$$
$$_6C^{13} + _1H^1 \rightarrow _7N^{14*} \text{ (7.5 Mev)}$$
$$_6C^{12} + _1H^2 \rightarrow _7N^{14*} \text{ (10.3 Mev)}$$
$$_6C^{11} + _1H^3 \rightarrow _7N^{14*} \text{ (22.7 Mev)}$$
$$_5B^{11} + _2He^3 \rightarrow _7N^{14*} \text{ (20.7 Mev)}$$
$$_5B^{10} + _2He^4 \rightarrow _7N^{14*} \text{ (11.6 Mev)}$$

of the original particles and the sum of their mass numbers. The compound nucleus has no "memory" of how it was formed, since its nucleons are mixed together regardless of origin and the energy brought into it by the incident particle is shared among all of them. A given compound nucleus may therefore be formed in a variety of ways. To illustrate this, Table 24.1 shows six reactions whose product is the compound nucleus $_7N^{14*}$. (The asterisk signifies an excited state; compound nuclei are invariably excited by amounts equal to at least the binding energies of the incident particles in them.) While $_7N^{13}$ and $_6C^{11}$ are beta-radioactive with such short half lives as to preclude the detailed study of their reactions to form $_7N^{14*}$, there is no doubt that these reactions can occur.

Compound nuclei have lifetimes of the order of 10^{-16} sec or so, which, while so short as to prevent actually observing such nuclei directly, are nevertheless long relative to the 10^{-21} sec or so required for a nuclear particle with an energy of several Mev to pass through a nucleus. A given compound nucleus may decay in one or more different ways, depending upon its excitation energy. Thus $_7N^{14*}$ with an excitation energy of, say, 12 Mev can decay via the reactions

$$_7N^{14*} \rightarrow _7N^{13} + _0n^1$$
$$_7N^{14*} \rightarrow _6C^{13} + _1H^1$$
$$_7N^{14*} \rightarrow _6C^{12} + _1H^2$$
$$_7N^{14*} \rightarrow _5B^{10} + _2He_4$$

or simply emit one or more gamma rays whose energies total 12 Mev, but it *cannot* decay by the emission of a triton ($_1H^3$) or a helium-3 ($_2He^3$) particle since it does not have enough energy to liberate them. Usually a particular decay mode is favored by a compound nucleus in a specific excited state.

The formation and decay of a compound nucleus has an interesting interpretation on the basis of the liquid-drop nuclear model described in Chap. 22. In terms of this model, an excited nucleus is analogous to a drop of hot liquid, with the binding energy of the emitted particles corresponding to the heat of vaporization of the liquid molecules. Such a drop of liquid will eventually evaporate one or more molecules, thereby cooling down. The evaporation process occurs when statistical fluctuations in the energy distribution within the drop cause a particular molecule to have enough energy for escape. Similarly, a compound nucleus persists in its excited state until a particular nucleon or group of nucleons momentarily happens to have a sufficiently large fraction of the excitation energy to leave the nucleus. The time interval between the formation and decay of a compound nucleus fits in nicely with this picture.

A large class of nuclear reactions may be represented by

24.20
$$a + X \rightarrow Y + b$$

This formula signifies that the particle a interacts with the nucleus X to yield the nucleus Y and the particle b. Equation 24.20 is usually abbreviated

$$X(a,b)\,Y$$

so that $_6C^{12}$ (d,n) $_7N^{13}$, for example, stands for a reaction between an incident deuteron $(d = {}_1H^2)$ and a $_6C^{12}$ nucleus to produce a $_7N^{13}$ nucleus with the emission of a neutron.

24.5 Excited States

Information about the excited states of nuclei can be gained from nuclear reactions as well as from radioactive decay. The presence of an excited state may be detected by a peak in the cross section versus energy curve of a particular reaction, as in the neutron-capture reaction of Fig. 24.8. Such a peak is called a *resonance* by analogy with ordinary acoustic or ac circuit resonances: a compound nucleus is more likely to be formed when the excitation energy provided exactly matches one of its energy levels than if the excitation energy has some other value.

The $_{48}Cd^{113}$ (n,γ) $_{48}Cd^{114}$ reaction of Fig. 24.8 has a resonance at 0.176 ev whose width (at half-maximum) is $\Gamma = 0.115$ ev. The uncertainty principle in the form $\Delta E\,\Delta t \geqslant \hbar$ enables us to relate the level width Γ of an excited state with the mean lifetime τ of the state. The width Γ evidently corresponds to the uncertainty ΔE in the energy of the state, and the mean lifetime τ corresponds to the uncertainty Δt in the time when the state will decay, in the present example by the emission of a gamma ray. The mean lifetime of an excited state is defined in general as

24.21
$$\tau = \frac{\hbar}{\Gamma}$$
Mean lifetime of excited state

In the case of the (n, γ) reaction in cadmium, the level width of 0.115 ev implies a mean lifetime for the compound nucleus of

$$\tau = \frac{1.054 \times 10^{-34} \text{ joule-sec}}{0.115 \text{ ev} \times 1.60 \times 10^{-19} \text{ joule/ev}}$$
$$= 5.73 \times 10^{-15} \text{ sec}$$

Excited States

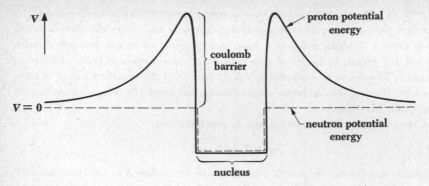

coulomb barrier

proton potential energy

$V = 0$

neutron potential energy

nucleus

FIGURE 24.9 Proton and neutron potential energy near a nucleus.

24.6 The Coulomb Barrier

As we saw in the previous chapter, every nucleus is surrounded by an electrostatic potential barrier that opposes both the entry and escape of positively charged particles. Neutrons, which have no charge, are not faced with this *coulomb barrier* and accordingly are more readily absorbed and emitted by nuclei than are protons, deuterons, and alpha particles. Figure 24.9 is a plot of the potential energies of a proton and a neutron near a nucleus. The coulomb barrier is about 3 Mev in height for carbon nuclei, 13 Mev for silver nuclei, and 20 Mev for lead nuclei.

Evidently a neutron of any energy can enter a nucleus by "falling into" its potential well, just as a rolling ball falls into a hole in the ground, while an approaching proton is faced with a potential hill. Classically the proton would have to possess at least as much energy as the height of the barrier in order to enter the nucleus. Quantum-mechanically, of course, the proton *can* get in with less than this amount of energy, but the probability of its doing so is small. Hence the height of the coulomb barrier represents an effective threshold energy for nuclear reactions initiated by such particles as protons, deuterons, and alpha particles, while no such threshold is present for incoming neutrons. Figure 24.10 shows how the cross sections for comparable neutron- and proton-induced nuclear reactions vary with energy. The neutron cross section *decreases* with increasing energy, because the likelihood that a neutron be captured depends upon how much time it spends near a particular nucleus, which is inversely proportional to its speed. The proton cross section *increases* with increasing energy because of the presence of the coulomb barrier.

The coulomb barrier also acts to impede the emission of a charged particle from a nucleus. A suitably excited nucleus is therefore usually more likely to eject a neutron than a proton, and reactions of the type (d,n), for example, are more common than those of the type (d,p).

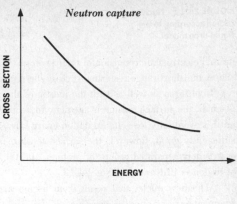

Neutron capture

CROSS SECTION

ENERGY

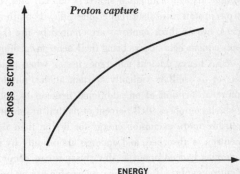

Proton capture

CROSS SECTION

ENERGY

FIGURE 24.10 Neutron and proton capture cross sections vary differently with particle energy.

24.7 Nuclear Fission

Another type of nuclear-reaction phenomenon that can be analyzed with the help of the liquid-drop model is *fission*, in which a heavy nucleus ($A > \sim 230$) splits into two lighter ones. When a liquid drop is suitably excited, it may oscillate in a variety of ways. A simple one is shown in Fig. 24.11: the drop successively becomes a prolate spheroid, a sphere, an oblate spheroid, a sphere, a prolate spheroid again, and so on. The restoring force of its surface tension always returns the drop to spherical shape, but the inertia of the moving liquid molecules causes the drop to overshoot sphericity and go to the opposite extreme of distortion.

While nuclei may be regarded as exhibiting surface tension, and so can vibrate like a liquid drop when in an excited state, they also are subject to disruptive forces due to the

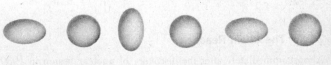

FIGURE 24.11 The oscillations of a liquid drop.

time ————→

FIGURE 24.12 Nuclear fission according to the liquid-drop model.

mutual electrostatic repulsion of their protons. When a nucleus is distorted from a spherical shape, the short-range restoring force of surface tension must cope with the latter long-range repulsive force as well as with the inertia of the nuclear matter. If the degree of distortion is small, the surface tension is adequate to do both, and the nucleus vibrates back and forth until it eventually loses its excitation energy by gamma decay. If the degree of distortion is sufficiently great, however, the surface tension is not adequate to bring back together the now widely separated groups of protons, and the nucleus splits into two parts. This picture of fission is illustrated in Fig. 24.12.

The new nuclei that result from fission are called *fission fragments*. Usually fission fragments are of unequal size (Fig. 24.13) and, because heavy nuclei have a greater neutron/proton ratio than lighter ones, they contain an excess of neutrons. To reduce this excess, two or three neutrons are emitted by the fragments as soon as they are formed, and subsequent beta decays bring their neutron/proton ratios to stable values.

A heavy nucleus undergoes fission when it acquires enough excitation energy (5 Mev or so) to oscillate violently. Certain nuclei, notably $_{92}U^{235}$, are sufficiently excited by the mere absorption of an additional neutron to split in two. Other nuclei, notably $_{92}U^{238}$ (which composes 99.3 percent of natural uranium, with $_{92}U^{235}$ composing the remainder), require more excitation energy for fission than the binding energy released when another neutron is absorbed, and undergo fission only by reaction with fast neutrons whose kinetic energies exceed about 1 Mev. Fission can occur after excitation by other means besides neutron capture, for instance, by gamma-ray or proton bombardment. Some nuclides are so unstable as to be capable of spontaneous fission, but they are more likely to undergo alpha decay before this takes place.

The most striking aspect of nuclear fission is the magnitude of the energy evolved. This energy is readily computed. The heavy fissionable nuclides, whose mass numbers are about 240, have binding energies of ~7.6 Mev/nucleon, while fission fragments, whose mass numbers are near 120, have binding energies of ~8.5 Mev/nucleon. Hence 0.9 Mev/nucleon is released during fission—over 200 Mev for the 240 or so nucleons involved! Ordinary chemical reactions, such as those that participate in the combustion of coal and oil, liberate only a few electron volts per individual reaction, and even nuclear reactions (other than fission) liberate no more than several million electron volts. Most of the energy that is released during fission goes into the kinetic energy of the fission fragments: the emitted neutrons, beta and gamma rays, and neutrinos carry off perhaps 20 percent of the total energy.

24.8 The Chain Reaction

Almost immediately after the discovery of nuclear fission in 1939 it was recognized that, because a neutron can induce fission in a suitable nucleus with the consequent evolution of additional neutrons, a self-sustaining sequence of fissions is, in principle, possible. The condition for such a chain reaction to occur in an assembly of fissionable material is simple: at least one neutron produced during each fission must, on the average, initiate another fission. If too few neutrons initiate fissions, the reaction will slow down and stop; if precisely one neutron per fission causes another fission, energy will be released at a constant rate; and if the frequency of fissions increases, the energy release will be so rapid that an explosion will occur. These situations are respectively called *subcritical*, *critical*, and *supercritical*.

Let us briefly examine the basic problems involved in the design of a *nuclear reactor*, which is a device for producing controlled power from nuclear fission. We shall use as an

FIGURE 24.13   The distribution of mass numbers in the fragments from the fission of $_{92}U^{235}$.

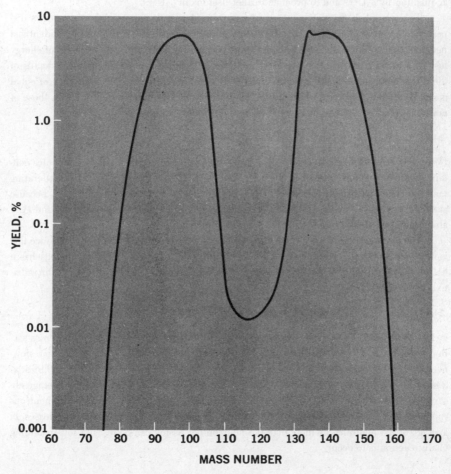

example a reactor fueled with natural uranium, similar to the one built in 1942 by Fermi in the first demonstration of the feasibility of chain reactions.

First we shall consider the loss of neutrons through the reactor surface and by non-fission-inducing absorption. The former problem can be met simply by increasing the reactor size, since a large object has less surface in proportion to its volume than a small one. The second problem is more difficult, since natural uranium contains only 0.7 percent of the fissionable isotope $_{92}U^{235}$. The more abundant isotope $_{92}U^{238}$ readily captures fast neutrons, but it usually rids itself of the resulting excitation energy by merely emitting a gamma ray rather than by undergoing fission. However, $_{92}U^{238}$ has only a small cross section for the capture of *slow* neutrons, while the cross section of $_{92}U^{235}$ for slow-neutron-induced fission is 550 barns—many times larger than its geometrical cross section. Hence it is necessary to rapidly slow down the neutrons that are liberated in fission both to prevent their acquisition by $_{92}U^{238}$ and to promote further fissions in $_{92}U^{235}$.

To accomplish the slowing down of fission neutrons, the uranium in a reactor is dispersed in a matrix of a *moderator*, a substance whose nuclei absorb energy from incident fast neutrons that collide with them without capture occurring. While the exact amount of energy lost by a moving particle that elastically collides with another depends upon the details of the interaction, in general the energy transfer is greatest when the participants are of equal mass. Hydrogen, however, whose proton nuclei have masses almost identical with those of neutrons, readily captures neutrons to form deuterons:

24.22
$$_1H^1 + _0n^1 \rightarrow _1H^2 + \gamma$$

Deuterons are less likely to react with neutrons, and so *heavy water*, whose molecules contain deuterium atoms instead of ordinary hydrogen atoms, makes a suitable moderating material. The first reactor employed carbon in the form of graphite as a moderator, because graphite was more readily available than heavy water in the necessary quantities and $_6C^{12}$ also has only a small cross section for neutron capture.

The cross section of $_{92}U^{235}$ for fission by slow neutrons is 550 barns so that, since this isotope composes 0.7 percent of natural uranium, the "average" uranium nucleus has a fission cross section of 0.007×550, or 3.9 barns. The cross section of $_{92}U^{235}$ for radiative neutron capture,

24.23
$$_{92}U^{235} + _0n^1 \rightarrow _{92}U^{236*} \rightarrow _{92}U^{236} + \gamma$$

is 101 barns, and the average uranium nucleus accordingly has for this process a cross section of 0.007×101, or 0.7 barns. The 99.3 percent abundant $_{92}U^{238}$ has a radiative slow-neutron-capture cross section of 2.8 barns, for an average uranium cross section of 0.993×2.8, or 2.8 barns. Hence the average fission cross section is 3.9 barns while the average absorption cross section is $0.7 + 2.8 = 3.5$ barns, and so only about half the slow neutrons captured in a block of uranium induce fissions. Because each fission in $_{92}U^{235}$ releases an average of 2.5 neutrons, no more than 0.5 neutron per fission can be lost if a self-sustaining chain reaction is to occur.

The actual operation of a reactor begins when a sufficiently large amount of fissionable material is brought together in the presence of a moderator. A single stray neutron—from the cosmic radiation, perhaps, or from a spontaneous fission—strikes a $_{92}U^{235}$ nucleus and causes it to split, releasing two or three additional neutrons. These neutrons are slowed down from energies of several Mev to energies of less than 1 ev by collisions with moderator nuclei, and then proceed to induce further fissions. The period of time between the release of a fission neutron and its later absorption is under a millisecond (0.001 sec). It is necessary to provide a means for controlling the chain-reaction rate. This is accomplished by means of rods made of a material, such as cadmium or boron, which readily absorbs slow neutrons; as these rods are inserted into the reactor, the reaction rate is progressively damped. The energy generated by a nuclear reactor is manifested as heat, and it can be extracted by circulating a suitable liquid or gaseous coolant through the reactor interior.

24.9   Transuranic Elements

Elements of atomic number greater than 92, which is that of uranium, have such short half lives that, had they been formed when the universe came into being, they would have disappeared long ago. Such *transuranic elements* may be produced in the laboratory by the bombardment of certain heavy nuclides with neutrons. Thus $_{92}U^{238}$ may absorb a neutron to become $_{92}U^{239}$, which beta-decays ($T_{1/2} = 23$ min) into $_{93}Np^{239}$, an isotope of the transuranic element *neptunium:*

$$_{92}U^{238} + {}_0n^1 \rightarrow {}_{92}U^{239}$$

$$_{92}U^{239} \rightarrow {}_{93}Np^{239} + e^-$$

This neptunium isotope is itself radioactive, undergoing beta decay with a half life of 2.3 days into an isotope of the transuranic element *plutonium:*

$$_{93}Np^{239} \rightarrow {}_{94}Pu^{239} + e^-$$

Plutonium alpha-decays into $_{92}U^{235}$ with a half life of 24,000 years:

$$_{94}Pu^{239} \rightarrow {}_{92}U^{235} + {}_2He^4$$

It is interesting to note that $_{94}Pu^{239}$, like $_{92}U^{235}$, is fissionable and can be used in nuclear reactors and weapons. Plutonium is chemically different from uranium; its separation from the remaining $_{92}U^{238}$ after neutron irradiation is more easily accomplished than the separation of $_{92}U^{235}$ from the much more abundant $_{92}U^{238}$ in natural uranium.

Transuranic elements past einsteinium ($Z = 99$) have half lives too short for their isolation in weighable quantities, though they can be identified by various means. The transuranic element of the highest atomic number yet discovered is lawrencium, $Z = 103$.

24.10   Thermonuclear Energy

The basic exothermic reaction in stars—and hence the source of nearly all of the energy in the universe—is the fusion of hydrogen nuclei into helium nuclei. This can take place under stellar conditions in two different series of processes. In one of them, the *proton-proton cycle*, direct collisions of protons result in the formation of heavier nuclei whose collisions in turn yield helium nuclei. The other, the *carbon cycle*, is a series of steps in which carbon nuclei absorb a succession of protons until they ultimately disgorge alpha particles to become carbon nuclei once more.

The initial reaction in the proton-proton cycle is

$$_1H^1 + {_1}H^1 \rightarrow {_1}H^2 + e^+ + \nu$$

the formation of deuterons by the direct combination of two protons accompanied by the emission of a positron. A deuteron may then join with a proton to form a $_2He^3$ nucleus:

$$_1H^1 + {_1}H^2 \rightarrow {_2}He^3 + \gamma$$

Finally two $_2He^3$ nuclei react to produce a $_2He^4$ nucleus plus two protons:

$$_2He^3 + {_2}He^3 \rightarrow {_2}He^4 + {_1}H^1 + {_1}H^1$$

The total evolved energy is $(\Delta m)c^2$, where Δm is the difference between the mass of four protons and the mass of an alpha particle plus two positrons; it turns out to be 24.7 Mev. Figure 24.14 shows the entire sequence.

The carbon cycle proceeds in the following way:

$$_1H^1 + {_6}C^{12} \rightarrow {_7}N^{13}$$
$$_7N^{13} \rightarrow {_6}C^{13} + e^+ + \nu$$
$$_1H^1 + {_6}C^{13} \rightarrow {_7}N^{14} + \gamma$$
$$_1H^1 + {_7}N^{14} \rightarrow {_8}O^{15} + \gamma$$
$$_8O^{15} \rightarrow {_7}N^{15} + e^+ + \nu$$
$$_1H^1 + {_7}N^{15} \rightarrow {_6}C^{12} + {_2}He^4$$

Carbon cycle

The net result again is the formation of an alpha particle and two positrons from four protons, with the evolution of 24.7 Mev; the initial $_6C^{12}$ acts as a kind of catalyst for the process, since it reappears at its end (Fig. 24.15).

Self-sustaining fusion reactions can occur only under conditions of extreme temperature and pressure, to ensure that the participating nuclei have enough energy to react despite their mutual electrostatic repulsion and that reactions occur frequently enough to counterbalance losses of energy to the surroundings. Stellar interiors meet these specifications. In the sun, whose interior temperature is estimated to be 2×10^6 °K, the proton-proton cycle has the greater probability for occurrence. In general, the carbon cycle is more efficient at high temperatures, while the proton-proton cycle is more efficient at low temperatures.

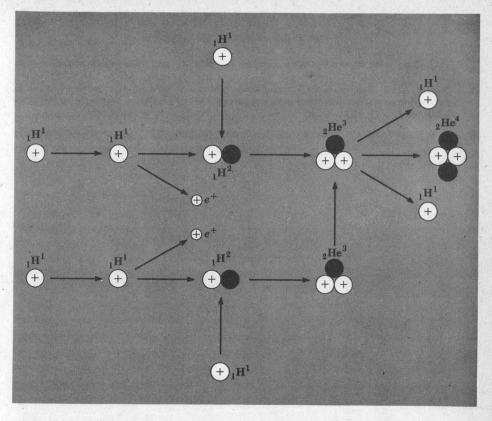

FIGURE 24.14 The proton-proton cycle. This is one of the two nuclear reaction sequences that take place in the sun and that involve the combination of four hydrogen nuclei to form a helium nucleus with the evolution of energy.

Hence stars hotter than the sun obtain their energy largely from the former cycle, while those cooler than the sun obtain the greater part of their energy from the latter cycle.

The energy liberated in the fusion of light nuclei into heavier ones is often called *thermonuclear energy*, particularly when the fusion takes place under man's control. On the earth neither the proton-proton nor carbon cycle offers any hope of practical application, since their several steps require a great deal of time. The fusion reactions that seem most promising as terrestrial energy sources are the direct combination of two deuterons in either of the following ways:

$$_1H^2 + {}_1H^2 \rightarrow {}_2He^3 + {}_0n^1 + 3.3 \text{ Mev}$$
$$_1H^2 + {}_1H^2 \rightarrow {}_1H^3 + {}_1H^1 + 4.0 \text{ Mev}$$

Thermonuclear Energy

and the direct combination of a deuteron and a triton to form an alpha particle,

$$_1H^3 + {}_1H^2 \rightarrow {}_2He^4 + {}_0n^1 + 17.6 \text{ Mev}$$

Capitalizing upon the above reactions requires an abundant, cheap source of deu-

FIGURE 24.15 The carbon cycle also involves the combination of four hydrogen nuclei to form a helium nucleus with the evolution of energy. The $_6C^{12}$ nucleus is unchanged by the series of reactions.

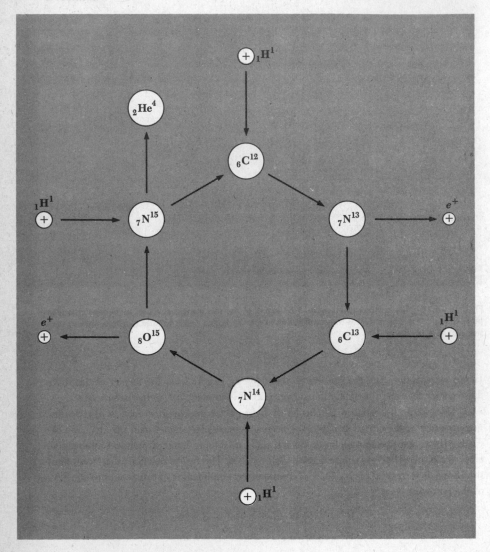

terium. Such a source is the oceans and seas of the world, which contain about 0.015 percent deuterium—a total of perhaps 10^{15} tons! In addition, a more efficient means of promoting fusion reactions than merely bombarding a target with fast particles from an accelerator is required, since the operation of an accelerator consumes far more power than can be evolved by the relatively few reactions that occur in the target. Current approaches to this problem all involve very hot fully ionized gases containing deuterium or deuterium-tritium mixtures which are contained by strong magnetic fields. The purpose of high temperature is to ensure that the individual $_1H^2$ and $_1H^3$ nuclei have enough energy to come together and react despite their electrostatic repulsion. A magnetic field is used as a container to keep the reactive gas from contacting any other material which might cool it down or contaminate it; there is little likelihood that the wall will melt since the gas, though at a temperature of several million degrees K, actually does not have a high energy density. While nuclear-fusion reactors present more severe practical difficulties than fission reactors, there is little doubt that they will eventually become a reality.

Problems

(The masses in u of neutral atoms of nuclides mentioned below are: $_1H^1$, 1.007825; $_1H^3$, 3.016050; $_2He^3$, 3.016030; $_2He^4$, 4.002603; $_5B^{10}$, 10.0129; $_6C^{12}$, 12.0000; $_6C^{13}$, 13.0034; $_7N^{14}$, 14.0031; $_8O^{16}$, 15.9949; $_8O^{17}$, 16.9994. The neutron mass is 1.008665 u. Atomic masses of the elements are listed in Table 10.1.)

1. Find the minimum energy in the laboratory system that an alpha particle must have in order to initiate the reaction

$$_2He^4 + {}_7N^{14} + 1.18 \text{ Mev} \rightarrow {}_8O^{17} + {}_1H^1$$

2. Find the minimum energy in the laboratory system that a neutron must have in order to initiate the reaction

$$_0n^1 + {}_8O^{16} + 2.20 \text{ Mev} \rightarrow {}_6C^{13} + {}_2He^4$$

3. Find the minimum energy in the laboratory system that a proton must have in order to initiate the reaction

$$p + d + 2.22 \text{ Mev} \rightarrow p + p + n$$

4. A particle of mass m_1 collides with a particle of mass m_2 which is at rest. Show that, if $m_1 > m_2$, the maximum angle in the laboratory system through which m_1 can be scattered is given by $\theta_{max} = \sin^{-1} m_2/m_1$.

5. There are approximately 6×10^{28} atoms/m^3 in solid aluminum. A beam of 0.5-Mev neutrons is directed at an aluminum foil 0.1 mm thick. If the capture cross section for neu-

trons of this energy in aluminum is 2×10^{-31} m^2, find the fraction of incident neutrons that are captured.

6. The density of $_5\text{B}^{10}$ is 2.5×10^3 kg/m^3. The capture cross section of $_5\text{B}^{10}$ is about 4,000 barns for "thermal" neutrons, that is, neutrons in thermal equilibrium with matter at room temperature. How thick a layer of $_5\text{B}^{10}$ is required to absorb 99 percent of an incident beam of thermal neutrons?

7. The density of iron is about 8×10^3 kg/m^3. The neutron-capture cross section of iron is about 2.5 barns. What fraction of an incident beam of neutrons is absorbed by a sheet of iron 1 cm thick?

8. The cross section of iron for neutron capture is 2.5 barns. What is the mean free path of neutrons in iron?

9. How much energy must a photon have if it is to split an alpha particle into a triton ($_1\text{H}^3$) and a proton?

10. How much energy must a photon have if it is to split an alpha particle into a $_2\text{He}^3$ nucleus and a neutron?

11. Complete the following nuclear reactions:

$$_{17}\text{Cl}^{35} + ? \rightarrow {}_{16}\text{S}^{32} + {}_2\text{He}^4$$
$$_5\text{B}^{10} + ? \rightarrow {}_3\text{Li}^7 + {}_2\text{He}^4$$
$$_3\text{Li}^6 + ? \rightarrow {}_4\text{Be}^7 + {}_0n^1$$

12. Complete the following nuclear reactions:

$$_{11}\text{Na}^{23} + {}_1\text{H}^1 \rightarrow {}_{10}\text{Ne}^{20} + ?$$
$$_{11}\text{Na}^{23} + {}_1\text{H}^2 \rightarrow {}_{12}\text{Mg}^{24} + ?$$
$$_{11}\text{Na}^{23} + {}_1\text{H}^2 \rightarrow {}_{11}\text{Na}^{24} + ?$$

13. The fission of $_{92}\text{U}^{235}$ releases approximately 200 Mev. What percentage of the original mass of $_{29}\text{U}^{235} + n$ disappears?

14. When a neutron is absorbed by a target nucleus, the resulting compound nucleus is usually more likely to emit a gamma ray than a proton, deuteron, or alpha particle. Why?

15. Certain stars obtain part of their energy by the fusion of three alpha particles to form a $_6\text{C}^{12}$ nucleus. How much energy does each such reaction evolve?

Chapter 25
Elementary
Particles

While nuclei are apparently composed solely of protons and neutrons, several score other elementary particles have been observed to be emitted by nuclei under appropriate circumstances. These particles, christened "strange particles" soon after their discovery about two decades ago, bring the total number merely of relatively stable elementary particles to over 30. To discern order in this multiplicity of particles has not proved to be an easy task. While certain regularities in elementary-particle properties have been established, and while such particles as the electron, the neutrino, and the π meson are relatively well-understood, no comprehensive theory of elementary particles has yet found wide acceptance. It is fitting to conclude our survey of modern physics with this topic, then, as a reminder that there remains much to be learned about the natural world.

25.1 The Theory of the Electron

The electron is the only elementary particle for which a satisfactory theory is known. This theory was developed in 1928 by P. A. M. Dirac, who obtained a wave equation for a charged particle in an electromagnetic field that incorporated the results of special relativity. When the observed mass and charge of the electron are inserted in the appropriate solutions of this equation, the intrinsic angular momentum of the electron is found to be ½ $\hbar$ (that is, spin ½) and its magnetic moment is found to be $e\hbar/2m$, one Bohr magneton. These predictions agree with experiment, and the agreement is strong evidence for the correctness of the Dirac theory.

Perhaps the most striking result of the Dirac theory is its prediction that electrons can exist in *negative* energy states. Dirac found that *both* the positive and negative roots of the relativistic energy equation

25.1
$$E = \pm \sqrt{m_0^2 c^2 + p^2 c^2}$$

are equally correct. The positive root permits the total energy E of an electron to have any value from $m_0 c^2$, its rest energy, to ∞. The negative root permits E to have any value from $-m_0 c^2$ to $-\infty$. There is nothing in the theory to prevent an electron in a positive energy state from undergoing a transition to a negative energy state by radiating a photon of

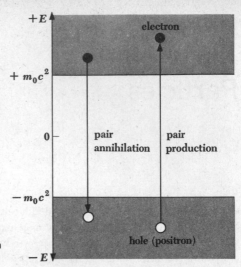

$+E$

electron

$+m_0c^2$

0 — pair annihilation   pair production

$-m_0c^2$

hole (positron)

FIGURE 25.1 Pair production and annihilation according to the Dirac theory.

$-E$

appropriate energy, or to prevent an electron in a negative energy state from falling to a still more negative state in the same way. Since systems always tend to evolve toward configurations of minimum energy, all the electrons in the universe should ultimately have energies of $E = -\infty$! There being no evidence that this peculiar destiny is indeed forthcoming, Dirac proposed that all negative-energy electron states are normally occupied. The exclusion principle therefore prevents positive-energy electrons from undergoing transitions into any of these states. Dirac further proposed that the "sea" of negative-energy electrons is not observable directly.

While positive-energy electrons are prevented from falling into negative energy states, since the latter are all occupied, the reverse process can occur. Given enough energy in some way, an electron of negative total energy can become one of positive total energy and thus become observable. This process is shown schematically in Fig. 25.1. The removal of an electron from the negative-energy electron sea leaves behind a "hole." Such a hole has interesting properties. Since it represents the absence of a particle of negative mass and kinetic energy, it behaves as though it is a particle of positive mass and kinetic energy. Moreover, a hole responds to electric and magnetic fields precisely as if it has a *positive* charge, as in the case of a hole in the electron structure of a crystal. Thus a hole has all the characteristics of an ordinary electron except that its charge is $+e$.

While Dirac's reasoning had no obvious flaws, his notion of holes in an infinite sea of negative-energy electrons was too strange to find acceptance when proposed. In 1932, however, positive electrons were actually detected in the flux of cosmic radiation at the earth's surface, completely verifying the hole hypothesis. Positive electrons are usually called *positrons*.

The formation of a positron requires a minimum energy of $2m_0c^2$ (1.02 Mev), since this amount of energy is needed to bring an electron in a state whose energy is $-m_0c^2$ to a state whose energy is $+m_0c^2$. When a positron is formed, an electron simultaneously appears, since it is the absence of this electron from the unobservable negative-energy sea which constitutes the positron. If we wish, we can regard the process of electron-positron pair creation as one involving the *materialization* of matter from energy, since $2m_0c^2$ of energy disappears whenever such a pair comes into being. Any available energy in excess of $2m_0c^2$ goes into the kinetic energy of the electron and positron. Experimentally electron-positron pairs are found to be produced when gamma rays of $hv > 1.02$ Mev pass near nuclei (Fig. 25.2); the presence of a relatively heavy nucleus is required in order that momentum as well as energy be conserved in the process.

The reverse of pair creation occurs when an electron of positive energy falls into a vacant negative energy state (Fig. 25.1). Since the latter is observed as a positron, we may regard the process as one in which an electron and a positron *annihilate* each other. The simultaneous disappearance of an electron and a positron liberates $2m_0c^2$ of energy. No nucleus or other particle is needed for annihilation to take place, since the energy evolved appears as two gamma rays whose directions are such as to conserve both momentum and energy.

Three processes in all are therefore responsible for the absorption of X and gamma rays in matter. At low photon energies Compton scattering is the sole mechanism, since there are definite thresholds for both the photoelectric effect (several ev) and electron pair production (1.02 ev). Both Compton scattering and the photoelectric effect decrease in importance with increasing energy, as shown in Fig. 25.3 for the case of a lead absorber, while the likelihood of pair production increases. At high photon energies the dominant mechanism of energy loss is pair production. The curve representing the total absorption in lead has its minimum at about 2 Mev. The ordinate of the graph is the linear absorption coefficient μ, which is equal to the ratio between the fractional decrease in radiation intensity $-dI/I$ and the absorber thickness dx. That is,

$$\frac{dI}{I} = -\mu \, dx$$

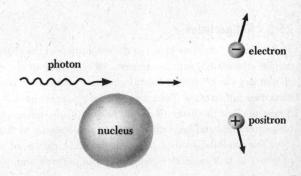

FIGURE 25.2   Pair production.

The Theory of the Electron

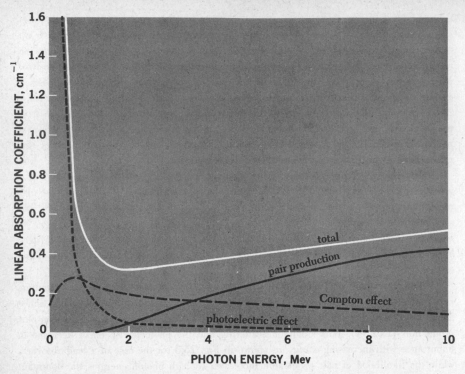

FIGURE 25.3 Linear absorption coefficients for photons in lead. These curves refer to energy absorption, not to the likelihood of interactions in the medium.

whose solution is

$$I = I_0 e^{-\mu x}$$

The intensity of the radiation decreases exponentially with the thickness of the absorber.

25.2 Antiparticles

The positron is often spoken of as the *antiparticle* of the electron, since it is able to undergo mutual annihilation with an electron. All other known elementary particles except for the photon and the π^0 and η^0 mesons also have antiparticle counterparts: the latter constitute their own antiparticles. The antiparticle of a particle has the same mass, spin, and lifetime if unstable, but its charge (if any) has the opposite sign and the alignment or antialignment between its spin and magnetic moment is also opposite to that of the particle.

The annihilation of a particle-antiparticle pair need not always result in a pair of photons, as it does in the case of electron-positron annihilation. When an antiproton is annihilated with a proton or neutron, for example, or an antineutron with a neutron or proton, several neutral and charged π mesons are usually produced. This is further evidence that π mesons may be regarded as quanta of the nuclear force field in the same sense that photons are quanta of the electromagnetic field.

The distinction between the neutrino and the antineutrino is a particularly interesting one. The spin of the neutrino is opposite in direction to the direction of its motion; viewed from behind, as in Fig. 25.4, the neutrino spins counterclockwise. The spin of the antineutrino, on the other hand, is in the same direction as its direction of motion; viewed from behind, it spins clockwise. Thus the neutrino moves through space in the manner of a left-handed screw, while the antineutrino does so in the manner of a right-handed screw.

Prior to 1956 it had been universally assumed that neutrinos could be either left-handed or right-handed, implying that, since no difference was possible between them except one of spin direction, the neutrino and antineutrino are identical. This assumption had roots going all the way back to Leibniz, Newton's contemporary and an independent inventor of calculus. The argument may be stated as follows: if we observe an object or a physical process of some kind both directly and in a mirror, we cannot ideally distinguish which object or process is being viewed directly and which by reflection. By definition, distinctions in physical reality must be capable of discernment or they are meaningless. Now the only difference between something seen directly and the same thing seen in a mirror is the interchange of right and left, and so *all* objects and processes must occur with equal probability with right and left interchanged. This plausible doctrine is indeed experimentally valid for nuclear and electromagnetic interactions, but until 1956 its applicability to neutrinos had never been actually tested. In that year T. D. Lee and C. N. Yang suggested

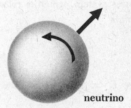

neutrino

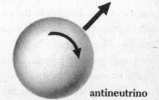

antineutrino

FIGURE 25.4 Neutrinos and antineutrinos have opposite directions of spin.

that several serious theoretical discrepancies would be removed if neutrinos and antineutrinos have different handedness, even though it meant that neither particle could therefore be reflected in a mirror. Experiments performed soon after their proposal showed unequivocally that neutrinos and antineutrinos are distinguishable, having left-handed and right-handed spins respectively. We might note that the absence of right-left symmetry in neutrinos can occur only if the neutrino mass is exactly zero, thereby resolving what had been the very difficult experimental problem of measuring the neutrino mass.

The significance of the handedness of neutrinos is explored further in Sec. 25.10.

25.3 π Mesons

As we saw in Chap. 22, the meson theory of nuclear forces is based upon the presumed exchange of particles called π mesons (or *pions*) between interacting nucleons. Such exchanges violate the law of conservation of energy, so that a limit is set by the uncertainty principle to the time available for them to take place. This limit leads to a predicted π-meson mass of at least 200 m_e. Twelve years after the meson theory was formulated, particles with the required properties were actually found to exist outside nuclei.

Two factors contributed to the belated discovery of the free π meson. First, enough energy must be supplied to a nucleon so that its emission of a π meson conserves energy. Thus at least $m_\pi c^2$ of energy, about 140 Mev, is required. To furnish a nucleon with this much energy in a collision, the incident particle must have considerably more kinetic energy than $m_\pi c^2$ in order that momentum as well as energy be conserved. Let us suppose that we wish to create a π meson by the impact of a fast proton on a stationary proton. In this case the amount of energy available for dissipation is the kinetic energy T' in the center-of-mass system, which is related to the kinetic energy T in the laboratory system by the formula

$$T' = \left(\frac{m_t}{m_i + m_t}\right) T$$

Here $T' = 140$ Mev and the masses of the incident and target particles are the same, so that

$$T = 2T' = 280 \text{ Mev}$$

Particles with kinetic energies of several hundred Mev are therefore required to produce free π mesons, and such particles are found in nature only in the diffuse stream of cosmic radiation that bombards the earth. Hence the discovery of the π meson had to await the development of sufficiently sensitive and precise methods of investigating cosmic-ray interactions. More recently high-energy accelerators were placed in operation; they yielded the necessary particle energies, and the profusion of mesons that were created with their help could be studied readily.

The second reason for the lag between the prediction and experimental discovery of the π meson is its instability: the half life of the charged π meson is only 1.8×10^{-8} sec, and that of the neutral π meson is 7×10^{-17} sec. The lifetime of the π^0 meson is so short, in fact, that its existence was not established until 1950. Charged π mesons almost invariably decay into lighter mesons called μ *mesons* and neutrinos:

$$\pi^+ \rightarrow \mu^+ + \nu_\mu$$
$$\pi^- \rightarrow \mu^- + \bar{\nu}_\mu$$

These neutrinos are not the same as those involved in beta decay, which is why their symbols are ν_μ and $\bar{\nu}_\mu$. The existence of two classes of neutrino was established in 1962. A metal target was bombarded with high-energy protons, and π mesons were created in profusion. Inverse reactions traceable to the neutrinos from the decay of these mesons produced μ mesons only, and no electrons. Hence these neutrinos must be somehow different from those associated with beta decay. The neutral π^0 meson decays into a pair of gamma rays:

$$\pi^0 \rightarrow \gamma + \gamma$$

The π^+ and π^- mesons have rest masses of 273 m_e, while that of the π^0 mesons is slightly less, 264 m_e. The π^- meson is the antiparticle of the π^+ meson, and the π^0 meson is its own antiparticle, a distinction it shares only with the photon and the η^0 meson.

25.4 μ Mesons

While the existence of π mesons is so readily understandable that they were predicted many years before their actual discovery, μ mesons (or *muons*) even today represent something of a puzzle. Their physical properties are known quite accurately. Positive and negative μ mesons have the same rest mass, $207 m_e$, and the same spin, ½. Both decay with a half life of 1.5×10^{-6} sec into electrons and neutrino-antineutrino pairs:

$$\mu^+ \rightarrow e^+ + \nu + \bar{\nu}$$
$$\mu^- \rightarrow e^- + \nu + \bar{\nu}$$

As with electrons, the positive-charge state of the μ meson represents the antiparticle. There is no neutral μ meson.

Unlike the case of π mesons, which, as we would expect, interact strongly with nuclei, the only interaction between μ mesons and matter is an electrostatic one. Accordingly μ mesons readily penetrate considerable amounts of matter before being absorbed. The majority of cosmic-ray particles at sea level are μ mesons from the decay of π mesons created in nuclear collisions caused by fast primary cosmic-ray atomic nuclei, since nearly all of the other particles in the cosmic-ray stream either decay or lose energy rapidly and are absorbed far above the earth's surface.

The mysterious aspect of the μ meson is its function—or, rather, its apparent lack of any function. Only in its mass and instability does the μ meson differ significantly from the electron, leading to the hypothesis that the μ meson is merely a kind of "heavy electron" rather than a unique entity. Other evidence, which we shall examine later in this chapter, is less unflattering to the μ meson, although it is still not wholly clear why, for instance, π mesons should preferentially decay into μ mesons rather than directly into electrons; only about 0.01 percent of π mesons decay directly into electrons and neutrinos.

25.5 K Mesons

The π and μ mesons do not exhaust the list of known particles with masses intermediate between those of the electron and proton. A third class of mesons, called *K mesons* (or *kaons*), has been discovered whose members may decay in a variety of ways. Charged K mesons have rest masses of $966m_e$, spins of 0, and half lives of 8×10^{-9} sec. The following decay schemes are possible for K^+ mesons:

$$K^+ \rightarrow \pi^+ + \pi^+ + \pi^-$$
$$\rightarrow \pi^+ + \pi^0 + \pi^0$$
$$\rightarrow \pi^+ + \pi^0$$
$$\rightarrow \mu^+ + \pi^0 + \nu$$
$$\rightarrow \mu^+ + \nu$$
$$\rightarrow e^+ + \pi^0 + \nu$$

The possible decays of K^- mesons, which are classed as antiparticles, follow a similar pattern.

There are apparently two distinct varieties of neutral K mesons, the K_1^0 and K_2^0. Both have rest masses of 974 m_e and spins of 0, but the former has a half life of about 7×10^{-11} sec, while that of the latter is about 4×10^{-8} sec. The following decay modes are known for neutral K mesons:

$$K_1^0 \rightarrow \pi^+ + \pi^-$$
$$\rightarrow \pi^0 + \pi^0$$
$$K_2^0 \rightarrow \pi^+ + \pi^- + \pi^0$$
$$\rightarrow \pi^0 + \pi^0 + \pi^0$$
$$\rightarrow \pi^- + \mu^+ + \nu$$
$$\rightarrow \pi^+ + \mu^- + \bar{\nu}$$
$$\rightarrow \pi^- + e^+ + \nu$$
$$\rightarrow \pi^+ + e^- + \bar{\nu}$$

In addition to their electromagnetic interaction with matter through which they pass, K mesons exhibit varying degrees of specifically nuclear interactions. The K^+ and K^0

mesons interact only weakly with nuclei, while their antiparticle counterparts are readily scattered and absorbed by nuclei in their paths.

25.6 Hyperons

Elementary particles heavier than protons are called *hyperons*. The known hyperons fall into four classes, Λ, Σ, Ξ, and Ω hyperons, in order of increasing mass. (Λ, Σ, Ξ, and Ω are, respectively, the Greek capital letters *lambda*, *sigma*, *xi*, and *omega*.) All are unstable with extremely brief mean lifetimes. The spin of all hyperons is ½ except that of the Ω hyperon, which is 3⁄2. The masses, half lives, and decay schemes of various hyperons are given in Table 25.1.

Hyperons exhibit definite interactions with nuclei. The Λ^0 hyperon is even able to act as a nuclear constituent. A nucleus containing a bound Λ^0 hyperon is called a *hyperfragment*; eventually the Λ^0 decays, of course, with the resulting nucleon and π meson either reacting with the parent nucleus or emerging from it entirely.

25.7 Systematics of Elementary Particles

Despite the multiplicity of elementary particles and the diversity of their properties, it is possible to discern an underlying order in their behavior. The fact of this order does not constitute a theory of elementary particles, however, any more than the order found in atomic spectra constitutes a theory of the atom, but it does provide hope that there may indeed be a single theoretical picture that can encompass elementary-particle phenomena in the manner that the quantum theory encompasses atomic phenomena. Thus far no such picture has emerged, although some intriguing lines of approach have been proposed. In

TABLE 25.1. HYPERON PROPERTIES

Particle	Mass, m_e	Half life, sec	Decay
Λ^0	2,182	1.7×10^{-10}	$\Lambda^0 \rightarrow p + \pi^-$
			$\rightarrow n + \pi^0$
			$\rightarrow p + e^- + \bar{\nu}$
Σ^+	2,328	0.6×10^{-10}	$\Sigma^+ \rightarrow p + \pi^0$
			$\rightarrow n + \pi^+$
Σ^-	2,341	1.2×10^{-10}	$\Sigma^- \rightarrow n + \pi^-$
Σ^0	2,332	$<10^{-12}$	$\Sigma^0 \rightarrow \Lambda + \gamma$
Ξ^-	2,583	0.9×10^{-10}	$\Xi^- \rightarrow \Lambda + \pi^-$
Ξ^0	2,571	1.0×10^{-10}	$\Xi^0 \rightarrow \Lambda + \pi^0$
Ω^-	3,290	$\sim 10^{-10}$	$\Omega^- \rightarrow \Lambda + K^-$
			$\rightarrow \Xi^0 + \pi^-$
			$\rightarrow \Xi^- + \pi^0$

the remainder of this chapter we shall examine the regularities observed in elementary particles and their apparent significance.

Table 25.2 is a listing in order of rest mass of the relatively stable elementary particles we have thus far mentioned plus the η meson, which we shall discuss shortly. By relatively stable is meant that the half lives of the particles all greatly exceed the time required for light to travel a distance equal to the "diameter" of an elementary particle. This diameter is probably a little over 10^{-15} m, and the characteristic time required to traverse it at the speed of light is therefore of the order of magnitude of 10^{-23} sec. Thus the particles in Table 25.2 are almost all capable of traveling through space as distinct entities along paths of measurable length in such devices as bubble chambers.

A considerable body of experimental evidence also points to the existence of many different "particles" whose lifetimes against decay are only about 10^{-23} sec. What can be meant by a particle which exists for so brief an interval? Indeed, how can a time of 10^{-23} sec even be measured? Such particles cannot be detected by observing their formation and subsequent decay in a bubble chamber or other instrument, but instead appear as resonant states in the interaction of more stable (and hence more readily observable) particles. Resonant states occur in atoms as energy levels; in Chap. 6 we reviewed the Franck-Hertz experiment, which showed the existence of atomic energy levels through the occurrence of inelastic electron scattering from atoms at certain energies only. An atom in a specific excited state is not the same as that atom in its ground state or in another excited state, but we do not usually speak of such an excited atom as though it were a member of a special species only because the interaction that gives rise to the excited state—the electromagnetic interaction —is well understood. A rather different situation holds in the case of elementary particles, where the various interactions involved are, except for the electromagnetic one, only partially understood, and much of our information comes from the properties of the resonances.

Let us see what is involved in a resonance in the case of elementary particles. An experiment is performed, for instance the bombardment of protons by energetic π^+ mesons, and a certain reaction is studied, for instance

$$\pi^+ + p \rightarrow \pi^+ + p + \pi^+ + \pi^- + \pi^0$$

The effect of the interaction of the π^+ meson and the proton is the creation of three new π mesons. In each such reaction the new mesons have a certain total energy that consists of their rest energies plus their kinetic energies relative to their center of mass. If we plot the number of events observed versus the total energy of the new mesons in each event, we obtain a graph like that of Fig. 25.5. Evidently there is a strong tendency for the total meson energy to be 785 Mev and a somewhat weaker tendency for it to be 548 Mev. We can say that the reaction exhibits resonances at 548 and 785 Mev or, equivalently, we can say that this reaction proceeds via the creation of an intermediate particle which can be either one whose mass is 548 Mev or one whose mass is 785 Mev. From the graph we can even estimate the mean lifetimes of these intermediate particles, which are known as the η and ω mesons, respectively. According to the uncertainty principle, the uncertainty in decay time of an un-

TABLE 25.2. ELEMENTARY PARTICLES STABLE AGAINST DECAY BY THE STRONG NUCLEAR INTERACTION

Class	Name	Particle +e	Particle 0	Particle −e	Antiparticle +e	Antiparticle 0	Antiparticle −e	Spin	Rest mass, m_e	Rest mass, Mev	Half life, sec	L	M	B	S	Y	I
												(Antiparticles have opposite signs)					
PHOTON	photon		γ			(γ)		1	0	0	stable	0	0				0
LEPTON	e-neutrino		ν_e			$\bar{\nu}_e$		1/2	0	0	stable	+1	0				0
LEPTON	μ-neutrino		ν_μ			$\bar{\nu}_\mu$		1/2	0	0	stable	0	+1				0
LEPTON	electron			e^-	e^+			1/2	1	0.51	stable	+1	0				0
LEPTON	μ meson			μ^-	μ^+			1/2	207	106	1.5×10^{-6}	0	+1				0
MESON	π meson		π^0			(π^0)		0	264	135	7×10^{-17}	0	0				1
MESON	π meson	π^+					π^-	0	273	140	1.8×10^{-8}	0	0				1
MESON	K meson	K^+					K^-	0	966	494	8×10^{-9}	0	0		+1	+1	1/2
MESON	K meson		K^0			$\bar{K}^0$		0	974	498	$7 \times 10^{-11};\ 4 \times 10^{-8}$	0	0		+1	+1	1/2
MESON	η meson		η^0			(η^0)		0	1,073	548	$\sim 10^{-18}$	0	0				0
BARYON	nucleon proton	p					$\bar{p}$	1/2	1,836	938	stable	0	0	+1	0	+1	1/2
BARYON	nucleon neutron		n			$\bar{n}$		1/2	1,839	940	6.5×10^2	0	0	+1	0	+1	1/2
BARYON	Λ hyperon		Λ^0			$\bar{\Lambda^0}$		1/2	2,182	1,115	1.7×10^{-10}	0	0	+1	−1	0	0
BARYON	Σ hyperon	Σ^+					$\bar{\Sigma}^+$	1/2	2,328	1,192	0.6×10^{-10}	0	0	+1	−1	0	1
BARYON	Σ hyperon		Σ^0			$\bar{\Sigma}^0$		1/2	2,332	1,194	$<10^{-12}$	0	0	+1	−1	0	1
BARYON	Σ hyperon			Σ^-	$\bar{\Sigma}^-$			1/2	2,341	1,197	1.2×10^{-10}	0	0	+1	−1	0	1
BARYON	Ξ hyperon		Ξ^0			$\bar{\Xi^0}$		1/2	2,571	1,310	1.0×10^{-10}	0	0	+1	−2	−1	1/2
BARYON	Ξ hyperon			Ξ^-	$\bar{\Xi}^+$			1/2	2,583	1,320	0.9×10^{-10}	0	0	+1	−2	−1	1/2
BARYON	Ω hyperon			Ω^-	$\bar{\Omega}^+$			3/2	3,290	1,676	$\sim 10^{-10}$	0	0	+1	−3	−2	0

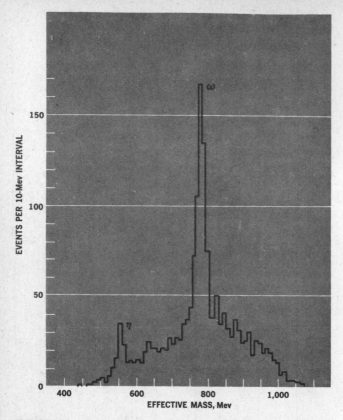

FIGURE 25.5 Resonant states in the reaction $\pi^+ + p \rightarrow \pi^+ + p + \pi^+ + \pi^- + \pi^0$ occur at effective masses of 548 and 785 Mev. By effective mass is meant the total energy, including mass energy, of the three new mesons relative to their center of mass.

stable particle—which is its mean lifetime τ—will give rise to an uncertainty in the determination of its energy—which is the width ΔE at half maximum of the corresponding peak in Fig. 25.4—whose relationship is

$$\tau \, \Delta E \approx \hbar$$

Hence the lifetimes of the resonances, or, just as well, the lifetimes of the η and ω mesons, can be established. The η lifetime is sufficiently long for it to be regarded as a relatively stable particle and it is included in Table 25.2, while the ω lifetime is too short by many orders of magnitude. We shall return to the resonance particles later in this chapter.

The particles in Table 25.2 seem to fall naturally into four general categories. In a class by itself is the photon, a stable particle with zero rest mass and unit spin. If there is a *graviton,* a particle that is the quantum of the gravitational field in the same sense that the photon is the quantum of the electromagnetic field or the π meson the quantum of the nuclear force field, it would be another member of this class. The graviton, as yet undetected, should be massless and stable, and should have a spin of 2. Its interaction with matter would be extremely weak, and it is unlikely that present techniques are capable of verifying its existence. (The zero mass of the graviton can be inferred from the unlimited range of gravitational forces. As we saw in Sec. 22.4, the mutual forces between two bodies can be regarded as transmitted by the exchange of particles between them. If energy conservation is to be preserved, the uncertainty principle requires that the range of the force be inversely proportional to the mass of the exchanging particles, so gravitational forces can have an infinite range only if the graviton mass is zero. A similar argument holds for the photon mass.)

After the photon in Table 25.2 come the e-neutrino and μ-neutrino, the electron, and the μ meson, all with spins of $\frac{1}{2}$. These particles are jointly called *leptons*. The π, K, and η mesons, all with spins of 0, are classed as *mesons*. (Despite its name, the μ meson has more in common with the other leptons than with the π, K, and η mesons.) The heaviest particles, namely the nucleons and hyperons, comprise the *baryons*.

While this grouping is reasonable on the basis of mass and spin alone, there is further evidence in its favor. Let us introduce three new quantum numbers, L, M, and B as follows. We assign the number $L = 1$ to the electron and the e-neutrino, and $L = -1$ to their antiparticles; all other particles have $L = 0$. We assign the number $M = 1$ to the μ meson and its neutrino, and $M = -1$ to their antiparticles; all other particles have $M = 0$. Finally, we assign $B = 1$ to all baryons, and $B = -1$ to all antibaryons; all other particles have $B = 0$. The significance of these numbers is that, in every process of whatever kind that involves elementary particles, the total values of L, M, and B remain constant. The classical conservation laws of energy, momentum, angular momentum, and electric charge plus the new conservation laws of L, M, and B help us to determine whether any given process is capable of taking place or not. An example is the decay of the neutron,

$$n^0 \rightarrow p^+ + e^- + \bar{\nu}_e$$

While $L = 0$ for the neutron and proton, $L = 1$ for the electron and -1 for the antineutrino, so the total value of L before and after the decay is 0. Similarly, $B = 1$ for both neutron and proton, so that the total value of B before and after the decay is 1. The stability of the proton is a consequence of energy and baryon-number conservation: There are no baryons of smaller mass than the proton, and so the proton cannot decay.

25.8 Strangeness Number

Despite the introduction of the quantum numbers L, M, and B, certain aspects of elementary-particle behavior still defied explanation. For instance, it was hard to see why certain heavy particles decay into lighter ones together with the emission of a gamma ray while others do not undergo apparently equally permissible decays. Thus the Σ^0 baryon decays into a Λ^0 baryon and a gamma ray,

$$\Sigma^0 \rightarrow \Lambda^0 + \gamma$$

while the Σ^+ baryon is never observed to decay into a proton and a gamma ray,

$$\Sigma^+ \nrightarrow p^+ + \gamma$$

Another peculiarity is based upon the general observation that physical processes in nature that release large amounts of energy take place more rapidly than processes that release small amounts. However, many strange particles whose decay releases considerable energy have relatively long lifetimes, well over a billion times longer than theortical calculations predict. A third odd feature is that strange particles are never created singly, but always two or more at a time. These and still other considerations led to the introduction of a quantity called *strangeness number S*. Table 25.2 shows the values of S that are assigned to the various elementary particles. We note that L, B, and S are 0 for the photon and π^0 and η^0 mesons. Since these particles are also uncharged, there is no way to distinguish between them and their antiparticles. For this reason the photon and π^0 and η^0 mesons are regarded as their own antiparticles. Before we consider the interpretation of the strangeness number, we shall have to examine the various kinds of particle interaction.

There are apparently four types of interaction between elementary particles that, in principle, give rise to all of the physical processes in the universe. The feeblest of these is the gravitational interaction. Next is the so-called weak interaction that is present between leptons and other leptons, mesons, or baryons in addition to any electromagnetic forces that may exist. The weak interaction is responsible for particle decays in which neutrinos are involved, notably beta decays. Stronger than gravitational and weak interactions are the electromagnetic interactions between all charged particles and also those with electric or magnetic moments. Finally, strongest of all are the nuclear forces (usually called simply strong forces when elementary particles are being discussed) that are found between mesons, baryons, and mesons and baryons.

The relative strengths of the strong, electromagnetic, weak, and gravitational interactions are in the ratios $1:15 \times 10^{-4}:3 \times 10^{-15}:10^{-40}$. Of course, the distances through which the corresponding forces act are very different. While the strong force between nearby nucleons is many powers of 10 greater than the gravitational force between them, when they are a meter apart the proportion is the other way. The structure of nuclei is determined by the properties of the strong interaction, while the structure of atoms is determined by those of the electromagnetic interaction. Matter in bulk is electrically neutral, and the strong and weak forces are severely limited in their range. Hence the gravitational interaction, utterly insignificant on a small scale, becomes the dominant one on a large scale. The role of the weak force in the structure of matter is apparently that of a minor perturbation that sees to it that nuclei with inappropriate neutron/proton ratios undergo corrective beta decays.

Let us now return to the strangeness number S. It is found that in all processes involving strong and electromagnetic interactions the strangeness number is conserved. The decay

$$\Sigma^0 \rightarrow \Lambda^0 + \gamma$$
$$S = -1 \quad -1 \quad 0$$

conserves S and is observed to occur, while the superficially similar decay

$$\Sigma^+ \nrightarrow p^+ + \gamma$$
$$S = -1 \quad 0 \quad 0$$

does not conserve S and has never been observed. Strange particles are created in high-energy nuclear collisions which involve strong interactions, and their multiple appearance results from the necessity of conserving S. The relative slowness with which all unstable elementary particles save the π^0 meson and η^0 meson decay is accounted for if we assume that weak interactions are also characteristic of mesons and baryons as well as leptons, though normally dominated by strong or electromagnetic interactions. With strong or electromagnetic processes impossible except in the above cases owing to the lack of conservation of S, only the weak interaction is available for processes in which the total value of S changes. Events governed by weak interactions take place slowly, as borne out by experiment. Even the weak interaction, however, is unable to permit S to change by more than $+1$ or -1 in a decay. Thus the Ξ^- hyperon does not decay directly into a neutron since

$$\Xi^- \nrightarrow n^0 + \pi^-$$
$$S = -2 \quad 0 \quad 0$$

but instead via the two steps

$$\Xi^- \rightarrow \Lambda^0 + \pi^-$$
$$S = -2 \quad -1 \quad 0$$
$$\Lambda^0 \rightarrow n^0 + \pi^0$$
$$S = -1 \quad 0 \quad 0$$

A quantity called *hypercharge*, Y, has also been found useful in characterizing particle families; it is conserved in strong interactions. Hypercharge is equal to the sum of the strangeness and baryon numbers of the particle families:

25.2 $$Y = S + B$$

For mesons the hypercharge is equal to the strangeness. The various hypercharge assignments are listed in Table 25.2.

25.9 Isotopic Spin

It is obvious from Table 25.2 that there are a number of particle families each of whose members has essentially the same mass and interaction properties but different charge. These families are called *multiplets*, and it is natural to think of the members of a multiplet as representing different charge states of a single fundamental entity. It has proved useful to categorize each multiplet according to the number of charge states it exhibits by a number I such that the multiplicity of the state is given by $2I + 1$. Thus the nucleon multiplet is assigned $I = \frac{1}{2}$, and its $2 \cdot \frac{1}{2} + 1 = 2$ states are the neutron and the proton. The π meson multiplet has $I = 1$, and its $2 \cdot 1 + 1 = 3$ states are the π^+, π^-, and π^0 mesons. The η meson has $I = 0$ since it occurs in only a single state and $2 \cdot 0 + 1 = 1$. There is evidently an analogy here with the splitting of an angular-momentum state of quantum number l into $2l + 1$ substates, and this has led to the somewhat misleading name of *isotopic spin quantum number* for I.

Pursuing the analogy with angular momentum, isotopic spin can be represented by a vector $\mathbf{I}$ in "isotopic spin space" whose component in a certain direction is governed by a quantum number customarily denoted I_3. The possible values of I_3 are restricted to $I, I - 1$, $\ldots, 0, \ldots, -(I - 1), -I$, so that I_3 is half-integral if I is half-integral and integral or zero if I is integral. The isotopic spin of the nucleon is $I = \frac{1}{2}$, which means that I_3 can be either $\frac{1}{2}$ or $-\frac{1}{2}$; the former is taken to represent the proton and the latter the neutron. In the case of the π meson, $I = 1$ and $I_3 = 1$ corresponds to the π^+ meson, $I_3 = 0$ to the π^0 meson, and $I_3 = -1$ to the π^- meson. The values of I_3 for the other mesons and baryons are assigned in a similar way.

The charge of a meson or baryon is related to its baryon number B, its strangeness number S, and the component I_3 of its isotopic spin by the formula

25.3
$$q = e \left(I_3 + \frac{B}{2} + \frac{S}{2} \right)$$

Each allowed orientation of the isotopic spin vector $\mathbf{I}$ hence is directly connected to the charge of the particle thus represented. In the case of the nucleon multiplet, the proton has $I_3 = \frac{1}{2}$, $B = 1$, and $S = 0$, so that $q = e$, while the neutron has $I_3 = -\frac{1}{2}$, $B = 1$, and $S = 0$, so that $q = 0$. In the case of the π meson multiplet, $B = S = 0$ and the three values of I_3 of 1, 0, and -1 respectively yield $q = e$, 0, and $-e$. Charge and baryon number B are conserved in all interactions. Thus I_3 must be conserved whenever S is conserved, namely in strong and electromagnetic interactions. Only in weak interactions does the total I_3 change.

An additional conservation law is suggested by the observed charge independence of nuclear forces, which result from the strong interaction. Such properties of a nucleus as its binding energy and pattern of energy levels change when a neutron is substituted for a proton or vice versa only by amounts that follow from purely electromagnetic considerations, implying that the strong interaction itself does not depend upon electric charge. Now the difference between a proton and a neutron in isotopic spin space lies only in the orientation of their isotopic spin vectors, so we can say that the charge independence of the strong interaction means that this interaction is independent of orientation in isotopic spin space. Angular momentum is likewise independent of orientation in real space and it is conserved in all interactions, which might lead us to surmise that isotopic spin is conserved in strong interactions. This surmise happens to be correct, and as a result the isotopic spin quantum number I is conserved in strong, but not in weak or in electromagnetic, interactions. We shall return to the relation between conservation principles and invariance with respect to symmetry operations in the next section.

We note that, although I_3 is conserved in electromagnetic interactions, I itself need not be. An example of a process in which I changes while I_3 does not is the decay of the π^0 meson into two photons:

$$\pi^0 \to \gamma + \gamma$$

A π^0 meson has $I = 1$ and $I_3 = 0$, while I is not defined for photons; there is no component I_3 of isotopic spin on either side of the equation, which is consistent with its conservation, although I has changed.

25.10 Symmetries and Conservation Principles

In Sec. 25.9 the charge independence of the strong interaction was expressed in terms of the isotropy of isotopic spin space, so that isotopic spin is independent of orientation in this space. By analogy with angular momentum, this symmetry was said to imply the conservation of isotopic spin in strong interactions. It is a remarkable fact that *all* known symmetries in the physical world lead directly to conservation laws, so the relationship between symmetry under rotations of **I** and the conservation of **I** is wholly plausible. Let us survey some of these symmetry-conservation relationships before continuing our discussion of elementary particles.

What is meant by a "symmetry"? Formally, if rather vaguely, we might say that a symmetry of a particular kind exists when a certain operation leaves something unchanged. A candle is symmetric about a vertical axis because it can be rotated about that axis without changing in appearance or any other feature; it is also symmetric with respect to reflection in a mirror. Table 25.3 lists the principal symmetry operations which leave the laws of physics unchanged under some or all circumstances. The simplest symmetry operation is translation in space, which means that the laws of physics do not depend upon where we choose the origin of our coordinate system to be. By more advanced methods than we are employing in this book, it is possible to show that the invariance of the description of nature to translations in space has as a consequence the conservation of linear momentum! Another simple symmetry operation is translation in time, which means that the laws of physics do not depend upon when we choose $t = 0$ to be, and this invariance has as a consequence the conservation of energy. Invariance under rotations in space, which means that the laws of physics do not depend upon the orientation of the coordinate system in which they are expressed, has as a consequence the conservation of angular momentum.

Conservation of electric charge is related to gauge transformations, which are shifts in the zeros of the scalar and vector electromagnetic potentials V and **A**. (As elaborated in electromagnetic theory, the electromagnetic field can be described in terms of the potentials

TABLE 25.3. SOME SYMMETRY OPERATIONS AND THEIR ASSOCIATED CONSERVATION PRINCIPLES

Symmetry operation	Conserved quantity
All interactions are independent of:	
Translation in space	Linear momentum **p**
Translation in time	Energy E
Rotation in space	Angular momentum **L**
Electromagnetic gauge transformation	Electric charge q
Lorentz transformation	Velocity of center of mass **V**
Interchange of identical particles	Type of statistical behavior
Inversion of space, time, and charge	Product of charge parity, space parity, and time parity CPT
?	Baryon number B
?	Lepton number L
?	Lepton number M
The strong and electromagnetic interactions only are independent of:	
Inversion of space	Parity P
Reflection of charge	Charge parity C, isotopic spin component I_3, and strangeness S
The strong interaction only is independent of:	
Charge	Isotopic spin I

V and **A** instead of in terms of **E** and **B**, where the two descriptions are related by the vector calculus formulas $\mathbf{E} = -\nabla V$ and $\mathbf{B} = \nabla \times \mathbf{A}$.) Gauge transformations leave **E** and **B** unaffected since they are obtained by differentiating the potentials, and this invariance leads to charge conservation.

Invariance under a Lorentz transformation signifies the isotropy of space-time and leads to the conservation of the velocity of the center of mass of an isolated system. In Chap. 2 we saw an indication of this relationship when $E = mc^2$ was derived starting from the latter conservation principle.

The interchange of identical particles in a system is a type of symmetry operation which leads to the preservation of the character of the wave function of a system. The wave function may be symmetric under such an interchange, in which case the particles do not obey the exclusion principle and the system follows Bose-Einstein statistics, or it may be antisymmetric, in which case the particles obey the exclusion principle and the system follows Fermi-Dirac statistics. Conservation of statistics (or, equivalently, of wave-function symmetry or antisymmetry) signifies that no process occurring within an isolated system can change the statistical behavior of that system. A system exhibiting Bose-Einstein behavior cannot spontaneously alter itself to exhibit Fermi-Dirac statistical behavior or vice versa. This conservation principle has applications in nuclear physics, where it is found that nuclei that contain an odd number of nucleons (odd mass number A) obey Fermi-Dirac statistics while those with even A obey Bose-Einstein statistics; conservation of statistics is thus a further condition a nuclear reaction must observe.

The conservations of the baryon number B and the lepton numbers L and M are alone among the principal conservation principles in having no known symmetries associated with them.

Apart from the charge independence of the strong interaction and its associated conservation of isotopic spin, which we have already mentioned, the remaining symmetry operations in Table 25.3 all involve *parities* of one kind or another. The term parity with no qualification refers to the behavior of a wave function under an inversion in space. By inversion in space is meant the reflection of spatial coordinates through the origin, with $-x$ replacing x, $-y$ replacing y, and $-z$ replacing z. If the sign of the wave function ψ does not change under such an inversion,

$$\psi(x, y, z) = \psi(-x, -y, -z)$$

and ψ is said to have *even* parity. If the sign of ψ changes,

$$\psi(x, y, z) = -\psi(-x, -y, -z)$$

and ψ is said to have *odd* parity. Thus the function $\cos x$ has even parity since $\cos x = \cos (-x)$, while $\sin x$ has odd parity since $\sin x = -\sin (-x)$.

If we write

$$\psi(x, y, z) = P\psi(-x, -y, -z)$$

we can regard P as a quantum number characterizing ψ whose possible values are $+1$ (even parity) and -1 (odd parity). Every elementary particle has a certain parity associated with it, and the parity of a system such as an atom or a nucleus is the product of the parity of the wave function that describes the coordinates of its constituent particles and the intrinsic parities of the particles themselves. Since $|\psi|^2$ is independent of P, the parity of a system is not a quantity that has an obvious physical consequence. However, it *is* found that the initial parity of an isolated system does not change during whatever events occur within it, which can be ascertained by comparing the parities of known final states of a reaction or transformation with the parities of equally plausible final states that are not observed to occur. A system of even parity retains even parity, a system of odd parity retains odd parity; this principle is known as conservation of parity.

The conservation of parity is an expression of the inversion symmetry of space, that is, of the lack of dependence of the laws of physics upon whether a left-handed or a right-handed coordinate system is used to describe phenomena. In Sec. 25.2 it was noted that the neutrino has a left-handed spin and the antineutrino a right-handed spin, so that there is a profound difference between the mirror image of either particle and the particle itself. This asymmetry implies that interactions in which neutrinos and antineutrinos participate—the weak interactions—need not conserve parity, and indeed parity conservation is found to hold true only in the strong and electromagnetic interactions. Historically the fact that spatial inversion is not invariably a symmetry operation was suggested by the failure of parity conservation in the decay of the K^+ meson, and was later confirmed by experiments showing the specific handedness of ν and $\bar{\nu}$.

Two other parities occur in Table 25.3, time parity T and charge parity C, which respectively describe the behavior of a wave function when t is replaced by $-t$ and when q is replaced by $-q$. The symmetry operation that corresponds to the conservation of time parity is *time reversal*. Time reversal symmetry implies that the direction of increasing time is not significant, so that the reverse of any process that can occur is also a process that can occur. In other words, if symmetry under time reversal holds, it is impossible to establish by viewing it whether a motion picture of an event is being run forwards or backwards. Although time parity T was long considered to be conserved in every interaction, it was discovered in 1964 that the K_2^0 meson can decay into a π^+ and a π^- meson, which violates the conservation of T. The symmetry of phenomena under time reversal thus has an ambiguous status at present. The symmetry operation that corresponds to the conservation of charge parity C is *charge conjugation*, which is the replacement of every particle in a system by its antiparticle. Charge parity C, like space parity P, is not conserved in weak interactions. However, despite the limited validities of the conservation of C, P, and T, there are good theoretical reasons for believing that the product CPT of the charge, space, and time parities of a system is invariably conserved. The conservation of CPT means that for every process there is an antimatter mirror-image counterpart that takes place in reverse, and this particular symmetry seems to hold even though its component symmetries sometimes fail individually.

25.11 Theories of Elementary Particles

In addition to the particles listed in Table 25.2 there are, as mentioned earlier, a great many "particles" of extremely brief lifetimes whose existence is revealed by resonances in interactions involving their longer-lived brethren. These resonant states are characterized by definite values of mass, charge, angular momentum, isotopic spin, parity, strangeness, and so on, and it is no more logical to disqualify them as legitimate particles because their existences are so transient than it is to consider the neutron, say, as merely an unstable state of the proton. Of course, it is possible to make out an excellent case for supposing the latter, and then to go on to generalize that *all* the various "elementary" particles are actually excited states of a very few truly elementary particles, as yet unidentified; this constitutes one line of attack toward a comprehensive theory of elementary particles. On the other hand, if we accept the particles of Table 25.2 as legitimate, then it is consistent to include the resonant states as well, and to seek a theoretical framework that embraces the entire collection of more than a hundred entities.

The latter point of view seems especially applicable to the strongly interacting particles, each of which may be regarded in a sense as a composite of the others. To see what this notion means, let us consider the strange case of the π^0 meson, which cannot participate in electromagnetic interactions since it has neither charge nor magnetic moment and yet which is observed to decay into a pair of photons, which are electromagnetic quanta. Apparently what happens is that the π^0 first undergoes a virtual transition to a nucleon-antinucleon pair, which the uncertainty principle permits to exist for a very short time even though energy is not conserved, and this nucleon-antinucleon pair can interact electromagnetically to produce two photons whose energies are consistent with the π^0 mass. This situation can be described by saying that the π^0 meson can communicate with otherwise inaccessible particle states through the intermediacy of the nucleon-antinucleon pair, with which it communicates via the strong interaction. Since the π^0 cannot be regarded as being itself (so to speak) at all times in view of its anomalous decay, it must be regarded as a composite of all the particle states with which it can communicate via the strong interaction. Similar arguments can be given for all the other strongly interacting particles, and this approach to a theory of elementary particles is under active study. To be sure, these incestuous relationships are confined to particles linked by the strong interaction, namely the mesons and baryons; the ground state of an atom, for instance, can be understood without reference to the properties of its excited states, although the proton, the "ground state" of the baryon family in the sense that it alone is stable in space, *cannot* be understood without bringing the other baryons into the description as well.

Several interesting and suggestive classification schemes have been devised for the strongly interacting particles based upon the abstract theory of groups. One of these schemes, the so-called *eightfold way*, collects isotopic spin multiplets into supermultiplets whose members have the same spin and parity but differ in charge and hypercharge (Figs. 25.6 and 25.7). The scheme prescribes the number of members each particular supermul-

FIGURE 25.6 Supermultiplets of spin-1/2 baryons and spin-0 mesons stable against decay by the strong nuclear interaction. Arrows indicate possible transformations according to the eightfold way.

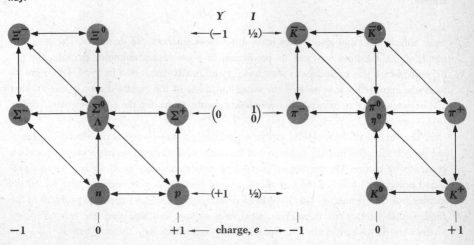

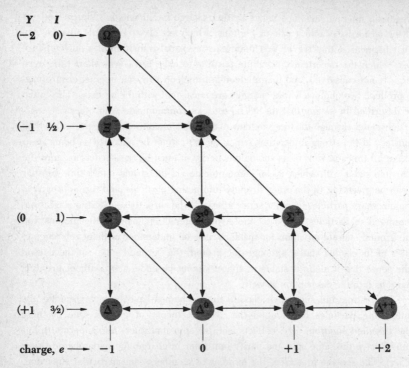

Y I

$(-2$ $0) \longrightarrow$ Ω^-

$(-1$ $\tfrac{1}{2}) \longrightarrow$ Ξ^- Ξ^0

$(0$ $1) \longrightarrow$ Σ^- Σ^0 Σ^+

$(+1$ $\tfrac{3}{2}) \longrightarrow$ Δ^- Δ^0 Δ^+ Δ^{++}

charge, $e \longrightarrow$ -1 0 $+1$ $+2$

FIGURE 25.7 Baryon supermultiplet whose members have spin 3/2 and (except Ω^-) are short-lived resonance particles; the Ξ and Σ particles here are heavier and have different spins from the ones in Table 25.2. Arrows indicate possible transformations according to the eightfold way. The Ω^- particle was predicted from this scheme.

tiplet should have and also relates mass differences among these members. The great triumph of the eightfold way was its prediction of a previously unknown particle, the Ω^- hyperon, which was subsequently searched for and finally discovered in 1964. Other group-theoretic approaches have related the supermultiplets of the eightfold way to one another and have attempted to incorporate relativistic considerations into the comprehensive picture that is emerging.

The success of the eightfold way in organizing our knowledge of the strongly interacting particles implies that the symmetry of its mathematical structure has a counterpart in a symmetry in nature. The further we probe into nature, the more hints we receive of a profound order that underlies the complications and confusions of experience. But for all the elegance of the symmetries that have been revealed, there still remains the problem of the fundamental interactions themselves, what they signify, and how they are related to one another and to the properties of the particles through which they are manifested.

Problems

1. Find the maximum kinetic energy of the electron emitted in the beta decay of the free neutron.

2. How much energy must an antineutrino have if it is to react with a proton to produce a neutron and a positron?

3. What is the energy of each of the gamma rays produced when a π^0 meson decays?

4. How much energy must a gamma-ray photon have if it is to materialize into a neutron-antineutron pair?

5. A positron collides head-on with an electron and both are annihilated. Each particle had a kinetic energy of 1 Mev. Find the wavelength of each of the resulting photons.

6. A π^0 meson whose kinetic energy is equal to its rest energy decays in flight. Find the angle between the two gamma-ray photons that are produced.

7. Why does a free neutron not decay into an electron and a positron?

8. Why are photons and π mesons the only particles multiply produced in collisions between elementary particles?

9. Why does a Λ hyperon not decay into a π^+ and a π^- meson?

10. Trace the decay of the Ξ^0 particle into stable particles.

11. A μ^- meson collides with a proton, and a neutron plus another particle are created. What is the other particle?

12. Does the reaction $\pi^- + p \rightarrow \pi^0 + n$ violate any conservation principles?

13. A proton of kinetic energy T_0 collides with a stationary proton, and a proton-antiproton pair is produced. If the momentum of the bombarding proton is shared equally by the four particles that emerge from the collision, find the minimum value of T_0.

14. How long does the uncertainty principle permit a π^0 meson to exist as a nucleon-antinucleon pair?

Answers to Odd-numbered Problems

CHAPTER 1
1. No, because it is accelerated as a consequence of the earth's rotation on its axis and revolution around the sun.
5. 4.2×10^7 m/sec
7. 6.3×10^4 years
9. 213 m

CHAPTER 2
1. $0.987c$
3. 4.2×10^7 m/sec
5. 1.87×10^8 m/sec; 1.64×10^8 m/sec
7. $39°$
9. 200 Mev
11. 8.9×10^{-28} kg
13. 3.6×10^{17} joules
15. 2.7×10^{11} kg
17. $1,960\ m_e$
19. 0.294 Mev
21. $1,630$ Mev$/c$
23. $F = \dfrac{m_0\ dv/dt}{(1 - v^2/c^2)^{3/2}}$
25. $T = 2\pi \sqrt{\dfrac{M_{iner}}{M_{grav}} \dfrac{L}{g}}$

CHAPTER 3
1. $1,800$ A
3. $5,400$ A; 3.9 ev
5. 2.83×10^{-19} joule
7. 1.71×10^{30} photons/sec
9. (a) 6.82×10^{-10} lb/in.^{2}
   (b) 4.24×10^{22} photons/m^2-sec
   (c) 3.9×10^{28} watts; 1.2×10^{48} photons/sec
   (d) 1.41×10^{14} photons/m^3
11. 1.24×10^5 volts
13. 3.14 A
15. 5×10^{18} cycles/sec
17. 0.015 A
19. 2.4×10^{19} cycles/sec
21. 0.565 A
23. (a) 2×10^{-3} ev
    (b) 2×10^{-25} ev
    (c) 3.5×10^{18} cycles/sec; 7.7×10^3 cycles/sec

CHAPTER 4
3. 0.10 A
5. 6.62×10^{-34} m
7. $\lambda = \dfrac{12.27}{V^{1/2}} \left(\dfrac{eV}{2m_0c^2} + 1 \right)^{-1/2}$
11. $u = w/2$
13. 6.2 percent
15. 7.3×10^5 m/sec; 4.0×10^2 m/sec
17. 2.7×10^{-6} percent

CHAPTER 5
1. $10°$
3. 0.876
7. 1.14×10^{-13} m
9. $m/m_0 = 1.002$
11. $f = \dfrac{1}{2\pi} \sqrt{\dfrac{Qe}{4\pi\varepsilon_0 mR^3}}$. For the hydrogen atom, $f = 6.6 \times 10^{15}$ sec^{-1}, which is comparable with the highest frequencies in the hydrogen spectrum.

CHAPTER 6
1. $18,750$ A
3. 920 A
5. 12 volts
7. 1.04 Mev
9. 8.2×10^6 rev
11. 1.05×10^5 °K
13. 2.4 A
15. They are the same.

CHAPTER 7
5. Hint: Since $x = r \sin \theta \cos \phi$, $y = r \sin \theta \sin \phi$, and $z = r \cos \theta$, we have
$$\frac{\partial}{\partial \phi} = \frac{\partial x}{\partial \phi} \frac{\partial}{\partial x} + \frac{\partial y}{\partial \phi} \frac{\partial}{\partial y} = x \frac{\partial}{\partial y} - y \frac{\partial}{\partial x}$$
7. Hint: Integrate by parts.

CHAPTER 8
1. 2.1 Mev
3. 3, 6, 9, 11, and $12 \times \pi^2 \hbar^2 / 2mL^2$. All but the lowest and highest of these levels are three-fold degenerate.
5. 2.94×10^4 °K

9. Classically $\bar{T} = \bar{V} = E/2$ where $\bar{T}$ and $\bar{V}$ are averages over an entire period of oscillation.
11. Hint: It is necessary to integrate by parts.

CHAPTER 9
7. 68 percent; 25 percent

CHAPTER 10
1. 182

3.
state	S	L	J
$^1S_0$	0	0	0
$^3P_2$	1	1	2
$^2D_{3/2}$	½	2	3/2
$^5F_5$	2	3	5
$^6H_{5/2}$	5/2	5	5/2

5. There are no other allowed states.
7. $^2P_{1/2}$
9. 1.39×10^{-4} ev
11. 0.0283 A
13. 20 webers/m^2
15. $5\mu_B$; 3

CHAPTER 11
3. The transitions that give rise to X-ray spectra are the same in all elements, since these transitions involve only inner, closed-shell electrons. The transitions that give rise to optical spectra, however, involve the outermost electrons, and the possible transitions depend critically upon the degree of occupancy of the outermost shell.
5. Cobalt $(Z = 27)$; molybdenum $(Z = 42)$

CHAPTER 12
1. The ionization energy of H_2 is 15.67 ev, while that of H is 13.6 ev. The ionization energy of H_2 is greater because each electron in H_2 is more strongly bound by the two protons, despite the presence of the other electron, than the electron in H is bound by its single proton.
3. Sodium ions have closed shells, while sodium atoms each have a single outer electron.
5. Both have one electron outside a closed shell.
7. 0.5 ev; 2.9×10^{-9} m
9. 1.9 ev; 7.6×10^{-10} m

CHAPTER 13
1. See text.
3. CO: 1 σ bond, 2 π bonds, for a total of 3
   CO+: ½ σ bond, 2 π bonds, for a total of 2½
   Hence CO is more strongly bound.
5. F_2: $\sigma_s{}^2\sigma_s^*{}^2\pi_{x,y}{}^4\pi_{x,y}^{*4}\sigma_z{}^2 = 1$ net bond
   $F_2{}^+$: $\sigma_s{}^2\sigma_s^*{}^2\pi_{x,y}{}^4\pi_{x,y}^*{}^3\sigma_z{}^2 = 1½$ net bonds
   $F_2{}^-$: $\sigma_s{}^2\sigma_s^*{}^2\pi_{x,y}{}^4\pi_{x,y}^*{}^4\sigma_z{}^2\sigma_z^* = ½$ net bond
   Hence $F_2{}^+$ has the highest bond energy and $F_2{}^-$ the lowest.

CHAPTER 14
1. C^{13}
3. 1.27 A
5. 2.23 A
7. No.
9. 1.24×10^{14} cycles/sec

CHAPTER 15
1. $e^{13.1 \times 10^6}$
5. $2/\sqrt{\pi} v_p$
7. 1.6×10^9 neutrons/m^3
9. (a) $1.00 : 1.68 : 0.89 : 0.22 : 0.027$
   (b) yes; 1533° K
11. $1.00 : 2.3 \times 10^{-10} : 6.2 \times 10^{-12} : 2.3 \times 10^{-12}$

CHAPTER 16
3. 5800°K
5. A fermion gas will exert the greatest pressure because the Fermi distribution has a larger proportion of high-energy particles than the other distributions; a boson gas will exert the least pressure because the Bose distribution has a larger proportion of low-energy particles than the others.
7. 18.70 ev; 16.84 ev. The He atoms are needed to maintain an inverted energy population in the Ne atoms by transferring energy to them in collisions, which supplements the direct excitation of Ne atoms by electron impact.

CHAPTER 17
1. Hydrogen atoms join into separate molecules because they are more stable in this form than as a continuous lattice of H atoms. Solid hydrogen consists of H_2 molecules held together by van der Waals bonds.

3. 7.29 ev; $n = 8.4$
5. The heat lost by the expanding gas is equal to the work done against the attractive van der Waals forces between its molecules.

CHAPTER 18
3. 1.52 A, 1.24 A, 1.37 A. Lithium is the largest atom because it has only a single electron in its outer shell, so that the effective nuclear charge acting on this electron is just $+e$.
7. (a) one jump every 10^3 sec
    (b) 7.2×10^5 jumps/sec
9. 203°C
11. Interstitial locations

CHAPTER 19
1. 0.35 A; 12 percent
3. 327°C
5. 8.5×10^{12} sec^{-1}; 7.8×10^{12} sec^{-1}; the vibrational frequency of aluminum atoms is 6.4×10^{12} sec^{-1}, as mentioned in the text, and the preceding figures are consistent with this, considering the approximate character of the various ways of calculating v_m.
7. 3.3 ev; 2.56×10^4 °K; 1.08×10^6 m/sec
9. 11 ev
11. 10.4 ev
13. There are no free electrons in nonmetals.
15. Hint: Since only positive values of n_x, n_y, n_z are involved, it is necessary to find the volume of an octant of radius n_F in n space.

CHAPTER 20
1. p-type
3. (a) In a metal, valence electrons can find unoccupied excited energy states in the conduction band for any excitation energy, however small.
    (b) The energy gap in semiconductors is small ($\leqslant$ 1.5 ev), and so photons of visible light can excite valence electrons to the conduction band although photons of infrared light have insufficient energy for this purpose.
    (c) The energy gap is so large that photons of visible light cannot provide enough excitation energy from electrons in the valence band to reach the conduction band.
5. 2

7. 50 A; the ionization energy of the electron is 0.009 ev, which is much smaller than the energy gap and not very far from the 0.025 ev value of kT at 20°C.
9. Because $m^*/m = 1.01$ in copper.

CHAPTER 21
1. 8.83 cm
3. 19 percent B^{10}, 81 percent B^{11}
5. 34.97 u
7. 7.98 Mev
9. 15.6 Mev

CHAPTER 22
1. Nuclear forces cannot be strongly charge-dependent.
3. The nucleon kinetic energy that corresponds to the momentum uncertainty implied by an uncertainty in position of 2 f is 5.2 Mev, which is entirely consistent with a potential well 35 Mev deep.
5. $\sim$0.6 Mev
7. $u_F = \dfrac{(3/\pi)^{4/3}h^2}{32mR_0{}^2}$; 29 Mev

CHAPTER 23
1. ¼
3. 1,620 years
5. 1.23×10^4 sec^{-1}
7. 3.37
9. A positron
11. $_{82}Pb^{206}$
13. $_{47}Hg^{107}$, $_{18}A^{38}$, $_{50}Sn^{120}$
15. The mass of $_4Be^7$ is not sufficiently larger than that of $_3Li^7$ to permit the creation of a positron.

CHAPTER 24
1. 1.52 Mev
3. 3.33 Mev
5. 1.2×10^{-6}
7. 0.21
9. 18.8 Mev
11. $_1H^1$, $_0n^1$, $_1H^2$
13. 0.1 percent
15. 7.28 Mev

CHAPTER 25

1. 0.78 Mev
3. 68 Mev
5. 8.6×10^{-13} m
7. This decay does not conserve baryon number or spin.
9. This decay does not conserve baryon number or spin.
11. A neutrino
13. 5,630 Mev

Index

Dirac, P. A. M., 232, 571
Dislocation, edge, 432
  screw, 432
Donor level in semiconductor, 465
Doppler effect in light, 29
Doublet state, 248
Duane-Hunt relationship, 63
Dulong-Petit law, 445

Edge dislocation, 432
Effective mass of electron in crystal, 482
  table of, 487
Eigenfunctions, 167
  orthogonality of, 194
Eigenvalue equation, 167
Eigenvalues, 167
Eightfold way, 591
Einstein, A., 8, 56, 232
Einstein theory of specific heat, 445
Elastic collison, 551
Electromagnetic field, vacuum fluctuations in, 389
Electromagnetic interaction, 584
Electromagnetic radiation, momentum of, 45
Electron, Dirac theory of, 571
  gyromagnetic ratio of, 226
Electron affinity, 286
Electron capture by nucleus, 545
Electron configurations of the elements, 238
Electron gas in a metal, 404
  statistical behavior of, 452
Electron sharing, 267
Electron shells in an atom, 232
Electron spin, 224
  quantum number of, 225
Electron subshells in an atom, 234
  order of filling, 240
Electronegativity, 301
  table of, 302
Electronic specific heat, 458
Electrons, delocalized in molecule, 315
  nuclear, 96, 492
Elementary particles, 571
  systematics of, 579
  theories of, 590
Elements, actinide, 234
  electron configurations of, 238
  electronegativities of, 302

Elements, ionization energies of, 285
  lanthanide, 234
  periodic table of, 235
  transition, 234
  transuranic, 565
Emission spectra, 125
Energy, binding, of atomic electrons, 233
    nuclear, 502, 516
  in center-of-mass system, 551
  conservation of, 23
  kinetic, operator, 162
  quantization of (see Quantization, of energy)
  rest, 40
  thermonuclear, 567
  total, 40
    operator, 162
Energy band in crystal, 406, 461
Energy levels, degenerate, 193
  of harmonic oscillator, 186
  of hydrogen atom, 132, 258
  of helium atom, 260
  of mercury atom, 262
  of particle in a box, 172, 193
  of rotating molecule, 316, 320
  of sodium atom, 259
  statistical weights of, 364
  of vibrating molecule, 327, 333
  and X-ray spectra, 263
Equivalence, principle of, 46
η meson, 580
Ether, 1
Ethylene molecule, 310
Exchange force in nucleus, 511
Exchange integral, 279
Excited state of atom, 132
Exclusion principle, 230
  repulsion due to, 267, 287
Expectation value, 159, 160
  of operator, 164

Face-centered cubic crystal, 394, 421
Fermi, E., 232
Fermi, the, 501
Fermi-Dirac statistics, 382, 453, 588
Fermi energy, 385, 406, 454, 486
  table of, 457
Fermi gas model of nucleus, 522

Fermi surface, 481
Fermion (Fermi particle), 232, 356
Feynman-Hellmann theorem, 269
Fine structure in spectral lines, 223, 228, 257
Fission, nuclear, 503, 561
Fission fragments, 562
Fluorescence, 345
Fluorite crystal, 423
Forbidden band in crystal, 463, 475
Forbidden transition, 254
Four-vector, 24
Frame of reference, 7
    center-of-mass, 549
    inertial, 8
Franck-Condon principle, 342
Franck-Hertz experiment, 137
Frank-Read mechanism, 435
Frenckel defect, 430
Fusion, nuclear, 503, 566

Galilean transformation, 10
Gamma decay, 546
Geiger-Marsden experiment, 103
General relativity, 46
Goudsmit, S. A., 224
Graphite, 408
Gravitational field, effect on clock, 48
Gravitational interaction, 584
Gravitational red shift, 74
Graviton, 583
Gravity in general relativity, 47
Gray tin, 406
Ground state of atom, 132
Group in periodic table, 234
Group velocity of waves, 81, 86
    in crystal, 482
Gyromagnetic ratio of electron spin, 226
    of orbital electron, 207

H_α line, 127, 143, 257
Half life of radioisotope, 523
Hall effect, 470
Hamiltonian operator, 165
Harmonic oscillator, 180
    average energy of, 379, 443
    energy levels of, 186

Harmonic oscillator, Schrödinger's equation for, 182
    selection rules for transitions in, 335
    wave functions of, 186
    zero-point energy of, 186
Heisenberg, W., 94, 147, 512
Helium atom, energy levels of, 260
    wave behavior of, 90
Hermite polynomial, 187
Heteronuclear diatomic molecules, properties of, 304
Hexagonal crystal, 418
    close-packed, 419
Hole, in electron sea, 572
    in a metal, 481
    in a semiconductor, 465
Homonuclear diatomic molecules, properties of, 292, 300
Hooke's law, 180
Hund's rule, 241, 291
Hybrid orbital, 307
Hydrogen atom, Bohr theory of, 128
    classical dynamics of, 118
    eigenvalue equations of, 214
    electron probability density in, 214, 220
    energy levels of, 132, 258
    quantum theory of, 195
    wave functions of, 197, 202
Hydrogen bond, 402
Hydrogen fluoride molecule, 303, 304
Hydrogen molecular ion, 269
Hydrogen molecule, 282
Hydrogen spectrum, 126, 135
Hydrogenic atom, 144
Hypercharge, 585
Hyperfine structure in spectral lines, 224, 501
Hyperfragment, 579
Hyperon, 579

Ice crystal, 404
Impact parameter, 110
Impurity semiconductor, 464
Induced emission of radiation, 386
Inelastic collision, 552
Inertial frame of reference, 8
Insulator, 464, 478
Internal conversion, 547

X-ray spectra, 62, 261
X-ray spectrometer, 67
X-ray wavelengths, 60
X rays, 59
    polarization of, 60

Yang, C. N., 575

Yukawa, H., 512

Zartman-Ko experiment, 363
Zeeman effect, 210, 223
    anomalous, 244
Zero-point energy, 186
Zinc blende crystal, 422

Selected Figures
and Tables

Selected Figures

Selected Tables

The Electromagnetic Spectrum

CLASS	GAMMA RAYS	X RAYS	ULTRAVIOLET RADIATION	VISIBLE LIGHT	INFRARED RADIATION	MICROWAVES	FM AND TV BROADCASTING	AM BROADCASTING

WAVELENGTH, m

10^{-12}	10^{-11}	10^{-10}	10^{-9}	10^{-8}	10^{-7}	10^{-6}	10^{-5}	10^{-4}	10^{-3}	10^{-2}	10^{-1}	1	10	10^{2}	10^{3}	10^{4}	10^{5}	10^{6}	10^{7}
0.01 A	0.1 A	1 A	10 A 1 mμ	100 A 10 mμ	1,000 A 100 mμ	10,000 A 1 μ	0.1 mm 10 μ	1 mm 100 μ	1 cm	10 cm					1 km	10 km	100 km	1,000 km	

FREQUENCY, c/sec

10^{21}	10^{20}	10^{19}	10^{18}	10^{17}	10^{16}	10^{15}	10^{14}	10^{13}	10^{12}	10^{11}	10^{10}	10^{9}	10^{8}	10^{7}	10^{6}	10^{5}	10^{4}	10^{3}	10^{2}
										10^4 mc/s	10^3 mc/s	100 mc/s	10 mc/s	1 mc/s	100 kc/s	10 kc/s	1 kc/s		

QUANTUM ENERGY, $h\nu$, ev

10^{7}	10^{6}	10^{5}	10^{4}	10^{3}	10^{2}	10	1	10^{-1}	10^{-2}	10^{-3}	10^{-4}	10^{-5}	10^{-6}
10 Mev	1 Mev	100 kev	10 kev	1 kev									

CHARACTERISTIC TEMPERATURE, $h\nu/k$, °K

10^{10}	10^{9}	10^{8}	10^{7}	10^{6}	10^{5}	10^{4}	10^{3}	10^{2}	10	1	10^{-1}

TYPE OF TRANSITION

Nuclear	Inner Electron	Outer Electron	Molecular Vibration	Molecular Rotation	Electron Spin	Nuclear Spin

ARTIFICIAL PRODUCTION

Betatron	X-ray Tube	Gas Discharge, Arcs, Sparks	Hot Filaments	Magnetron, Klystron, Traveling-wave Tube	Electrical Circuit	AC Generator

DETECTION

Geiger and Scintillation Counters, Ionization Chamber	Photoelectric and Photo-multiplier Cells	Bolometer, Thermopile	Crystal	Electrical Circuit